International Conference on Non-Perturbative/Background Independent Quantum Gravity

(LOOPS'11)

AA002786

Journal of Physics: Conference Series Volume 360

Madrid, Spain
23-28 May 2011

Editors:

Guillermo A Mena Marugán
J Fernando Barbero G
Luis J Garay
Eduardo J S Villaseñor
Javier Olmedo

ISBN: 978-1-62276-777-9
ISSN: 1742-6588

Printed from e-media with permission by:

Curran Associates, Inc.
57 Morehouse Lane
Red Hook, NY 12571

Some format issues inherent in the e-media version may also appear in this print version.

Printed by Curran Associates, Inc. (2013)

Institute of Physics
Dirac House, Temple Back
Bristol BS1 6BE UK

Phone: 44 1 17 929 7481
Fax: 44 1 17 920 0979

techtracking@iop.org

Additional copies of this publication are available from:

Curran Associates, Inc.
57 Morehouse Lane
Red Hook, NY 12571 USA
Phone: 845-758-0400
Fax: 845-758-2634
Email: curran@proceedings.com
Web: www.proceedings.com

TABLE OF CONTENTS

PLENARY CONTRIBUTIONS

012001 Some Recent Advances in Loop Quantum Cosmology ... 1
A Ashtekar

012002 Ten Questions on Group Field Theory (and Their Tentative Answers) ... 10
A Baratin, D Oriti

012003 Chiral Vacuum Fluctuations in Quantum Gravity ... 20
L Bethke, J Magueijo

012004 Towards Computational Insights Into the Large-scale Structure of Spin Foams ... 29
B Dittrich, F C Eckert

012005 Quantum Scalar Field in Quantum Gravity with Spherical Symmetry ... 39
R Gambini, J Pullin, S Rastgoo

012006 Extension of Loop Quantum Gravity to Metric Theories Beyond General Relativity ... 42
Y Ma

012007 New Insights in Quantum Geometry ... 52
H Sahlmann

012008 Bianchi-I Spacetimes in Loop Quantum Cosmology: Physics of Singularity Resolution ... 62
P Singh

012009 The Diffeomorphism Constraint Operator in Loop Quantum Gravity ... 72
M Varadarajan

FOUNDATIONS OF QUANTUM GRAVITY

012010 3D Gravity and Quantum Deformations: A Drinfel'd Double Approach ... 81
A Ballesteros, F J Herranz, C Meusburger

012011 Perturbative Running of the Immirzi Parameter ... 85
D Benedetti, S Speziale

012012 S-Matrix for AdS from General Boundary QFT ... 89
D Colosi, M Dohse, R Oeckl

012013 A Systematic Construction of Curved Phase Space: A Gravitational Gauge Theory with Symplectic Form ... 93
J S Hazboun, J T Wheeler

012014 Effective Relational Dynamics ... 97
P A Höhn

012015 A View on the Problems of Quantum Gravity ... 101
T P Shestakova

012016 Loop Quantum Effect and the Fate of Tachyon Field Collapse ... 105
Y Tavakoli, P V Moniz, J Marto, A H Ziaie

012017 The Geometric Role of Symmetry Breaking in Gravity ... 109
D K Wise

HAMILTONIAN LOOP QUANTUM GRAVITY

012018 Complete Loop Quantization of a Dimension 1+2 Lorentzian Gravity Theory ... 113
R M S Barbosa, C P Constantinidis, Z Oporto, O Piguet

012019 Dynamics for a Simple Graph Using the U(N) Framework for Loop Quantum Gravity ... 117
E F Borja, J Díaz-Polo, L Freidel, I Garay, E R Livine

012020 SU(2) Gauge Symmetry in Gravity Phase Space ... 122
F Cianfrani

012021 Boundary Terms and the 3 + 1 Decomposition of the Holst Action ... 126
A Corichi, J D Reyes

012022 On Ashtekar's Formulation of General Relativity ... 130
C Fleischhack

012023 Loop Gravity in Terms of Spinors ... 140
E Livine, J Tambornino

012024 SU(2) Gauge Theory of Gravity with Topological Invariants 145
S Sengupta
012025 Perfect Discretization of Path Integrals 149
S Steinhaus

LOOP QUANTUM COSMOLOGY

012026 Quantum Back-reaction in a Universe with Positive Cosmological Constant 153
D Brizuela
012027 Inflationary Spectra and Observations in Loop Quantum Cosmology 157
G Calcagni
012028 Consistent Probabilities in sLQC 161
D A Craig, P Singh
012029 2-point Functions in Quantum Cosmology 165
S Gielen
012030 Plane Gravitational Waves and Loop Quantization 169
F Hinterleitner, S Major
012031 Loop Quantum Cosmology of the Bianchi I Model: Complete Quantization 173
M Martín-Benito, L J Garay, G A Mena Marugán, E Wilson-Ewing
012032 Inhomogenous Loop Quantum Cosmology with Matter 177
D Martín-De Bias, M Martín-Benito, G A Mena Marugán
012033 Inflation and Inhomogeneities: A Hybrid Quantization 181
J Olmedo, M Fernández-Méndez, G A Mena Marugán
012034 Palatini Approach to Bouncing Cosmologies and DSR-like Effects 185
G J Olmo

BLACK HOLE PHYSICS AND DYNAMICAL TRIANGULATIONS

012035 Black Hole Entropy in Loop Quantum Gravity 189
I Agulló, J F Barbero G, E F Borja, J Díaz-Polo, E J S Villaseñor
012036 Hawking Radiation from Loop Black Holes 193
E Alesci, L Modesto
012037 Entropy of Generic Quantum Isolated Horizons 197
C Beetle, J Engle
012038 The Effective Kinetic Term in CDT 201
T G Budd
012039 Curved Geometry and Graphs 205
F Caravelli
012040 Non-commutative Holonomies in 2 + 1 LQG and Kauffman's Brackets 209
K Noui, A Perez, D Pranzetti

COVARIANT LOOP QUANTUM GRAVITY AND SPINFOAMS

012041 Regularized Hamiltonians and Spinfoams 213
E Alesci
012042 Operator Spin Foams: Holonomy Formulation and Coarse Graining 217
B Bahr
012043 Scalar-tensor Theory As a Singular Subsector of $\Lambda(\Phi)$ Plebanski Gravity 222
D Beke
012044 The Time-oriented Boundary States and the Lorentzian-Spinfoam Correlation Functions 227
E Bianchi, Y Ding
012045 Singular Topologies in the Boulatov Model 231
S Carrozza
012046 Holomorphic Simplicity Constraints for 4D Riemannian Spinfoam Models 235
M Dupuis, E R Livine
012047 Canonical Formalism for Simplicial Gravity 239
P A Höhn
012048 Dynamics of Loop Quantum Gravity Encoded in Graphs 243
M Kisielowski

012049 Effective Action for EPRL/FK Spin Foam Models 247
A Mikovic, M Vojinovic
012050 Einstein-Regge Equations in Spinfoams 251
C Perini
012051 Shape of Spinfoams Encoded in Graphs 255
J Puchta
012052 Non-commutative Dual Representation for Quantum Systems on Lie Groups 260
M Raasakka
012053 Spinfoams: Summing Or Refining? 264
M Smerlak
012054 Local Gauge Theory and Coarse Graining 267
J A Zapata
012055 Loop Quantum Brans-Dicke Theory 272
X Zhang, Y Ma

SHAPE DYNAMICS AND PHENOMENOLOGY DERIVED FROM QUANTUM GRAVITY

012056 Braided Topology and the Emergence of Matter 276
S Bilson-Thompson
012057 Observable Consequences of Quantum Gravity: Can Light Fermions Exist? 280
A Eichhorn
012058 The Coupling of Shape Dynamics to Matter 284
H Gomes
012059 Shape Dynamics 289
T Koslowski
012060 Longitudinal and Transverse Relativity of Spacetime Locality in Planck-scale-deformed Phase Spaces 294
N Loret, L Barcaroli, G Rosati
012061 Quantum Geometry Phenomenology: Angle and Semiclassical States 298
S A Major
012062 Shape Dynamics and AdS/CFT 302
F Mercati
Author Index

Preface

The international conference LOOPS'11 took place in Madrid from the 23–28 May 2011. It was hosted by the Instituto de Estructura de la Materia (IEM), which belongs to the Consejo Superior de Investigaciones Científicas (CSIC). Like previous editions of the LOOPS meetings, it dealt with a wealth of state-of-the-art topics on Quantum Gravity, with special emphasis on non-perturbative background-independent approaches to spacetime quantization.

The main topics addressed at the conference ranged from the foundations of Quantum Gravity to its phenomenological aspects. They encompassed different approaches to Loop Quantum Gravity and Cosmology, Polymer Quantization, Quantum Field Theory, Black Holes, and discrete approaches such as Dynamical Triangulations, amongst others. In addition, this edition celebrated the 25th anniversary of the introduction of the now well-known Ashtekar variables and the Wednesday morning session was devoted to this silver jubilee.

The structure of the conference was designed to reflect the current state and future prospects of research on the different topics mentioned above. Plenary lectures that provided general background and the 'big picture' took place during the mornings, and the more specialised talks were distributed in parallel sessions during the evenings. To be more specific, Monday evening was devoted to Shape Dynamics and Phenomenology Derived from Quantum Gravity in Parallel Session A, and to Covariant Loop Quantum Gravity and Spin foams in Parallel Session B. Tuesday's three Parallel Sessions dealt with Black Hole Physics and Dynamical Triangulations (Session A), the continuation of Monday's session on Covariant Loop Quantum Gravity and Spin foams (Session B) and Foundations of Quantum Gravity (Session C). Finally, Thursday and Friday evenings were devoted to Loop Quantum Cosmology (Session A) and to Hamiltonian Loop Quantum Gravity (Session B).

The result of the conference was very satisfactory and enlightening. Not only was it a showroom for the research currently being carried out by many groups throughout the world, but there was also a permanent look towards the future. During these days, the CSIC Campus witnessed many scientific conversations triggered by the interaction amongst the people and groups that participated in LOOPS'11 Madrid and which, in many cases, will crystallise into new results and advances in the field.

The conference would not have been possible without the generous help of a number of national and international institutions. The organizers would like to acknowledge the financial support provided by the Spanish Ministry of Science and Innovation (Ministerio de Ciencia e Innovación), the Spanish Research Council, CSIC (Consejo Superior de Investigaciones Científicas), The BBVA Foundation (Fundación BBVA), The CONSOLIDER-CPAN project, the Spanish Society for Gravitation and Relativity (SEGRE),

The Universidad Carlos III of Madrid (UC3M), and the European Science Foundation (ESF).

The ESF, through the Quantum Gravity and Quantum Geometry network, provided full support for a number of young participants that have contributed to these proceedings: Dario Benedetti (Albert Einstein Institute, Potsdam, Germany), Norbert Bodendorfer (Institute for Theoretical Physics III, FAU Erlangen Nürnberg, Germany), Mariam Bouhmadi López (CENTRA, Centro Multidisciplinar de Astrofísica, Lisbon), Timothy Budd (Institute for Theoretical Physics, Utrecht University, The Netherlands), Miguel Campiglia (Institute for Gravitation and the Cosmos, Penn State University, USA), Gianluca Delfino (School of Mathematical Sciences, University of Nottingham, UK), Maite Dupuis (Institute for Theoretical Physics III, FAU Erlangen Nürnberg, Germany), MichałDziendzikowski (Institute of Theoretical Physics, Warsaw University, Poland), Muxin Han (Centre de Physique Théorique de Luminy, Marseille, France), Philipp Höhn (Institute for Theoretical Physics, Utrecht University, The Netherlands), Jacek Puchta (Centre de Physique Théorique de Luminy, Marseille, France), James Ryan (Albert Einstein Institute, Potsdam, Germany), Lorenzo Sindoni (Albert Einstein Institute, Golm, Germany), David Sloan (Institute for Theoretical Physics, Utrecht University, The Netherlands), Johannes Tambornino (Laboratoire de Physique, ENS Lyon, France), Andreas Thurn (Institute for Theoretical Physics III, FAU Erlangen Nürnberg, Germany), Francesca Vidotto (Laboratoire de Physique Subatomique et de Cosmologie, Grenoble, France), and Matteo Smerlak (Albert Einstein Institute, Golm, Germany).

We would like to conclude this preamble by thanking all the attendants of the conference for their high and enthusiastic participation. The presence of a large number of past and present Loop Quantum Gravity practitioners, as well as a significant number of top researchers in other approaches to quantum gravity, provided ample opportunities for fruitful scientific exchanges and a very lively atmosphere. It is encouraging to see that, 25 years after the inception of Loop Quantum Gravity, there is a vibrant young community of researchers entering the field. Let us hope that, with their help, the quantization of general relativity can be successfully accomplished in the near future.

The Editors

Journal of Physics: Conference Series **360** (2012) 011002 doi:10.1088/1742-6596/360/1/011002

Committees and Sponsors

Local Organizing Committee

J Fernando Barbero González (IEM, CSIC)
Laura Castelló Gomar (Univ. Complutense, Madrid)
Mikel Fernández Méndez (IEM, CSIC)
Iñaki Garay Elizondo (Univ. País Vasco)
Luis J Garay Elizondo (Univ. Complutense, Madrid)
Mercedes Martín-Benito (Perimeter Institute, Canada)
Daniel Martín de Blas (IEM, CSIC)
Guillermo A Mena Marugán (IEM, CSIC)
Javier Olmedo Nieto (IEM, CSIC)
Gonzalo Olmo Alba (IFIC, CSIC)
Tomasz Pawlowski (Warsaw University, Poland)
Eduardo J Sánchez Villaseñor (Univ. Carlos III, Madrid)

Scientific International Committee

Abhay Ashtekar (Penn State University, USA)
J Fernando Barbero González (IEM, CSIC, Spain)
John Barrett (University of Nottingham, UK)
José Manuel Cidade Mourão (Instituto Superior Técnico de Lisboa, Portugal)
Laurent Freidel (Perimeter Institute, Canada)
Jerzy Lewandowski (Warsaw University, Poland)
Guillermo A Mena Marugán (IEM, CSIC, Spain)
Jorge Pullin (Louisiana State University, USA)
Carlo Rovelli (Univ. Méditerranée, Marseille, France)
Thomas Thiemann (Univ. Erlangen-Nürnberg, Germany)

Sponsors

Journal of Physics: Conference Series **360** (2012) 011003 doi:10.1088/1742-6596/360/1/011003

Peer review statement

All papers published in this volume of *Journal of Physics: Conference Series* have been peer reviewed through processes administered by the proceedings Editors. Reviews were conducted by expert referees to the professional and scientific standards expected of a proceedings journal published by IOP Publishing.

Journal of Physics: Conference Series **360** (2012) 012001 doi:10.1088/1742-6596/360/1/012001

Some Recent Advances in Loop Quantum Cosmology

Abhay Ashtekar

E-mail: ashtekar@gravity.psu.edu
Institute for Gravitation and the Cosmos & Physics Department, The Pennsylvania State University, University Park PA 16802, USA

Abstract. In my talk I discussed three recent advances in loop quantum cosmology: i) Path integral formulation and its WKB approximation; 2) Cosmological spin foams and lessons they provide; and 3) Probability of a slow roll inflationary phase compatible with the 7 year WMAP data. In addition to presenting an overview, this discussion also provides the necessary background for a number of talks in the parallel sessions.

1. Path Integrals and the WKB approximation in LQC

The key difference between the Wheeler DeWitt theory (WDW) theory and Loop Quantum Cosmology (LQC) is that, thanks to the quantum geometry effects inherited from Loop Quantum Gravity, LQC has a novel, in-built repulsive force. While it is completely negligible when curvature is less than, say, 1% of the Planck scale, it grows dramatically once curvature becomes stronger, overwhelming the classical gravitational attraction and causing a quantum bounce that resolves the big-bang singularity. (For details, see e.g. [1]). From a path integral viewpoint (see, e.g., [2]), on the other hand, this stark departure from classical solutions seems rather surprising at first. For, in the path integral formulation quantum effects usually become important when the action is small, comparable to the Planck's constant $\hbar$, while the Einstein-Hilbert action along classical trajectories that originate in the big-bang is generically very large. Thus, there is an apparent conceptual tension. Recent detailed analyses [3, 4] have resolved this issue in the k=0 and k=1 Friedman Lemaitre Robertson Walker (FLRW) model. I will summarize the main ideas and results using the k=0 model.

1.1. Strategy

Since LQC uses a Hilbert space framework, it is most natural to return to the original derivation of path integrals, where Feynman began with the expressions of transition amplitudes in the Hamiltonian theory and *reformulated* them as an integral over all kinematically allowed paths [5]. But in non-perturbative quantum gravity, there is a twist: at a fundamental level, one deals with a constrained system without external time whence the notion of a transition amplitude does not have an a priori meaning. It is replaced by an *extraction amplitude* —a Green's function which extracts physical quantum states from kinematical ones and also provides the physical inner product between them. (For a discussion in the cosmological context, see, e.g., [6, 7, 8, 9]). If the theory can be deparameterized it inherits a relational time variable, and then the extraction amplitude should reduce to the standard transition amplitude with respect to that time. In LQC with a massless scalar field, a natural deparametrization is indeed available [10, 11, 12, 1]

Journal of Physics: Conference Series **360** (2012) 012001 doi:10.1088/1742-6596/360/1/012001

and this expectation is borne out [7]. However, more generally —e.g. if one were to introduce a potential for the scalar field— it is difficult to find a *global* time variable. Since the conceptual tension between the LQC results and the path integral intuition is generic, it is best not to have to rely heavily on deparametrization. Therefore, in [3] the comparison was carried out in the timeless framework. This is also the setting of spin foams, the path integral approach to full LQG; see e.g. [13]. The idea then is to start with the expression of the extraction amplitude in the Hilbert space framework of LQC, cast it as a path integral, and re-examine the tension between the path integral intuition and the singularity resolution.

Recall that solutions to the quantum constraint equation, as well as inner product between them, can be obtained through a group averaging procedure [14, 15, 16]. Let us work in the basis $|\nu, \phi\rangle$ in which the volume and the scalar field operators, $\hat{V}$ and $\hat{\phi}$, on the kinematical Hilbert space $\mathcal{H}_{\rm kin}^{\rm total}$ are diagonal. The extraction amplitude $\mathcal{E}(\nu, \phi; \nu', \phi')$ is a Green's function that results from this group averaging:

$$\mathcal{E}(\nu_f, \phi_f; \nu_i, \phi_i) := \textstyle\int_{-\infty}^{\infty} \mathrm{d}\alpha\ \langle \nu_f, \phi_f|\, e^{\frac{i}{\hbar}\alpha\hat{C}}\, |\nu_i, \phi_i\rangle\ , \tag{1}$$

where $\hat{C} = -\hbar^2(\partial_\phi^2 + \hat{C}_{\rm grav})$ is the full Hamiltonian constraint and α is a parameter (with dimensions $[L^{-2}]$). The integral averages the ket (or the bra) over the group generated by the constraint. Since the integrated operator is heuristically '$\delta(\hat{C})$,' the amplitude $\mathcal{E}(\nu_f, \phi_f; \nu_i, \phi_i)$ satisfies the Hamiltonian constraint in both of its arguments. Consequently, it serves as a Green's function that extracts physical states $\Psi_{\rm phys}(\nu, \phi)$ in $\mathcal{H}_{\rm phy}$ from kinematical states $\Psi_{\rm kin}(\nu, \phi)$ in $\mathcal{H}_{\rm kin}^{\rm total}$ through a convolution

$$\Psi_{\rm phys}(\nu, \phi) = \sum_{\nu'} \textstyle\int \mathrm{d}\phi'\ \mathcal{E}(\nu, \phi; \nu', \phi')\ \Psi_{\rm kin}(\nu', \phi'). \tag{2}$$

This is why $\mathcal{E}$ is referred to as the *extraction amplitude*. Since it is the group averaging Green's function, $\mathcal{E}$ also enables us to write the physical inner product in terms of the kinematical:

$$(\Phi_{\rm phys},\ \Psi_{\rm phys}) := \sum_{\nu, \nu'} \textstyle\int \mathrm{d}\phi\, \mathrm{d}\phi'\ \bar{\Phi}_{\rm kin}(\nu, \phi)\, \mathcal{E}(\nu, \phi; \nu', \phi')\Psi_{\rm kin}(\nu',\ \phi'). \tag{3}$$

Thus, in the 'timeless' framework without any deparametrization, all the information in the physical sector of the quantum theory is neatly encoded in the extraction amplitude $\mathcal{E}(\nu, \phi; \nu', \phi')$. Therefore to relate the Hilbert space and path integral frameworks, it suffices to recast the expression (1) of this amplitude as a path integral.

1.2. Path integral for the extraction amplitude

Let us begin by recalling the procedure Feynman used to arrive at the path integral expression of the transition amplitude in quantum mechanics. He began with the Hamiltonian framework, wrote the unitary evolution as a composition of N infinitesimal ones, inserted a complete basis between these infinitesimal evolution operators to arrive at a 'discrete time' path integral, and finally took the limit $N \to \infty$. In the timeless framework, we need to adapt this procedure to the extraction amplitude $\mathcal{E}(\nu_f, \phi_f; \nu_i, \phi_i)$, using $e^{\frac{i}{\hbar}\alpha\hat{C}}$ in the integrand of Eq. (1) in place of the 'evolution' operator, and then performing the α integration in a final step.

More precisely, the integrand of (1) can be thought of as a matrix element of a *fictitious* evolution operator $e^{\frac{i}{\hbar}\alpha\hat{C}}$. One can regard $\alpha\hat{C}$ as playing the role of a (purely mathematical) Hamiltonian, the evolution over an unit time interval. We can then decompose this fictitious evolution into N evolutions of length $\epsilon = 1/N$ and insert a complete basis at suitable intermediate steps to write the extraction amplitude as a 'skeletonized' *phase space* path integral and finally take the limit $N \to \infty$ (or $\epsilon \to 0$). As is standard in the path integral literature, the result can be expressed as a formal infinite dimensional integral,

$$\mathcal{E}(\nu_f, \phi_f; \nu_i, \phi_i) = \int \mathrm{d}\alpha \int [\mathcal{D}\nu(T)]\ [\mathcal{D}\mathrm{b}(T)]\ [\mathcal{D}p_{(\phi)}(T)]\ [\mathcal{D}\phi(T)]\ e^{\frac{i}{\hbar}S} \tag{4}$$

where b and $p_{(\phi)}$ are momenta conjugate to ν and ϕ respectively. A detailed calculation [3] shows that the action S in this expression is given by

$$S = \int_0^1 \mathrm{d}T \left(p_{(\phi)}\phi' - \frac{1}{2}\mathrm{b}\nu' - \alpha \left(p_{(\phi)}^2 - 3\pi G \nu^2 \frac{\sin^2 \lambda \mathrm{b}}{\lambda^2} \right) \right) \tag{5}$$

where the prime denotes derivative with respect to the (fictitious) time T and $\lambda^2 == 4\sqrt{3}\pi\gamma\ell_{\mathrm{Pl}}^2$ is the area gap of LQC. Note that the final integration is over *all* paths in the classical phase space, *including the ones which go through the point* $\nu = 0$, *normally regarded as the singularity.* Therefore, the tension between the Hamiltonian LQC and the path integral formulation is brought to forefront: How can we see the singularity resolution in the path integral setting provided by (4)? The answer is that the paths are not weighted by the standard Einstein-Hilbert action but by a 'polymerized' version (5) of it which still retains the memory of the quantum geometry underlying the Hamiltonian theory. As we will see, this action is such that a path going through the classical singularity has negligible contribution whereas bouncing trajectories give the dominant contribution.

Note that the new action (5) retains memory of quantum geometry through the area gap of LQC which depends on $\hbar$. This means that the Einstein-Hilbert action itself has received quantum corrections. This may seem surprising at first. However, this occurs also in some familiar examples if one systematically arrives at the path integral starting from the Hilbert space framework. Perhaps the simplest such example is that of a non-relativistic particle on a curved Riemannian manifold for which the standard Hamiltonian operator is simply $\hat{H} = -(\hbar^2/2m)g^{ab}\nabla_a\nabla_b$. When the quantum dynamics generated by this $\hat{H}$ is recast as a path integral, the integrand turns out to be $e^{iS_{\mathrm{corr}}}$ where $S_{\mathrm{corr}} = S_{\mathrm{cl}} + \int \mathrm{d}t(\frac{\hbar^2}{12m}R)$ is a *quantum corrected* action [17]. Note that the extrema of this action are not the geodesics one obtains in the classical theory but rather particle trajectories in a $\hbar$-dependent potential; the two can be qualitative different. Thus, this elementary example shows that one *cannot* always assume that the correct quantum theory will result from the path integral, where each path is weighted just by $e^{iS_{\mathrm{cl}}}$.

1.3. The steepest descent approximation

We can now use the steepest descent approximation to understand the singularity resolution in LQC from a path integral perspective. However there are two subtleties that need to be addressed. First, the standard WKB analysis refers to unconstrained systems and we have to adapt it to the constrained one, replacing the Schrödinger equation with the quantum constraint. This is not difficult (see, e.g., sections 3.2 & 5.2 of [18], or Appendix A in [3]). However, as in the standard WKB approximation for the transition amplitude, the procedure assumes that the action that features in the path integral has no explicit $\hbar$ dependence. The second subtlety arises because, in our case, the action does depend on $\hbar$ through $\lambda^2 \sim \gamma^3\hbar$. Therefore to explore the correct semi-classical regime of the theory now one has to take the limit $\hbar \to 0$, keeping $\gamma^3\hbar$ fixed [3]. One can then evaluate the action in the saddle point approximation. One finds that it is precisely the bouncing solutions —rather than the classical FLRW metrics— that dominate the WKB approximation.

To summarize, in the timeless framework all the physical information is contained in the extraction amplitude which reduces to the standard transition amplitude if a global deparametrization can be found. Following Feynman, one can start with the Hilbert space expression of the extraction amplitude and recast it as a phase space path integral. Quantum geometry effects of LQC leave their trace on the weight associated with each path: The action functional is modified. This in turn implies that, in the WKB approximation, the extraction amplitude is dominated by universes that undergo a bounce. Thus, from the LQC perspective, it would be incorrect to simply define the theory starting with the Einstein Hilbert action because this procedure completely ignores the quantum nature of the underlying Riemannian geometry. For a satisfactory treatment of ultraviolet issues such as the singularity resolution, it is crucial that the calculation retains appropriate memory of its quantum nature. Indeed, this is why in the spin foam models one sums over quantum geometries, not smooth metrics. For further details, see [1, 3, 8]

2. Cosmological Spin foams

2.1. General Setting

Spin foams were discussed by several speakers at this conference. Let us recall the key ideas. This is a sum over histories approach whose distinguishing feature relative to the older path integral approaches such as those of [2] is that now the sum is over specific *quantum* geometries that originate in loop quantum gravity. In the simplest version one can consider a simplicial decomposition of a 4-manifold of interest and consider the 2-complex dual to it. Each vertex of the 2-complex lies in a simplex, each face is labeled by a spin and each edge by an 'intertwiner'. Thus, a 2-complex can be thought of as the 'time evolution' of spin a network, with a genuine change/evolution occurring at the vertices. As explained in section 1, the basic object determining dynamics is the extraction amplitude which, in full quantum gravity, is written out as a sum over partial amplitudes, each associated with a 2-complex. Since the 2-complex joining the 'in' and the 'out' spin network is characterized by the number of vertices it contains, this formal expression of the extraction amplitude is called the *vertex expansion*.

Over the last three years, there have been significant advances in the area of spin foams (see, e.g. [13, 19]). However, key issues still remain: Can the vertex expansion be summed to yield a meaningful answer? Does it provide the full extraction amplitude or does one have to take a 'continuum limit' of the 2-complex as in the standard Regge approach? Is the vertex amplitude necessarily local or can there be Planck scale non-localities that become invisible on coarse graining? What should the summand/integrand reduce to in the *classical* limit? There is a debate on whether the integrand should yield e^{is} where S is the Einstein-Hilbert action or $\cos S$. This is related to the question: What paths should one sum over? Should they be restricted to be suitably time oriented [20]?

2.2. Insights from LQC

LQC models provide a particularly convenient setting to probe these conceptual issues [6, 7, 8, 9, 21]. Because the Hamiltonian theory in these models is fully under control, as in section 1, we can again use the procedure followed by Feynman [5] to rewrite the well-defined extraction amplitude as a sum over histories. To obtain a sum analogous to that used in spin foams, however, now one keeps it as a sum over paths in the *configuration space*, without introducing a further sum over momentum basis in the intermediate steps as was done in section 1.

For simplicity let us consider the FLRW model where, in the timeless framework, the configuration variables are volume ν of the fiducial cell and the scalar field ϕ. (For details on the Bianchi I model, see [21]). Then the extraction amplitude $\mathcal{E}(\nu_f, \phi_f; \nu_i, \phi_i)$ can be expressed as

Journal of Physics: Conference Series **360** (2012) 012001 doi:10.1088/1742-6596/360/1/012001

$$\begin{aligned}\mathcal{E}(\nu_f, \phi_f;\, \nu_i, \phi_i) &= \int \mathrm{d}\alpha\, \langle \nu_f, \phi_f |\, e^{i\alpha \hat{C}}\, | \nu_i, \phi_i \rangle \\ &= \int \mathrm{d}\alpha\, \langle \nu_f, |e^{i\alpha \hat{C}_{\mathrm{grav}}} | \nu_i, \rangle\, \langle \phi_f\, | e^{i\alpha \hat{C}_{\mathrm{matt}}} |\, \phi_i \rangle \\ &= \int \mathrm{d}\alpha\, A_{\mathrm{grav}}(v_f, v_i;\, \alpha)\;\, A_\phi(\phi_f, \phi_i;\, \alpha) \end{aligned} \tag{6}$$

It is completely straightforward to calculate the matter amplitude A_ϕ. Mathematically, we can regard the gravitational amplitude A_{grav} as the 'transition amplitude' between $|v_i\rangle$ and $|v_f\rangle$ in a 'unit time interval', caused by a 'Hamiltonian operator' $\alpha \hat{C}_{\mathrm{grav}}$. Although 'time' and the 'Hamiltonian' are mathematical constructs without direct physical significance, this strategy lets us follow, step by step, Feynman's construction [5] to express A_{grav} as a path integral. The resulting sum can be rearranged by grouping together all paths in which there are precisely M volume transitions between ν_i and ν_f in 'unit time interval'. Finally, the integral over α can be carried out explicitly to express the full extraction amplitude $\mathcal{E}(\nu_f, \phi_f;\, \nu_i, \phi_i)$ as a sum over amplitudes $A(v_M, \ldots, \nu_0;\, \phi_f, \phi_i)$ associated with quantum histories in which the universe makes ordered transitions[1] $\nu_i \to \nu_1 \to \nu_2 \to \ldots \to \nu_{M-1} \to \nu_f$ at increasing but otherwise arbitrary values of $\phi \in (\phi_f,\, \phi_i)$:

$$\mathcal{E}(\nu_f, \phi_f;\, \nu_i, \phi_i) = \sum_{M=0}^{\infty} \Big[\sum_{\nu_{M-1}, \ldots, \nu_1} A(v_M, \ldots, \nu_0;\, \phi_f, \phi_i) \Big] = \sum_{M=0}^{\infty} A(M)\,. \tag{7}$$

Here in the last step $A(M)$ denotes the contribution to the extraction amplitude obtained by summing over all quantum histories in which there are *precisely* M volume transitions between the initial and the final configurations. This is the analog of the Mth term in the spin foam vertex expansion obtained by summing over all colorings for a given 2-complex with precisely M vertices. (The analogy becomes even closer when anisotropies are included [21, 1].) These partial amplitudes are explicitly given by matrix elements of the gravitational part $\hat{C}_{\mathrm{grav}}$ of the quantum Hamiltonian constraint.

2.3. Lessons

This analysis in the cosmological models has shed light on several issues that have been open in the spin-foam models. First, since the Hamiltonian theory is fully under control, in these models one can start with a well-defined expression of the extraction amplitude $\mathcal{E}(\nu_f, \phi_f;\, \nu_i, \phi_i)$ which we know to be an exact (i.e. not just a formal) expression from the Hamiltonian theory. Thus we are simply asking if this correct expression can be recast as a 'vertex expansion'. One finds that this is possible and —modulo one precise assumption that an infinite sum can be interchanged with an integral— the sum is convergent. In particular, since the histories summed over are 'quantum', the allowed values of ν are discrete and remain discrete in the final sum; a continuum limit is not necessary. This provides a strong support for the general paradigm underlying the spin foam program. Of course in the cosmological models one freezes out all but a finite number of degrees of freedom whence new difficult issues will arise in testing convergence in the full theory. But the analysis supports the underlying *conceptual* viewpoint that, thanks to quantum geometry, a continuum limit is not necessary.

There are also a number of other insights. Since the scalar field serves as a good clock in the cosmological models, it can be regarded as a relational time variable to deparameterize the classical as well as the quantum theory. In the deparameterized quantum description, we have a Hamiltonian operator $\hat{H}$ generating evolution. Therefore, we can construct a path integral following Feynman [5] not just mathematically but physically. The result is the *transition*

[1] Dynamics of LQC implies that the relevant part of the spectrum of the volume operator is discrete; all ν that feature in these transitions are given by $4n\,\nu_o$ where ν_o is a fixed number.

Journal of Physics: Conference Series **360** (2012) 012001 doi:10.1088/1742-6596/360/1/012001

amplitude $\langle\nu_f|\,e^{i\hat{H}(\phi_f-\phi_i)}\,|\nu_i\rangle$ for the universe to pass from a initial state $|\nu_i\rangle$ at time ϕ_i to the final state $|\nu_f\rangle$ at time ϕ_f. This transition amplitude in the deparameterized theory can be shown to be exactly equal to the physical inner product $\langle[\nu_f,\phi_f]|[\nu_i,\phi_i]\rangle$ between physical states $[\nu_f,\phi_f]$ and $[\nu_i,\phi_i]\rangle$ [7]. Thus, one of the implicit assumptions on the relation between the sum over histories in the timeless and deparameterized frameworks is borne out in detail in these models. Secondly, the analysis has also provided a perturbative expansion for the extraction amplitude a la group field theory [22] and shown in detail that an early suggestion of Oriti that the coupling constant in the group field theory may be related to the cosmological constant is realized in a precise fashion [7]. Furthermore, it is possible to spell out the precise meaning of the approximation in which one truncates the sum just to a finite number of terms.

Finally, the analysis also suggests directions for future work in spin foams. First, in the cosmological context, the correct extraction amplitude of the Hamiltonian theory is recovered in the spin foam type expansion *only if* one restricts the sum to paths which satisfy an appropriate positive frequency requirement. When this is done, in the *classical limit* the amplitude is given by e^{iS} rather than $\cos S$. This suggests that it may be necessary to restrict the sum in the general spin foam program to quantum histories that are 'time oriented' in a suitable sense. For further details, see [20, 7, 23].

3. Probability of inflation

3.1. The question

The inflationary paradigm has been extremely successful in accounting for the observed inhomogeneities in the CMB which serve as seeds for the subsequent formation of the large scale structure in the universe. Consequently, it is widely regarded as the leading candidate to describe the very early universe. However, it faces two conceptual issues. First, Borde, Guth and Vilenkin [24] have shown that the inflationary space-times are necessarily past incomplete; even with eternal inflation one cannot avoid the initial big-bang. In LQC, on the other hand, thanks to the quantum geometry effects, the singularity is resolved and inflationary space-times are past complete.

The second issue is that of 'naturalness.' While a given theory —such as general relativity— may admit solutions with an appropriate inflationary phase, is the occurrence of such a phase generic, or, does it require a careful fine tuning? For definiteness, let us suppose we have an inflaton ϕ with a quadratic potential, $V(\phi) = (1/2)m^2\phi^2$. Then the WMAP data [25] provides us with a remarkably detailed picture of the conditions at the onset of inflation [26]. More precisely, the data refers to the time $t(k_\star)$ at which a reference mode $k_\star$ used by WMAP exited the Hubble horizon in the early universe. (Today, this mode has a wave-length about 12% of the radius of the Hubble horizon.) Within error bars of less than 4.5%, the data tells us that the field configurations were:

$$\phi(t(k_\star)) = \pm 3.15\, m_{\rm Pl}, \qquad \dot\phi(t(k_\star)) = \mp 1.98\times 10^{-7}\, m_{\rm Pl}^2, \qquad H(t(k_\star)) = 7.83\times 10^{-6}\, m_{\rm Pl}\,, \quad (8)$$

where, the dot denotes derivative with respect to the proper (or cosmic) time and $H = \dot a/a$ is the Hubble parameter [26]. Thus, for the quadratic potential, the 7 year WMAP data requires the dynamical trajectory of the universe to have entered a *tiny* neighborhood of the phase-space point given by (8). One's first reaction would be that this condition would be met only by a *very* small fraction of all dynamical trajectories. If so, a priori the required inflationary phase would seem very implausible and the theory would be left with a heavy burden of 'explaining' why it actually occurred.

3.2. The Liouville measure

Issue of 'naturalness' has, of necessity, a subjective element and one could just say that our universe simply happened to pass through this tiny region of phase space and, since we have only one universe, the issue of likelihood is irrelevant. However, the broader community did not adopt this viewpoint. Rather, it has sought to sharpen the question of naturalness by introducing a measure on the space $\mathbb{S}$ of solutions of the given theory: the required probability is then be given by the *fractional* volume occupied by those solutions which do pass through configurations specified by the WMAP data at *some* time during their evolution.

To find the measure, the following general strategy was introduced over twenty years ago [27, 28, 29]. Recall that solutions to the field equations are in 1-1 correspondence with phase space trajectories, and the natural Liouville measure $d\mu_L$ on the phase space is preserved by the dynamical evolution. On the one hand, this measure is natural because it is constructed using *just* the phase space structure. On the other hand, precisely for the same reason, it does not encode additional information that may be important for the physics of the specific system. Therefore, it only provides *a priori* probabilities, i.e. 'bare' estimates. Further physical input can and should be used to provide sharper probability estimates and a more reliable likelihood. However, a priori probabilities themselves can be directly useful if they are very low or very high. In these cases, it would be an especially heavy burden on the fundamental theory to come up with the physical input that significantly alters them.

The question of probability of inflation along these lines has received a considerable attention in the literature (see for eg. [32, 33, 30, 31, 34, 36, 26, 35]). In the general relativity literature, conclusions have been vastly different: For the slow roll associated with configurations (8)), they range from the probability being close to 1 [30] to its being exponentially suppressed [31] (by factors of $\sim e^{-195}$. It turns out that these vast differences arise because procedure is intrinsically ambiguous because of the big bang singularity. As I will now sketch the ambiguity can be naturally resolved in LQC by working with the initial data at the big bounce [26, 35].

3.3. General relativity versus effective LQC

The subtleties involved can be summarized as follows. It turns out that the Liouville measure of the full phase space is infinite. Recall that, to calculate a priori probabilities one has to compute the *relative* volumes occupied by regions of interest in the phase space. If one were to proceed directly, this amounts to comparing one infinity with another, whence the answer is ambiguous. However, this infinity is physically spurious because it is associated with the *gauge freedom* $a \to \lambda a$ in rescaling the scale factor by a positive number λ. Thus, given a solution $(a(t), \phi(t))$ to the field equations, the solution $(\lambda a(t), \phi(t))$ is mathematically distinct but cannot be physically distinguished from the original one. Under this rescaling, both the symplectic structure and hence the Liouville volume is rescaled (but so is the Hamiltonian with the consequence that the Hamiltonian vector field is left invariant). Therefore, the Liouville measure does not project down to the quotient space. To remove this spurious infinity, then, one has to gauge fix. In the space-time picture, this can be interpreted as working with the initial data $(\phi, \dot\phi)$ at a fixed time instant (fixing a, say $a = 1$, there). In general relativity, the only natural 'instant' is the big bang where everything diverges. Thus, there is no natural gauge fixing. In LQC, by contrast, one can fix gauge by working at the big bounce where all fields are regular. Then it is possible to ask and answer a well-defined and physically interesting question: What is the fractional volume of the initial data *at the big bounce* which, when evolved, yields a dynamical trajectory that will eventually passe through the tiny region of the phase space selected by the WMAP data? To answer it, one needed not only analytical calculations that address the conceptual subtleties correctly, but also non-trivial numerical simulations, carried out by Sloan [26]. They showed that the probability is *very* close to one; greater than $1 - 3 \times 10^{-6}$. In this precise sense, *provided there is a scalar field with a suitable potential,* the desired inflationary phase occurs almost always in LQC. The analysis also provides detailed information on dynamics between the bounce and the onset of the slow roll compatible with WMAP which is now being used to investigate the seeds of structure formation within LQC.

Let us summarize. The recent investigations [26, 35] asked a sharp question: given a quadratic potential, what is the a priori probability of obtaining slow roll inflation that is compatible with the 7 year WMAP data? This is a much more refined than asking just for a the probability of obtaining a certain number of slow roll e-foldings [30, 31]. To address it unambiguously, one has to take into account certain subtleties associated with the Liouville measure: The procedure requires one to fix a time instant, consider all permissible initial data at that time, and ask for the fractional $d\tilde{\mu}_L$-volume of the subset of these configurations which, upon evolution, enter the small region of phase space singled out by WMAP. In general relativity, the only natural choice of the time instant is the big bang where fields of interest diverge. In LQC, there is a natural choice of the initial time —the bounce time. During the passage from the big bounce to the onset of the WMAP slow roll, the density decreases by some 12 orders of magnitude. The detailed analysis showed that, in this passage, the trajectories starting out at essentially any initial data at the bounce get funneled into the tiny region of the phase space allowed by the WMAP data. This impressive focussing of the effective LQC trajectories is much more than the qualitative statement that inflationary trajectories are attractors in general relativity.

For further details, see [1, 26, 35].

Acknowledgments

It is a pleasure to thank the Madrid gravity group at CSIC for organizing a stimulating conference. Results reported in this article are largely based on the joint work with Miguel Campiglia, Adam Henderson and David Sloan and the summary presented here closely follows the much more exhaustive recent review [1]. This work was supported in part by the NSF grants PHY0854743 and PHY1068743 and the Eberly research funds of Penn State.

[1] A. Ashtekar and P. Singh, Loop quantum cosmology: A status report, Class. Quant. Grav. **28** 213001 (2011)
[2] S. W. Hawking, The path integral approach to quantum gravity, in *General Relativity, An Einstein Centenary Survey,* (Cambridge University Press, London (1979))
[3] A. Ashtekar, M. Campiglia and A. Henderson, Path Integrals and the WKB approximation in loop quantum cosmology, Phys. Rev. D**82** 124043 (2010)
[4] H. Huang, Y. Ma and L. Qin, Path integral and effective Hamiltonian in loop quantum cosmology, `arXiv:1102.4755`
[5] R. P. Feynman, Space-time approach to non-relativistic quantum mechanics, Rev. Mod. Phys. **20** 367-387 (1948)
[6] A. Ashtekar, M. Campiglia and A. Henderson, Loop quantum cosmology and spin foams, Phys. Lett. **B681** 347-352 (2009)
[7] A. Ashtekar, M. Campiglia and A. Henderson, Casting loop quantum cosmology in the spin foam paradigm, Class. Quant. Grav. **27** 135020 (2010)
[8] A. Henderson, *Loop Quantum Cosmology: A Window into the Path Integral Representation of Quantum Gravity* Ph.D. Dissertation, The Pennsylvania State University (2011)
[9] A. Henderson, C. Rovelli, F. Vidotto and E. Wilson-Ewing, Local spinfoam expansion in loop quantum cosmology, Class. Quant. Grav. **28** 025003 (2011)
[10] A. Ashtekar, T. Pawlowski and P. Singh, Quantum nature of the big bang, Phys. Rev. Lett. **96** 141301 (2006)
[11] A. Ashtekar, T. Pawlowski and P. Singh, Quantum nature of the big bang: An analytical and numerical investigation I, Phys. Rev. **D73** 124038 (2006)
[12] A. Ashtekar, T. Pawlowski and P. Singh, Quantum nature of the big bang: Improved dynamics, Phys. Rev. D**74** 084003 (2006)
[13] C. Rovelli, A new look at loop quantum gravity, `arXiv:1004.1780`
[14] D. Marolf, Refined algebraic quantization: Systems with a single constraint. `arXive: gr-qc/9508015`; Quantum observables and recollapsing dynamics. Class. Quant. Grav. **12** 1199–1220 (1994)
[15] A. Ashtekar, J. Lewandowski, D. Marolf, J. Mourão and T. Thiemann, Quantization of diffeomorphism invariant theories of connections with local degrees of freedom. J. Math. Phys. **36** 6456–6493 (1995)
[16] A. Ashtekar, L. Bombelli and A. Corichi, Semiclassical states for constrained systems, Phys. Rev. **D72** 025008 (2005)
[17] B. S. DeWitt, Dynamical theory in curved space-times. I. A review of the classical and quantum action principles, Rev. Mod. Phys. **29** 377-397 (1957)
[18] C. Rovelli,*Quantum Gravity* (Cambridge University Press, Cambridge (2004))

[19] A. Ashtekar, Introduction to loop quantum gravity, in: *The Proceedings of the Third School on Quantum Geometry and Quantum Gravity* Pos/QGQG2011-001, edited by F. Hellmann et al (at press); also available at arXiv: 1201.4598
[20] D. Oriti, The Feynman propagator for spin foam quantum gravity, Phys. Rev. Lett. **94** 111301 (2005)
[21] M. Campiglia, A. Henderson and W. Nelson, Vertex Expansion for the Bianchi I model, Phys. Rev. **D82** 064036 (2010)
[22] D. Oriti, The group field theory approach to quantum gravity, arXiv:gr-qc/0607032
[23] G. Calcagni, S. Gielen and D. Oriti, Two-point functions in (loop) quantum cosmology, Class. Quantum Grav. **28** 125014 (2011)
[24] A. Borde, A. Guth and A. Vilenkin, Inflationary space-times are not past-complete, Phys. Rev. Lett. **90** 151301 (2003)
[25] E. Komatsu et al, Seven-Year Wilkinson micowave anisotropy probe (WMAP) observations: Cosmological interpreation, arXiv:1001.4538
[26] A. Ashtekar and D. Sloan, Probability of inflation in loop quantum cosmology, Gen. Rel. Grav. **43** 3619-3656 (2011)
[27] G. W. Gibbons, S.W. Hawking and J. Stewart, A Natural Measure on the set Of all universes, Nucl. Phys. **B281** 736 (1987)
[28] D.N. Page, Probability of R^2 inflation, Phys. Rev. D. **36** 1607 (1987)
[29] S. W. Hawking and D. N. Page, How probable is inflation?, Nucl. Phys. **B298** 789 (1988)
[30] L. Kofman, A. Linde, V. Mukhanov, Inflationary theory and alternative cosmology, JHEP **0210** 057 (2002)
[31] G.W. Gibbons, N. Turok, The measure problem in cosmology, Phys. Rev. D **77** 063516 (2008)
[32] S. W. Hawking, T. Hertog, Why does inflation start at the top of the hill?, Phys. Rev. D **66** (2002) 123509.
[33] S. Hollands, R. M. Wald, An alternative to inflation, Gen. Rel. Grav. **34** 2043 (2002)
[34] C. Germani, W. Nelson, M. Sakellariadou, On the onset of inflation in loop quantum cosmology, Phys. Rev. D **76** 043529 (2007)
[35] A. Corichi and A. Karami, On the measure problem in slow roll inflation and loop quantum cosmology, Phys. Rev. D**83** 104006 (2011)
[36] A. Ashtekar and D. Sloan, Loop quantum cosmology and slow roll inflation, Phys. Lett. B**694** 108-112 (2010)

Loops 11: Non-Perturbative / Background Independent Quantum Gravity IOP Publishing
Journal of Physics: Conference Series **360** (2012) 012002 doi:10.1088/1742-6596/360/1/012002

Ten questions on Group Field Theory (and their tentative answers)

Aristide Baratin, Daniele Oriti

Max Planck Institute for Gravitational Physics (Albert Einstein Institute), Am Mühlenberg 1, D-14476 Golm, Germany, EU

E-mail: aristide.baratin@aei.mpg.de, daniele.oriti@aei.mpg.de

Abstract. We provide a short and non-technical summary of our current knowledge and some possible perspectives on the group field theory formalism for quantum gravity, in the form of a (partial) FAQ (with answers). Some of the questions and answers relate to aspects of the formalism that concern loop quantum gravity. This summary also aims at giving a brief, rough guide to the recent literature on group field theory (and tensor models).

1. Introduction

The aim of this paper is to provide a short summary of what we know about the group field theory (GFT) formalism for quantum gravity and its longer term goals. For greater agility of presentation and reading, the format is that of a FAQ: we outline tentative and provisional answers to some basic questions about the GFT formalism, which aim at reflecting the current understanding of the subject, from the strict point of view of the authors.

Having clarified our aims, let us clarify what this paper is *not* meant to be. First, it is not meant to be a technical paper dwelling into any of the details of the formalism, or explaining in detail any of the recent results. For this we will refer to the relevant literature. Second, it is not a pedagogical, extensive introduction to GFTs. For early introductions, the reader should go to [2, 3], and for the general idea of tensor models one should look at the early papers [4] and at the literature on matrix models [5]. For a more extensive, and almost up to date introduction to GFTs for quantum gravity, we refer to [6]. Third, it is not a proper review of the subject nor of its recent developments, which are many and important. For many of the recent results on (colored) tensor models, a complete and up-to-date (and beautifully written) review is [7]. For many of the recent results that relate to the geometry of group field theories, we refer to the review [8]. For a discussion of spin foam models, in particular the recent ones, and thus some important aspects of the corresponding group field theories, we refer to the many reviews on spin foams [9]. Many other results, concerning for example GFT and tensor model renormalization can unfortunately only be found in the original papers.

2. Ten questions, with their tentative answers

(i) What is a group field theory?

Group field theories are a specific class of tensor quantum field theories which generalize matrix models to higher dimensions. The fundamental variables are in fact higher rank *tensors* formally denoted by $T_{i_1 \dots i_d}$ where the indices label points in the direct product of d domain spaces. The domain space could be a manifold or a finite set. The dynamics of tensor models is governed by a classical action characterized by a peculiar pairing of the indices in the interaction term:

$$S_{int}(T) = \lambda \sum_{\vec{i^j}} T_{\vec{i^1}} \cdots T_{\vec{i^{d+1}}}$$

where $\vec{i^j} = (i^{jj-1}, \dots, i^{j0}, i^{jd}, \dots, i^{jj+1})$ is the set of indices of the j copy of the tensor in the monomial of degree $d+1$, with $i^{jk} = i^{kj}$. For example the action for the independent identically distributed 3-tensor model is

$$S(T) = \frac{1}{2} \sum_{i,j,k} T_{ijk} T_{kji} - \frac{\lambda}{4!} \sum_{ijklmn} T_{ijk} T_{klm} T_{mjn} T_{nli} \tag{1}$$

A tensor can be graphically represented as a d-valent node with an index on each line, or dually as a (d-1)-simplex with an index on each (d-2)-face. The interaction term pairs the indices along the links of a network with d+1 d-valent nodes; dually it patterns the glueing of d+1 (d-1)-simplices along common faces to form a d-simplex. The kinetric term describes instead the identification of two nodes or (d-1)-simplices. Different choices of kinetic and interaction terms, are possible and define different models in the same class. The peculiar interaction combinatorics of any tensor model, shows up in the perturbative expansion of the theory in Feynman diagrams

$$Z = \int \mathcal{D}T \, e^{-S[T]} = \sum_{\Gamma} \frac{\lambda^V}{sym[\Gamma]} Z(\Gamma),$$

where V is the number of interaction vertices in the Feynman graph Γ, $sym[\Gamma]$ is a symmetry factor for the graph and $Z(\Gamma)$ the corresponding Feynman amplitude. By construction the Feynman diagrams will be represented as stranded diagrams dual to cellular (simplicial) complexes of arbitrary topology, obtained by arbitrary glueings of d-simplices along (d-1)-faces.

Tensors models are thus a straightforward combinatorial generalization of matrix models for 2d quantum gravity. In matrix models, the Feynman amplitudes associated to each diagram can be re-interpreted in terms of a discretization of 2d general relativity on an equilateral triangulation, and the overall sum as a generalization of a lattice path integral for 2d gravity to arbitrary triangulations and arbitrary topologies. Among the key results in matrix models [5] are the large N topological expansion dominated by trivial topologies (where N is the dimension of the matrix) and the identification of critical behaviour for some value of the coupling constant. This critical behaviour defines the continuum limit, shown by various methods to match the effective continuum dynamics to 2d Liouville quantum gravity. Tensors models try to export these ideas and results to higher dimensions. However, while even simple tensor models of the type (1) have an interest for quantum gravity, as they relate directly to the approach of dynamical triangulations, both the richness of spacetime geometry in higher dimensions (especially in 4 dimensions) and the results obtained by other approaches to quantum gravity suggest that a richer set of data and symmetries should be added to the tensors, and that a more involved dynamics should be chosen, to have a better chance to describe quantum geometry and gravity.

Group field theories (GFT) is the name used, in fact, for those tensor models where the domain space is chosen to be the local gauge group G of gravity (i.e. the Lorentz group or its euclidean counterpart), so that tensors thus turn into fields $\varphi \in L^2(G^d)$. An additional symmetry is invoked which captures local gauge invariance:

$$\varphi(hg_1, \cdots hg_d) = \varphi(g_1 \cdots g_d) \quad \forall h \in G,$$

and, possibly, kinetic and interaction terms are characterized by non-trivial kernels. In turn, alternative representations in terms of group representations or as non-commutative quantum field theories over $\mathbb{R}^{|G|}$, where $|G|$ is the dimension of G, become available, thanks to the group structure and tools from representation theory and non-commutative geometry. An example is the Boulatov model for 3d Riemanian quantum gravity [10], where $G = \mathrm{SU}(2)$ and with action:

$$S_{3d}[\varphi] = \frac{1}{2} \int [\mathrm{d}g]^3 \, \varphi_{123}\varphi_{321} - \frac{\lambda}{4!} \int [\mathrm{d}g]^6 \, \varphi_{123}\varphi_{345}\varphi_{526}\varphi_{641} \qquad (2)$$

where φ_{ijk} is a short notation for $\varphi(g_i, g_j, g_k)$ and $\mathrm{d}g$ is the Haar measure on the gauge group.

The GFT formalism thus attempts to *define* a theory of quantum spacetime as a superposition of discrete spaces, each generated as a possible interaction process of fundamental building blocks, tentative quanta of space, by incorporating into the tensor model framework lessons and insights of other approaches to quantum geometry – quantum Regge calculus, quantum simplicial geometry and Loop Quantum Gravity (LQG) [11], as well as mathematical tools taken from non-commutative geometry:

- The insight from quantum simplicial geometry suggests the variables needed to describe a 'quantum geometric (d-1)-simplex' which has to become the fundamental atom of quantum space, the basic 'quantum' in our enriched tensor model.
- A characterization of the geometry and of the whole kinematical phase space to be associated to a simplex is also provided by the analysis of discretized gravity actions in various dimensions and in various formulations. In particular, 1st order formulations of gravity in terms of connection and (conjugate) vielbein variables suggest to use group and Lie algebra manifolds to define the phase space of such simplices, and the domain spaces of tensor models, leading then to group field theories.
- Loop Quantum Gravity provides further justification for the kinematical structures used for defining the Hilbert space of quantum states of these models, several insights on their interpretation, and a main source of inspiration for constructing models and specifying the correct dynamics.
- Simplicial gravity path integrals, and in particular quantum Regge calculus, provide a template to analyze the corresponding Feynman amplitudes, and to evaluate to what extent they capture correctly the dynamics of discrete geometry that can be associated to each Feynman diagram/simplicial complex.
- Tools from non-commutative geometry then provide the means to define different representations of the same models, and to analyze their properties.

(ii) **How does the GFT formalism relate to loop quantum gravity?**

Polynomial gauge invariant observables of the group field theory correspond to functions of a finite number of gauge group elements associated to the links of a (d-valent) graph (plus possible additional variables) satisfying a gauge invariance condition at each node. A basis of such observables is written in terms of so-called spin networks functionals, generically labelled by a graph $(\Gamma, j_e, \imath_v)$ with edges e and vertices v respectively group representations j_e and invariant tensors $\imath_v$. GFT n-point functions can thus be expressed in terms of LQG-like states. The exact matching holds for some models, at least in representation space (and as far as the spectrum of some quantum geometric operators is concerned).

Intuitively, GFTs provide a 'second quantized' framework for LQG: spin network vertices become the quanta of a GFT field, and the LQG wave function associated to an individual vertex is turned to a field operator that creates/annihilates spin network vertices from/to a 'no-space' vacuum, where no geometrical nor topological structure is present. It is difficult, however, to go beyond this intuitive picture and construct a proper Fock space of spin network vertices: to do so, one has to face the issue of choosing a statistics for them, which is both a very difficult and very exotic question that has never been considered in the LQG literature.

As far as the quantum dynamics of LQG is concerned, the main link lies in the structure of the GFT Feynman amplitudes $I_{\mathcal{G}}$. These can indeed be written in terms of *spin foam models* [9], which show up in LQG as a natural arena to encode the Hamiltonian evolution of spin-networks as a sum over histories. A 'spin foam' $(\mathcal{J}, j_f, \imath_e)$ is a higher dimensional analogue of a spin-network (a 'history' of spin-network evolution): it is a system of branching surfaces $\mathcal{J}$ taking the form of a 2-complex, with polygonal faces labeled by group representations j_f and edges labeled by intertwiners $\imath_e$. To each GFT Feynman diagram $\mathcal{G}$ corresponds a 2-complex $\mathcal{J}$: vertices and edges of $\mathcal{G}$ corresponds to vertices and edges of $\mathcal{J}$ and the loops of strands on $\mathcal{G}$ corresponds to the 2d faces of $\mathcal{J}$. One can show that the GFT amplitudes (in a specific representation) takes the generic form of 'spin foam amplitudes':

$$I_{\mathcal{G}} = \sum_{j_f, \imath_e} \prod_f A_f(j_f) \prod_e A_e(j_f, \imath_e) \prod_v A_v(j_f, \imath_e) \tag{3}$$

characterized by a choice of local amplitudes A_f, A_e and A_v assigned to the faces, edges and vertices of the 2-complex $\mathcal{J}$ defined by $\mathcal{G}$. For example, the Feynman amplitudes of the Boulatov model (2) give the Ponzano-Regge model for 3d quantum gravity. This duality between GFT and spin foam amplitudes is general: the amplitudes of any spin foam model can be obtained as the Feynman amplitude of a GFT [12].

In a covariant, sum-over-histories formulation of loop quantum gravity, both the combinatorial and algebraic data that specify a quantum state of geometry are randomized with a quantum probability amplitude in order to define the full quantum theory. GFTs thus provide a definition of this quantum dynamics, in the form of a quantum field theory expansion in possible interaction processes for the same building blocks of quantum geometry, which includes a sum over all topologies.

The main object of interest for the LQG dynamics is the physical inner product between states, projecting onto the kernel of the Hamiltonian constraint. Indications of where this object should be found in the GFT formalism stem from the interpretation of GFT as a second quantized formalism for spin network vertices. First, one would expect that some form of Hamiltonian constraint acting on spin network vertices can already be identified at the level of the GFT classical action and equations of motion. Second, as we mentioned, the dynamics encoded into a single evolution history of a spin network state can be found in all its details in the form of a GFT Feynman amplitude. Obviously, from the field theory perspective, this is but a tiny corner of the true quantum dynamics, which has to be looked for in the Schwinger-Dyson equations (SDE) for the n-point functions of the GFT model [13, 2]. In other words, one would expect that the SDE for polynomial field correlations, that is spin network functionals, would admit (at least in a regime suppressing topology change) an interpretation and a transcription as Hamiltonian and diffeomorphism constraints of a quantum gravity theory, and thus provide the definition of the physical inner product. This is indeed what happens, in the appropriate limit, in the simple case of matrix models. In particular, this should be true in the continuum limit of the GFT model, which is necessarily a non-perturbative domain. The same continuum dynamics can be also looked for at the level of effective actions in this continuum limit, at the level of symmetries satisfied by the model and, if a phase transition is part of the picture, in the nature of the critical exponents for the scaling of interesting observables.

(iii) **Why is the GFT formalism useful to LQG?**

This way of defining the quantum dynamics is advantageous, from a purely LQG perspective, for a variety of reasons. The main one is that a field theory framework is the most convenient one for dealing with the infinite number of degrees of freedom which should ultimately be present in the full continuum theory: a generic continuum geometry will in fact be captured by a highly complicated (superposition of) spin network state(s), that is, in GFT language, a hugely populated many-particle state. Furthermore, contrary to direct attempts at the definition of the continuum gravitational dynamics from the canonical quantum gravity perspective, the GFT formalism offers a compact, but in principle complete definition of the theory: it is given by the path integral of the GFT model. The difficulty is of course to go beyond the formal definition, prove its well-posedness, and extract interesting physics from it.

The GFT definition also opens the door to the application of more or less standard quantum and statistical field theory tools (and ideas) to the study of the same dynamics, a very powerful machinery indeed. Last, it permits to LQG structures and results to force their way out of the "canonical quantization of continuum gravity cage ", which is not only technically extremely challenging, but also suspicious for a variety of reasons. Several arguments indeed (very different in nature, each not conclusive, and still each very reasonable) can be put forward suggesting that general relativity (GR) is but an effective field theory, and that it should not be quantized as such, but only emerge in some corner of a fundamental theory defined in altogether different terms. GFTs allow in principle to test and, if realized, to describe in detail this scenario.

(iv) **How to define the right quantum dynamics of geometry as a GFT?**

Given the ingredients entering the very definition of a GFT model, the most natural way of proceeding to its construction relies mainly on quantum simplicial geometry. It amounts to answering three questions:

a) what is a quantum geometric (d-1)-simplex?
b) what is a quantum geometric d-simplex, or, equivalently, how to glue d+1 (d-1)-simplices to form a d-simplex?
c) how does the geometric information of one d-simplex propagate to a neighboring one?

The answer to a) dictates the nature and properties of the GFT field and its variables. The answer to b) indicates what interaction kernel one should use to define the classical action of the corresponding GFT. The answer to c) gives the form of the kinetic term of the action, that is, the propagator of the field theory. Having all the ingredients at hand, one can write down the GFT action, and thus the full model, in perturbative expansion.

A first source of inspiration for answering these questions comes from ideas of geometric quantization, also at the roots of spin foam models [14]. Let us see this in the case of the Boulatov model (2): upon Peter-Weyl expansion into irreducible SU(2) representations labelled by $j_i \in \frac{1}{2}\mathbb{N}$, the gauge invariant field is expressed in terms of three valent-invariant tensors

$$\iota^{j_1,j_2,j_3} \in \mathrm{Inv}\left[\mathcal{H}_{j_1} \otimes \mathcal{H}_{j_2} \otimes \mathcal{H}_{j_3}\right]$$

In this picture, the Boulatov field represents a quantum triangle with fixed lengths: each $\mathcal{H}_{j_i}$ represents the Hilbert space of states of a quantized vector of length j_i, upon geometric quantization that promotes the classical edge vector $\vec{x}_i$ into the SU(2) generators $\widehat{X}_i$; the classical closure condition $\vec{x}_1 + \vec{x}_2 + \vec{x}_3 = 0$ for a triangle then projects into the subspace of invariant tensors. Geometric quantization here also gives us the glueing rules for triangles as traces of tensors. In such a representation of group field theory, the Feynman amplitudes take the form of spin foam models. The challenge has been to export these ideas to four dimensions for the construction of realistic quantum gravity models.

An elegant, complementary approach [15], which has the advantage of making the geometry content of GFT more manifest, has recently been proved fruitful [16]. In this approach, the field variables are data encoding classical simplicial geometry (edge vectors in 3d, area bivectors in 4d...), and their 'quantization' results in a non-commutative star-product on the space of fields. This non-commutative structure rests on the phase space structure of the discrete classical theories these GFT models are set to quantize, usually given by the cotangent bundle of a group manifold, with configuration space being a Lie group and the dual space being the corresponding Lie algebra; it thus encodes the classical non-commutativity of the 'momentum' space of the classical theory. In the case of the model (2), upon so-called group Fourier transform [17]

$$\widehat{\varphi}(x_1, \cdots x_3) := \int [\mathrm{d}g_i]^3 \, \varphi(g_1, \cdots g_3) \, \mathrm{e}^{i \,\mathrm{Tr}\, x_1 g_1} \cdots \mathrm{e}^{i \,\mathrm{Tr}\, x_3 g_3} \tag{4}$$

the model is expressed in terms of fields on copies of $\mathfrak{su}(2) \sim \mathbb{R}^3$ endowed with a non-commutative star-product (dual to group convolution). Gauge invariance translates into a closure condition $x_1 + x_2 + x_3 = 0$ for the variables, thus naturally interpreted as the edge vectors of a (non-commutative) triangle. Glueing rules for the triangles are then simply dictated by the star product, in a way that identifies the edge vectors of the common edges by means of non-commutative $\star$-delta functions. This result in the following expression for the action (2):

$$S[\widehat{\varphi}] = \frac{1}{2} \int [d^3 x_i]^3 \, \widehat{\varphi}_{123} \star \widehat{\varphi}_{321} - \frac{\lambda}{4!} \int [d^3 x_i]^6 \, \widehat{\varphi}_{123} \star \widehat{\varphi}_{345} \star \widehat{\varphi}_{526} \star \widehat{\varphi}_{641} \tag{5}$$

where the star-product relates pairwise the variables x_i with the same index i. In such a representation of group field theory, the Feynman amplitudes take the form of simplicial gravity path integrals. Remarkably, this approach easily extends to 4d gravity models.

(v) **What are the symmetries of a GFT model and what are their consequences?**

We start entering a less explored territory, despite the great importance of the issue. Symmetries are in fact crucial, in a field theory context, to constrain model building and understanding the true nature of the models, to define appropriate observables, to identify universality classes of critical behaviour, to distinguish different phases of the system, and so on.

Up to now, only a few symmetries of GFT models for topological and, even less, gravity 4d models are understood and under some control [10, 18, 19, 20, 21]. The GFT implementation of the symmetry which becomes, at the level of the GFT Feynman amplitudes, the local gauge invariance of simplicial path integrals is well-understood. Recently, also the GFT symmetry corresponding to 3d simplicial diffeos (after gauge invariance is imposed) and to translation invariance in BF models has been uncovered [20]. Interestingly, all these transformations, which are local symmetries in the simplicial path integral, are global transformations of the GFT field, from a field-theoretic point of view. Moreover, diffeos/translations are implemented as *quantum group* symmetries. These are clearly at least partially broken in 4d gravity models, but the details of this breaking have not been elucidated yet.

Combinatorial symmetries with respect to permutations of the arguments of the GFT field have also been considered, but their implications beyond those on the structure of the Feynman diagrams have not been explored. A more careful study of GFT symmetries and of their consequences in GFT models, the 4d gravity ones in particular, is certainly needed. One class of transformations that would be interesting to identify is the GFT and tensor models analogue of the unitary matrix transformations that allow the re-writing of matrix models in terms of matrix eigenvalues (the true degrees of freedom). This re-writing is crucial for many of the key results in matrix models and one could expect a similar role in tensor models and GFTs.

(vi) **Can one control the sum over diagrams and topologies?**

The perturbative sum is in fact pretty wild, including both simplicial manifolds of arbitrary topology and pseudo-manifolds [22]. A number of works have been devoted to a better understanding of the properties of the generated complexes. An important development was the definition of a GFT analogue of the large-N expansion of matrix models, dominated by a particular class of triangulations of the sphere, for simple tensor models and topological GFTs in any dimension [23]. The GFT analogue of the size N of the tensors is a cut-off in the momenta (group representations). This expansion is yet to be understood in its physical significance and in its subdominant terms, but it opens the way for several applications, both in quantum gravity and statistical physics. Mathematically, it is complemented by a variety of results on the scaling behaviour with the cut-off of the amplitudes of topological BF models (and some more limited results on 4d gravity models), including rigorous power counting results [24]. For some models, it was also possible to prove the Borel summability of the whole perturbative series [26].

(vii) **Is the GFT perturbative expansion physically useful?**

From the GFT point of view, the discrete data appearing in the GFT action and amplitudes are physical, in the sense that they correspond to true degrees of freedom of the theory, and not mere discretization artifacts. However, they are obviously very different from the variables we are used to adopt for describing continuum spacetime physics. From simplicial gravity, we know that these data correspond to (piecewise-flat) singular continuum geometries, which in turn can be used to reconstruct smooth geometries in the limit in which large numbers of the same data are considered (at least at the classical level). This points to the non-perturbative sector of the GFT formalism as the locus of continuum geometric physics (as indeed shown explicitly in the simple case of matrix models). However, being physical, even the data associated to finite (sums of) simplicial complexes (e.g. those obtained truncating the GFT perturbative sum to finite order) could in principle be used to describe physical phenomena, in the same way in which a finite number of particles could be used, in some limited circumstances, to mimic the behaviour of a fluid. What is hard, however, is to pinpoint what these circumstances are, exactly. In general, being simplicial gravity path integrals, one would say that the Feynman amplitudes of GFTs describe correctly gravitational phenomena to the extent in which piecewise-flat Regge geometries can be used to approximate them. This means we could take advantage of the expertise developed, in the use of such approximations, in numerical relativity. However, one could expect that the perturbative GFT expansion is more and more physically relevant, and convenient as a mathematical language, the closer we are, physically, to the spacetime configuration represented by the GFT perturbative vacuum, which is unfortunately a very degenerate configuration in which no geometry and no space is actually present.

(viii) **Why does one want to have a renormalizable GFT? is it even possible?**

Being a field theory, and aiming at a fundamental description of space and time down to the Planck scale, a GFT model should be renormalizable. Non-renormalizability would instead implies that the GFT model considered can at best be an effective description of a more fundamental theory. The renormalization group flow of a given GFT will tell us what type of interaction (thus, which dynamics) is relevant at various scales.

The question becomes then whether one can hope to obtain such a renormalizable GFT [27]. The main source of doubt probably comes from the non-renormalizability of gravity around Minkowski spacetime. However GFTs are background independent formulations of gravity (hopefully), not limited to fluctuations of the metric around a given continuum spacetime geometry, and their perturbative expansion is formulated around a very different vacuum, one that corresponds to no-space at all. Being formulated in terms of fundamental structures that are either non-geometric in nature or anyway only associated to discrete geometries, it is not obvious what effective continuum degrees of freedom it captures. While this is obviously a difficulty in relating GFTs to continuum physics, it also warns us against the direct application of conventional wisdom about continuum perturbative quantum gravity.

Studies of tensor models and GFT renormalization are just beginning [25], but already one example of a fully renormalizable non-trivial tensor model in 4 dimensions, though much simpler than full-blown GFTs for 4d gravity, has been recently exhibited [28].

(ix) **Where should one look for continuum gravity?**

The dynamics of smooth geometries should emerge as an effective dynamics for many degrees of freedom, i..e from large boundary graph-based states and the re-summation of infinite Feynman diagrams. In other words, continuum gravitational physics resides in the non-perturbative sector of the current definition of GFTs. This means looking for different vacua of the fundamental GFT and for the dynamics around them, possibly accompanied by 'large-scale' and semi-classical approximations. Notice, however, that not only the semi-classical limit is a distinct limit from the continuum one (the latter being naturally a limit of increasing number of degrees of freedom, regardless of their quantum or classical nature), but the appropriate continuum limit could even emerge because of the quantum properties of the fundamental degrees of freedom, which should not be washed out beforehand. The example that comes to mind is that of quantum liquids. Obviously, this is *terra incognita*, at present, so it is hard to do more than speculations and some educated guesses.

If more and more interacting degrees of freedom are to be taken into account on the way to the continuum limit, and if the theory has to be trusted over a wide range of scales (as a fundamental theory of quantum gravity should be), a phase transition should occur. Physically, the occurrence of such a phase transition switching from microscopic excitations to collective ones, would correspond to a 'geometrogenesis' [29], the true birth of continuum spacetime and geometry from a fundamental non-geometric phase. This scenario has been proposed repeatedly in the GFT context [30, 31, 27] and outside it [32]. We are gaining experience of phase transitions following from the resummation of the perturbative series in the large-N limit, in simple [33] and less simple [34] tensor models. But more work is needed to establish such scenario in more involved GFT models, and to understand its physical significance.

(x) **Can we do physics with GFTs?**

In light of the above scenario, the answer is unfortunately 'not yet', given the current status of our understanding of the subject. The task is to improve our control over the non-perturbative sector of GFT and our tools for studying phase transitions in GFTs. In the meantime, the exploration of approximation methods for GFT has started, to study [35, 36, 37, 38] the effective dynamics for fluctuations around non-trivial GFT field configurations by expanding the GFT action around them, interpreted as non-trivial quantum spacetime backgrounds. The results (including the derivation of effective geometrodynamic equations, effective Hamiltonian constraint operators, emergent non-commutative matter field theories) are very interesting, but there is clearly still a long way to go. In parallel, the construction of simplified GFT-like models has also started, in particular models with possible cosmological interpretation and application [39]. The idea is that such simplified models could either be derived from or at least guide the mean field theory treatment of fundamental GFT models. Again, this is just a beginning.

3. Conclusions

We have answered, to the best of our current understanding of the subject, some questions about the GFT approach to quantum gravity, its relation with LQG and other approaches, and its present status. In the process, we have mentioned many of the important results obtained recently in this context, and outlined some direction for future work. We hope to have well represented the fast growth of this area of research, and convinced the reader of its promise. At the same time, we hope it stands clear that much more work is needed to address many of these questions fully, and in particular those whose answer will really establish GFTs as a solid candidate formalism for quantum gravity.

References

[1] D. Oriti, ed., *Approaches to Quantum Gravity*, Cambridge University Press, Cambridge (2009)

[2] L. Freidel, Int.J.Phys. **44**, 1769-1783, (2005) arXiv: hep-th/0505016.

[3] D. Oriti, in [1], [arXiv:gr-qc/0607032]; D. Oriti, arXiv:0912.2441 [hep-th]; D. Oriti, in *Quantum Gravity*, B. Fauser, J. Tolksdorf, E. Zeidler, eds., Birkhaeuser, Basel, (2007), [arXiv: gr-qc/0512103]

[4] M. Gross, Nucl. Phys. Proc. Suppl. **25A**, 144-149, (1992); J. Ambjorn, B. Durhuus, T. Jonsson, Mod. Phys. Lett. **A6**, 1133-1146, (1991); N. Sasakura, Mod.Phys.Lett. A6 (1991) 2613-2624

[5] F. David, Nucl. Phys. B257, **45** (1985); P. Ginsparg, arXiv: hep-th/9112013; P. Di Francesco, P. Ginsparg, J. Zinn-Justin, Phys. Rept. 254 (1995) 1-133,hep-th/9306153; P. Ginsparg, G. Moore, hep-th/9304011

[6] D. Oriti, in G. Ellis, J. Murugan, A. Weltman (eds), *Foundations of space and time*, Cambridge University Press, Cambridge (2011), arXiv:1110.5606 [hep-th]

[7] R. Gurau, J. Ryan, SIGMA, arXiv:1109.4812 [hep-th]

[8] A. Baratin, D. Oriti, SIGMA, to appear

[9] A. Perez, Class. Quant. Grav. **20**, R43, (2003), [arXiv: gr-qc/0301113]; D. Oriti, gr-qc/0311066; J. C. Baez, Class. Quant. Grav. 15, 1827 (1998) [arXiv:gr-qc/9709052],

[10] D. V Boulatov, Mod. Phys. Lett. A**7**:1629-1646 (1992), arXiv:hep-th/9202074.

[11] C. Rovelli, *Quantum Gravity*, Cambridge University Press, Cambridge, (2006).

[12] M. P. Reisenberger and C. Rovelli, Class. Quant. Grav. 18, 121 (2001) [arXiv:gr-qc/0002095].

[13] H. Ooguri, Prog.Theor.Phys.89:1-22,1993, arXiv:hep-th/9210028.

[14] J. W. Barrett, L. Crane, J. Math. Phys. 39, 3296 (1998), arXiv:gr-qc/9709028.

[15] A. Baratin, D. Oriti, Phys. Rev. Lett. **105** (2010) 221302, arXiv:1002.4723 [hep-th].

[16] A. Baratin, B. Dittrich, D. Oriti, J. Tambornino, Class. Quant. Grav. 28 (2011) 175011, arXiv:1004.3450 [hep-th], A. Baratin, D. Oriti, to appear in New J. Phys., arXiv:1108.1178 [gr-qc]. A. Baratin, D. Oriti, arXiv:1111.5842[hep-th].

[17] L. Freidel, E. Livine, Class. Quant. Grav. **23**, 2021(2006), arXiv:hep-th/0502106 . S. Majid, *Foundations of Quantum Group Theory*, Cambridge University Press (1995).

[18] H. Ooguri, Mod. Phys. Lett. A **7**, 2799 (1992), hep-th/9205090.

[19] F. Girelli, E. Livine, Class. Quant. Grav. 27, 245018 (2010), arXiv:1001.2919.

[20] A. Baratin, F. Girelli, D. Oriti, Phys. Rev. D **83** 104051 (2011), arXiv:1101.0590 [hep-th].

[21] J. Ben Geloun, arXiv:1107.3122.

[22] R. De Pietri and C. Petronio, J. Math. Phys. 41, 6671 (2000) [arXiv:gr-qc/0004045]. R. Gurau, Class. Quant. Grav. 27, 235023 (2010), arXiv:1006.0714 [hep-th].

[23] R. Gurau, Annales Henri Poincare 12:829-847 (2011), arXiv:1011.2726 [gr-qc]. R. Gurau and V. Rivasseau, Europhys.Lett.95:50004,201, arXiv:1101.4182[gr-qc]

[24] L.Freidel, R. Gurau, D. Oriti, Phys.Rev. D80:044007 (2009), arXiv:0905.3772[hep-th]. J. Ben Geloun, T. Krajewski, J. Magnen, V. Rivasseau, arXiv:1002.3592[hep-th]. T. Krajewski, J. Magnen, V. Rivasseau, A. Tanasa, P. Vitale, Phys. Rev. D82:124069 (2010), arXiv:1007.3150 [gr-qc].

[25] J. Ben Geloun, V. Bonzom, Int.J.Theor.Phys.50:2819-2841 (2011), arXiv:1101.4294[hep-th].

[26] L. Freidel, D. Louapre, Phys. Rev. D68 (2003) 104004, arXiv:hep-th/0211026[hep-th].

[27] V. Rivasseau, PoS CNCFG2010 (2010) 004, arXiv:1103.1900 [gr-qc].

[28] J. Ben Geloun, V. Rivasseau arXiv:1111.4997 [hep-th].

[29] T. Konopka, F. Markopoulou, L. Smolin, hep-th/0611197

[30] D. Oriti, J.Phys.Conf.Ser. 67 (2007) 012052, hep-th/0612301

[31] D. Oriti, Proceedings of Science, [arXiv:0710.3276]; L. Sindoni, arXiv:1105.5687 [gr-qc]; D. Oriti, arXiv:1107.4534 [physics.pop-ph]

[32] N. Seiberg, hep-th/0601234

[33] V. Bonzom, R. Gurau, A. Riello, V. Rivasseau, Nucl.Phys. B853 (2011) 174-195, arXiv:1105.3122 [hep-th]

[34] D. Benedetti, R. Gurau, Nucl.Phys. B855 (2012) 420-437, arXiv:1108.5389 [hep-th]

[35] D. Oriti, L. Sindoni, New J.Phys. 13 (2011) 025006, arXiv:1010.5149 [gr-qc]

[36] W. Fairbairn and E. Livine, Class. Quant. Grav. **24**, 5277 (2007) [arXiv:gr-qc/0702125]

[37] F. Girelli, E. Livine and D. Oriti, Phys. Rev. D81, 024015 (2010), [arXiv: 0903.3475 [gr-qc]]

[38] E.Livine, D. Oriti, J. Ryan, to appear in Class.Quant.Grav., arXiv:1104.5509 [gr-qc]

[39] G. Calcagni, S. Gielen, D. Oriti, to appear

Journal of Physics: Conference Series **360** (2012) 012003 doi:10.1088/1742-6596/360/1/012003

Chiral vacuum fluctuations in quantum gravity

Laura Bethke and João Magueijo

Theoretical Physics, Blackett Laboratory, Imperial College, London, SW7 2BZ, United Kingdom

E-mail: j.magueijo@imperial.ac.uk

Abstract.
In this paper we investigate cosmological tensor modes in terms of the Ashtekar variables of loop quantum gravity, for complex values of the Immirzi parameter. While, on-shell, the classical Hamiltonian reduces to the usual expression found in cosmological perturbation theory, the quantum Hamiltonian displays significant differences. We can find a physical Fourier space Hamiltonian in terms of graviton creation and annihilation operators, after selecting out the non-physical modes through the inner product which itself is determined by the reality conditions. We are left with the usual graviton modes but with a chiral asymmetry in the the vacuum energy and fluctuations. The latter depends on γ (in particular it vanishes for purely real γ) and on the ordering of the 2-point function. Such an effect would leave a distinctive imprint in the polarisation of the cosmic microwave background, thus finally engaging quantum gravity in meaningful experimental test.

1. Introduction

Loop Quantum Gravity (see e.g. [1, 2, 3, 4]) is a promising framework for quantising gravity. It has a rigorous mathematical structure and a lot of progress has recently been made on specifying important field theoretical concepts such as the graviton propagator. However, as for any other theory of quantum gravity, currently missing is a distinct experimental test that could verify predictions made by LQG as well as distinguish it from other approaches. In a series of recent publications [5, 6, 7], we have provided such an experimental test by recasting the standard inflationary calculation for tensor perturbations in terms of Ashtekar variables. In these proceedings we will summarise this work.

In LQG, the Holst action [8] (which includes a topological term, dependent on the Immirzi parameter γ, that vanishes classically) instead of the standard Palatini-Kibble formalism is used and the connection is treated as the central gravitational variable. Using this approach and applying standard cosmological perturbation theory [9, 10], it turns out that the vacuum fluctuations of inflationary tensor modes, gravitational waves, exhibit a chirality, i.e. depend on the graviton polarisation. This is a complete novelty compared to the normal second order formalism [11]. Classically, the two formulations are equivalent, as they should be, with novelties only appearing in the quantum formulation. The observed chirality will be shown to depend on ordering prescriptions and in particular on the value of the Immirzi parameter γ (with the values $\gamma = \pm i$ which render the connection self-dual(SD)/anti-self-dual(ASD) being special cases) which is taken to be complex in general.

In Section 2 we set up the calculation by identifying the Ashtekar variables for tensor perturbations around an inflationary de Sitter background, deriving the perturbed Hamiltonian, and stating the classical solution. In Section 3 several issues that have to be taken into account to obtain the right theory are highlighted, and other novelties of the formalism are discussed. We will then set up the quantum theory in Section 4: we define reality conditions, commutation relations and a quantum Hamiltonian in terms of graviton creation and annihilation operators (of which only half are physical before reality conditions are imposed). Section 6 will show how this framework leads to a chirality in the gravitational vacuum fluctuations. We will summarise our results and outline plans for future work in the conclusion.

Throughout this paper we shall use units for which $\hbar = c = 1$ and we parameterise the strength of gravity with $l_P^2 = 8\pi G$. We'll be concerned with the real world, so the metric will invariably be Lorentzian; its signature is taken to be $-+++$.

2. Classical solution

In this section we provide all the ingredients for the calculation by defining the Ashtekar variables in cosmological perturbation theory, their Fourier expansions and the Hamiltonian formulation. We will show that the classical solution is the one we expect from the second order formalism. In Section 3, certain aspects of the approach used are explained in more detail and some steps that are needed to obtain the right results are highlighted.

2.1. *Canonical variables*

To describe inflationary tensor fluctuations, we consider perturbations around de Sitter space-time in the flat slicing:

$$ds^2 = a^2[-d\eta^2 + (\delta_{ab} + h_{ab})dx^a dx^b] \,, \tag{1}$$

where h_{ab} is a symmetric TT tensor, $a = -1/H\eta$, $H^2 = \Lambda/3$ and $\eta < 0$. Using the convention $\Gamma^i = -\frac{1}{2}\epsilon^{ijk}\Gamma^{jk}$ (where Γ^{ik} is the spin connection), the Ashtekar-Immirzi-Barbero connection is given by $A^i_a = \Gamma^i_a + \gamma\Gamma^{0i}_a$ (where $\gamma = \pm i$ corresponds to a SD/ASD connection, complex γ to the Immirzi connection and real γ to the Barbero connection). Its canonical conjugate is the densitised inverse triad, $E^a_i = \det\left(e^j_b\right) e^a_i$, where indices $i = 1,2,3$ are spatial Lorentz algebra indices and $a = 1,2,3$ are spatial indices for the base manifold. Making use of the torsion free condition

$$T^i = de^i + \Gamma^{ij} \wedge e^j = 0 \tag{2}$$

for the zeroth order solution, the canonical variables can be expressed as:

$$A^i_a = \gamma H a \delta^i_a + \frac{a^i_a}{a} \tag{3}$$

$$E^a_i = a^2\delta^a_i - a\delta e^a_i \,. \tag{4}$$

To make contact with cosmological perturbation theory and standard perturbative quantum field theory we need to expand the perturbed variables in terms of Fourier modes:

$$\begin{aligned}
\delta e_{ij} &= \int \frac{d^3k}{(2\pi)^{\frac{3}{2}}} \sum_r \epsilon^r_{ij}(\mathbf{k})\tilde{e}_{r+}(\mathbf{k},\eta)e^{i\mathbf{k}\cdot\mathbf{x}} + \epsilon^{r\star}_{ij}(\mathbf{k})\tilde{e}^\dagger_{r-}(\mathbf{k},\eta)e^{-i\mathbf{k}\cdot\mathbf{x}} \\
a_{ij} &= \int \frac{d^3k}{(2\pi)^{\frac{3}{2}}} \sum_r \epsilon^r_{ij}(\mathbf{k})\tilde{a}_{r+}(\mathbf{k},\eta)e^{i\mathbf{k}\cdot\mathbf{x}} + \epsilon^{r\star}_{ij}(\mathbf{k})\tilde{a}^\dagger_{r-}(\mathbf{k},\eta)e^{-i\mathbf{k}\cdot\mathbf{x}}
\end{aligned} \tag{5}$$

where $\tilde{e}_{rp}(\mathbf{k},\eta) = e_{rp}(\mathbf{k})\Psi_e(k,\eta)$ and $\tilde{a}_{rp}(\mathbf{k},\eta) = a_{rp}(\mathbf{k})\Psi^{rp}_a(k,\eta)$, and ϵ^r_{ij} are polarization tensors. Amplitudes $\tilde{a}_{rp}(\mathbf{k})$ and $\tilde{e}_{rp}(\mathbf{k})$ have two indices (contrasting with previous literature, e.g. [13, 14]): $r = \pm 1$ for right and left helicities, and p for graviton ($p = 1$) and anti-graviton ($p = -1$) modes. The a_{rp} and e_{rp} can be chosen so as *not* to carry any time-dependence, and for simplicity we will assume that they are equal. After imposing on-shell conditions we'll find that functions $\Psi_a(k,\eta)$ must then carry an r and p dependence.

2.2. *Hamiltonian and Hamilton's equations*

In the Ashtekar formalism, the gravitational Hamiltonian constraint is given by

$$\mathcal{H} = \frac{1}{2l_P^2}\int d^3x N E_i^a E_j^b\left[\epsilon_{ijk}(F_{ab}^k + H^2\epsilon_{abc}E_k^c) - 2(1+\gamma^2)K_{[a}^i K_{b]}^j\right] \tag{6}$$

where $K_a^i = \frac{A_a^i - \Gamma_a^i(E)}{\gamma}$ is the extrinsic curvature of the spatial surfaces. The canonical variables have symplectic structure $\{A_a^i(\mathbf{x}), E_j^b(\mathbf{y})\} = \gamma l_P^2 \delta_a^b \delta_j^i \delta(\mathbf{x}-\mathbf{y})$. We can derive Hamilton's equations and by expanding them to first order get equations of motion for the perturbations in metric and connection variables. Combining them, we obtain

$$\delta e_{ij}'' - \left(\partial^2 + \frac{2}{\eta^2}\right)\delta e_{ij} = 0\,. \tag{7}$$

This is exactly the same equation as obtained for the "v" variable used in cosmology to describe metric tensor perturbations. Classically, the Ashtekar approach therefore reduces to the second order formalism.

Our aim is to identify perturbative graviton states within a Fourier space Hamiltonian. Obviously gravitons should have dynamics, therefore we need to identify the part of the Hamiltonian that is not constrained to be zero. This is given by ${}^2_1\mathcal{H}$, the second order Hamiltonian in terms of products of first order variables. This is non-zero as only the full second order Hamiltonian, ${}^2\mathcal{H} = {}^2_1\mathcal{H} + {}^2_2\mathcal{H}$, must vanish on shell. The first term provides a candidate for the Hamiltonian to be identified with that of the second quantized QFT. The second contains the backreaction or compensation resulting from the non-linearity of the gravitational field, ensuring that the Hamiltonian constraint is satisfied. For general γ, the dynamical Hamiltonian to second order is given by (where we are being careful with the ordering with an eye on quantisation)

$$\begin{aligned}\mathcal{H}_{eff} = \frac{1}{2l_P^2}\int d^3x \quad &\left[\frac{1}{\gamma^2}a_{ij}a_{ij} - 2H^2a^2\delta e_{ij}\delta e_{ij} + \left(1-\frac{1}{\gamma^2}\right)\epsilon_{ikl}(\partial_k\delta e_{lj})a_{ij}\right.\\ &\left. -\left(1+\frac{1}{\gamma^2}\right)\epsilon_{ikl}a_{ij}(\partial_k\delta e_{lj}) + \left(1+\frac{1}{\gamma^2}\right)\epsilon_{ikl}\epsilon_{jmn}(\partial_k\delta e_{lj})(\partial_m\delta e_{ni})\right]\,.\end{aligned} \tag{8}$$

2.3. *Classical solution*

As opposed to the second order formalism, we treat connection and metric as independent variables initially. On-shell, i.e. when the equations of motion are satisfied, they are related by the torsion-free condition (2). Using this condition, we can find an expression for the first order connection on shell in terms of first order metric variables:

$$a_{ij} = \epsilon_{ikl}\partial_k\delta e_{lj} + \gamma\delta e_{ij}' \tag{9}$$

We can plug this into decomposition (5) to find the corresponding expression in Fourier space, $\Psi_a^{r+} = \gamma\Psi_e' + rk\Psi_e$ and $\Psi_a^{r-} = \gamma^*\Psi_e' + rk\Psi_e$. As the metric satisfies equation (7) on-shell, the functions $\Psi_e(k,\eta)$ will satisfy the corresponding Fourier space equation of motion, $\Psi_e'' + \left(k^2 - \frac{2}{\eta^2}\right)\Psi_e = 0$, which has solution $\Psi_e = \frac{e^{-ik\eta}}{2\sqrt{k}}\left(1-\frac{i}{k\eta}\right)$. We will be concerned with modes that were inside the horizon at the onset of inflation (for which $|k\eta| \gg 1$) as these are the modes we observe in CMB experiments now. On-shell, they satisfy

$$\Psi_a^{rp} = \Psi_e k\left(r - i\gamma_R + p\gamma_I\right)\,, \tag{10}$$

so for a γ with an imaginary part the connection is p-dependent and therefore made up of graviton and anti-graviton states.

3. Issues to be addressed

Classically, the Ashtekar formailsm is equivalent to the standard second order calculation. We should therefore rediscover all known classical results for the tensor perturbations when using the Ashtekar variables. This provides us with a good test of whether the formalism is consistent. It turned out that there were quite a few subtleties that had to be taken into account before the right result was reobtained. Furthermore, there have been a few misconceptions in previous literature, in particular related to Fourier expansions, that need to be addressed.

In this section we will summarise the main issues that have to be considered when doing cosmological perturbation theory in the Ashtekar framework.

- **Helicity and duality** There has been a belief (although proven to be wrong in [12]) that helicity (right and left handed) and duality (SD and ASD) states of the graviton should align. However, it should be obvious that this cannot be true: the two types of states can never align because helicity states are real whereas duality states must be complex for a Lorentzian space. Reality conditions therefore relate SD and ASD states; but they can never impose a constraint upon helicity states. The cause of the confusion lies in the Fourier expansion of the canonical variables: they should include positive and negative frequency states. This was missed, for example, in the works of [13, 14]. This can be easily seen by looking at the on-shell condition (10) which vanishes for certain r and p if $\gamma_I = \pm 1$. The surviving contributions are summarised in table 1. In the case $\gamma = \pm i$, the connection is therefore made up of the right (left) graviton and the left (right) anti-graviton.

Table 1. Helicity and Duality

	$r = +$ [R]	$r = -$ [L]
$p = +$ $[G]$	SD	ASD
$p = -$ $[\overline{G}]$	ASD	SD

- **Fourier space expansion** The expansions (5) are chosen to fulfil the following criteria:
  - As reality conditions are yet to be enforced there must be graviton and anti-graviton modes, so it's essential not to forget the negative frequencies in all expansions, and ensure that they are initially independent of the positive frequencies.
  - For a clearer physical picture, it is convenient to use the quantum field theory convention stipulating that for free modes the spatial vector $\mathbf{k}$ points in the direction of propagation *for both positive and negative frequencies.*

  If the above is employed, the physical Hamiltonian will not contain couplings between $\mathbf{k}$ and $-\mathbf{k}$ modes inside the horizon. The presence of such couplings in the formalism [13, 14] merely reflects not having properly identified the direction of propagation (and thus the polarisation). As the modes leave the horizon, couplings between $\mathbf{k}$ and $-\mathbf{k}$ may appear, and represent the production of particle pairs by the gravitational field (where the particles in each pair move in opposite directions) [15].
- **Other constraints** In Hamiltonian GR, the full Hamiltonian constraint contains two further constraints, Gauss and diffeomorphism. Whereas the diffeomorphism constraint is automatically satisfied to all orders considered, solving for the perturbations to first order allows for a non-vanishing Gauss constraint to second order (in the form of expressions quadratic in linear perturbation variables). This is also true for the torsion. To use the notations defined in the text, ${}^2_1G_i \neq 0$ and ${}^2_1T^a \neq 0$. As with the Hamiltonian constraint,

it is the second order (or backreaction) terms ${}^2_2G_i$ and ${}^2_2T^a$, that enforce these constraints to second order. In deriving the Ashtekar Hamiltonian from the ADM formalism [4] these two constraints are used, and therefore ${}^2_1\mathcal{H}$ acquires extra terms. However, these turn out to be irrelevant full divergences.

- **Perturbation variables as a canonical transformation** If one expands the full Ashtekar variables in terms of their perturbations and naively evaluates ${}^2_1\mathcal{H}$, then, contrary to expectations, one does *not* obtain the standard cosmological Hamiltonian. The reason is that the second order Hamiltonian needs to acquire an additional term: the perturbative expansions (3) and (4) can be regarded as a canonical transformation into new variables, a^i_a and δe^a_i, which happen to be "small". As this transformation is time dependent, as in standard Hamiltonian mechanics [16], an additional term has to be added to ${}^2_1\mathcal{H}$ which is given by the time derivative of the generating function of the transformation. Not only does this procedure ensure to give us the correct Hamiltonian, it also gives a more rigorous meaning to the perturbative quantisation procedure. We are not quantising the fluctuations whilst "freezing" the quantum mechanics of the background; we are merely quantising the full theory in new variables, which happen to be "small" in some circumstances.
- **Boundary term** Boundary terms [17, 18, 19] are often ignored in the literature, usually by invoking suitable fall-off conditions [13, 19]. However, it turns out that this leads to the wrong result for the cosmological Hamiltonian. The reason is that planes waves, the central tool of cosmological perturbation theory, do not satisfy the fall-off conditions, say, in a deSitter background. Therefore the boundary term has to be included to the right order in the expansion in order to obtain the correct Hamiltonian to be employed in quantising the graviton modes.
- **Hamiltonian is complex off shell** Whereas on shell the Hamiltonian is weakly zero and therefore real, off-shell, for a general γ, we have an intrinsically complex Hamiltonian. In particular this is true for its perturbative expression. Note, however, that the Hamiltonian is still Hermitian *with respect to the inner product used*, i.e. when taking the reality conditions into account. This looks very similar to the non-Hermitian Hamiltonians studied by Bender and collaborators [20], which are initially complex but become Hermitian with respect to a non-trivial inner product.

4. The Quantum Hamiltonian

4.1. Reality conditions and commutation relations

To set up the quantum theory, we need to determine the reality conditions for the Fourier space variables (which are present as γ, and therefore the connection, is complex in general). These will let us determine the physical inner product. The reality of the metric translates to the simple condition $e_{r+}(\mathbf{k}) = e_{r-}(\mathbf{k})$ in terms of modes. For the connection in real space, we have $\Re A^i = \Gamma^i + \gamma_R \Gamma^{0i}$ and $\Im A^i = \gamma_I \Gamma^{0i}$. We can write the spin connection in terms of the metric by again imposing the torsion-free condition (2). The reality conditions are supposed to embody the non-dynamical part of the torsion-free condition, and, in terms of modes, turn out to be

$$i\gamma^* \tilde{a}_{r+}(\mathbf{k},\eta) - i\gamma \tilde{a}_{r-}(\mathbf{k},\eta) = 2rk\gamma_I \tilde{e}_{r+}(\mathbf{k},\eta) \tag{11}$$

$$-i\gamma \tilde{a}^\dagger_{r+}(\mathbf{k},\eta) + i\gamma^* \tilde{a}^\dagger_{r-}(\mathbf{k},\eta) = 2rk\gamma_I \tilde{e}^\dagger_{r-}(\mathbf{k},\eta)\ . \tag{12}$$

Note that for a purely real γ the connection is real and therefore there are no reality conditions (instead $a_{r+} = a_{r-}$); though the torsion-free condition is still needed on shell to relate metric and connection. We also need to derive the commutation relations between the metric and connection in Fourier space, starting from the Poisson bracket of the real theory. They are given by [6, 7]

$$[\tilde{a}_{rp}(\mathbf{k}), \tilde{e}^{\dagger}_{sq}(\mathbf{k}')] = -i(\gamma_R + pi\gamma_I)\frac{l_P^2}{2}\delta_{rs}\delta_{p\bar{q}}\delta(\mathbf{k} - \mathbf{k}') , \tag{13}$$

where $\bar{q} = -q$. All other commutators are zero.

4.2. The Hamiltonian

We can find the quantum Hamiltonian by plugging the mode expansion (5) into the second order dynamical Hamiltonian (8). We will consider modes inside the horizon, $|k\eta| \gg 1$, so terms containing Ha can be neglected. We can then write $\mathcal{H}$ in terms of graviton creation and annihilation operators:

$$\begin{aligned}\mathcal{H}_{eff} = \frac{1}{2l_P^2}\int d^3k \sum_r &-(1+i\gamma r)G_{r\mathcal{P}_+}(\mathbf{k})G_{r\mathcal{P}_-}(-\mathbf{k}) - (1-i\gamma r)G_{r\mathcal{P}_-}(\mathbf{k})G_{r\mathcal{P}_+}(-\mathbf{k})\\ &+(1+i\gamma r)G_{r\mathcal{P}_+}(\mathbf{k})G^{\dagger}_{r\mathcal{P}_+}(\mathbf{k}) + (1-i\gamma r)G^{\dagger}_{r\mathcal{P}_+}(\mathbf{k})G_{r\mathcal{P}_+}(\mathbf{k}) + (1-i\gamma r)G_{r\mathcal{P}_-}(\mathbf{k})G^{\dagger}_{r\mathcal{P}_-}(\mathbf{k})\\ &+(1+i\gamma r)G^{\dagger}_{r\mathcal{P}_-}(\mathbf{k})G_{r\mathcal{P}_-}(\mathbf{k}) - (1-i\gamma r)G^{\dagger}_{r\mathcal{P}_+}(\mathbf{k})G^{\dagger}_{r\mathcal{P}_-}(-\mathbf{k}) - (1+i\gamma r)G^{\dagger}_{r\mathcal{P}_-}(\mathbf{k})G^{\dagger}_{r\mathcal{P}_+}(-\mathbf{k})\end{aligned} \tag{14}$$

Note that this contains physical, donated by $\mathcal{P}_+$, and unphysical graviton states donated by $\mathcal{P}_-$. They are summarised in table 2. Using the reality conditions, we can see that $G_{r\mathcal{P}}$ and $G^{\dagger}_{r\mathcal{P}}$ are indeed conjugates of each other (this is trivially true for a real γ). The unphysical modes can be seen to have negative energy from their commutators, and they vanish classically, i.e. when imposing the on-shell condition (10). This means that on-shell, we don't have any spurious $\mathbf{k}$, $-\mathbf{k}$ couplings that would signify particle production (which does not occur in the inside horizon limit $H \to 0$, which corresponds to Minkowski space). Quantum mechanically, the unphysical modes can be removed by determining the inner product through the reality conditions. We choose a holomorphic representation diagonalising $G^{\dagger}_{r\mathcal{P}}$,

Table 2. Physical and unphysical graviton modes

Physical $\mathcal{P} = \mathcal{P}_+ = 1$	Unphysical $\mathcal{P} = \mathcal{P}_- = -1$
$G_{r\mathcal{P}_+} = \frac{-r}{i\gamma}(\tilde{a}_{r+} - k(r+i\gamma)\tilde{e}_{r+})$	$G_{r\mathcal{P}_-} = \frac{-r}{i\gamma}(\tilde{a}_{r+} - k(r-i\gamma)\tilde{e}_{r+})$
$G^{\dagger}_{r\mathcal{P}_+} = \frac{r}{i\gamma}(\tilde{a}^{\dagger}_{r-} - k(r-i\gamma)\tilde{e}^{\dagger}_{r-})$	$G^{\dagger}_{r\mathcal{P}_-} = \frac{r}{i\gamma}(\tilde{a}^{\dagger}_{r-} - k(r+i\gamma)\tilde{e}^{\dagger}_{r-})$
$\left[G_{r\mathcal{P}_+}(\mathbf{k}), G^{\dagger}_{s\mathcal{P}_+}(\mathbf{k}')\right] = kl_P^2\delta_{rs}\delta(\mathbf{k}-\mathbf{k}')$	$\left[G_{r\mathcal{P}_-}(\mathbf{k}), G^{\dagger}_{s\mathcal{P}_-}(\mathbf{k}')\right] = -kl_P^2\delta_{rs}\delta(\mathbf{k}-\mathbf{k}')$

$$G^{\dagger}_{r\mathcal{P}}\Phi(z) = z_{r\mathcal{P}}\Phi(z) \quad \Rightarrow \quad G_{r\mathcal{P}}\Phi = \mathcal{P}kl_P^2\frac{\partial\Phi}{\partial z_{r\mathcal{P}}} \tag{15}$$

(using the commutation relations to determine the action of $G_{r\mathcal{P}}$) and we can impose the reality conditions through $\langle\Phi_1|G^{\dagger}_{r\mathcal{P}}|\Phi_2\rangle = \overline{\langle\Phi_2|G_{r\mathcal{P}}|\Phi_1\rangle}$, fixing the inner product [2, 13, 21]. The measure μ of a general inner product $\langle\Phi_1|\Phi_2\rangle = \int dz d\bar{z} e^{\mu(z,\bar{z})}\bar{\Phi}_1(\bar{z})\Phi_2(z)$ can then be determined as

$$\mu(z,\bar{z}) = \int d\mathbf{k} \sum_{r\mathcal{P}} \frac{-\mathcal{P}}{kl_P^2} z_{r\mathcal{P}}(\mathbf{k})\bar{z}_{r\mathcal{P}}(\mathbf{k}) . \tag{16}$$

From this we see that the variables corresponding to the unphysical modes $\mathcal{P}_- = -1$ actually lead to a divergent measure and should therefore be removed from the physical Hilbert space. We can also use this representation to determine a perturbative ground state (just a constant in this case) as well as particle states (which are monomials in the physical z variable).

5. Vacuum fluctuations

From the analysis of the last section, we know that the physical Hamiltonian needed to determine the vacuum fluctuations is given by

$$\mathcal{H}^{ph}_{eff} \approx \frac{1}{2l_P^2}\int d\mathbf{k}\sum_r [G_r^{ph}G_r^{ph\dagger}(1+ir\gamma) + G_r^{ph\dagger}G_r^{ph}(1-ir\gamma)] \tag{17}$$

where $G_r^{ph} = G_{r\mathcal{P}_+}$. When we normal order this expression, we get a γ-independent graviton spectrum which is symmetric for right and left particles, however we are left with a chiral vacuum energy $V_r = 1 + i\gamma r$. Note that this contains imaginary terms in general, in particular even for a purely real γ (although the total vacuum energy, $V_R + V_L = 2$ is γ independent). This could be traced back to the fact that the physical Hamiltonian is in fact only Hermitian (under the chosen inner product) if γ is purely imaginary. It turns out that if we choose a symmetric ordering of the Hamiltonian, the vacuum energy is real and symmetric for both right and left gravitons. Whether this is a hint towards a preferred ordering of the Hamiltonian is up to debate.

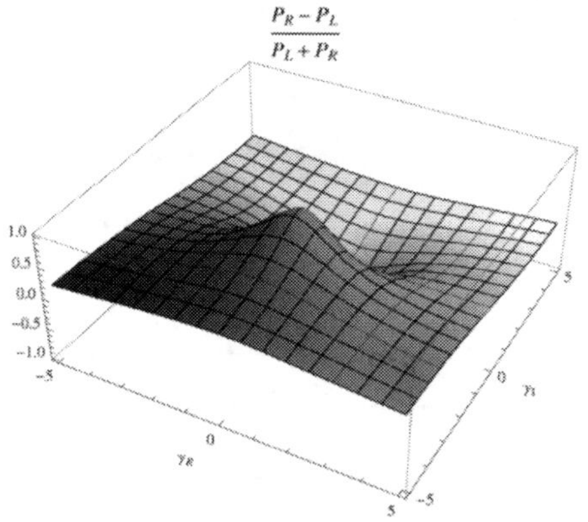

Figure 1. Power spectrum asymmetry as a function of a generally complex Immirzi parameter γ.

Whereas the vacuum energy is not physically measurable, the vacuum fluctuations are. These are given by the 2-point functions of the connection, $\langle 0|A_r^\dagger(\mathbf{k})A_r(\mathbf{k}')|0\rangle = P_r(k)\delta(\mathbf{k}-\mathbf{k}')$, where $A_r(\mathbf{k})$ includes the Fourier components of the connection, $A_r(\mathbf{k}) = a_{r+}(\mathbf{k})e^{-ik\cdot x} + a^\dagger_{r-}(\mathbf{k})e^{ik\cdot x}$. We can use our on-shell condition to express this 2-point function in terms of physical graviton states only,

$$\langle 0|A_r^{ph\dagger}(\mathbf{k})A_r^{ph}(\mathbf{k}')|0\rangle = \frac{(r+i\gamma)(r-i\gamma^*)}{4}\langle 0|G_{r\mathcal{P}_+}(\mathbf{k})G^\dagger_{r\mathcal{P}_+}(\mathbf{k}')|0\rangle\ , \tag{18}$$

which lets us determine the power spectrum asymmetry between right and left-handed states:

$$\frac{P_R - P_L}{P_R + P_L} = -\frac{2\gamma_I}{1+|\gamma|^2}\ . \tag{19}$$

Equation (19) shows that the chiral asymmetry vanishes for a purely real γ, but is present in general. This is in contrast with the second order theory, where right and left gravitons are completely symmetric. While the ordering of the Hamiltonian does not affect the 2-point function, the ordering of the connection components does. If we use $A_r A_r^\dagger$ instead of $A_r^\dagger A_r$, expression (19) flips sign, and for a symmetric ordering we get no chirality at all. The power spectrum asymmetry is shown in figure 1. It shows that the maximum chirality occurs at the SD/ASD values $\gamma = \pm i$ and that the chirality vanishes for $\gamma_I = 0$ as well as in the limit $|\gamma| \to \infty$, representing the Palatini-Kibble theory.

Conclusion

We have shown that using the Ashtekar formulation of canonical GR leads to a chiral gravitational wave background. The effect depends on the value of the free parameter of the theory, the Immirzi parameter γ: chirality vanishes for a purely real γ and is maximised for $\gamma = \pm i$, corresponding to a SD/ASD connection. Additionally, it also depends on the ordering of the 2-point function. In [22] it was shown that parity violation, which leads to a non-vanishing TB (temperature and magnetic mode) correlator, would render the detection of gravitational waves much easier. In [6] we argued that we can expect a siginificant TB signal for a range of $\frac{1}{800} < |\gamma| < 800$. Upcoming CMB polarisation experiments like PLANCK will therefore potentially be able to measure this chiral effect, assuming there is a sufficiently large tensor to scalar ratio.

Other mechanisms to produce gravitational chirality have been proposed (e.g. [23, 24, 25]), however the one pointed out here is by far the simplest. It would be useful to bridge our result to the one of [26], where a chirality in the graviton propagator was observed. However, as their approach relies on a Euclidean signature and real γ, the relationship is not obvious.

In addition to comparing different methods of producing gravitational chirality, there are a few other interesting lines of investigation to follow. Instead of using a holomorphic representation for the graviton operators (dependent on metric and connection), a representation purely in terms of the connection can be found. This will allow us to analyse the perturbative Kodama state [27] (a non-perturbative solution to the gravitational constraints) which turns out (as shown in [28]) *not* to be normalisable in terms of our inner product, and therefore cannot describe gravitons. Furthermore, it would be interesting to investigate the process of horizon crossing and the behaviour of modes outside the horizon. The dynamics are then fully driven by the metric, and we expect a complete reduction to second order theory. Understanding this transition might give us insight into the problem of decoherence in Cosmology and the nature of squeezed states.

References

[1] Ashtekar A 1988 *New Perspectives in Canonical Gravity* (Naples: Bibliopolis)
Ashtekar A and Tate R S 1991 *Lectures on Non-Perturbative Canonical Gravity* (Singapore: World Scientific)
[2] Gambini R and Pullin J 1996 *Loops, Knots, Gauge Theories and Quantum Gravity* (Cambridge: CUP)
[3] Rovelli C 2004 *Quantum Gravity* (Cambridge: CUP)
[4] Thiemann T 2007 *Modern Canonical Quantum General Relativity* (Cambridge: CUP)
[5] Magueijo J and Benincasa D 2011 *Phys. Rev. Lett.* **106** 121302
[6] Bethke L and Magueijo 2011 *Phys. Rev.* D **84** 024014
[7] Bethke L and Magueijo J 2011 *Preprint* arXiv:1108.0816
[8] Holst S 1996 *Phys. Rev.* D **53** 5966
[9] Liddle A and Lyth D 2000 *Cosmological Inflation and Large-scale Structure* (Cambridge: CUP)
[10] Mukhanov V 2005 *Physical Foundations of cosmology* (Cambridge: CUP)
[11] Arnowitt R, Deser S and Misner C W 1960 *Phys. Rev.* **117** 1595
[12] Ashtekar A 1986 *J. Math.Phys.* **27**, 824
[13] Ashtekar A, Rovelli C and Smolin L 1991 *Phys. Rev.* D **44** 1740
[14] Freidel L and Smolin L 2004 *Class. Quant. Grav.* **21** 3831-44

[15] Grishchuk L P and Sidorov Y V 1990 *Phys. Rev.* D **42** 3413
Grishchuk L P 2010 *Preprint* arXiv:0707.3319v4
[16] Goldstein H 1980 *Classical mechanics* (Boston: Addison-Wesley)
[17] Gibbons G W and Hawking S W 1977 *Phys. Rev.* D **15** 2752
[18] Regge T and Teitelboim C 1974 *Ann. Phys.* **88** 236
[19] Ashtekar A, Eagle J and Sloan D 2008 *Class. Quant. Grav.* bf 25 095020
[20] Bender C M 2007 *Rep. Prog. Phys.* **70** 947
[21] Ashtekar A and Tate R S 1994 *J. Math. Phys.* **35** 6434
[22] Contaldi C, Magueijo J and Smolin L 2008 *Phys.Rev.Lett.* **101** 141101
[23] Alexander S and Calcagni G 2008 *Found.Phys.* **38** 1148-84
Alexander S and Calcagni G 2009 *Phys. Lett.* B **672** 386
[24] Alexander S 2007 *Preprint* arXiv:0706.4481
[25] S. Mercuri 2011 *Preprint* arXiv:1007.3732v4
[26] Rovelli C 2006 *Phys. Rev. Lett.* **97** 151301
Bianchi E *et al.* 2006 *Class. Quant. Grav.* **23** 6989
Bianchi E, Magliaro E and Perini C 2009 *Nuc. Phys.* B **822** 245
[27] Kodama H 1990 *Phys. Rev.* D **42** 2548
[28] Benincasa D, Bethke L and Magueijo J (unpublished)

Loops 11: Non-Perturbative / Background Independent Quantum Gravity IOP Publishing
Journal of Physics: Conference Series **360** (2012) 012004 doi:10.1088/1742-6596/360/1/012004

Towards computational insights into the large-scale structure of spin foams

Bianca Dittrich and Frank C Eckert

MPI f. Gravitational Physics, Albert Einstein Institute,
Am Mühlenberg 1, D-14476 Potsdam, Germany

E-mail: `dittrich@aei.mpg.de`, `frank.eckert@aei.mpg.de`

Abstract.
Understanding the large-scale physics is crucial for the spin foam approach to quantum gravity. We tackle this challenge from a statistical physics perspective using simplified, yet feature-rich models. In particular, this allows us to explicitly answer whether broken symmetries will be restored by renormalization: We observe a weak phase transition in both Migdal-Kadanoff and tensor network renormalization. In this work we give a concise presentation of the concepts, results and promises of this new direction of research.

1. Will broken symmetries be restored in the large-scale limit?

The investigation of the large-scale structure of spin foam models is crucial. There are three reasons for this.

First, spin foam models can be understood as giving a meaning to the path integral for quantum gravity [1; 2] and, in order to do so in a well defined way, spin foams introduce a regularization. The regularization is based on a discretization by means of a lattice, triangulation or two-complex and this will generically break (global) Lorentz and (local) diffeomorphism invariance [3]. Will these symmetries be restored in the large-scale limit[1] [5]? As diffeomorphisms constitute the fundamental symmetry of general relativity this question is essential to demonstrate the viability of spin foams as a quantum theory of gravity.

Second, it is not clear whether such a statistical limit will give a space-time modelled by a smooth four-dimensional manifold. It is imperative, however, to realize a smooth continuum limit in order to validate any approach to quantum gravity and to make predictions of measurable physical effects. Studies in (causal) dynamical triangulations have shown [6] that degenerate configurations may appear and are relevant. It is therefore necessary to understand whether, when and how pathological structures appear and dominate the state sum. In other words, it is necessary to understand the phase structure of spin foams.

Third, even though spin foams currently enjoy much attention little progress has been made in probing the large-scale limit. The question is important and other approaches to quantum gravity have progressed in this direction. Therefore, understanding the large-scale structure of spin foams is now one of the most pressing problems in the field.

[1] The auxiliary discretization and its large-scale limit is not to be confused with another source of discreteness (Planck scale discreteness) in the theory [4], namely in the spectra of observables, and its corresponding $\hbar \to 0$ limit. The relationship between the two is an open issue.

The reason why our understanding of the large-scale physics of spin foams is stagnating is simple: the models are very complex and mathematically involved, making it difficult to perform even simple calculations, let alone to use the computer as a computational tool.

We approach this issue by radically simplifying the mature spin foam models. This will facilitate the development of renormalization tools, methods and most importantly numerical algorithms. Furthermore, by studying the renormalization behavior of increasingly more complicated models, it is possible to deduce properties for the full models. In this work we start with the study of the simplest models: Instead of non-Abelian Lie groups we use finite Abelian groups. We also confine our models to regular lattices (instead of general two-complexes), which allows to apply effective approximation schemes. As is argued in [5] a restoration of diffeomorphism symmetry should also ensure independence from the chosen underlying lattice or triangulation. Hence the restriction to regular lattices is justified. Finally, we consider dimensionally reduced versions of the four-dimensional models, so called *spin net models*, which in fact will be the focus of this article. Interestingly, spin net models in two dimensions share many statistical properties with spin foam models in four dimensions [7].

Simplification as put forward here is not only done so that the resulting models become computationally manageable. Rather, we choose to simplify in such a way that important features are conserved and meaningful problems can be studied. In particular, important (gauge) symmetries are preserved, the model space includes a topological (BF like) theory and we introduce certain *Abelian cutoff models*, which break the symmetries in a controlled way and can be used to study the question of whether broken symmetries are being restored under renormalization.

In doing so we connect with statistical and condensed matter physics. We can make use of the power of computers and of highly developed methods, which we refine and adapt to suit the questions we are asking. Ultimately, this enables us do calculations, to study the large-scale physics of simplified spin foam and spin net models and to explicitly answer the question of symmetry restoration. While interesting in itself, we see promise in this as a first step towards computational insights into the large-scale structure of mature spin foam models. Furthermore, as we will see, results obtained across a range of finite groups can inform about the analogue and more complicated situation with the corresponding continuous group.

This article is a concise presentation and discussion of work published in [8] to which we refer the reader for details. The outline is as follows. We first construct a setting in which we have simple models and a gauge symmetry which can be broken, simple enough to analyze it computationally. We then discuss two numerical renormalization schemes, Migdal-Kadanoff and tensor network renormalization, and the weak phase transition we observe in both cases. Finally, we outline the next steps to be done.

2. Spin net models: controlled simplification retaining symmetries

Spin net models are dimensionally reduced versions of spin foam models [2; 8]. They are a class of lattice theories that include the Ising model. In this article we will solely focus on simple 2D spin net models with finite Abelian groups. A more general and detailed treatment can be found in [8].

Given a regular, oriented graph and an element g of the finite Abelian group $\mathbb{Z}_q$ a spin net model is defined via its partition function

$$Z = \frac{1}{q^{\sharp v}} \sum_{\{g_v\}} \prod_e w(g_{s_e} g_{t_e}^{-1}), \tag{1}$$

where $\sharp v$ denotes the number of vertices in the graph, $\sum_{\{g_v\}} = \prod_v \sum_{g_v}$ is the sum over all assignments of group elements to the graph's vertices and $\prod_e$ is the product over all edges.

Journal of Physics: Conference Series **360** (2012) 012004 doi:10.1088/1742-6596/360/1/012004

Here, the index s_e (resp. t_e) gives the source (resp. target) vertex and w is the weight function. In analogy to spin foam models we call this way of writing the state sum the *holonomy representation*.

The *spin net representation*, in contrast, is obtained by expressing the state sum in terms of irreducible representations of the group. This is done via a group Fourier transform which is given by the character decomposition,

$$w(g) = \sum_{k=0}^{q-1} \tilde{w}_k \, \chi_k(g) \qquad \tilde{w}_k = q^{-1} \sum_{g=0}^{q-1} w(g) \overline{\chi}_k(g) \,. \tag{2}$$

For $\mathbb{Z}_q$, the irreducible representations are one dimensional, they are labeled by $k \in \mathbb{Z}_q$ and the characters, given by $\chi_k(g) = \exp((2\pi i/q)k \cdot g)$ for $g \in \mathbb{Z}_q$, obey $\chi_k(g_1 \cdot g_2) = \chi_k(g_1) \cdot \chi_k(g_2)$ and $\chi_k(g^{-1}) = \chi_k^{-1}(g) = \overline{\chi}_k(g)$. Furthermore, the weight function with trivial Fourier coefficients gives the q-periodic delta function over the group, $\delta^{(q)} = \sum_k \chi_k(g)$, satisfying $q^{-1} \sum_g \delta^{(q)}(g) f(g) = f(0)$. With this, the state sum in the spin net representation takes the form

$$Z = \sum_{\{k_e\}} \prod_e \tilde{w}_{k_e} \prod_v \tilde{P}^v(\{k_e\}_{e \supset v}), \qquad \tilde{P}^v(\{k_e\}_{e \supset v}) := \delta^{(q)}\Big(\sum_{e \supset v} \varepsilon_v^e k_e\Big), \tag{3}$$

where we have introduced an orientation coefficient ε_v^e which equals $+1$ (resp. -1) if v is e's source (resp. target).

In equation (3) the vertex projector $\tilde{P}^v$ imposes Gauß constraints at each vertex. More mathematically, this corresponds to a projection onto the irreducible subrepresentation in the tensor product of (dual) representations associated to (incoming) outgoing edges. The projection can be restricted even further in order to construct more complicated models. We will exemplify one such case here.

A choice of $\tilde{w}(k) \equiv 1$ in the spin net representation (corresponding to $w(g) = \delta^{(q)}(g)$ in the holonomy representation) defines a topological model without local degrees of freedom, also called BF theory.[2] From a statistical physics perspective, it is a 'frozen' model at zero temperature: the group elements of each connected component of the graph have to agree. This model is invariant under a gauge transformation, also known as *translation symmetry*, given by

$$k_e \mapsto k_e' = k_e + \sum_{f \supset e} \varepsilon_e^f k_f, \tag{4}$$

where the sum is over faces bounding the respective edge with corresponding parameters $k_f \in \mathbb{Z}_q$. ε_e^f is again an orientation factor.

The implementation of a certain non-trivial projector then leads to *Abelian cutoff models* which break this gauge symmetry. The name stems from the fact that the effect of the desired projector restriction can also be achieved by a restriction on weights $\tilde{w}(k)$, namely

$$\tilde{w}(k) = \begin{cases} 1, & \text{for } |k| \leq K \\ 0, & \text{for } |k| > K \end{cases} \quad , -\frac{q}{2} < k \leq \frac{q}{2} \tag{5}$$

effectively *cutting off* the sum that defines the delta. Defined in this way the Abelian cutoff models are also symmetric, $\tilde{w}(k) = \tilde{w}(-k) \, \forall k \in \mathbb{Z}_q$, and thus, as desired, independent of edge orientations and certain subdivisions [9].

[2] The name BF theory origins from the spin foam models. There BF theory models are distinguished by the appearance of translation symmetries, equation (4), from other spin foam models. As these symmetries also appear in the corresponding spin net models, we will apply this name for these spin nets.

The setup we now have includes a BF model whose gauge symmetry is an analogue of diffeomorphism symmetry found in 3D BF with group $SO(3)$ and cutoff models which implement non-trivial projectors and thereby break the gauge symmetry. Yet the models are much simpler to handle and this sets the stage for an explicit analysis of the problem whether the broken symmetries will be restored under renormalization.

3. Simple, non-refinable approximation: Migdal-Kadanoff

In order to determine the behaviour of cutoff models under renormalization a coarse graining procedure has to be implemented. However, coarse graining in general leads to a growing number of (non-local) couplings at each step and this makes it mandatory to apply an approximation scheme.

The Migdal-Kadanoff (MK) approximation [10; 11] is a simple method to restrict the renormalized Hamiltonian to local couplings. This is done by moving bonds (links between vertices) away from vertices until the valency is effectively reduced to at most two, mirroring the one-dimensional situation in which no non-local couplings appear during coarse graining (see figure 1).

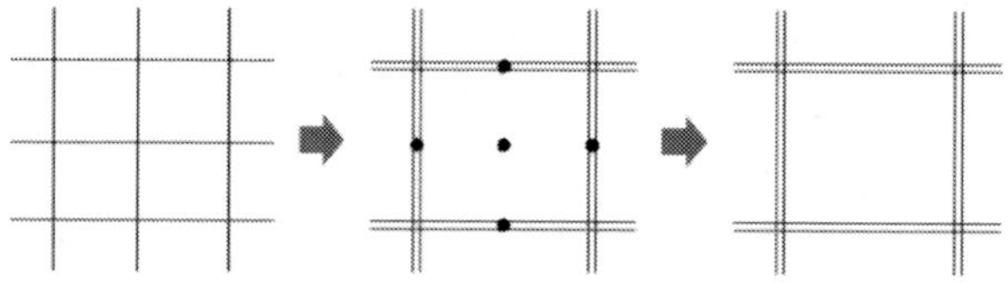

Figure 1. The derivation of MK recursion relations for 2D spin net models involves two steps: moving bonds and subsequently integrating out spins.

Kadanoff derived a lower bound on the free energy for this approximation [11]. Further justification for this otherwise ad hoc method is drawn from fairly good predictions of fixed points and phase transitions [12] and from comparison with Monte Carlo simulations [13]. The order of phase transitions and some fixed point of Kosterlitz-Thouless type [7], however, are not correctly reproduced. The MK approximation is thus a simple and computationally efficient tool to probe the large-scale limit. It, however, does not offer means for systematic improvement, for instance by taking non–local couplings into account.

The (isotropic) version of the MK recursion relation used here have the interesting property to coincide for 2D (Abelian) spin net models and 4D (Abelian) spin foam models [7]. Thus 2D spin nets can be seen as a statistical toy models for 4D spin foams.

At the core of the MK approximation, the derivation of the recursion relations is based on the aforementioned bond-moving: In a systematic way, some terms (bonds) in the state sum are dropped while others are strengthened which leads to convolution of Fourier coefficients,

$$w(g_{e'}) \to w^2(g_{e'}) = \sum_k \left(\sum_j \tilde{w}(k-j)\tilde{w}(j) \right) \chi_k(g_{e'}). \tag{6}$$

Convolution is a central feature of the recursion relations which, after normalization, $Q(k) = \tilde{w}(k)/\tilde{w}(0)$, take the form

$$Q_{n+1}(k) = \left(\frac{\sum_{j=0}^{q-1} Q_n(k-j)Q_n(j)}{\sum_{j=0}^{q-1} Q_n(-j)Q_n(j)} \right)^2 \tag{7}$$

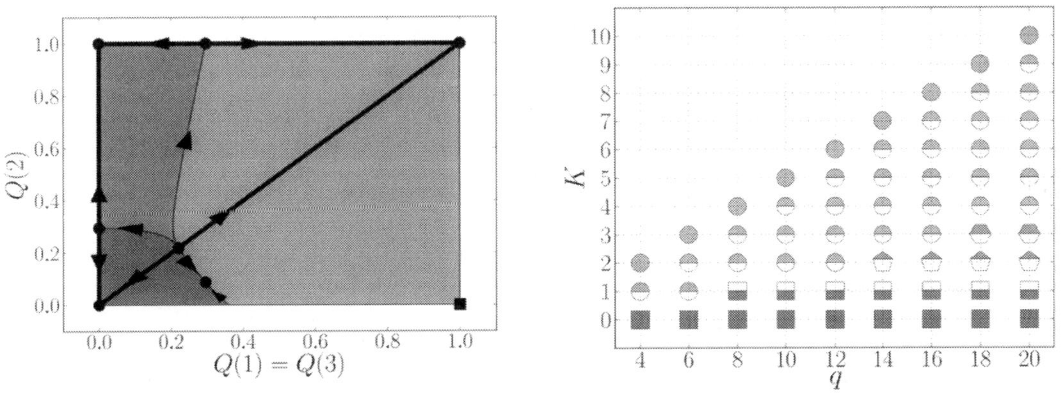

(a) fixed point structure for $\mathbb{Z}_4$

(b) $q - K$ diagram showing weak phase transition

Figure 2. (a) Parameter space of $\mathbb{Z}_4$ symmetric models and their renormalization flow. The three regions correspond to configurations flowing to HTF $(0,0)$, LTF / BF $(1,1)$ and the 'cyclic fixed point' $(0,1)$. (b) renormalization behaviour of cutoff models parametrized by q and K. Markers half-filled at the top flow to BF/LTF (diagonal models, $K = q/2$), markers half-filled at the bottom flow to HTF (horizontal modesl with $K = 0$). Pentagonal and hexagonal markers denote unstable fixed points.

and feature a rich fixed point structure exemplified in figure 2a. From these relations it can be seen that the space of models they act on depends on q. We further restrict this space to symmetric models ($\tilde{w}(k) = \tilde{w}(-k)\,\forall k \in \mathbb{Z}_q$) which are preserved by (7). This includes the low and high temperature as well as the cutoff models. Due to small numerical errors, we have to artificially project onto symmetric configurations after each coarse graining step. The recursion relations feature two competing effects: convolution levels out differences in the Fourier coefficients and draws the model towards the low temperature fixed point (LTF). The overall exponent, on the other hand, dampens all but $Q(0)$, drawing the configuration towards the high temperature fixed point (HTF).

We now turn to the behaviour of cutoff models under MK renormalization. The results displayed in figure 2b show the dependence of the result on q and the phase transition between models flowing to LTF (or BF phase) and those flowing to HTF. For most models, the dampening is too weak to stop them from flowing to BF, they do so quickly (5-15 iterations). Only a series of models in immediate proximity to HTF ($K = 1, q \geq 8$) actually flow to HTF (after about 60 iterations). Thus we can conclude that the translation symmetry is restored for a range of cutoff models for finite q.

However, the phase transition we observe is weak and becomes weaker with growing q. This can be seen in rising iteration numbers for fixed K and growing q (e.g. we observe $8, 15, 60$ iterations for $K = 2$ and $q = 8, 10, 12$ respectively). Starting with $q = 14$ we also observe the appearance of quasi fixed points which are stable over large numbers of iterations.

These results are in accordance with similar considerations of MK relations for $q \to \infty$, i.e. for $U(1)$. The Mermin-Wagner theorem states that that the continuous symmetry cannot be spontaneously broken at finite temperature. There, however, is a Kosterlitz-Thouless phase transition of infinite order present, even though it is not detected by the MK approximation [7]. Similarly, the generalized Ising fixed point between HTF and LTF is no longer present in the limit of large q [8].

Similarly to the 2D models there is also a phase transition present in 4D $U(1)$ lattice gauge theory [14]. For 4D non–Abelian lattice gauge theories, however, the confinement conjecture states the absence of such a transition. All configurations would thus flow back to the HTF and in this case the translational symmetries would not be restored. Note that the HTF corresponds to a geometrically degenerate phase for the gravitational spin foam models. An essential question for future research will be whether this behavior is altered if non–trivial projectors are taken into account for the non–Abelian models. Such non–trivial projectors would no longer allow the model to be rewritten in standard lattice gauge theory form.

In 3D the gravitational model coincides with the BF (i.e. LTF) phase, i.e. the projector in (3) is the one on the maximal subspace of trivial representations. Here, for the finite Abelian group models the tendency to flow to HTF is even stronger than in the 2D spin net and 4D spin foam models. For $q \to \infty$ there is no phase transition within the MK framework [15], which also holds for the non–Abelian Lie groups [16]. Hence a restoration of translational symmetries does not take place for the gravitational models 3D based on $SO(3)$. A question for future research will be, whether a phase transition takes place for the gravitational models with cosmological constant, which are based on quantum groups [17]. These might resemble the finite group case more closely.

4. Refinable, computationally intensive renormalization with tensor networks

The tensor network renormalization (TNR) scheme [18] is more general than the MK approximation discussed so far. It is also refinable. However, it has so far only been successfully applied to 2D models.

Spin net models as in (3) can easily be rewritten in *tensor network representation* which takes the general form of a tensor trace over tensors attached to vertices with indices contracted along edges,

$$Z = \sum_{a,b,c,d\ldots} T^{abcd} T^{aefg} T^{bhij} \ldots \quad . \tag{8}$$

In order to arrive at this form it is only necessary to split the Fourier weights in (3) attached to the edges in two, $u(a) := \sqrt{\tilde{w}(a)}$, and to pull each half into the definition of the tensor at the vertex

$$T^{abcd} = u(a)u(b)u(c)u(d)\tilde{P}(\{a,b,c,d\}) = u(a)u(b)u(c)u(d)\delta^{(q)}(a+b-c-d). \tag{9}$$

The naive approach to coarse graining or blocking such a system will collect several tensors into one new tensor with several edges pointing in each direction or equivalently, with one edge with an exponentially growing index range. Here, the approximation comes in: it limits the index range appropriately. The renormalization group flow then depends on the choice of cutoff D_c. This is in contrast to the MK method where the cutoff would be determined by the size of the group $\mathbb{Z}_q$. Indeed we find that in the TNR method the flow of Abelian cutoff models coincides for different choices of Abelian groups $\mathbb{Z}_q$, as long as the relation $q \geq 4K + 2$ holds. The cutoff D_c can be increased in order to refine the accuracy of predictions. In the case of the Ising model, this can indeed be shown explicitly, i.e. the accuracy of the critical temperatures found in the different approximations determined by D_c increases with the cutoff D_c [8].

The TNR method we follow here takes the state sum and reinterprets it as trace over matrices. As such, the state sum implicitly contains a sum over singular values of these matrices via a singular value decomposition. The approximation then proceeds by cutting off the sum over

Journal of Physics: Conference Series **360** (2012) 012004 doi:10.1088/1742-6596/360/1/012004

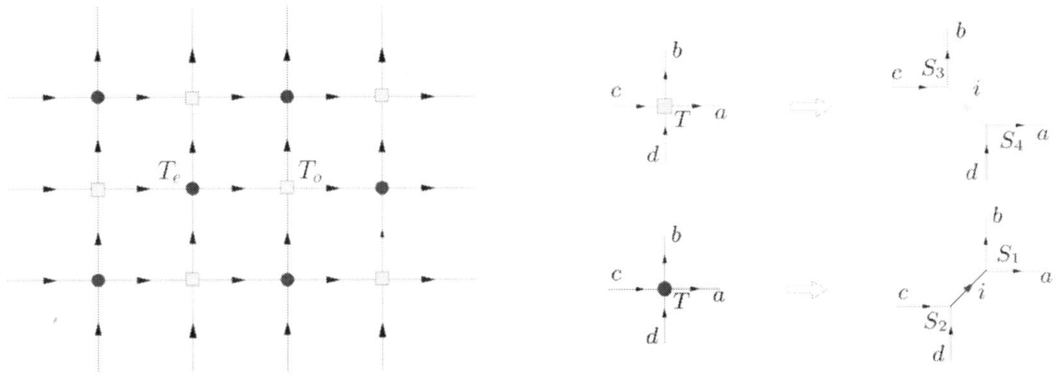

(a) square lattice (b) splitting of vertices

Figure 3. The square lattice with even (blue circles) and odd (green squares) vertices.

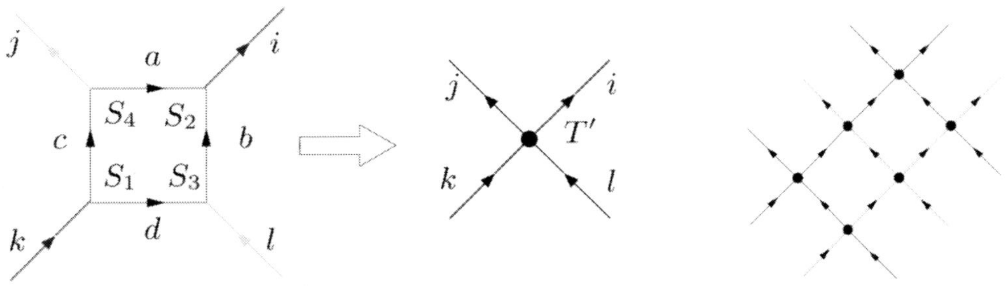

(a) contraction (b) coarse grained lattice

Figure 4. *(a)* Contraction of the four S tensor to the new T' tensor. *(b)* The coarse grained lattice.

singular values, including only the D_c largest among them:

$$T^{abcd} = M_1^{ab,cd} = \sum_{i=0}^{q^2-1} U_1^{ab,i}\lambda_i (V_1^\dagger)^{i,cd} \approx \sum_{i=0}^{D_c-1} (U_1^{ab,i}\sqrt{\lambda_i})(\sqrt{\lambda_i}(V_1^\dagger)^{i,cd}) = \sum_i S_1^{ab,i} S_2^{cd,i} \tag{10}$$

$$T^{abcd} = M_1^{cb,ad} = \sum_{i=0}^{q^2-1} U_1^{cb,i}\lambda_i (V_1^\dagger)^{i,ad} \approx \sum_{i=0}^{D_c-1} (U_1^{cb,i}\sqrt{\lambda_i})(\sqrt{\lambda_i}(V_1^\dagger)^{i,ad}) = \sum_i S_3^{cb,i} S_4^{ad,i} \tag{11}$$

The blocking can then be performed by contracting four tensors,

$$T'^{ijkl} = \sum_{a,b,c,d} S_2^{ab,i} S_4^{ac,j} S_1^{dc,k} S_3^{db,l} \quad , \tag{12}$$

yielding a new, coarse grained tensor network, tilted by 45°. The splitting and contraction of vertices is illustrated in the figures 3 and 4 respectively.

It is possible to implement the TNR method in such a way that the Gauß constraints are explicitly preserved at each coarse graining step [8; 19]. This is desirable because it preserves this

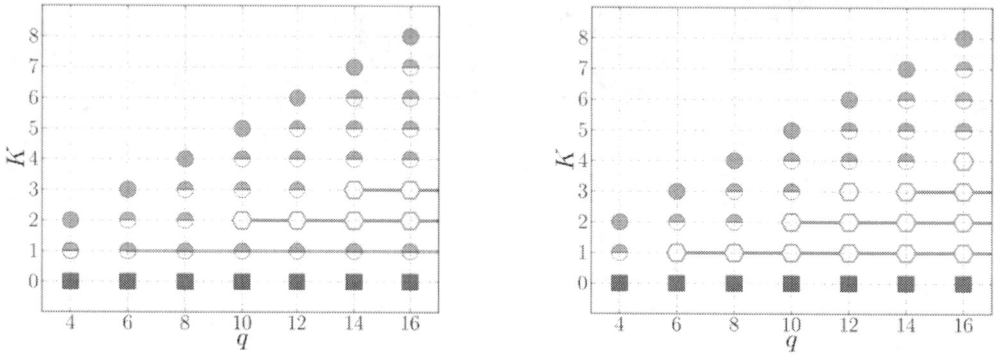

(a) $D_c = 16$ (b) $D_c = 25$

Figure 5. Flow behaviour of different cutoff models, labelled by K and q. Markers half-filled at the top flow to LTF/LTF×CDL (diagonal models, $K = q/2$), empty hexagonal markers indicate a quasi fixed point or oscillating behaviour. Connected markers illustrate equivalent models.

piece of geometrical interpretation in the otherwise very abstract framework of tensor networks. It also increases both speed and stability of the numerical algorithm.

Similar to the MK approximation, we also observe quasi fixed points in the tensor network renormalization of cutoff models. Moreover, because of the larger number of parameters compared to MK, the TNR exhibits a richer fixed point structure, including non-isolated fixed points, given e.g. by Corner Double Line (CDL) tensors [8; 18].

In our TNR simulations, the overall results of the MK analysis of cutoff models is confirmed (see figure 5). However, the simulations also show the unique features of the TNR approach: additional CDL fixed point structures for some LTF points and equivalence of models.

Similar to our MK studies, we also observe some instabilities which lead to oscillating singular values and breaking of the equivalence of models. On the one hand, these effects occur when the models are in the weak phase transition regime and therefore particularly sensitive. On the other hand, numerical instabilities and the rigid imposition of the cutoff lead to asymmetric configurations. Further research is devoted to understanding the instabilities and refine the algorithm accordingly. Thus it may be possible to decide with these refined algorithms whether for instance the Kosterlitz-Thouless transition, which is missed by the MK method, can be observed with the TNR method.

5. Next steps include refined algorithms and less simplification

We have argued that accessing and understanding the large-scale structure is crucial for the spin foam approach to quantum gravity. The work presented here proposes a new way of tackling the issue, starting out from simplified, yet feature-rich models and a statistical physics mindset. The simplified models are an ideal test bed for developing coarse graining and renormalization methods. Furthermore, studies with increasingly more complicated models will make it possible to put forward and test conjectures about the gravitational spin foam models.

We hope that this line of research will significantly advance our grasp on the large-scale dynamics encoded in loop quantum gravity and spin foam models. This will be key to construct improved models, even with the potential to obtain uniqueness results or falsifying the approach.

The first steps we considered here opens up numerous opportunities for further research. The observed instabilities in the TNR scheme pose an immediate challenge we would like to tackle. Possible refinements of the algorithm include choosing a symmetric parametrization for the models and measures to ensure symmetry in the framework we use so far. In particular, an adaptive or block-dependent cutoff could be implemented, along with a projection onto symmetric configurations in order to eliminate numerical errors.

Then, in order to make a connection with the mature spin foam models, less simplified models have to be constructed that are still statistically analyzable. This makes it necessary, in particular, to move to non-Abelian groups and, in the TNR scheme, to higher dimensions. A class of models with finite groups that resemble the gravitational spin foam models [20] is in the making [21]. Here the main question will be whether a HTF phase, corresponding to a geometrically degenerate phase, will continue to dominate the behavior in the limit of infinite group size, as is the case for standard lattice gauge theories. Current hopes, however, rely on a flow towards a BF like fixed point [22]. For such a fixed point the restoration of diffeomorphism symmetries can also be expected [5].

In addition, the application of TNR methods in a canonical formalism is promising [23]. Here, as an improvement of the mean field approach, the TNR methods would provide a variational ansatz for the physical wave function and an efficient way to compute expectation values of observables, such as the master constraint. These methods can thus help in the master constraint [24] or uniform discretization program [25] to find wave functions minimizing the violation of the master constraint and thus providing an approximation to physical (vacuum) states.

Further interesting questions comprise the generalization to random rather than regular lattices [26], the relationship between degenerate phases and choice of projectors as well as the relationship between the phase structures of fixed lattice and sum over lattices models [27].

The plethora of relevant research opportunities encourages us to advocate the statistical approach to understanding the large-scale structure of spin foams. We see great benefits in this connection between quantum gravity and statistical / condensed matter physics and hope to contribute to cross-fertilization in both communities.

Acknowledgments
We would like to thank Mercedes Martin-Benito for collaboration on [8].

References

[1] Perez A 2003 *Class.Quant.Grav.* **20** R43 (*Preprint* gr-qc/0301113)
Perez A 2012 *Liv.Rev.Rel. To appear*

[2] Bahr B, Dittrich B and Ryan J P 2011 (*Preprint* 1103.6264)

[3] Dittrich B 2009 *Adv.Sci.Lett.* **2** 151–164 (*Preprint* 0810.3594)
Bahr B and Dittrich B 2009 *Class.Quant.Grav.* **26** 225011 (*Preprint* 0905.1670)
Bahr B and Dittrich B 2009 *AIP Conference Proceedings* **1196** 10–17 (*Preprint* 0909.5688)

[4] Rovelli C and Smolin L 1995 *Nucl.Phys.* **B442** 593–622 (*Preprint* gr-qc/9411005)
Ashtekar A and Lewandowski J 1997 *Class.Quant.Grav.* **14** A55–A82 (*Preprint* gr-qc/9602046)
Dittrich B and Thiemann T 2009 *J.Math.Phys.* **50** 012503 (*Preprint* 0708.1721)
Rovelli C 2007 (*Preprint* 0708.2481)

[5] Bahr B and Dittrich B 2009 *Phys.Rev.* **D80** 124030 (*Preprint* 0907.4323)
Bahr B, Dittrich B and He S 2011 *New J.Phys.* **13** 045009 (*Preprint* 1011.3667)
Bahr B, Dittrich B and Steinhaus S 2011 *Phys.Rev.* **D83** 105026 (*Preprint* 1101.4775)

[6] Ambjorn J 1995 *Class.Quant.Grav.* **12** 2079–2134
Ambjorn J, Jurkiewicz J and Loll R 2006 *Contemp.Phys.* **47** 103–117 (*Preprint* hep-th/0509010)
Ambjorn J, Gorlich A, Jurkiewicz J, Loll R, Gizbert-Studnicki J *et al.* 2011 *Nucl.Phys.* **B849** 144–165 (*Preprint* 1102.3929)

[7] Ito K R 1985 *Phys.Rev.Lett.* **54** 2383

[8] Dittrich B, Eckert F C and Martin-Benito M 2011 (*Preprint* 1109.4927)

[9] Bojowald M and Perez A 2010 *Gen.Rel.Grav.* **42** 877–907 (*Preprint* gr-qc/0303026)
Bahr B 2011 *Class.Quant.Grav.* **28** 045002 (*Preprint* 1006.0700)

[10] Migdal A A 1975 *Sov.Phys.JETP* **42** 413

[11] Kadanoff L 1976 *Annals Phys.* **100** 359–394

[12] Creutz M and Roberts L 1983 *Nucl.Phys.* **B215** 447–457

[13] Creutz M, Jacobs L and Rebbi C 1979 *Phys.Rev.Lett.* **42** 1390
Creutz M, Jacobs L and Rebbi C 1979 *Phys.Rev.* **D20** 1915

[14] Polyakov A M 1975 *Phys.Lett.* **B59** 82–84
Banks T, Myerson R and Kogut J B 1977 *Nucl.Phys.* **B129** 493

[15] Ito K 1984 *Commun.Math.Phys.* **95** 247

[16] Muller V and Schiemann J 1985 *Commun.Math.Phys.* **97** 605

[17] Turaev V and Viro O 1992 *Topology* **31** 865–902

[18] Levin M and Nave C P 2007 *Physical Review Letters* **99** 120601 (*Preprint* arXiv:cond-mat/0611687)
Gu Z C and Wen X G 2009 *Phys.Rev.* **B80** 155131 (*Preprint* 0903.1069)

[19] Singh S, Pfeifer R N and Vidal G 2010 *Phys.Rev.* **A82** 050301 (*Preprint* 0907.2994)

[20] Engle J, Livine E, Pereira R and Rovelli C 2008 *Nucl.Phys.* **B799** 136–149 (*Preprint* 0711.0146)

[21] Bahr B, Dittrich B, Hellmann F and Kaminski W *To appear*

[22] Rovelli C 2011 (*Preprint* 1107.2310)

[23] Vidal G 2007 *Physical Review Letters* **99** 220405 (*Preprint* arXiv:cond-mat/0512165)
Gu Z C, Levin M and Wen X G 2008 (*Preprint* 0807.2010)
Verstraete F, Cirac J and Murg V 2009 (*Preprint* 0907.2796)

[24] Thiemann T 2006 *Class.Quant.Grav.* **23** 2211–2248 (*Preprint* gr-qc/0305080)
Dittrich B and Thiemann T 2006 *Class.Quant.Grav.* **23** 1067–1088 (*Preprint* gr-qc/0411139)
Dittrich B and Thiemann T 2006 *Class.Quant.Grav.* **23** 1025–1066 (*Preprint* gr-qc/0411138)

[25] Campiglia M, Di Bartolo C, Gambini R and Pullin J 2006 *Phys.Rev.* **D74** 124012 (*Preprint* gr-qc/0610023)
Gambini R, Pullin J and Rastgoo S 2009 *Class.Quant.Grav.* **26** 215011 (*Preprint* 0906.1774)

[26] Markopoulou F 2000 (*Preprint* hep-th/0006199)
Markopoulou F 2003 *Class.Quant.Grav.* **20** 777–800 (*Preprint* gr-qc/0203036)

[27] Bonzom V, Gurau R, Riello A and Rivasseau V 2011 *Nucl.Phys.* **B853** 174–195 (*Preprint* 1105.3122)
Benedetti D and Gurau R 2011 (*Preprint* 1108.5389)
Bonzom V, Gurau R and Rivasseau V 2011 (*Preprint* 1108.6269)
Carrozza S and Oriti D 2011 (*Preprint* 1104.5158)
Rovelli C and Smerlak M 2010 (*Preprint* 1010.5437)

Journal of Physics: Conference Series **360** (2012) 012005 doi:10.1088/1742-6596/360/1/012005

Quantum scalar field in quantum gravity with spherical symmetry

Rodolfo Gambini[1], **Jorge Pullin**[2], **Saeed Rastgoo**[1]

1. Instituto de Física, Facultad de Ciencias, Iguá 4225, esq. Mataojo, Montevideo, Uruguay.
2. Department of Physics and Astronomy, Louisiana State University, Baton Rouge, LA 70803-4001

E-mail: `rgambini@fisica.edu.uy,saeed@fisica.edu.uy,pullin@lsu.edu`

Abstract. This is a summary of the talk presented by one of us in Loops 2011. We discuss the application of the uniform discretization approach to spherically symmetric gravity coupled to a spherically symmetric scalar field.

1. Introduction

Canonical loop quantum gravity is being developed in detail in scenarios of increasing complexity. The first detailed calculations were carried out for homogeneous cosmologies [1], constituting what is know called loop quantum cosmology. Further work has been developed in spherical symmetry in vacuum [2], Gowdy cosmologies [3] and parameterized field theory [4]. For various reasons in all those examples one could ignore the "problem of dynamics" of canonical quantum gravity: the fact that the constraint algebra is not a Lie algebra. In homogeneity there is only one constraint and a trivial algebra. In vacuum spherical symmetry one can choose special gauges where the problem is not present. The Gowdy case has been tackled through a hybrid quantization that bypasses the problem. The parameterized field theory has special properties that also allows to ignore the problem.

However, even in relatively simple model like spherically symmetric gravity coupled to a spherically symmetric scalar field there is currently no known method of avoiding having to deal with a constraint algebra that has structure functions. As such, we do not know how to treat the problem using the Dirac quantization procedure. To handle the situation we decided to use the uniform discretization approach [5]. In it, one discretizes the theory one wishes to study and constructs the master constraint [6] and studies its quantum spectrum. If the quantum spectrum contains the zero eigenvalue, then the continuum limit is achieved. If it does not, then one has a theory with a fundamental level of discreteness that will approximate the theory of interest at wavelengths long compared to the size of the minimum eigenvalue.

2. Polymerized spherical gravity coupled to an unpolymerized scalar field

We started our study by polymerizing and doing a loop quantization for the gravitational part of the variables whereas we quantized the scalar field in a traditional fashion [7]. We did this for simplicity and because we wanted to probe the regime close to flat space, where we expect the usual Fock treatment for the scalar field should emerge. It should be noted that the model

Journal of Physics: Conference Series **360** (2012) 012005 doi:10.1088/1742-6596/360/1/012005

we are studying can in principle exhibit a very rich dynamics including the formation of black holes, critical phenomena in gravitational collapse, quasinormal ringing, etc. Our goals initially are very modest: we would be happy to complete the quantization in a regime where the fields are very weak and space-time is close to flat.

Unfortunately, even in such a regime, carrying out the uniform discretization program for a theory as complex as the one we wish to study here in closed form does not appear feasible at present. The master constraint is a quite complicated operator to work out its spectrum. To deal with this shortcoming we decided to pursue a variational technique. We posited a trial state dependent on parameters. The choice we made was to consider a state that was a product of Gaussians in the gravitational variables at each lattice site, centered around flat space, times a Fock-like vacuum for the scalar field. We polymerized the gravitational variables but not the scalar field in a first study. We then proceeded to find the minimum for the master constraint in terms of the free parameters, the widths of the Gaussians and the lattice spacing. As such we are computing a "ground state". It turns out the master constraint's minimum occurs for a lattice spacing different from zero. This is not surprising, the model we are considering has a scalar field that contributes a zero point field whose energy diverges when the lattice spacing goes to zero. So one has to live with a lattice that will be larger than Planck scale, but is very small compared to, say, particle physics lengths. The minimum we find for the master constraint is non-zero. We are therefore in a situation where the discrete quantum theory we find will not approximate in the semiclassical regime general relativity at all scales, but only at scales large

3. Polymerizing the field

3.1. Validating the approximation

We then went on to remove the restriction we made on the scalar field and polymerize it [8]. In turn, one would like to validate that the calculations described in the previous section were justified. To this end we took the quantum master constraint with all variables polymerized and we computed its expectation value on the ground state we discussed above. We found that the expectation value differs very little from the one computed before when we had not polymerized the field. Given that the approximation is validated, we continue to use the same ground state to study the polymerized theory.

3.2. Effects of the polymerization of the field

We decided to study the polymerized scalar field as an interacting quantum field theory treating the polymerization parameter as a "coupling". That is, in the limit in which the polymerization parameter goes to zero one has a "free" theory which is just an ordinary massless scalar field. The effect of the addition of the polymerization is viewed as the addition of an "interaction" for the field with itself, and can be studied with standard Feynman diagrammatics.

There is a choice one faces in this example. Because in the lattice in $1+1$ dimensions both the scalar field and its canonical momentum are scalars one can choose to polymerize the field or polymerize the momentum. We analyzed both cases and proceeded to compute the interacting propagator to lowest order in perturbation theory. The result is that one recovers the usual propagator for the scalar field on a lattice. In the case in which the field is polymerized the former acquires a mass proportional to the square of the polymerization parameter. The resulting propagators violate Lorentz invariance due to the presence of the lattice.

3.3. Lorentz invariance

Since one is dealing with a field theory with spatial degrees of freedom we can probe if the final result makes contact with Lorentz invariant quantum field theory. As we mentioned in the previous subsection, the resulting propagators violate Lorentz invariance. This runs the danger of falling in the class of examples considered by Collins *et al.* [9] in which they argue that the

Lorentz violations generate catastrophic experimental signatures when one considers interacting quantum field theory. We have however argued [10, 8] that the situation here differs because one is dealing with a generally covariant theory and therefore one should ask questions about Dirac observables. Propagators can be turned into Dirac observables if one introduces real clocks and rods to label the points of space-time. But real clocks and rods come with fundamental limitations in the accuracy of measurement. The limitations are significantly larger than the Planck scale (or in our case, of the lattice spacing). This has the practical effect in calculations of cutting off frequencies considerably lower than Planck's. The net result is that the resulting Lorentz violations in terms of propagators that are observable are suppressed [8].

4. Summary

We have studied quantum spherically symmetric gravity coupled to a spherical quantum scalar field using the uniform discretization technique. We minimized the master constraint using a variational technique. We studied the polymerized scalar field treating its polymerization parameter as a coupling constant and computed the physical propagators. We argued that the resulting theory violates Lorentz invariant but in magnitudes that are too small to be experimentally detectable.

5. Acknowledgments

This work was supported in part by grant NSF-PHY-0650715, funds of the Hearne Institute for Theoretical Physics, FQXi, CCT-LSU, Pedeciba and ANII PDT63/076. This publication was made possible through the support of a grant from the John Templeton Foundation. The opinions expressed in this publication are those of the author(s) and do not necessarily reflect the views of the John Templeton Foundation.

[1] See A. Ashtekar, "Singularity Resolution in Loop Quantum Cosmology: A Brief Overview," arXiv:0812.4703 [gr-qc] for a recent review.
[2] R. Gambini and J. Pullin, Phys. Rev. Lett. **101**, 161301 (2008) [arXiv:0805.1187 [gr-qc]].
[3] L. J. Garay, M. Martin-Benito, G. A. Mena Marugan, Phys. Rev. **D82**, 044048 (2010). [arXiv:1005.5654 [gr-qc]].
[4] A. Laddha, M. Varadarajan, Phys. Rev. **D83**, 025019 (2011). [arXiv:1011.2463 [gr-qc]].
[5] M. Campiglia, C. Di Bartolo, R. Gambini and J. Pullin, Phys. Rev. D **74**, 124012 (2006) [arXiv:gr-qc/0610023].
[6] T. Thiemann, Class. Quant. Grav. **23**, 2249 (2006) [arXiv:gr-qc/0510011] and references therein.
[7] R. Gambini, J. Pullin, S. Rastgoo, Class. Quant. Grav. **26**, 215011 (2009). [arXiv:0906.1774 [gr-qc]].
[8] R. Gambini, J. Pullin, S. Rastgoo, "Quantum scalar field in quantum gravity: the propagator and Lorentz invariance in the spherically symmetric case," [arXiv:1105.0667 [gr-qc]].
[9] J. Collins, A. Perez, D. Sudarsky, L. Urrutia and H. Vucetich, Phys. Rev. Lett. **93**, 191301 (2004) [arXiv:gr-qc/0403053].
[10] R. Gambini, S. Rastgoo, J. Pullin, Class. Quant. Grav. **28**, 155005 (2011). [arXiv:1106.1417 [gr-qc]].

Journal of Physics: Conference Series **360** (2012) 012006 doi:10.1088/1742-6596/360/1/012006

Extension of Loop Quantum Gravity to Metric Theories beyond General Relativity

Yongge Ma

Department of Physics, Beijing Normal University, Beijing 100875, China

E-mail: mayg@bnu.edu.cn

Abstract. The successful background-independent quantization of Loop Quantum Gravity relies on the key observation that classical General Relativity can be cast into the connection-dynamical formalism with the structure group of $SU(2)$. Due to this particular formalism, Loop Quantum Gravity was generally considered as a quantization scheme that applies only to General Relativity. However, we will show that the nonperturbative quantization procedure of Loop Quantum Gravity can be extended to a rather general class of metric theories of gravity, which have received increased attention recently due to motivations coming form cosmology and astrophysics. In particular, we will first introduce how to reformulate the 4-dimensional metric $f(R)$ theories of gravity, as well as Brans-Dicke theory, into connection-dynamical formalism with real $SU(2)$ connections as configuration variables. Through these formalisms, we then outline the nonpertubative canonical quantization of the $f(R)$ theories and Brans-Dicke theory by extending the loop quantization scheme of General Relativity.

1. Introduction

In the recent 25 years, loop quantum gravity(LQG), a background independent approach to quantize general relativity (GR), has been widely investigated [1, 2, 3, 4]. It is remarkable that, as a non-renormalizable theory, GR can be non-perturbatively quantized by the loop quantization procedure. This background-independent quantization relies on the key observation that classical GR can be cast into the connection-dynamical formalism with the structure group of $SU(2)$[5, 6]. Thus one is naturally led to ask the following questions. Does the non-perturbative quantization scheme of LQG apply only to GR? What is the applicable scope of LQG? On this Loops 11 Conference, we just learned from the talks by Thiemann et al that LQG is applicable to GR in arbitrary dimensions, since higher dimensional GR could also be cast into connection dynamics with certain compact structure group. We are going to answer the question: whether GR is a unique relativistic theory of gravity with the desired connection-dynamical character?

In fact, modified gravity theories have recently received increased attention due to motivations coming form cosmology and astrophysics. A series of independent observations, including type Ia supernova, weak lens, cosmic microwave background anisotropy, baryon oscillation, etc, implied that our universe is currently undergoing a period of accelerated expansion[7]. This result has carried the "dark energy" problem in GR. A large number of phenomenological models for dark energy have been proposed, such as quintessence, phantom, Chaplygin gas, K-essence etc. In all these models, certain scalar fields are added by hand, and their origins are not understood. Moreover, there are indirect evidences suggesting that the bulk of the matter of the universe

Journal of Physics: Conference Series **360** (2012) 012006 doi:10.1088/1742-6596/360/1/012006

is invisible or dark (see e.g. [8]). The strongest evidence comes from the rotational velocity of the isolated stars or hydrogen cloud on the outskirts of galaxies. Although a few candidates for dark matter are proposed, such as sterile neutrinos, neutralinos etc, none of them are satisfying. To explain the accelerated expansion of the universe, as well as dark matter, from fundamental physics is now a great challenge. It is reasonable to consider the possibility that GR is not a valid theory of gravity on a galactic or cosmological scale. Historically, Einstein's GR is the simplest relativistic theory of gravity with correct Newtonian limit. It is worth pursuing all alternatives, which provide a high chance to new physics.

A large variety of models of $f(\mathcal{R})$ modified gravity have been proposed to explain the "dark universe" without recourse to dark energy and dark matter [9]. The action of metric $f(\mathcal{R})$ theories reads:

$$S[g]=\frac{1}{2}\int d^4x\sqrt{-g}f(\mathcal{R}) \tag{1}$$

where f is a general function of the scalar curvature $\mathcal{R}$, and we set $8\pi G=1$. Besides $f(\mathcal{R})$ theories, a well-known competing relativistic theory of gravity was proposed by Brans and Dicke in 1961 [10], which is apparently compatible with Mach's principle. To represent a varying "gravitational constant", a scalar field is non-minimally coupled to the metric as

$$S[g,\phi]=\int d^4x\sqrt{-g}(\phi\mathcal{R}-\frac{\omega}{\phi}(\partial_\mu\phi)\partial^\mu\phi). \tag{2}$$

The scalar field in Brans-Dicke theory (BDT) of gravity is then expected to account for "dark energy". It can naturally lead to cosmological acceleration if certain potential term of the scalar is added in the original action (2). It is also possible to account for the dark matter problem. Furthermore, a large part of the non-trivial tests on gravity theory is related to Einstein's equivalence principle (EEP) [11], which contains the following three elements. (i) The trajectory of a freely falling "test" body is independent of its internal structure and composition. (ii) The outcome of any non-gravitational experiments is independent of the velocity of the freely falling reference frame in which it is performed. (iii) The outcome of any non-gravitational experiments is independent of where and when in the universe it is performed. There exist many local experiments in solar-system supporting EEP, which implies the metric theories of gravity. In a metric theory of gravity, spacetime is endowed with a metric, and the trajectories of freely falling "test" bodies are geodesics of that metric. Moreover, in local freely falling reference frames, the non-gravitational laws of physics are those written in the language of Special Relativity. Actually, besides GR, both $f(\mathcal{R})$ theories and BDT belong to metric theories of gravity. For metric theories, gravity is still geometry with diffeomorphism invariance as in GR. The differences between them are just reflected in dynamical equations and additional variables. Hence, a background-independent and non-perturbative quantization for metric theories of gravity is preferable. Since metric $f(\mathcal{R})$ theories, as well as BDT, are a class of representative metric theories, which have received most attention, we will take them as examples to carry out the extension of LQG to metric theories. Throughout the paper, we use Greek alphabet for spacetime indices, Latin alphabet a,b,c,..., for spatial indices, and i,j,k,..., for internal indices.

2. Connection Dynamics of $f(\mathcal{R})$ and Brans-Dicke Theories

By introducing an independent variable s and a Lagrange multiplier ϕ, an action equivalent to (1) of $f(\mathcal{R})$ theories is proposed as [12]

$$S[g,\phi,s]=\frac{1}{2}\int d^4x\sqrt{-g}(f(s)-\phi(s-\mathcal{R})). \tag{3}$$

Journal of Physics: Conference Series **360** (2012) 012006 doi:10.1088/1742-6596/360/1/012006

The variation of (3) with respect to s yields $\phi = df(s)/ds \equiv f'(s)$. Assuming s could be resolved from the above equation, action (3) is reduced to

$$S[g,\phi] = \frac{1}{2}\int d^4x\sqrt{-g}(\phi\mathcal{R} - \xi(\phi)) \equiv \int d^4x\mathcal{L} \tag{4}$$

where $\xi(\phi) \equiv \phi s - f(s)$. Comparing action (4) with (2), it is obvious that metric $f(\mathcal{R})$ theories could be regarded as a particular kind of generalized BDT with certain potentials of ϕ and a vanishing coupling parameter ω. The virtue of (4) is that it admits a treatable Hamiltonian analysis [12]. By doing 3+1 decomposition and Legendre transformation, the Hamiltonian of metric $f(\mathcal{R})$ gravity can be derived as a liner combination of first-class constraints as:

$$H_{total} = \int_\Sigma N^a V_a + NH,$$

where N and N^a are the lapse function and shift vector respectively, and the diffeomorphism and Hamiltonian constraints read respectively

$$\begin{aligned} V_a &= -2D^b(p_{ab}) + \pi\partial_a\phi, \\ H &= \frac{2}{\sqrt{h}}(\frac{p_{ab}p^{ab} - \frac{1}{3}p^2}{\phi} + \frac{1}{6}\phi\pi^2 - \frac{1}{3}p\pi) + \frac{1}{2}\sqrt{h}(\xi(\phi) - \phi R + 2D_aD^a\phi), \end{aligned}$$

where p^{ab} and π are the momentum respectively conjugate to the spatial 3-metric h_{ab} and scalar field ϕ on the spatial manifold Σ. Although the above Hamiltonian analysis is started with the action (4) where a non-minimally coupled scalar field is introduced, we can show that the resulted Hamiltonian formalism is equivalent to the Lagrangian formalism. The symplectic structure reads

$$\{h_{ab}(x), p^{cd}(y)\} = \delta_a^{(c}\delta_b^{d)}\delta^3(x,y), \qquad \{\phi(x), \pi(y)\} = \delta^3(x,y). \tag{5}$$

To achieve the connection dynamics of metric $f(\mathcal{R})$ modified gravity, we let

$$\tilde{K}^{ab} \equiv \phi K^{ab} + \frac{h^{ab}}{2N}(\dot{\phi} - N^c\partial_c\phi),$$

where K_{ab} is the extrinsic curvature of Σ, and introduce

$$E_i^a \equiv \sqrt{h}e_i^a, \qquad \tilde{K}_a^j \equiv \tilde{K}_{ab}e^{bj},$$

where e_i^a is the triad such that $h_{ab}e_i^a e_j^b = \delta_{ij}$. We first extend the phase space of geometry to the space consisting of pairs $(E_i^a, \tilde{K}_a^i)$. It is then easy to see that the symplectic structure (5) can be derived from the following Poisson brackets:

$$\{E_j^a(x), E_k^b(y)\} = \{\tilde{K}_a^j(x), \tilde{K}_b^k(y)\} = 0, \qquad \{\tilde{K}_a^j(x), E_k^b(y)\} = \delta_a^b\delta_k^j\delta(x,y). \tag{6}$$

Thus there is a symplectic reduction from the extended phase space to the original one. Since $\tilde{K}^{ab} = \tilde{K}^{ba}$, we have an additional constraint:

$$G_{jk} \equiv \tilde{K}_{a[j}E_{k]}^a = 0. \tag{7}$$

For second step, we make a canonical transformation by defining

$$A_a^i = \Gamma_a^i + \gamma\tilde{K}_a^i,$$

where Γ_a^i is the spin connection determined by E_i^a and γ is a nonzero real number. Then the Poisson brackets among the new variables read

$$\{A_a^j(x), E_k^b(y)\} = \gamma\delta_a^b\delta_k^j\delta(x,y), \quad \{A_a^i(x), A_b^j(y)\} = 0.$$

Now, the phase space consists of conjugate pairs (A_a^i, E_j^b) and (ϕ, π). Combining Eq.(7) with the compatibility condition: $\partial_a E_i^a + \epsilon_{ijk}\Gamma_a^j E^{ak} = 0$, we obtain the standard Gaussian constraint

$$\mathcal{G}_i = \mathcal{D}_a E_i^a \equiv \partial_a E_i^a + \epsilon_{ijk}A_a^j E^{ak}, \tag{8}$$

which justifies A_a^i as an Ashtekar-Barbero $su(2)$-connection. Had we let $\gamma = \pm i$, the (anti-)self-dual complex connection formalism would be obtained. The original diffeomorphism constraint can be expressed in terms of new variables up to Gaussian constraint as

$$\mathcal{V}_a = \frac{1}{\gamma}F_{ab}^i E_i^b + \pi\partial_a\phi,$$

where $F_{ab}^i \equiv 2\partial_{[a}A_{b]}^i + \epsilon_{kl}^i A_a^k A_b^l$ is the curvature of A_a^i. The original Hamiltonian constraint can be written up to Gaussian constraint as

$$\begin{aligned} H_{fR} &= \frac{\phi}{2}[F_{ab}^j - (\gamma^2 + \frac{1}{\phi^2})\varepsilon_{jmn}\tilde{K}_a^m\tilde{K}_b^n]\frac{\varepsilon_{jkl}E_k^aE_l^b}{\sqrt{h}} \\ &+ \frac{1}{2}(\frac{2}{3\phi}\frac{(\tilde{K}_a^iE_i^a)^2}{\sqrt{h}} + \frac{4}{3}\frac{(\tilde{K}_a^iE_i^a)\pi}{\sqrt{h}} + \frac{2}{3}\frac{\pi^2\phi}{\sqrt{h}} + \sqrt{h}\xi(\phi)) + \sqrt{h}D_aD^a\phi. \end{aligned} \tag{9}$$

Similar to GR, the constraints are of first class. Thus the total Hamiltonian is a linear combination of the constraints. We have obtained the real $su(2)$-connection dynamical formalism of Lorentz $f(\mathcal{R})$ gravity [13, 14]. Note that a connection dynamical formalism of Euclidean $f(\mathcal{R})$ theories in Einstein frame was derived in [15]

We now turn to the BDT. By doing 3+1 decomposition and Legendre transformation, the Hamiltonian of BDT with $\omega \neq -3/2$ can be derived from the action (2) as a liner combination of first-class constraints as $H_{total} = \int_\Sigma N^a V_a + NH$, where the diffeomorphism constraint V_a has the same form as that of metric $f(\mathcal{R})$ theories, while the Hamiltonian constraint reads [16, 17]

$$H = \frac{2}{\sqrt{h}}\left(\frac{p_{ab}p^{ab} - \frac{1}{2}p^2}{\phi} + \frac{(p-\phi\pi)^2}{2\phi(3+2\omega)}\right) + \frac{1}{2}\sqrt{h}\left(-\phi R + \frac{\omega}{\phi}(D_a\phi)D^a\phi + 2D_aD^a\phi\right).$$

In comparison with metric $f(\mathcal{R})$ theories, there is a new kinetic term $\frac{\omega\sqrt{h}}{2\phi}(D_a\phi)D^a\phi$ of the scalar field in the above Hamiltonian constraint. However, this kinetic term will not affect the canonical transformations to connection dynamics. Following the same canonical transformations in $f(\mathcal{R})$ theories, we obtain new conjugate variables for the geometry as: $A_a^i = \Gamma_a^i + \gamma\tilde{K}_a^i$ and $E_i^a = \sqrt{h}e_i^a$, as well as the Gaussian constraint (8). In terms of the new variables, the Hamiltonian constraint reads

$$\begin{aligned} H_{BD} &= \frac{\phi}{2}\left(F_{ab}^j - (\gamma^2 + \frac{1}{\phi^2})\varepsilon_{jmn}\tilde{K}_a^m\tilde{K}_b^n\right)\frac{\varepsilon^{jkl}E_k^aE_l^b}{\sqrt{h}} \\ &+ \frac{1}{3+2\omega}\left(\frac{(\tilde{K}_a^iE_i^a)^2}{\phi\sqrt{h}} + 2\frac{(\tilde{K}_a^iE_i^a)\pi}{\sqrt{h}} + \frac{\pi^2\phi}{\sqrt{h}}\right) + \frac{\omega}{2\phi}\sqrt{h}(D_a\phi)D^a\phi + \sqrt{h}D_aD^a\phi. \end{aligned} \tag{10}$$

The special case of $\omega = -3/2$ can also be dealt with [17]. Hence both $f(\mathcal{R})$ theories and BDT can be cast into connection dynamics with $su(2)$-connection as one of configuration variables.

Journal of Physics: Conference Series **360** (2012) 012006 doi:10.1088/1742-6596/360/1/012006

3. Loop Quantization of $f(\mathcal{R})$ and Brans-Dicke Theories

Based on the connection dynamical formalisms, the nonperturbative loop quantization procedure can be straightforwardly extended to both $f(\mathcal{R})$ theories and BDT. Since the scalar field in $f(\mathcal{R})$ theories and BDT still reflects gravity, it is natural to employ the polymer-like representations for both the scalar field and the connection. The following quantum kinematical structure is valid for both $f(\mathcal{R})$ theories and BDT.

For the geometry sector, we have the unique diffeomorphism and internal gauge invariant representation for the quantum holonomy-flux algebra[1]. The kinematical Hilbert space of geometry reads

$$\mathcal{H}^{geo}_{kin} = L^2(\bar{\mathcal{A}}, d\mu_{AL}),$$

with the so-called spin-network basis $T_\alpha(A) \equiv T_{\alpha,j,m,n}(\bar{A})$. The spatial geometric operators of LQG, such as the area [18], the volume [19] and the length operators [20, 21] are still valid here. Hence, the important physical result that both the area and the volume are discrete at quantum kinematical level remains true for loop quantum $f(\mathcal{R})$ gravity as well as loop quantum Brans-Dicke gravity.

For the polymer-like quantization of the scalar field [22], one extends the space $\mathcal{U}$ of smooth scalar fields to the quantum configuration space $\bar{\mathcal{U}}$. A simple element $U \in \bar{\mathcal{U}}$ may be thought as a point holonomy: $U_\lambda = \exp(i\lambda\phi(x))$ at point $x \in \Sigma$, where λ is a real number. By GNS structure, there is a natural diffeomorphism invariant measure $d\mu$ on $\bar{\mathcal{U}}$. Thus the kinematical Hilbert space of scalar field reads

$$\mathcal{H}^{sca}_{kin} = L^2(\bar{\mathcal{U}}, d\mu),$$

with the scalar-network basis

$$T_X(\phi) \equiv \prod_{x_j \in X} U_{\lambda_j}(\phi(x_j)),$$

where $X = \{x_1, \ldots, x_n\}$ is an arbitrary given set of finite number of points in Σ. The total kinematical Hilbert space for $f(\mathcal{R})$ and Brans-Dicke gravity reads $\mathcal{H}_{kin} := \mathcal{H}^{geo}_{kin} \otimes \mathcal{H}^{sca}_{kin}$, with an orthonormal basis

$$T_{\alpha,X}(A,\phi) \equiv T_\alpha(A) \otimes T_X(\phi).$$

The holonomy $h_e(A) = \mathcal{P}\exp\int_e A_a$, flux $E(S,f) := \int_S \epsilon_{abc}E^a_i f^i$, point holonomy U_λ, and smeared momentum $\pi(R) := \int_R d^3x\pi(x)$ of scalar field become basic operators in $\mathcal{H}_{kin}$. Their actions read respectively

$$\begin{aligned}\hat{h}_e(A)\Psi(A,\phi) &= h_e(A)\Psi(A,\phi), \quad \hat{E}(S,f)\Psi(A,\phi) = i\hbar\{E(S,f),\Psi(A,\phi)\},\\ \hat{U}_\lambda(\phi(x))\Psi(A,\phi) &= \exp(i\lambda\phi(x))\Psi(A,\phi), \quad \hat{\pi}(R)\Psi(A,\phi) = i\hbar\{\pi(R),\Psi(A,\phi)\}.\end{aligned}$$

As in LQG, it is straightforward to promote the Gaussian constraint $\mathcal{G}(\Lambda)$ to a well-defined operator. The kernel of $\hat{\mathcal{G}}(\Lambda)$ in $\mathcal{H}_{kin}$ is the internal gauge invariant Hilbert space $\mathcal{H}_G$, with gauge invariant spin-scalar-network basis: $T_{s,c}(A,\phi) \equiv T_{s=(\alpha,j,i)}(A) \otimes T_X(\phi)$. Since the diffeomorphisms of Σ act covariantly on the cylindrical functions in $\mathcal{H}_G$, the so-called group averaging technique can be employed to solve the diffeomorphism constraint [2, 3, 4]. Thus we can also obtain the desired diffeomorphism and gauge invariant Hilbert space $\mathcal{H}_{Diff}$ for $f(\mathcal{R})$ and Brans-Dicke gravity.

The quantum dynamics is a nontrivial issue. To implement the Hamiltonian constraint at quantum level, we need to deal with the metric $f(\mathcal{R})$ and Brans-Dicke theories separately. The smeared version of the Hamiltonian constraint of metric $f(\mathcal{R})$ theories consists of seven terms in the order of Eq.(9) as: $H_{fR}(N) = \sum_{i=1}^{7} H_i(N)$. Comparing it with that of GR in connection formalism, the new ingredients that we have to deal with are $\phi(x), \phi^{-1}(x), \xi(\phi)$ and the four

[1] See e.g. talks by Sahlmann, Giesel, Lewandowski on Loops 11 Conference in Madrid

Journal of Physics: Conference Series **360** (2012) 012006 doi:10.1088/1742-6596/360/1/012006

terms: $H_3(N), H_4(N), H_5(N), H_7(N)$. By introducing certain small constant λ_0, an operator corresponding to the scalar $\phi(x)$ can be defined as

$$\hat{\phi}(x) = \frac{1}{2i\lambda_0}(U_{\lambda_0}(\phi(x)) - U_{-\lambda_0}(\phi(x))).$$

The ambiguity of λ_0 is the price that we have to pay in order to represent field ϕ in the polymer-like representation. To further define an operator corresponding to $\phi^{-1}(x)$, we can use the classical identity

$$\phi^{-1}(x) = sgn[\phi]\left(l^{-1}sgn[\phi]\{|\phi(x)|^l, \pi(R)\}\right)^{\frac{1}{1-l}}, \tag{11}$$

for any rational number $l \in (0,1)$. For example, one may choose $l = \frac{1}{2}$ for positive $\phi(x)$ and replace the Poisson bracket by commutator to define

$$\hat{\phi}^{-1}(x) = (\frac{2}{i\hbar}[\sqrt{\hat{\phi}(x)}, \hat{\pi}(R)])^2.$$

Then all the functions $\xi(\phi)$ which can be expanded as powers of $\phi(x)$ have been quantized. For other non-trivial types of $\xi(\phi)$, we may replace the argument ϕ by $\hat{\phi}$, provided that no divergence would arise after the replacement. In the case where divergence does appear, there remain the possibilities to employ tricks similar to Eq.(11) to deal with it. Hence it is reasonable to believe that most physically interesting functions $\xi(\phi)$ can be quantized. Then it is straight-forward to quantize the term $H_6(N)$ in the Hamiltonian constraint as an operator acting on an basis vector $T_{\alpha,X}$ as

$$\hat{H}_6(N) \cdot T_{\alpha,X} = \frac{1}{2}\sum_{v \in V(\alpha)} N(v)\hat{\xi}(\phi(v))\hat{V}_v \cdot T_{\alpha,X}.$$

Moreover, by the regularization techniques developed for the Hamiltonian constraint operators of LQG and polymer-like scalar field [2, 23], all the terms $H_3(N), H_4(N), H_5(N), H_7(N)$ can be quantized as operators acting on cylindrical functions in $\mathcal{H}_{kin}$ in state-dependent ways. The regularization procedure involves re-expressing some variables by Thiemann's trick, point-splitting, triangulating Σ in adaptation to some graph $\alpha \cup X$ underling a cylindrical function, and replacing connections by holonomies. Replacing the classical variables and Poisson brackets by the corresponding operators and commutators respectively, we can obtain the regulated Hamitonian constraint operator $\hat{H}^{\epsilon}_{fR}(N)$. It turns out that the regulators of $\hat{H}^{\epsilon}_3(N), \hat{H}^{\epsilon}_4(N), \hat{H}^{\epsilon}_5(N)$ can be straightforwardly removed. However, the action of $\hat{H}^{\epsilon}_7(N)$ on a basis vector $T_{\alpha,X}$ is graph changing. It adds a finite number of vertices with representation λ_0 at $t(s_I(v)) = \epsilon$ for edges $e_I(t)$ starting from each high(≥ 3)-valent vertex of α. The family of operators $\hat{H}^{\epsilon}_7(N)$ fails to be weakly convergent when $\epsilon \to 0$. However, due to the diffeomorphism covariant properties of the triangulation, the limit operator can be well defined via the so-called uniform Rovelli-Smolin topology induced by diffeomorphism-invariant states. The operators corresponding to the four new terms in the Hamiltonian constraint act on a basis vector as follows.

$$\hat{H}_3(N) \cdot T_{\alpha,X} = \frac{4}{3\gamma^3(i\hbar)^4}\sum_{v \in V(\alpha)} N(v)\hat{\phi}^{-1}(v)[\hat{H}^{Euc}(1), \sqrt{\hat{V}_v}]\,[\hat{H}^{Euc}(1), \sqrt{\hat{V}_v}] \cdot T_{\alpha,X},$$

$$\begin{aligned}\hat{H}_4(N) \cdot T_{\alpha,X} = & -\sum_{v \in V(\alpha)\cap X} \frac{2^{20}N(v)}{3^6\gamma^6(i\hbar)^6E(v)^2}\hat{\pi}(v) \sum_{v(\Delta)=v(\Delta')=v} \varepsilon(s_L, s_M, s_N)\varepsilon(s_I, s_J, s_K)\epsilon^{LMN} \\ & \times Tr(\tau_i\hat{h}_{s_L(\Delta)}[\hat{h}^{-1}_{s_L(\Delta)}, \hat{\tilde{K}}])Tr(\tau_i\hat{h}_{s_M(\Delta)}[\hat{h}^{-1}_{s_M(\Delta)}, (\hat{V}_v)^{3/4}]\hat{h}_{s_N(\Delta)}[\hat{h}^{-1}_{s_N(\Delta)}, (\hat{V}_v)^{3/4}])\epsilon^{IJK} \\ & \times Tr(\hat{h}_{s_I(\Delta')}[\hat{h}^{-1}_{s_I(\Delta')}, (\hat{V}_v)^{1/2}]\hat{h}_{s_J(\Delta')}[\hat{h}^{-1}_{s_J(\Delta')}, (\hat{V}_v)^{1/2}]\hat{h}_{s_K(\Delta')}[\hat{h}^{-1}_{s_K(\Delta')}, (\hat{V}_v)^{1/2}]) \cdot T_{\alpha,X},\end{aligned}$$

$$\begin{aligned}\hat{H}_5(N)\cdot T_{\alpha,X} &= \sum_{v\in V(\alpha)\cap X}\frac{2^{18}N(v)}{3^5\gamma^6(i\hbar)^6E(v)^2}\hat{\pi}(v)\phi(v)\hat{\pi}(v)\sum_{v(\Delta)=v(\Delta')=v}\varepsilon(s_I,s_J,s_K)\varepsilon(s_L,s_M,s_N)\\ &\times\ \epsilon^{IJK}Tr(\hat{h}_{s_I(\Delta)}[\hat{h}^{-1}_{s_I(\Delta)},(\hat{V}_v)^{1/2}]\hat{h}_{s_J(\Delta)}[\hat{h}^{-1}_{s_J(\Delta)},(\hat{V}_v)^{1/2}]\hat{h}_{s_K(\Delta)}[\hat{h}^{-1}_{s_K(\Delta)},(\hat{V}_v)^{1/2}])\epsilon^{LMN}\\ &\times\ Tr(\hat{h}_{s_L(\Delta')}[\hat{h}^{-1}_{s_L(\Delta')},(\hat{V}_v)^{1/2}]\hat{h}_{s_M(\Delta')}[\hat{h}^{-1}_{s_M(\Delta')},(\hat{V}_v)^{1/2}]\hat{h}_{s_N(\Delta')}[\hat{h}^{-1}_{s_N(\Delta')},(\hat{V}_v)^{1/2}])\cdot T_{\alpha,X},\end{aligned}$$

$$\begin{aligned}\hat{H}_7(N)\cdot T_{\alpha,X} &= -\sum_{v\in V(\alpha)}\frac{2^7N(v)}{3\gamma^2 i\lambda_0(i\hbar)^2E(v)}\sum_{b(e)=v}\hat{X}^i_e\sum_{v(\Delta)=v}\varepsilon(s_I,s_J,s_K)\\ &\times\ \epsilon^{IJK}\hat{U}^{-1}_{\lambda_0}(\phi(s_{s_I(\Delta')}))\left(\hat{U}_{\lambda_0}(\phi(t_{s_I(\Delta')}))-\hat{U}_{\lambda_0}(\phi(s_{s_I(\Delta')}))\right)\\ &\times\ Tr(\tau_i\hat{h}_{s_J(\Delta)}[\hat{h}^{-1}_{s_J(\Delta)},(\hat{V}_v)^{1/2}]\hat{h}_{s_K(\Delta)}[\hat{h}^{-1}_{s_K(\Delta)},(\hat{V}_v)^{1/2}])\cdot T_{\alpha,X},\end{aligned}$$

where $\varepsilon(s_I,s_J,s_K):=\mathrm{sgn}(\det(\dot{s}_I\dot{s}_J\dot{s}_K)(v))$ takes the values $(+1,-1,0)$ if the tangents of the three segments s_I,s_J,s_K at v (in that sequence) form a matrix of positive, negative or vanishing determinant, and $E(v)\equiv\binom{n(v)}{3}$ with $n(v)$ denoting the valence of the vertex v. The whole Hamiltonian constraint operator $\hat{H}(N)=\sum_v\hat{H}(N)_v$ is internal gauge invariant and hence also well defined in the gauge invariant Hilbert space $\mathcal{H}_G$.

Although $\hat{H}(N)$ can dually act on the diffeomorphism invariant states, there is no guarantee for the resulted states to be still diffeomorphism invariant. Hence it is difficult to define a Hamiltonian constraint operator directly on the diffeomorphism invariant Hilbert space $\mathcal{H}_{Diff}$. The idea of resolution is to introduce the master constraint as in LQG [24]:

$$\mathbf{M}:=\frac{1}{2}\int_\Sigma d^3x\frac{|H(x)|^2}{\sqrt{|\det h(x)|}}.$$

One then gets the master constraint algebra as a Lie algebra:

$$\begin{aligned}\{\mathcal{V}(\vec{N}),\ \mathcal{V}(\vec{N}')\} &= \mathcal{V}([\vec{N},\vec{N}']),\\ \{\mathcal{V}(\vec{N}),\ \mathbf{M}\} &= 0,\\ \{\mathbf{M},\ \mathbf{M}\} &= 0,\end{aligned}$$

where the subalgebra of diffeomorphism constraints forms an ideal. So it is possible to define a corresponding master constraint operator on $\mathcal{H}_{Diff}$. The master constraint can be regulated via a point-splitting strategy as [25]:

$$\mathcal{M}^\epsilon=\frac{1}{2}\int_\Sigma d^3y\int_\Sigma d^3x\chi_\epsilon(x-y)\frac{H(x)}{\sqrt{V_{U^\epsilon_x}}}\frac{H(y)}{\sqrt{V_{U^\epsilon_y}}}.$$

Introducing a partition $\mathcal{P}$ of the 3-manifold Σ into cells C, we have an operator $\hat{H}^\varepsilon_{C,\alpha}$ acting on gauge invariant spin-scalar-network basis $T_{s,c}$ via a state-dependent triangulation as

$$\hat{H}^\varepsilon_{C,\alpha}\cdot T_{s,c}\equiv(\hat{H}^\varepsilon_{C,GR}+\sum_{i=3}^{7}\hat{H}^\varepsilon_{C,i})\cdot T_{s,c}=\sum_{v\in V(\alpha)}\chi_C(v)(\sum_{v(\Delta)=v}\hat{H}^{\varepsilon,\Delta}_{GR,v}+\sum_{i=3}^{7}\hat{H}^\varepsilon_{i,v})\cdot T_{s,c},$$

where α denotes the underlying graph of the spin-network state T_s. Here,

$$\begin{aligned}\hat{H}^{\varepsilon,\Delta}_{GR,v} &= \frac{32}{3E(v)}\varepsilon(s_I,s_J,s_K)\epsilon^{IJK}(\frac{1}{i\hbar\gamma}\hat{\phi}(v)Tr(\hat{h}_{\alpha_{IJ}(\Delta)}\hat{h}_{s_K(\Delta)}[\hat{h}^{-1}_{s_K(\Delta)},\sqrt{\hat{V}_{U^\epsilon_v}}])\\ &- \frac{1}{(i\hbar)^3\gamma^3}(\hat{\phi}^{-1}(v)+\gamma^2\hat{\phi}(v))Tr(\hat{h}_{s_I(\Delta)}[\hat{h}^{-1}_{s_I(\Delta)},\hat{\bar{K}}]\hat{h}_{s_J(\Delta)}[\hat{h}^{-1}_{s_J(\Delta)},\hat{\bar{K}}]\hat{h}_{s_K(\Delta)}[\hat{h}^{-1}_{s_K(\Delta)},\sqrt{\hat{V}_{U^\epsilon_v}}])),\end{aligned}$$

Journal of Physics: Conference Series **360** (2012) 012006 doi:10.1088/1742-6596/360/1/012006

$$\hat{H}^{\varepsilon}_{3,v} = \frac{16}{3\gamma^3(i\hbar)^4}\hat{\phi}^{-1}(v)[\hat{H}_{Euc}(1),(\hat{V}_{U^{\varepsilon}_v})^{1/4}][\hat{H}_{Euc}(1),(\hat{V}_{U^{\varepsilon}_v})^{1/4}],$$

and $\hat{H}^{\varepsilon}_{4,v}$, $\hat{H}^{\varepsilon}_{5,v}$, $\hat{H}^{\varepsilon}_{6,v}$, $\hat{H}^{\varepsilon}_{7,v}$ can be defined similarly [14]. The family of operators $\hat{H}^{\varepsilon}_{C}$ are cylindrically consistent up to diffeomorphisms. The inductive limit operator $\hat{H}_C$ can be well defined by the uniform Rovelli-Smolin topology. Thus we can define master constraint operator $\hat{\mathcal{M}}$ on diffeomorphism invariant states as

$$(\hat{\mathcal{M}}\Phi_{Diff})T_{s,c} = \lim_{\mathcal{P}\to\Sigma;\varepsilon,\varepsilon'\to 0}\Phi_{Diff}[\frac{1}{2}\sum_{C\in\mathcal{P}}\hat{H}^{\varepsilon}_{C}(\hat{H}^{\varepsilon'}_{C})^{\dagger}T_{s,c}].$$

It is obvious that $\hat{\mathcal{M}}$ is diffeomorphism invariant. For any given diffeomorphism invariant spin-scalar-network state $T_{[s,c]}$, the norm $\|\hat{\mathcal{M}}T_{[s,c]}\|_{Diff}$ is finite. So $\hat{\mathcal{M}}$ is densely defined in $\mathcal{H}_{Diff}$. Moreover, $\hat{\mathcal{M}}$ is positive and symmetric and hence admits a unique self-adjoint Friedrichs extension. It is then possible to obtain the physical Hilbert space of quantum $f(\mathcal{R})$ gravity by the direct integral decomposition of $\mathcal{H}_{Diff}$ with respect to $\hat{\mathcal{M}}$.

Now we turn to the quantum dynamics of BDT. Comparing the Hamiltonian constraint (2) of Brans-Dicke gravity with that of metric $f(\mathcal{R})$ gravity in connection formalism, the only new ingredient that we have to deal with is the kinetic term:

$$H_k(N) = \int_{\Sigma} d^3x N\frac{\omega}{2\phi}\sqrt{h}(D_a\phi)D^a\phi.$$

By the regularization techniques similar to those for $f(\mathcal{R})$ gravity, it can be quantized as a well-defined operator in $\mathcal{H}_{kin}$ as [16]

$$\begin{aligned}
\hat{H}_k\cdot T_{\alpha,X} \quad &= \quad \sum_{v\in V(\alpha)}\frac{2^{17}\omega N(v)}{3^6\gamma^4(i\lambda_0)^2(i\hbar)^4E(v)^2}\hat{\phi}^{-1}(v)\sum_{v(\Delta)=v(\Delta')=v}\varepsilon(s_I,s_J,s_K)\varepsilon(s_L,s_M,s_N)\\
&\times \quad \epsilon^{LMN}\hat{U}^{-1}_{\lambda_0}(\phi(s_{s_L(\Delta)}))[\hat{U}_{\lambda_0}(\phi(t_{s_L(\Delta)}))-\hat{U}_{\lambda_0}(\phi(s_{s_L(\Delta)}))]\\
&\times \quad Tr(\tau_i\hat{h}_{s_M(\Delta)}[\hat{h}^{-1}_{s_M(\Delta)},(\hat{V}_v)^{3/4}]\hat{h}_{s_N(\Delta)}[\hat{h}^{-1}_{s_N(\Delta)},(\hat{V}_v)^{3/4}])\\
&\times \quad \epsilon^{IJK}\hat{U}^{-1}_{\lambda_0}(\phi(s_{s_I(\Delta')}))[\hat{U}_{\lambda_0}(\phi(t_{s_I(\Delta')}))-\hat{U}_{\lambda_0}(\phi(s_{s_I(\Delta')}))]\\
&\times \quad Tr(\tau_i\hat{h}_{s_J(\Delta')}[\hat{h}^{-1}_{s_J(\Delta')},(\hat{V}_v)^{3/4}]\hat{h}_{s_K(\Delta')}[\hat{h}^{-1}_{s_K(\Delta')},(\hat{V}_v)^{3/4}])\cdot T_{\alpha,X}.
\end{aligned}$$

Since all the other terms in the Hamiltonian constraint are equal to the corresponding terms of metric $f(\mathcal{R})$ theories up to coefficients, the whole Hamiltonian constraint operator of BDT with $\omega\neq -3/2$ has been well defined. The quantum dynamics for the special case $\omega\neq -3/2$ of BDT can also be well defined [17]. For the same reason as in $f(\mathcal{R})$ gravity, we also wish to define a master constraint operator for Brans-Dicke gravity. The "square root" of the new term in the regulated master constraint can be quantized as

$$\begin{aligned}
\hat{H}^{\varepsilon}_{k,v} \quad &= \quad \sum_{v(\Delta)=v(\Delta')=v}\frac{2^{15}\omega}{3^4\gamma^4(i\lambda_0)^2(i\hbar)^4E(v)^2}\varepsilon(s_I,s_J,s_K)\varepsilon(s_L,s_M,s_N)\\
&\times \quad \hat{\phi}^{-1}(v)\epsilon^{LMN}\hat{U}^{-1}_{\lambda_0}(\phi(s_{s_L(\Delta)}))[\hat{U}_{\lambda_0}(\phi(t_{s_L(\Delta)}))-\hat{U}_{\lambda_0}(\phi(s_{s_L(\Delta)}))]\\
&\times \quad Tr(\tau_i\hat{h}_{s_M(\Delta)}[\hat{h}^{-1}_{s_M(\Delta)},(\hat{V}_v)^{1/2}]\hat{h}_{s_N(\Delta)}[\hat{h}^{-1}_{s_N(\Delta)},(\hat{V}_v)^{3/4}])\\
&\times \quad \epsilon^{IJK}\hat{U}^{-1}_{\lambda_0}(\phi(s_{s_I(\Delta')}))[\hat{U}_{\lambda_0}(\phi(t_{s_I(\Delta')}))-\hat{U}_{\lambda_0}(\phi(s_{s_I(\Delta')}))]\\
&\times \quad Tr(\tau_i\hat{h}_{s_J(\Delta')}[\hat{h}^{-1}_{s_J(\Delta')},(\hat{V}_v)^{1/2}]\hat{h}_{s_K(\Delta')}[\hat{h}^{-1}_{s_K(\Delta')},(\hat{V}_v)^{3/4}]),
\end{aligned}$$

from which a positive and self-adjoint master constraint operator can be defined on $\mathcal{H}_{Diff}$. Thus the master constraint program is also valid for BDT.

Journal of Physics: Conference Series **360** (2012) 012006 doi:10.1088/1742-6596/360/1/012006

4. Summary and Outlook

Based on the above loop quantization procedure of metric $f(\mathcal{R})$ theories and BDT, we can propose a general quantization scheme for 4-dimensional metric theories of gravity. The prerequisite is that the theories should have well-defined geometrical dynamics, which means a Hamiltonian formalism with 3-metric h_{ab} as one of configuration variables and a closed (first-class) constraint algebra (perhaps after solving some second-class constraints). Without loss of generality, we assume that the classical phase space of the theory consists of conjugate pairs (h_{ab}, p^{ab}) and (ϕ_A, π^A), where ϕ_A could be a rather arbitrary scalar, vector, tensor or spinor field. Then the quantization scheme consists of the following steps. (i) To obtain a connection dynamical formalism, we first enlarge the phase space by transforming to the triad formulation as

$$(h_{ab}, p^{ab}) \rightsquigarrow (E^a_j \equiv \sqrt{h}h^{ab}e_{aj}, \tilde{K}^j_a \equiv \tilde{K}_{ab}e^{bj}),$$

where $\tilde{K}_{ab} = \frac{2}{\sqrt{h}}(p_{ab} - \frac{1}{2}ph_{ab})$ and $\tilde{K}_{a[i}E^a_{j]} = 0$. Then we make a canonical transformation to connection formulation as:

$$(E^a_j, \tilde{K}^j_a) \rightsquigarrow (E^a_j, A^j_a \equiv \Gamma^j_a + \gamma\tilde{K}^j_a),$$

with the Gaussian constraint, $\mathcal{D}_a E^a_i \equiv \partial_a E^a_i + \epsilon_{ijk} A^j_a E^{ak} = 0$, appeared by construction. It is straightforward to write all the constraints in terms of the new variables. (ii) For loop quantization, we first find the polymer-like representation of the fields (ϕ_A, π^A), together with the LQG representation of the holonomy-flux algebra. Then the kinematical Hilbert space reads $\mathcal{H}_{kin} := \mathcal{H}^{geo}_{kin} \otimes \mathcal{H}^{\phi}_{kin}$, where the basic operators and geometrical operators could be well defined. By implementing the Gaussian and diffeomorphism constraints as in standard LQG, we could get the gauge and diffeomorphism invariant Hilbert space as: $\mathcal{H}_{kin} \rightsquigarrow \mathcal{H}_G \rightsquigarrow \mathcal{H}_{Diff}$. To implement quantum dynamics, one may first construct the Hamiltonian constraint operator at least in $\mathcal{H}_G$. Usually it could not be well defined on $\mathcal{H}_{Diff}$. Then we can construct master constraint operator in $\mathcal{H}_{Diff}$ by using the structure of the Hamiltonian operator. (iii) One may try to understand the physical Hilbert space by the direct integral decomposition of $\mathcal{H}_{Diff}$ with respect to the master constraint operator. (iv) One may also do certain semiclassical analysis in order to confirm the classical limits of the Hamiltonian and master constraint operators as well as the constraint algebra. The low energy physics is also expected in the analysis. (v) Finally, to complement above canonical approach, we can also try the covariant path integral (spin foam) quantization. It should be noted that in the present work we only finished steps (i) and (ii) for metric $f(\mathcal{R})$ theories and BDT.

To summarize, our main results are in two folds. First, the 4-dimensional connection dynamics of metric $f(\mathcal{R})$ theories, as well as Brans-Dicke theory, have been obtained by canonical transformations from their geometrical dynamics. Thus GR is not the unique theory of gravity with connection dynamical character. Second, due to the $su(2)$-connection dynamical formalism, the 4-dimensional metric $f(\mathcal{R})$ theories and Brans-Dicke theory have been nonperturbatively quantized by extending LQG scheme. Hence, the non-perturbative loop quantization procedure is not only valid for GR but also valid for a general class of metric theories of gravity. In fact, it is not difficult to extend LQG further to general scalar-tensor theories of gravity [17]. Moreover, since higher dimensional scalar-tensor theories of gravity have well-defined Hamiltonian geometrical dynamics, and the symplectic reduction of Bodendorfer-Thiemann-Thurn connection formalism to metric formalism does not depend on dynamics, LQG may also be extended to higher(> 4) dimensional scalar-tensor theories of gravity [26].

Of course, there are still many open issues on the extension of LQG. It is desirable to find suitable actions for the connection dynamics of $f(\mathcal{R})$ theories and that of Brans-Dicke theory.

We will explore the applications of loop quantum $f(\mathcal{R})$ and Brans-Dicke theories to cosmology and black holes in future work [27]. It is also desirable to quantize metric theories of gravity by the covariant spin foam approach [28]. To conclude, our conservative observation is that LQG could be applicable to metric theories of gravity (with well-defined geometrical dynamics) in arbitrary dimensions. However, a caution arises from the difference between the 4(or 3)-dimensional and higher-dimensional connection dynamical formulations of GR. In particular, in 4-dimensional case, we have both Ashtekar-Barbero connection dynamics and Bodendorfer-Thiemann-Thurn connection dynamics of GR. Are the quantum theories corresponding to them unitary equivalent to each other [2]? If the answer is negative, are there any theoretical criteria for judging them? Is Bodendorfer-Thiemann-Thurn connection dynamics only preferable in higher dimensions? Another interesting question is whether LQG can be extended to non-metric theories, e.g., metric-affine $f(\mathcal{R})$ theories. All these open issues are fascinating and deserve future investigating.

Acknowledgments

The author would like to thank Thomas Thiemann and Xiangdong Zhang for helpful discussion and fruitful collaboration and acknowledge the organizers of Loops 11 conference for the financial support. This work is supported in part by NSFC (Grant No.10975017) and the Fundamental Research Funds for the Central Universities.

References

[1] C. Rovelli, Quantum Gravity, (Cambridge University Press, 2004).
[2] T. Thiemann, Modern Canonical Quantum General Relativity, (Cambridge University Press, 2007).
[3] A. Ashtekar and J. Lewandowski 2004 Class. Quant. Grav. **21** R53
[4] M. Han, Y. Ma and W. Huang 2007 Int. J. Mod. Phys. D **16** 1397
[5] A. Ashtekar 1986 Phys. Rev. Lett. **57** 2244
[6] J. Barbero 1995 Phys. Rev. **D 51** 5507
[7] J. Friemann, M. Turner, D. Huterer 2008 Ann. Rev. Astron. Astrophys. **46** 385
[8] W. Freedman and M. Turner 2003 Rev. Mod. Phys. **75** 1433
[9] T. P. Sotiriou and V. Faraoni 2010 Rev. Mod. Phys. **82** 451
[10] C. Brans and R. H. Dicke 1961 Phys. Rev. **124** 925
[11] C. M. Will 2006 Living Rev. Rel. **9** 3
[12] N. Deruelle, Y. Sendouda, and A. Youssef 2009 Phys. Rev. D **80** 084032
[13] X. Zhang and Y. Ma 2011 Phys. Rev. Lett. **106** 171301
[14] X. Zhang and Y. Ma 2011 Phys. Rev. D **84** 064040
[15] L. Fatibene, M. Ferraris, M. Francaviglia 2010 Class. Quant. Grav. **27** 185016
[16] X. Zhang and Y. Ma 2011 Loop quantum Brans-Dicke theory (To appear in Journal of Physics: Conference Series).
[17] X. Zhang and Y. Ma 2011 Phys. Rev. D **84** 104045
[18] C. Rovelli and L. Smolin 1995 Nucl. Phys. B **442** 593
[19] A. Ashtekar and J. Lewandowski 1998 Adv. Theor. Math. Phys. **1** 388
[20] T. Thiemann 1998 J. Math. Phys. **39** 3372
[21] Y. Ma, C. Soo, J. Yang 2010 Phys. Rev. D **81** 124026
[22] A. Ashtekar, J. Lewandowski, H. Sahlmann 2003 Class. Quant. Grav. **20** L11
[23] M. Han and Y. Ma 2006 Class. Quant. Grav. **23** 2741
[24] T. Thiemann 2006 Class. Quant. Grav. **23** 2211
[25] M. Han and Y. Ma 2006 Phys. Lett. B **635** 225
[26] Y. Han, X. Zhang and Y. Ma 2011 (in preperation).
[27] H. Guo, X. Wu, X. Zhang and Y. Ma 2011 (in preperation).
[28] Z. Zhou, X. Zhang and Y. Ma 2011 (in preperation).

[2] See the talk by Thomas Thiemann on Loops 11 Conference in Madrid

Journal of Physics: Conference Series **360** (2012) 012007 doi:10.1088/1742-6596/360/1/012007

New insights in quantum geometry

Hanno Sahlmann
Asia Pacific Center for Theoretical Physics, Pohang (South Korea)
Physics Department, Pohang University of Science and Technology, Pohang (South Korea)
E-mail: sahlmann@apctp.org

Abstract. Quantum geometry, i.e., the quantum theory of intrinsic and extrinsic spatial geometry, is a cornerstone of loop quantum gravity. Recently, there have been many new ideas in this field, and I will review some of them. In particular, after a brief description of the main structures and results of quantum geometry, I review a new description of the quantized geometry in terms of polyhedra, new results on the volume operator, and a way to incorporate a classical background metric into the quantum description. Finally I describe a new type of exponentiated flux operator, and its application to Chern-Simons theory and black holes.

1. Introduction

Loop quantum geometry is based on a canonical quantization of an SU(2) connection A and its canonical conjugate E. E encodes the intrinsic spatial geometry of a spatial slice Σ, whereas A contains the extrinsic geometry of Σ in space-time. The quantum theory of these fields – before the implementation of the quantum version of Einstein's equations – is called *quantum geometry*. It is a cornerstone of loop quantum gravity (LQG). The basic properties of this quantum theory [1, 2, 3, 4, 5] have been studied a long time and are now well understood. Nevertheless, there has been a lot of activity in this area recently. While our basic understanding has not changed, new aspects have been discovered and there is a much refined understanding of many technical problems. Some of these developments have the potential to change the way we think about and approach the big challenges that confront LQG, such as a deeper understanding of the dynamics and the contact with observations.

I have picked four areas of development for a review, namely

(i) U(N) structure on intertwiner spaces and interpretation in terms of polygons and twisted geometries [6, 7, 8, 9, 10, 11].

(ii) Sign factors and volume spectra [12, 13].

(iii) Quantum geometry with a classical spatial background [14, 15].

(iv) New flux operators and their application to Chern-Simons theory and the description of black holes [16, 17, 18, 19].

These are not meant to be the most important ones. Rather, the idea was to give both examples of technical and of conceptual progress, which happened roughly speaking within the last two years. In the end, there is also a personal flavor to this selection, as I have worked on two of the topics myself. Thus there are important developments at least partially related to quantum geometry which I will not mention further in this report. For example, there are new results

Journal of Physics: Conference Series **360** (2012) 012007 doi:10.1088/1742-6596/360/1/012007

on the connection of the formalism of quantum geometry in symmetry reduced models to the full theory [20, 21]. Moreover, I will only mention the description of classical and quantum geometry in terms of SU(2) spinors [9, 22, 23, 24, 25] in passing. A very important topic that will be left out are the new insights regarding how the 3d and the 4d picture can come together [26, 27, 28],[29], [30, 31]. Similarly I will not consider the exciting developments on quantum geometry in higher dimensions [32].

2. Quantum geometry – a reminder

The basic variables that encode the quantized geometry are Wilson-loop and flux-like variables

$$W_\alpha^{(j)} = \mathrm{tr}\left(\pi_j\left(P\exp\oint_\alpha A\right)\right), \qquad E_{S,f} = \int_S f^I \Sigma_I, \qquad \Sigma_I = \epsilon_{abc} E_I^a \, dx^b \wedge dx^c. \tag{1}$$

α is an oriented loop j an SU(2) irrep, S an oriented surface, and f a smearing function taking values in su(2)*. These variables are chosen because of their simple transformation properties under spatial diffeomorphisms and gauge transformations. In the quantum theory they are promoted to operators. Their commutation relations turn the set of operators into an algebra $\mathfrak{H}\mathfrak{F}$, the *holonomy-flux algebra*. The product of the operators $W_\alpha^{(j)}$ is determined by the pointwise product of the corresponding functionals, and hence in part by SU(2) representation theory. For example

$$(W_\alpha^{(j)})^2 = \sum_{k=0}^{2j} W_\alpha^{(k)}. \tag{2}$$

A second class of relations in $\mathfrak{H}\mathfrak{F}$ encodes the canonical commutation relations between A and E. For example, for $\triangle$ and $\bigcirc$ having a single transverse intersection of positive orientation $\bigcirc\!\!\triangle$ at a point p,

$$[E_{\triangle,f}, W_{\bigcirc}^{(j)}] = 8\pi\beta\ell_{\mathrm{P}} f^I(p) W_{\bigcirc_I}^{(j)}, \tag{3}$$

where $W^{(j)}{}_{\bigcirc_I}$ is an element of $\mathfrak{H}\mathfrak{F}$ corresponding to an insertion to an element of su(2) in the functional (1). A third class of relations makes $\mathfrak{H}\mathfrak{F}$ into a Lie algebra. For example

$$[[E_{S,f}, E_{S',f'}], W_\alpha^{(j)}] = (8\pi\beta\ell_{\mathrm{P}})^2 f^I(p) f^J(p) W_{\alpha IJ}^{(j)}, \tag{4}$$

where we have assumed that S and S' intersect each other and α in a point p. Again $W_{\alpha IJ}^{(j)}$ is an element of $\mathfrak{H}\mathfrak{F}$ corresponding to an insertion of elements of su(2).

An important representation for $\mathfrak{H}\mathfrak{F}$, the *Ashtekar-Lewandowski representation*. It is the unique representation invariant under spatial diffeomorphisms. In it, the W operators act as creation operators, generating states as excitations over a ground state $|\,\rangle$. An orthonormal basis in the corresponding Hilbert space $\mathcal{H}_{\mathrm{AL}}$ can be labeled by *spin networks*, diagrams that consist of a graph in Σ and a labeling of graph edges with su(2) irreps and of graph vertices with invariants in the tensor product of the representations of edges attached to the respective vertex. There is a decomposition into orthogonal subspaces $\mathcal{H}_\gamma$ labeled by graphs γ. A finer decomposition into sub-Hilbert spaces can be made using the spin network diagrams. The rules are given in fig. 1. The full Hilbert space also has an analytic description, $\mathcal{H}_{\mathrm{AL}} = L^2(\overline{\mathcal{A}/\mathcal{G}}, \mathrm{d}\mu_{\mathrm{AL}})$, as a space of square-integrable distributional connections modulo gauge transformations.

In $\mathcal{H}_{\mathrm{AL}}$, one can find operators corresponding to geometric quantities such as area, volume, length, or angle, in terms of the operators J_e representing SU(2) at the end of the edges (see fig.1). As an example, consider the area operator. In a situation where a single simple loop intersects a surface, $\bigcirc\!\!\triangle$, one has

$$\widehat{A}_{\triangle} |\bigcirc_j\rangle = 8\pi\beta l_p^2 \sqrt{j(j+1)} |\bigcirc_j\rangle. \tag{5}$$

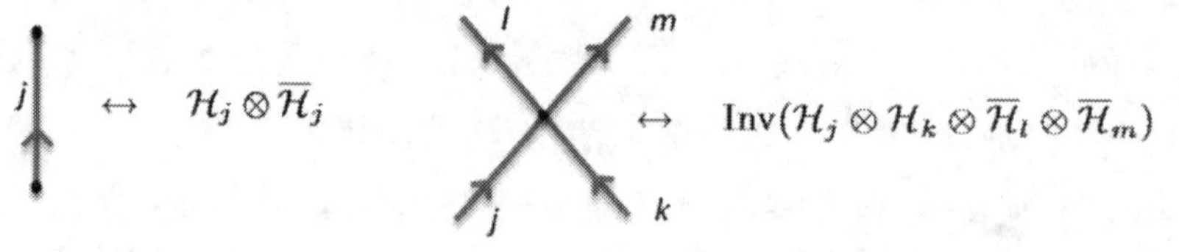

Figure 1. Decomposition in to invariant spaces wrt. the operators $\vec{J}_e$. $\mathcal{H}_j$ stands for the spin j representation of SU(2).

Volume operators act in the intertwiner spaces at the vertices. Their action is much more complicated and their spectrum is not known. At the same time, volume operators are a crucial ingredient in the quantum dynamics of LQG. In the next two sections we will present new results that also shed light on the action of the volume operators.

3. New developments

3.1. Polyhedral interpretation

It is an old idea to associate some kind of classical geometry to a spin network: Since the volume operator acts at the vertices, volume can be thought to reside there. The edges on the other hand carry area. Conventionally, the geometry is thought of as distributional, the metric being degenerate except for points on the spin network graph. But recently, a new geometric picture emerged, which is supported by some old result on the geometry of polyhedra. The idea is to envision the geometry associated to the spin network as given by a gluing of polyhedra, one for each vertex.

To describe the aspects of this new picture in more detail, we start by reminding the reader that given a graph γ the associated Hilbert space of spin network states has a decomposition

$$\mathcal{H}_\gamma = \bigoplus_{\{j_e|e\in\gamma\}} \bigotimes_{v\in\gamma} \mathrm{Inv}(j_e, e \text{ at } v). \tag{6}$$

Thus it is natural to investigate the geometric interpretation of the intertwiner spaces associated to the vertices. The first to describe the geometries associated to the states in this Hilbert space in detail were Livine and Freidel [7], based on earlier work by Livine and Girelli [6] which described an U(n) structure on the n-valent intertwiner space

$$\mathcal{H}_n = \bigoplus_{(j_1,j_2,\ldots j_n)} \mathrm{Inv}(j_1, j_2, \ldots j_n). \tag{7}$$

This action is obtained by expressing the generators of the SU(2) gauge rotations at a given vertex by two sets of annihilation and creation operators, the so-called *Schwinger representation.* For one copy of SU(2), this reads

$$J_z = \frac{1}{2}(a^\dagger a - b^\dagger b), \quad J_+ = a^\dagger b, \quad J_- = ab^\dagger. \tag{8}$$

Now, with n copies of SU(2) one has 2n sets of annihilation and creation operators, and the combination

$$E_{ij} := a_i^\dagger a_j + b_i^\dagger b_j \tag{9}$$

commutes with global SU(2) rotations and has the commutation relations of U(n) generators. Moreover, it preserves the subspace of intertwiners with a fixed sum $\sum_i j_i$ of incoming spins. this

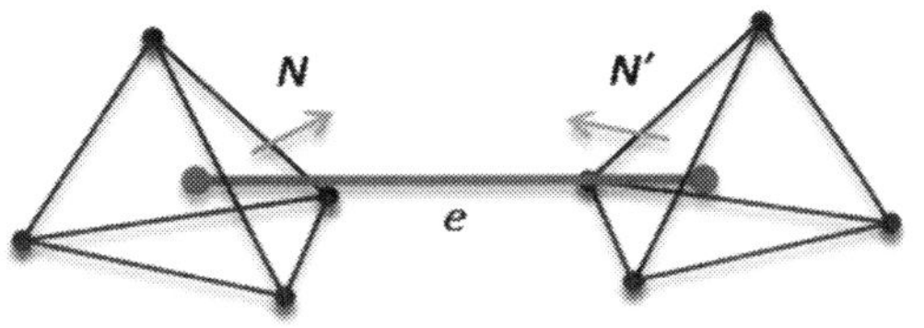

Figure 2. The gluing of two flat tetrahedra dual to two vertices connected by an edge.

can be interpreted as the *preservation of area*. Hence Livine and Freidel interpret the states in $\mathcal{H}_n$ as those of a quantized sphere, and the U(n) action as discretized action of diffeomorphisms. The U(n) structure is also useful for practical purposes, for example for determining the dimensions of various subspaces of the intertwiner space, and for constructing *intertwiner coherent states* [9], a new type of coherent states contained in $\mathcal{H}_n$. The latter arise from coherent states for the harmonic oscillators in the Schwinger representation, and a suitable group averaging. Using the U(n) structure, one can also introduce spinorial coordinates in the space of intertwiners which give a wholly new perspective [9, 22, 23, 24, 25].

Now one can also consider how to combine the chunks of quantized space. In [8] Freidel and Speziale call the result *twisted geometries*. $\mathcal{H}_e = L^2(SU(2))$ can be understood as carrier of a quantization of a phase space T*SU(2) associated to the edge. But what is its interpretation in terms of physical quantities? The classic answer [33] is directly in terms of the fields (A, E) used in LQG: Points in the phase space can be parametrized by $(h_e, \int_{S_e} E)$, where h is the holonomy along E and S_e is a fixed surface transversal to e. The surfaces around a vertex form a curved polyhedron. [8, 10] suggest to parametrize the phase space in a way that uses the geometry of two flat polyhedra glued together, possibly with extrinsic curvature. A point is given by (N, N', j, ξ), where N, N' are the unit normal vectors of the surfaces involved, and j describes their area, see fig. 2. The parameter ξ is carrying the remaining information about the parallel transport from one polyhedron to the next. An interesting point, corresponding to an earlier result [34], is that the geometries obtained in this way are twisted in the sense that the shapes of the glued triangles do not necessarily match.

The picture is rounded off by the following observations [11]: according to a theorem by Minkowski, there is a 1-1 correspondence between sets $\{a_i\}_{i=1\ldots n}$, $\{N_i\}_{i=1\ldots n}$ of positive real numbers and unit vectors in $\mathbb{R}^3$ satisfying the closure relation

$$\sum_i N_i a_i = 0, \tag{10}$$

and the equivalence classes of convex polyhedra under rotation. The a_i correspond to the areas of the faces, the N_i to their normal vectors under this correspondence. The space of this data carries a natural symplectic structure due to Kapovich and Millson, and its quantization can be precisely related to the structures at an n-valent vertex in LQG. To each edge e, one has a dual face whose area is given by the operator $|\vec{J}_e|$ and and whose product of area and normal vector corresponds to $\vec{J}_e$. The closure condition (10) is then just gauge invariance at the vertex, and the intertwiners acquire an interpretation as the quantum states of a convex polyhedron. This has very interesting consequences for the quantum black hole [35] and for the volume operator [36].

3.2. Sign factors and volume spectra

The volume operators are a crucial ingredient in the quantum dynamics of LQG, but their action and spectrum is very complicated. Given a vertex v one defines the operators

$$\widehat{q}_{e,e',e''} = \text{sign}(\det(\dot{e}, \dot{e}', \dot{e}''))\epsilon^{IJK} J_I^e J_J^{e'} J_K^{e''}, \qquad \widehat{q}_v = \sum_{e,e',e'' \text{ at } v} \widehat{q}_{e,e',e''} \tag{11}$$

The volume operator at the vertex is then given by

$$\widehat{V} = (|\sum_v \widehat{q}_v|)^{1/2} \quad \text{or} \quad \widehat{V} = (\sum_v |\widehat{q}_v|)^{1/2}, \tag{12}$$

depending on which version, Ashtekar-Lewandowski (first expression), or Rovelli-Smolin (second one) one considers. A particular complication of the first version version are the sign factors in the definition of $\widehat{q}_{e,e',e''}$, since these can have substantial influence on the spectrum for vertices of valence higher than three. While it may seem at first that arbitrary sign combinations may occur when letting e, e', e'' range over all the triples of edges at a given vertex, Brunnemann and Rideout [12] have observed that by no means all sign combinations can actually be realized by configurations of tangent vectors in $\mathbb{R}^3$. How to effectively characterize the realizable sign configurations has been a question that has hampered the numerical analysis of the volume spectrum for some time. Recently however, Brunnemann and Rideout have found in matroid theory the tool to answer all the questions pertaining to these signs [13]. Matroids are combinatorial structures generalizing linear dependencies in a real vector space, and [13] describes in detail how they can be used to describe the diffeomorphism invariance classes of tangential structure of vertices.

Using these results together with some analytic ones [37], Brunnemann and Rideout have expanded the numerical study of the spectrum of the AL volume operator for various vertices, see for example fig. 3. It can clearly be seen that the sign combinations have a drastic effect on the

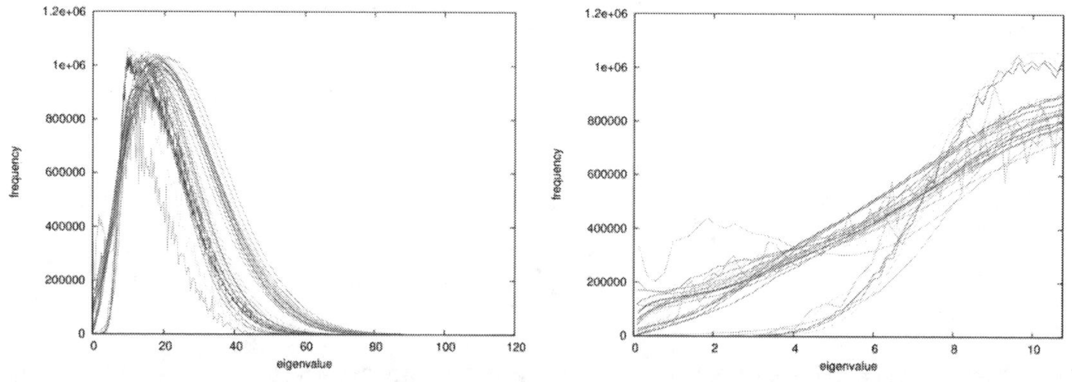

Figure 3. From [13] with kind permisson: Histograms for each sign configuration at the 6-vertex, with incoming spins up to $j_{\max} = 13/2$. On the right: The small eigenvalue region.

eigenvalue density, with some combinations showing a *volume gap*, whereas other combinations have high eigenvalue populations also for very small eigenvalues. These results will be quite important when the action of the Hamilton constraint (of which the AL volume operator is an important ingredient) is studied in detail. Additionally, as pointed out in [13], matroids can also be used to describe the combinatorics of edge composition in graphs, thus opening another potential application in LQG. This idea is extended even further in [38], where duality in the matroid, together with a duality operation on connections leads to a duality transformation in the whole theory.

Journal of Physics: Conference Series **360** (2012) 012007 doi:10.1088/1742-6596/360/1/012007

3.3. New representations

The AL representation of $\mathfrak{H}\mathfrak{F}$ is uniquely singled out by its properties with respect to diffeomorphisms. But at least for some applications in situations where one considers highly excited states of the gravitational field, it may be interesting to also consider representations that are not invariant under diffeomorphisms. One class of such representations has been studied in [14, 15]. Let $E^{(0)}$ be a classical triad field. Then we can define a *new* representation π' on the AL Hilbert space $\mathcal{H}_{\text{AL}}$, by changing the action of the fluxes:

$$\pi'(E_{S,r}) = X_{S,r} + E^{(0)}_{S,r}\,\text{id}, \qquad \text{with } E^{(0)}_{S,r} = \int_S \Sigma^{(0)I} r_I. \tag{13}$$

It is easily checked that this gives another representation of $\mathfrak{H}\mathfrak{F}$. We list some elementary properties:

(i) The spectra of the fluxes have changed. If λ is an eigenvalue of $X_{S,f}$ then $\lambda + E^{(0)}_{S,f}$ is an eigenvalue of $\pi'(E_{s,f})$. In particular, the AL vacuum $|\rangle$ is now an eigenstate of $\pi'(E_{s,f})$ with eigenvalue $E^{(0)}_{S,f}$, which is non-vanishing in general.

(ii) The new representation is still cyclic, with $|\rangle$ as cyclic vector.

(iii) The new representation is unitarily inequivalent to the AL representation.

(iv) The standard kinematic representation can be viewed as a special case of the new one, for $E^{(0)} = 0$.

To better interpret the new representations, it is very useful to consider the action of operators corresponding to geometric measures such as area and volume. It turns out that area and volume operators can be defined in the new representations with exactly the same regularization and quantization technique as in the standard representation, and the resulting operators have a very simple property:

$$V_R = V_R^{\text{vac}} + V_R^{(0)}\,\text{id}, \qquad A_S = A_S^{\text{vac}} + A_S^{(0)}\,\text{id}, \tag{14}$$

with V^{vac}, A^{vac} the geometric operators in the vacuum representation, and $V_R^{(0)}$, $A_S^{(0)}$ the classical values in the background geometry. This shows clearly that the new representations contain a background geometry given by the background field $E^{(0)}$.

Turn next to the gauge symmetries of LQG: The operators implementing diffeomorphisms and gauge transformations in the standard representation are still well defined and unitary in the new representation, but it can be easily checked that they do not implement the algebra-automorphisms anymore. The problem is that operators on the Hilbert space of π' can not change change the background geometry. This can be remedied by going over to a large direct sum of Hilbert spaces. For a given background $E^{(0)}$ one considers

$$\mathcal{H}_{G(E^{(0)})} := \bigoplus_{\widetilde{E}^{(0)} \in G(E^{(0)})} \mathcal{H}_{\widetilde{E}^{(0)}} \tag{15}$$

where $G(E^{(0)})$ is the equivalence class of $E^{(0)}$ with respect to diffeomorphisms and gauge transformations. The Hilbert space is thus effectively labeled by a spatial background metric modulo diffeomorphisms. The Hilbert space $\mathcal{H}_{G(E^{(0)})}$ carries a representation of $\mathfrak{H}\mathfrak{F}$, which is simply the direct sum of the representations on the individual spaces. But now one can implement the diffeomorphisms and gauge transformations unitarily [14, 15]. The unitary implementation of the symmetries can now be used to find invariant states and Hilbert spaces by *group averaging*, just as has been done for the vacuum representation [4]. For example, one

Journal of Physics: Conference Series **360** (2012) 012007 doi:10.1088/1742-6596/360/1/012007

obtains diffeomorphism invariant functionals $(T_\alpha, E^{(0)}|$ on cylindrical functions by averaging with respect to certain quotients of the diffeomorphism group. Similar results can be obtained for gauge transformations and, taking the semi-direct product of diffeomorphisms and gauge transformations, for bundle automorphisms. One thereby obtains the automorphism invariant Hilbert space $\mathcal{H}_{\text{aut}}$. Thus the quantum kinematics can be developed to exactly the same point for the new representations as for the AL representation. What is more, the latter appears simply as a special case of the constructions we sketched above.

I should stress that, while $\mathcal{H}_{G(E^{(0)})}$ is very large, this is partially remedied by the group averaging. Vectors $|1, E^{(0)}\rangle$, $|1, E'^{(0)}\rangle$ are mapped onto *the same* vector in $\mathcal{H}_{\text{aut}}$. More generally

$$(f, \phi \triangleright E| = (\phi^{-1} \triangleright f, E|. \tag{16}$$

I should also say that it seems that there is no problem with operators on $\mathcal{H}_{\text{aut}}$. As an example consider the operator V for the volume of the entire spatial slice. Also in the new representations, it commutes with all automorphisms. It thus defines an operator on $\mathcal{H}_{\text{aut}}$. Moreover, this operator acts in precisely the way one would expect. If f is an eigenstate of V^{vac}, with eigenvalue λ, then $|f, E^{(0)})$ is an eigenvector of V with eigenvalue $\lambda + V^{(0)}$.

3.4. Non-commuting fluxes, Chern-Simons theory and horizons

I start by reminding the reader that the flux operators are mutually non-commuting in quantum geometry, see (4). This means that there is some form of non-commutative geometry present in LQG. Recently, it turned out that this is not only a nuisance, but very useful in applications to Chern-Simons theory and black holes [16, 17, 18, 19]. The starting point are new gauge covariant exponentiated flux operators. Usually, the field E is integrated over surfaces in the quantum theory, but it turns out that even the non-integrated quantum field makes sense as an operator valued distribution $\widehat{E}$. Then given a surface S without self-intersections, and a system of paths that connect all points of the surface with some base-point on the edge of the surface, one defines

$$W_S = \mathbb{I}_2 + 8\pi i c \int_S \operatorname{ad}_{h_s}(\widehat{\Sigma}(s)) + (8\pi i c)^2 \int_{S^2} K_{s,s'} \operatorname{ad}_{h_s}(\widehat{\Sigma}(s)) \operatorname{ad}_{h_{s'}}(\widehat{\Sigma}(s')) + \ldots \tag{17}$$

Here, K is an integration kernel that implements a *surface ordering* and the holonomies h_s connect the point s on the surface with the base point via the path system. There are two problems with the above definition. The first is that consecutive actions of $\widehat{E}^a(p)$ give rise to delta distributions that are concentrated precisely at the boundary of integration enforced by surface ordering, and a prescription for the evaluation of these has to be adopted. This can be resolved via a standard regularization. The second problem is that the operators $\widehat{E}_I(p)$ at a fixed point p do not commute, therefore there is an ordering ambiguity inherent in the above definition. This ambiguity can be fixed using the *Duflo map*. This is a quantization map Υ from the free algebra of symbols $\{E_I\}_I$ with the Poisson bracket $\{E_I, E_J\} = f_{IJ}{}^K E_K$, ($f$ being the structure constants of a semisimple Lie-algebra $\mathfrak{g}$) into the universal enveloping algebra $U(\mathfrak{g})$, extending the map $E_I \mapsto X_I$ on generators. The defining property of Υ is that it is an *algebra isomorphism* between the invariant subspaces under the action of the corresponding Lie group G. Υ is an improved version of symmetric quantization χ,

$$\Upsilon = \chi \circ j^{\frac{1}{2}}(\partial), \tag{18}$$

where $j^{\frac{1}{2}}(\partial)$ is a differential operator that can be obtained by inserting derivatives ∂^I into the following function on $\mathfrak{g}$:

$$j^{\frac{1}{2}}(x) = \det{}^{\frac{1}{2}} \left(\frac{\sinh \frac{1}{2} \operatorname{ad} x}{\frac{1}{2} \operatorname{ad} x} \right) = 1 + \frac{1}{48} \|x\|^2 + \ldots, \tag{19}$$

with $\|x\|^2 = \mathrm{tr}(\mathrm{ad}_x^2)$ the square of the Cartan-Killing norm of G. For G=SU(2) on finds for example $\Upsilon(\|E\|^2) = \Delta_{\mathrm{SU}(2)} + \frac{1}{8}\mathbf{1}$, where $\Delta_{\mathrm{SU}(2)}$ is the Laplacian. I should also point out that the Duflo map has recently been used in [39] and it shows up in the context of the *flux representation* [40, 41, 42, 44]. Its use has also been advocated, in a slightly different context, in [43]. Using Υ to order powers of E the action of the operator W_S becomes well defined. The operators W_S satisfy some useful relations. For S, S_1, S_2 surfaces, $S_1 + S_2$ the disjoint union of the latter,

$$W_{S_1+S_2} = W_{S_1} W_{S_2}, \qquad W_S^\dagger = W_{-S}, \qquad U_g W_S U_g^{-1} = g(b) W_S g(b)^{-1}, \tag{20}$$

where $-S$ is S with opposite orientation, U_g is the unitary corresponding to a gauge transformation g, b is the base-point of S, and paths on S_1, S_2, $S_1 + S_2$ have been chosen in a compatible way. Moreover, for SU(2) and spin nets $|\,|_{j'}\rangle$ intersecting S transversally once with spin j, one finds

$$\mathrm{tr}_j\,(W_S)\,|\,|_{j'}\rangle = \frac{\sin\left[\pi c(2j+1)(2j'+1)\right]}{\sin\left[\pi c(2j'+1)\right]}|\,|_{j'}\rangle. \tag{21}$$

Finally, states $|\,|_j\rangle$ are not eigenstates of the operator valued matrices W_S. Rather,

$$W_S|\,|_j\rangle = c_1(j)\,\mathbb{I}_{2\times 2}|\,j\rangle + c_2(j)|\dashv_j\rangle, \qquad |\dashv_j\rangle = \mathrm{ad}_{h_p}(T_I)\kappa^{IJ}\widehat{E}_J(p)|\,|_j\rangle,$$

where $c_1(j), c_2(j)$ are constants that are complicated to compute explicitly, and $|\dashv_j\rangle$ carries a new vertex at p and a link between p and the base point of the loop along the path system on S.

3.4.1. Application to Chern Simons theory Since the work of Witten [45], it is well known that path integral expectation values of holonomy traces in CS theory are related to link invariants. For G =SU(2) and traces in the defining representation one obtains the *Kauffman bracket*, with the conventional variable A replaced by $q^{1/4}$, $q = \exp(2\pi i/k)$.[1] One can reproduce some of these results using the operators W_S. The key is that

$$-\frac{8\pi i}{k}\epsilon_{abc}\frac{\delta}{\delta A_c}e^{iS_{\mathrm{CS}}[A]} = F_{ab}e^{iS_{\mathrm{CS}}[A]} \tag{22}$$

for SU(2) CS theory with level k, and same, up to a numerical factor, for other gauge groups. The non-Abelian version of Stokes' theorem suggests that holonomy functionals can be replaced by the new flux operators under the path integral. To make this reasoning explicit, let S be a smooth, oriented, simply connected surface, ρ some representation of the structure group G, and and L be some functional of G-connections. Then formally

$$\begin{aligned}\langle L\,\mathrm{tr}_\rho(h_{\partial S})\rangle &= \int_{\mathcal{A}} L[A]\,\mathrm{tr}_\rho(h_{\partial S})[A]e^{iS_{\mathrm{CS}}[A]}\,\mathrm{d}\mu[A] = \int_{\mathcal{A}} L[A]\,\mathrm{tr}_\rho(W_S)e^{iS_{\mathrm{CS}}[A]}\,\mathrm{d}\mu[A]\\ &= \int_{\mathcal{A}} (\mathrm{tr}_\rho(W_{-S})L)[A]e^{iS_{\mathrm{CS}}[A]}\,\mathrm{d}\mu[A] = \langle(\mathrm{tr}_\rho(W_{-S})L)\rangle.\end{aligned} \tag{23}$$

While the manipulations under the path integral are formal, taking the first and the last line gives an equality in which all objects are defined, at least as long as S has no self-intersections. Consider G=SU(2) and chose $c = 1/k$ with k integer [47, 48] in the definition of W_S. Let

[1] In fact, with CFT methods on finds $q = \exp(2\pi i/(k+2))$ [45], but we do not see the shift of the level with our method.

$H^+(j_1, j_2) = \mathrm{tr}_{j_1}(h_{\alpha_1})\,\mathrm{tr}_{j_2}(h_{\alpha_2})$ denote the right handed Hopf link spin network. Using the above strategy and the eigenvalues (21), one finds

$$\langle H^+(j_1, j_2)\rangle_{\mathrm{CS}} = \mathrm{tr}_{j_1}(g_{j_2})\,\mathrm{tr}_{j_2}(g_0) = \frac{\sin\left[\frac{\pi}{k}(2j_1+1)(2j_2+1)\right]}{\sin\left[\frac{\pi}{k}\right]}, \tag{24}$$

thus reproducing the known values for the Kauffman bracket and its generalization [46]. Note that these numbers are important in related contexts: They are equivalently given by the trace of the square of the R-matrix of $U_q(\mathrm{su}(2))$ on $j_1 \otimes j_2$ or, up to normalization, by the Verlinde coefficients [49] in the $\mathrm{SU}(2)_k$ WZW model of conformal field theory. Similar results can be obtained for other gauge groups [17].

3.4.2. Application to Black holes The quantization of an isolated horizon (IH) is a remarkable success of LQG (see [50] for a recent overview). However, it is only an effective description, in the sense that it uses a number of elements that are not intrinsic to the formalism of LQG. This can be partially overcome using the operators W_S. For spherically symmetric IH, the pullback to the IH, of the dynamical fields satisfy the horizon condition [51]

$$\underleftarrow{F}(A) = -\frac{\pi(1-\beta^2)}{a_H}\underleftarrow{\Sigma}(E). \tag{25}$$

In the quantum theory, we thus call a surface H a type I quantum horizon, if the surface ordered exponential of the above equation, $\mathrm{tr}\, h_{\partial S}\Psi = \mathrm{tr}\, W_S\Psi$, is valid for all surfaces S lying entirely within H. Here, the operators W_S are evaluated with $c = -\pi\beta(1-\beta^2)\ell_{\mathrm{p}}^2/2a_H$. Solutions Ψ of this equation do not exist in the standard representation of LQG, but arguably in other representations. A spin network $|\psi\rangle$ determines puncture data $\mathcal{P} = \{(p_1, j_1, m_1), (p_2, j_2, m_2), \ldots (p_N, j_N, m_N)\}$ given by points p_i on H irreducible representations j_i and and magnetic quantum numbers m_i. $\mathcal{P}$ defines a functional on functions of traces of loops that encircle at most one puncture:

$$\mu(\prod_i \mathrm{tr}_{k_i}(h_{\alpha_i})) := \prod_i \mathrm{tr}_{k_i}(g_{j_i}). \tag{26}$$

This functional is positive and consistent with all the relations among these traces. It is however not yet clear whether it extends all holonomy functionals on H. The resulting horizon theory would seem to reproduce many of the results that have been obtained earlier [53, 54], up to the fact that the fluxes E, pulled back to the horizon are still a-priori independent degrees of freedom describing the intrinsic geometry of H.

4. Concluding remarks

I hope that it has become clear that quantum geometry still intrigues with new facets and discoveries. All of this is potentially relevant for physics: The polyhedral picture is already applied to give new forms of dynamics and for a better understanding of the quantum theory of black hole horizons. The detailed knowledge of spectrum and properties of the volume operator are relevant for the implementation of the Hamilton constraint, and the representations with classical background may be used to construct effective versions of the dynamics. The new flux operators have already proven useful in the description of black hole horizons. But since these are new developments, other applications are bound to emerge. I hope that this short overview can be helpful in this respect.

Acknowledgments

I thank the organizers of *Loops 11* for making it such a nice and productive event, and for the opportunity to speak. I gratefully acknowledge partial support through the Spanish MICINN

Project No. FIS2008-06078-C03-03.

References

[1] C. Rovelli and L. Smolin, Nucl. Phys. B **331**, 80 (1990).
[2] C. Rovelli and L. Smolin, Nucl. Phys. B **442**, 593 (1995) [arXiv:gr-qc/9411005].
[3] A. Ashtekar, J. Lewandowski, D. Marolf, J. Mourao and T. Thiemann, J. Math. Phys. **36**, 6456 (1995) [arXiv:gr-qc/9504018].
[4] A. Ashtekar and J. Lewandowski, Class. Quant. Grav. **14**, A55 (1997) [arXiv:gr-qc/9602046].
[5] J. Lewandowski, Class. Quant. Grav. **14**, 71 (1997) [arXiv:gr-qc/9602035].
[6] F. Girelli and E. R. Livine, Class. Quant. Grav. **22**, 3295 (2005) [gr-qc/0501075].
[7] L. Freidel and E. R. Livine, J. Math. Phys. **51**, 082502 (2010) [arXiv:0911.3553 [gr-qc]].
[8] L. Freidel and S. Speziale, Phys. Rev. D **82**, 084040 (2010) [arXiv:1001.2748 [gr-qc]].
[9] L. Freidel and E. R. Livine, J. Math. Phys. **52**, 052502 (2011) [arXiv:1005.2090 [gr-qc]].
[10] C. Rovelli and S. Speziale, Phys. Rev. D **82**, 044018 (2010) [arXiv:1005.2927 [gr-qc]].
[11] E. Bianchi, P. Dona and S. Speziale, Phys. Rev. D **83**, 044035 (2011) [arXiv:1009.3402 [gr-qc]].
[12] J. Brunnemann and D. Rideout, Class. Quant. Grav. **25**, 065001 (2008) [arXiv:0706.0469 [gr-qc]].
[13] J. Brunnemann and D. Rideout, Class. Quant. Grav. **27**, 205008 (2010) [arXiv:1003.2348 [gr-qc]].
[14] T. A. Koslowski, arXiv:0709.3465 [gr-qc].
[15] H. Sahlmann, Class. Quant. Grav. **27**, 225007 (2010) [arXiv:1006.0388 [gr-qc]].
[16] H. Sahlmann and T. Thiemann, J. Geom. Phys. **62**, 204 (2012) [arXiv:1004.2834 [gr-qc]].
[17] H. Sahlmann and T. Thiemann, J. Geom. Phys. **61**, 1104 (2011) [arXiv:1101.1690 [gr-qc]].
[18] H. Sahlmann, Phys. Rev. D **84**, 044049 (2011) [arXiv:1104.4691 [gr-qc]].
[19] H. Sahlmann and T. Thiemann, arXiv:1109.5793 [gr-qc].
[20] C. Fleischhack, arXiv:1010.0449 [math-ph].
[21] J. Brunnemann and T. A. Koslowski, arXiv:1012.0053 [gr-qc].
[22] E. F. Borja, L. Freidel, I. Garay and E. R. Livine, Class. Quant. Grav. **28**, 055005 (2011) [arXiv:1010.5451]
[23] E. R. Livine and J. Tambornino, arXiv:1105.3385 [gr-qc].
[24] E. R. Livine, S. Speziale and J. Tambornino, arXiv:1108.0369 [gr-qc].
[25] E. R. Livine and J. Tambornino, arXiv:1109.3572 [gr-qc].
[26] J. Engle, E. Livine, R. Pereira and C. Rovelli, Nucl. Phys. B **799**, 136 (2008) [arXiv:0711.0146 [gr-qc]].
[27] Y. Ding and C. Rovelli, Class. Quant. Grav. **27**, 165003 (2010) [arXiv:0911.0543 [gr-qc]].
[28] Y. Ding and C. Rovelli, Class. Quant. Grav. **27**, 205003 (2010) [arXiv:1006.1294 [gr-qc]].
[29] S. Alexandrov, Phys. Rev. D **82**, 024024 (2010) [arXiv:1004.2260 [gr-qc]].
[30] M. Geiller, M. Lachieze-Rey, K. Noui and F. Sardelli, SIGMA **7**, 083 (2011) [arXiv:1103.4057 [gr-qc]].
[31] M. Geiller, M. Lachieze-Rey and K. Noui, Phys. Rev. D **84**, 044002 (2011) [arXiv:1105.4194 [gr-qc]].
[32] N. Bodendorfer, T. Thiemann and A. Thurn, arXiv:1105.3703 [gr-qc], arXiv:1105.3704 [gr-qc], arXiv:1105.3705 [gr-qc], arXiv:1105.3708 [gr-qc].
[33] T. Thiemann, Class. Quant. Grav. **18**, 2025 (2001) [arXiv:hep-th/0005233].
[34] B. Dittrich and J. P. Ryan, Class. Quant. Grav. **28**, 065006 (2011) [arXiv:0807.2806 [gr-qc]].
[35] E. Bianchi, Class. Quant. Grav. **28**, 114006 (2011) [arXiv:1011.5628 [gr-qc]].
[36] E. Bianchi and H. M. Haggard, Phys. Rev. Lett. **107**, 011301 (2011) [arXiv:1102.5439 [gr-qc]].
[37] J. Brunnemann and T. Thiemann, Class. Quant. Grav. **23**, 1289 (2006) [arXiv:gr-qc/0405060].
[38] J. A. Nieto, arXiv:1003.4750 [hep-th].
[39] K. Noui, A. Perez and D. Pranzetti, JHEP **1110**, 036 (2011) [arXiv:1105.0439 [gr-qc]].
[40] L. Freidel and E. R. Livine, Class. Quant. Grav. **23**, 2021 (2006) [hep-th/0502106].
[41] L. Freidel and S. Majid, Class. Quant. Grav. **25**, 045006 (2008) [hep-th/0601004].
[42] A. Baratin, B. Dittrich, D. Oriti and J. Tambornino, Class. Quant. Grav. **28** (2011) 175011 [arXiv:1004.3450]
[43] A. Alekseev, A. P. Polychronakos and M. Smedback, Phys. Rev. D **71**, 067501 (2005) [hep-th/0405036].
[44] D. Oriti and M. Raasakka, Phys. Rev. D **84** (2011) 025003 [arXiv:1103.2098 [hep-th]].
[45] E. Witten, Commun. Math. Phys. **121**, 351 (1989).
[46] S. P. Martin, Nucl. Phys. B **338**, 244 (1990).
[47] S. Deser, R. Jackiw and S. Templeton, Phys. Rev. Lett. **48**, 975 (1982).
[48] S. Deser, R. Jackiw and S. Templeton, Annals Phys. **140**, 372 (1982)
[49] E. P. Verlinde, Nucl. Phys. B **300**, 360 (1988).
[50] J. Diaz-Polo and D. Pranzetti, arXiv:1112.0291 [gr-qc].
[51] J. Engle, K. Noui, A. Perez and D. Pranzetti, Phys. Rev. D **82**, 044050 (2010) [arXiv:1006.0634 [gr-qc]].
[52] H. Sahlmann, J. Math. Phys. **52**, 012503 (2011) [arXiv:gr-qc/0207112].
[53] J. Engle, A. Perez and K. Noui, Phys. Rev. Lett. **105**, 031302 (2010) [arXiv:0905.3168 [gr-qc]].
[54] J. Engle, K. Noui, A. Perez and D. Pranzetti, JHEP **1105**, 016 (2011) [arXiv:1103.2723 [gr-qc]].

Journal of Physics: Conference Series **360** (2012) 012008 doi:10.1088/1742-6596/360/1/012008

Bianchi-I spacetimes in loop quantum cosmology: physics of singularity resolution

Parampreet Singh

Department of Physics and Astronomy, Louisiana State University, Baton Rouge, Louisiana 70803, USA

E-mail: psingh@phys.lsu.edu

Abstract. Quantum geometry effects in loop quantum cosmology are known to resolve various singularities in the isotropic and anisotropic models. Previous investigations have shown that singularity resolution is tied to boundedness of physical quantities such as the energy density and the expansion and shear scalars in the spacetime. Using effective dynamics, we show that these do not necessarily lead to a generic boundedness of spacetime curvature in the Bianchi-I model and find the potential possibilities where the curvature invariants can diverge. These include cases where the pressure of the matter content diverges at a finite volume and energy density. By analyzing the strength of singularities, we show that such pressure singularities are weak. All known types of strong singularities of the classical theory are resolved. This analysis brings us closer to generalize the results on generic resolution of singularities in the isotropic models in loop quantum cosmology, to the anisotropic spacetimes.

1. Introduction

Bianchi-I spacetimes are one of the simplest spacetimes which admit a non-vanishing Weyl curvature. The presence of anisotropies, which lead to the Weyl part of the spacetime curvature, generate a very rich structure of singularities in the classical theory. These singularities have been widely studied in GR (see for eg. [1]). Depending on the behavior of different scale factors, singularities in Bianchi-I spacetimes have been classified as isotropic (point-like) singularities, barrel, cigar and pancake singularities [2, 3]. The occurrence of these singularities depend on the choice of matter component, and whether it has a vanishing (such as a perfect fluid) or non-vanishing anisotropic stress (such as in presence of magnetic fields). For all the singularities studied so far in the classical theory, at least one of the scale factors in the metric vanishes. At these singularities, directional Hubble rates, energy density, expansion and shear scalars diverge as the physical volume vanishes, which results in various curvature invariants becoming infinite. Analysis of the geodesic equations show that geodesic evolution breaks down at these singuilarities [4]. In terms of the conditions of Tipler [5] and Królak [6], these singularities are strong curvature type, which lead to a complete destruction of arbitrary detectors falling in the singularities. Existence of these singularities signals that the limit of the applicability of GR has been reached. It has been hoped that a quantum theory of gravity would lead to the resolution of space-like singularities. Our goal, here is to understand the way quantum properties of spacetime as understood in loop quantum cosmology (LQC) lead to the resolution of above singularities in the Bianchi-I spacetime sourced with a fluid with a vanishing anisotropic stress.

Journal of Physics: Conference Series **360** (2012) 012008 doi:10.1088/1742-6596/360/1/012008

LQC is a quantization of homogeneous spacetimes based on the techniques of loop quantum gravity (see Ref. [7] for an extensive review). In contrast to earlier attempts to quantize cosmological and symmetry reduced spacetimes, such as in Wheeler-DeWitt theory, a full quantization of several spacetimes has been accomplished and detailed physics has been analyzed in recent years. These include spatially flat isotropic models [8, 9, 10, 11, 12, 13, 14], spatially curved models [15, 16, 17, 18] and anisotropic models [19, 20, 21, 22]. Detailed physics of these models has been analyzed using full quantum evolution (see for eg. [9, 10, 16]) as well as effective Hamiltonian method (see Sec V of ref. [7] and references there in) using embedding method approach [23, 24, 25]. One of the main results of these investigations is that the cosmological singularities are resolved and replaced by a bounce in the Planck regime.[1] The occurrence of bounce (or multiple bounces in case of anisotropic spacetimes) is tied to the existence of upper bound on expansion scalar (θ) of the geodesics [27, 28, 29] which arises due to the underlying quantum geometric effects. In the isotropic models, these bounces can also be understood as arising due to an upper bound on the energy density (ρ) (see for eg. [30, 31, 10, 26]). However, for anisotropic models, due to a non-vanishing contribution by the Weyl curvature, this correspondence becomes invalid and one has to carefully include the role of shear scalar σ^2, which also turns out to be universally bounded for Bianchi-I model [32, 33] and under certain conditions for Bianchi-II and Bianchi-IX models [28].

The boundedness of ρ, θ and σ^2 signals positively for the generic resolution of singularities. Indeed, in the spatially flat isotropic models in LQC, using effective dynamics, one finds that the boundedness of expansion scalar leads to a resolution of all strong singularities and the spacetime is geodesically complete [27]. Investigations on the spatially curved models strongly indicate that these results are valid also for $k = \pm 1$ models [34]. These studies find that even though the ρ and θ are bounded in isotropic LQC, the spacetime curvature can still be unbounded (see [36] for the first phenomenological example of this kind), however in all such cases the singularity turns out be weak and is thus harmless [27]. Further, geodesics can be extended beyond such events [27]. In this work, we analyze the issue of boundedness of curvature invariants and the strength of the singularities in Bianchi-I model. Our analysis is based on the work in Ref. [29]. In the following we first briefly review the dynamics of Bianchi-I model in classical theory and the effective Hamiltonian approach and show the way bounds on ρ, θ and σ^2 arise in LQC. This is followed by an analysis of curvature invariants and necessary and sufficient conditions for singularities to be strong are investigated. From these conditions, we show that all known strong singularities of Bianchi-I model (all these occur at a vanishing physical volume) in classical theory are resolved. We also briefly comment on the conditions which lead to weak singularities. Details of the consequences for geodesic evolution, and a generic lack of strong singularities for perfect fluids and more general fluids are discussed in Ref. [29].

2. Dynamical equations in Bianchi-I spacetimes

We consider a homogeneous Bianchi-I anisotropic spacetime with a spacetime manifold $\mathbb{R}^4$. Due to its non-compact topology, one needs to introduce a fiducial cell – a regulator, to define the symplectic structure. We denote this cell by $\mathcal{V}$ with a fiducial volume $V_o = l_1 l_2 l_3$ with respect to the fiducial metric $\mathring{q}_{ab}$ endowed on the spatial manifold, and l_i refer to the coordinate lengths of the each side of the fiducial cell.

Imposing the underlying symmetries of a homogeneous spacetime, the phase space variables – the matrix valued Ashtekar-Barbero connection A_a^i and triad E_i^a, can be expressed in a simple form: symmetry reduced connection c_i and and triads p_i respectively (where $i = 1, 2, 3$) [35, 19]. These satisfy $\{c_i, p_j\} = 8\pi G\gamma\delta_{ij}$, where $\gamma = 0.2375$ is the Barbero-Immirzi parameter. The

[1] In the case of a spatially flat model sourced with a massless scalar field, it has been shown that the bounce occurs for *all* the states in a dense subspace of the physical Hilbert space [26].

triads are kinematically related to the directional scale factors a_i in the spacetime metric of the orthogonal Bianchi-I spacetime given by

$$ds^2 = -N^2\,dt^2 + a_1^2\,dx^2 + a_2^2 dy^2 + a_3^2 dz^2\ , \tag{1}$$

as

$$|p_1| = l_2 l_3\, a_2 a_3, \quad |p_2| = l_1 l_3\, a_1 a_3, \quad |p_3| = l_2 l_3\, a_2 a_3 \tag{2}$$

where the modulus sign arises because of orientation of the triads. Without any loss of generality, in the following we restrict to the positive orientation of the triads.

2.1. *Classical dynamics*

The Hamiltonian constraint for the Bianchi-I spacetime in classical GR, when expressed in terms of c_i and p_i takes the following form for $N = 1$:

$$\mathcal{H}_{\rm cl} = -\frac{N}{8\pi G\gamma^2 V}(c_1 p_1\, c_2 p_2 + c_3 p_3\, c_1 p_1 + c_2 p_2\, c_3 p_3) + \mathcal{H}_{\rm matt} \tag{3}$$

where $\mathcal{H}_{\rm matt}$ is the matter Hamiltonian. The vanishing of the Hamiltonian constraint: $\mathcal{H}_{\rm cl} \approx 0$ lead us to the physical solutions, and also to a relation between the energy density ρ and the directional Hubble rates H_i (defined as $H_i = \dot{a}_i/a_i$, where a dot denotes the derivative with respect to proper time):

$$\rho = \frac{1}{8\pi G}(H_1 H_2 + H_2 H_3 + H_3 H_1)\ . \tag{4}$$

Equations of motions for the phase space variables are obtained by solving Hamilton's equations:

$$\dot{p}_i = \{p_i, \mathcal{H}_{\rm cl}\} = -8\pi G\gamma\, \frac{\partial \mathcal{H}}{\partial c_i} \tag{5}$$

and

$$\dot{c}_i = \{c_i, \mathcal{H}_{\rm cl}\} = 8\pi G\gamma\, \frac{\partial \mathcal{H}}{\partial p_i}\ . \tag{6}$$

Using (3) and (5) we obtain the relation between connection components and the time derivative of scale factors,

$$c_i = \gamma l_i \dot{a}_i = \gamma l_i\, H_i a_i\ . \tag{7}$$

Dynamical equations for the matter components can be obtained in a similar way. It is interesting to note that c_i and p_i satisfy the following relation if one considers matter with a vanishing anisotropic stress, i.e. when $\rho(p_1, p_2, p_3) = \rho(V)$,

$$p_i c_i - p_j c_j = \gamma V_o \alpha_{ij} \tag{8}$$

where α_{ij} is a constant antisymmetric matrix [35]. Using (7), in the above equation, one obtains: $V(H_i - H_j) = \gamma V_o \alpha_{ij}$. This in turn implies that Σ^2, a measure of the anisotropic shear defined as $\Sigma^2 = \sigma^2 a^6/6$ is a constant of motion in GR. Here $a = (a_1 a_2 a_3)^{1/3} = V^{1/3}$ is the mean scale factor and σ^2 is the shear scalar $\sigma^2 \equiv \sigma_{ab}\sigma^{ab}$, where

$$\sigma_{ij} = H_{ij} - \frac{1}{3}(H_1 + H_2 + H_3)\delta_{ij}\ , \tag{9}$$

H_{ij} is the expansion matrix and θ is the expansion scalar $\theta = (H_1 + H_2 + H_3)/3$. The expression for the shear scalar in terms of the Hubble rates is given by

$$\begin{aligned} \sigma^2 &= \sum_{i=1}^{3}(H_i - \theta)^2 = \frac{1}{3}\left((H_1 - H_2)^2 + (H_2 - H_3)^2 + (H_3 - H_1)^2\right) \\ &= \frac{1}{3a^6}\left(\alpha_{12}^2 + \alpha_{23}^2 + \alpha_{31}^2\right)\ . \end{aligned} \tag{10}$$

Journal of Physics: Conference Series **360** (2012) 012008 doi:10.1088/1742-6596/360/1/012008

In the classical evolution if the mean scale factor vanishes or the directional Hubble rates diverge (in a finite time), then energy density, expansion and shear scalar also diverge. Due to their divergence, the geodesic equations break down and a singularity is reached in a finite time. We will later see that such a divergence also leads to a strong singularity where all in-falling detectors, irrespective of their construction, are destroyed by the tidal forces.

2.2. Effective dynamics

The effective Hamiltonian of the Bianchi-I model with lapse $N = 1$ can be written as:

$$\mathcal{H}_{\rm eff} = -\frac{1}{8\pi G\gamma^2 V}\left(\frac{\sin(\bar\mu_1 c_1)}{\bar\mu_1}\frac{\sin(\bar\mu_2 c_2)}{\bar\mu_2}p_1p_2 + \text{cyclic terms}\right) + \mathcal{H}_{\rm matt} \tag{11}$$

where

$$\bar\mu_1 = \lambda\sqrt{\frac{p_1}{p_2p_3}}, \quad \bar\mu_2 = \lambda\sqrt{\frac{p_2}{p_1p_3}}, \quad \text{and } \bar\mu_3 = \lambda\sqrt{\frac{p_3}{p_1p_2}} . \tag{12}$$

Here $\lambda^2 = 4\sqrt{3}\pi\gamma l_{\rm Pl}^2$ arises from the underlying quantum geometry. It is the minimum area of the closed loop over which holonomies of connection c_i are computed [37].

Using the above Hamiltonian, the Hamilton's equations yield

$$\dot p_1 = \frac{p_1}{\gamma\lambda}\cos(\bar\mu_1 c_1)\,(\sin(\bar\mu_2 c_2) + \sin(\bar\mu_2 c_2)), \tag{13}$$

$$\begin{aligned}\dot c_1 = \frac{1}{2p_1\gamma\lambda}\Bigg[& c_2p_2\cos(\bar\mu_2c_2)(\sin(\bar\mu_1c_1)+\sin(\bar\mu_3c_3)) + c_3p_3\cos(\bar\mu_3c_3)(\sin(\bar\mu_1c_1)+\sin(\bar\mu_2c_2)) \\ & - c_1p_1\cos(\bar\mu_1c_1)(\sin(\bar\mu_2c_2)+\sin(\bar\mu_3c_3)) - \bar\mu_1p_2p_3\Big[\sin(\bar\mu_2c_2)\sin(\bar\mu_3c_3) \\ & + \sin(\bar\mu_1c_1)\sin(\bar\mu_2c_2) + \sin(\bar\mu_3c_3)\sin(\bar\mu_1c_1)\Big]\Bigg] + 8\pi G\gamma\sqrt{\frac{p_2p_3}{p_1}}\left(\frac{\rho}{2} + p_1\frac{\partial\rho}{\partial p_1}\right),\end{aligned} \tag{14}$$

and similarly for the other components of the symmetry reduced connection and triad. Using these equations, it is straightforward to obtain the expressions for directional Hubble rates in LQC, such as

$$H_i = \frac{\dot a_i}{a_i} = \frac{1}{2\gamma\lambda}\left(\sin(\bar\mu_ic_i - \bar\mu_jc_j) + \sin(\bar\mu_ic_i - \bar\mu_kc_k) + \sin(\bar\mu_jc_j + \bar\mu_kc_k)\right) \quad \text{where } i\neq j\neq k\,. \tag{15}$$

Using this equation, one finds that $c_i \neq \gamma\dot a_i$. Thus, the relation between connection components and the time derivative of the scale factors is not as simple as in GR. As in the classical theory, if one considers matter with a vanishing anisotropic stress, $(p_ic_i - p_jc_j)$ is a constant in LQC. However, unlike in GR, this does not translate to the invariance of Σ^2 in the quantum evolution, since $p_ic_i - p_jc_j \neq V(H_i - H_j)$.

The behavior of the expansion and shear scalars in LQC is qualitatively different than in GR. Using (15), in the definitions of the expansion and shear scalars we get [32, 28]

$$\theta = \frac{1}{2\gamma\lambda}\left(\sin(\bar\mu_1c_1 + \bar\mu_2c_2) + \sin(\bar\mu_2c_2 + \bar\mu_3c_3) + \sin(\bar\mu_1c_1 + \bar\mu_3c_3)\right) \tag{16}$$

Journal of Physics: Conference Series **360** (2012) 012008 doi:10.1088/1742-6596/360/1/012008

and

$$\sigma^2 = \frac{1}{3\gamma^2\lambda^2}\Bigg[(\cos(\bar{\mu}_3c_3)(\sin(\bar{\mu}_1c_1)+\sin(\bar{\mu}_2c_2))-\cos(\bar{\mu}_2c_2)(\sin(\bar{\mu}_1c_1)+\sin(\bar{\mu}_3c_3)))^2$$
$$+(\cos(\bar{\mu}_3c_3)(\sin(\bar{\mu}_1c_1)+\sin(\bar{\mu}_2c_2))-\cos(\bar{\mu}_1c_1)(\sin(\bar{\mu}_2c_2)+\sin(\bar{\mu}_3c_3)))^2$$
$$+(\cos(\bar{\mu}_2c_2)(\sin(\bar{\mu}_1c_1)+\sin(\bar{\mu}_3c_3))-\cos(\bar{\mu}_1c_1)(\sin(\bar{\mu}_2c_2)+\sin(\bar{\mu}_3c_3)))^2\Bigg] .$$

Thus, we find that in the effective dynamics, expansion and shear scalars can be expressed as a combination of trignometric functions, and are bounded. Their maximum values are given by [28],

$$\theta_{\rm max} = \frac{3}{2\gamma\lambda} \approx \frac{2.77786}{l_{\rm Pl}} . \tag{17}$$

and

$$\sigma^2{}_{\rm max} = \frac{10.125}{3\gamma^2\lambda^2} = \frac{11.56995}{l_{\rm Pl}^2} . \tag{18}$$

Similarly, it is straightforward to show that the energy density ρ is also a bounded function in LQC. To show this, one solves for $\mathcal{H}_{\rm eff} \approx 0$ by using (12), which yields

$$\rho = \frac{1}{8\pi G\gamma^2\lambda^2}\left(\sin(\bar{\mu}_1c_1)\sin(\bar{\mu}_2c_2) + \text{cyclic terms}\right) . \tag{19}$$

The energy density has a maxima given by

$$\rho \leq \rho_{\rm max}, \quad \rho_{\rm max} = \frac{3}{8\pi G\lambda^2} \approx 0.41\rho_{\rm Pl} . \tag{20}$$

It is interesting to note that the upper bound for the energy density in the Bianchi-I anisotropic model is equal to maximum of ρ in the loop quantization of the spatially flat isotropic model[10].

The boundedness of Hubble rates, expansion and shear scalars and energy density in the loop quantization of Bianchi-I model is a direct consequence of the underlying quantum geoemtry. In the limit, where $\lambda \to 0$, above expressions lead to their classical analogs which are unbounded. These results show that classical singularities which are accompanied by a divergence of above physical quantities are avoided in LQC. If initial data leading to such singularities in the classical theory are evolved using effective dynamics in LQC, then physical solutions agree with their classical counterparts when spacetime curvature is small compared to the Planck scale. However, as the curvature grows, departures from GR become significant. As the Planck regime is approached, directional Hubble rates, given by eq.(15), attain their maximum values. In the subsequent evolution, Hubble rates vanish and the scale factors bounce or recollapse depending on whether they were initially contracting or expanding. The evolution of individual scale factors and Hubble rates is analogous to the dynamical evolution in the loop quantization of the spatially flat isotropic model. There in the backward evolution of an expanding branch, universe enters a regime of superinflation ($\dot{H} > 0$) where the isotropic Hubble rate attains a maximum, before a bounce occurs at $H = 0$. In the isotropic case, energy density attains its maximum values. In contrast, at bounces of directional scale factors in the Bianchi-I spacetime, ρ, θ and σ^2 do not necessarily attain their maximum values [32, 28].

As an example, we illustrate the way singularity resolution occurs for the case of Ekpyrotic/Cyclic model in the Bianchi-I spacetime. The Ekpyrotic/Cyclic model aims to provide a non-inflation based alternative explanation of the origin of structures in our universe, however it suffers from the past singularity problem which does not allow viability of cyclic evolution. Though the past singularity problem can be resolved in the effective dynamics of the isotropic LQC, the modified dynamics in isotropic LQC does not allow cyclic evolution for Ekpyrotic potential [31]. Interestingly, these problems can be successfully overcome in the Bianchi-I model (see Ref. [32] for details). In Fig. 1, we show the results from a simulation for this model using effective dynamical equations obtained in this section, which were analyzed in Ref. [32]. From the plots, we see that the mean scale factor ($a = (a_1a_2a_3)^{1/3}$) bounces and the shear scalar remains bounded during the evolution, resulting in that the singularity is resolved.

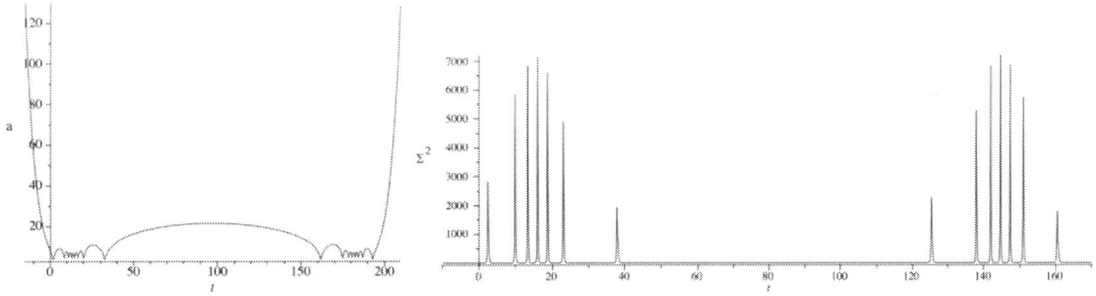

Figure 1. Plots of the non-singular evolution of the mean scale factor and Σ^2 are shown for an evolution with Ekpyrotic/Cyclic potential in Bianchi-I LQC. Multiple bounces of the mean scale factor occur and Σ^2 attains a constant value in the classical limit.

3. Behavior of curvature invariants and the absence of strong singularities

Dynamical equations obtained from the effective Hamiltonian (11) show that energy density and expansion and shear scalars are bounded by universal values in the loop quantization of Bianchi-I spacetime [32, 33, 28]. Boundedness of these quantities plays a crucial role in the resolution of various singularities in the Bianchi-I model. However, it is important to note that these do not necessarily imply that the spacetime curvature is *always* bounded in LQC. This is straightforward to understand from the expressions of the curvature invariants. For simplicity, we here discuss only the Ricci and the Kretschmann scalars, which for the metric (1), are given by

$$R = 2\left(H_1H_2 + H_2H_3 + H_3H_1 + \sum_{i=1}^{3} \frac{\ddot{a}_i}{a_i}\right) , \tag{21}$$

and

$$K = 4\left(H_1^2H_2^2 + H_1^2H_3^2 + H_2^2H_3^2 + \sum_{i=1}^{3} \frac{\ddot{a}_i^2}{a_i}\right) . \tag{22}$$

Other curvature scalars can be written in a similar way in terms of the directional Hubble rates and second order time derivatives of the scale factors. The pertinent question is whether curvature invariants are bounded functions as is the case for H_i, ρ, θ and σ^2 in the loop quantization of Bianchi-I model. To answer this, we have to analyze the second order time derivatives of scale factors. Using the Hamilton's equations for p_i and c_i, a straightforward calculation yields [29],

$$\frac{\ddot a_1}{a_1} = \frac{1}{2\gamma\lambda}\Bigg[\cos(\bar\mu_1 c_1 - \bar\mu_2 c_2)\,\frac{\mathrm{d}}{\mathrm{d}t}(\bar\mu_1 c_1 - \bar\mu_2 c_2) \;+\; \cos(\bar\mu_1 c_1 - \bar\mu_3 c_3)\,\frac{\mathrm{d}}{\mathrm{d}t}(\bar\mu_1 c_1 - \bar\mu_3 c_3) \tag{23}$$
$$+\cos(\bar\mu_2 c_2 + \bar\mu_3 c_3)\,\frac{\mathrm{d}}{\mathrm{d}t}(\bar\mu_2 c_2 + \bar\mu_3 c_3) + \frac{1}{2\gamma\lambda}\Big(\sin(\bar\mu_1 c_1 - \bar\mu_2 c_2) + \sin(\bar\mu_1 c_1 - \bar\mu_3 c_3) +$$
$$+\sin(\bar\mu_2 c_2 + \bar\mu_3 c_3)\Big)^2\Bigg]\,.$$

Equations for $\ddot a_2$ and $\ddot a_3$ can be obtained by cyclic permutations of the terms in the above equation. If we assume matter has a vanishing anisotropic stress, then we obtain

$$\frac{\ddot a_1}{a_1} = -4\pi G(\rho + P) + \frac{1}{4V}\Big(6(\cos(\bar\mu_1 c_1 - \bar\mu_2 c_2)\kappa_{21} + \cos(\bar\mu_1 c_1 - \bar\mu_3 c_3)\kappa_{31})H + (\kappa_{12} + \kappa_{13})\frac{\dot p_1}{p_1}$$
$$+\kappa_{23}\left(\frac{\dot p_2}{p_2} - \frac{\dot p_3}{p_3}\right)\Big) + \frac{1}{4\gamma^2\lambda^2}\left(\sin(\bar\mu_1 c_1 - \bar\mu_2 c_2) + \sin(\bar\mu_1 c_1 - \bar\mu_3 c_3) + \sin(\bar\mu_2 c_2 + \bar\mu_3 c_3)\right)^2\,.$$

Substituting eqs. (15) and (23), and similar equations for the time derivatives of a_2 and a_3 in the expression for the Ricci scalar (21), we obtain for matter with a vanishing anisotropic stress [29]:

$$R = -24\pi G(\rho + P) + \frac{1}{V}\left(\frac{\dot p_1}{p_1}(\kappa_{12} + \kappa_{13}) + \frac{\dot p_2}{p_2}(\kappa_{23} + \kappa_{21}) + \frac{\dot p_3}{p_3}(\kappa_{31} + \kappa_{32})\right)$$
$$+\frac{1}{2\gamma^2\lambda^2}\Big[3 + \cos^2(\bar\mu_1 c_1)\left(\sin^2(\bar\mu_3 c_3) + 4\sin(\bar\mu_2 c_2)\sin(\bar\mu_3 c_3) - \cos(2\bar\mu_2 c_2)\right)$$
$$+\cos^2(\bar\mu_2 c_2)\left(\sin^2(\bar\mu_1 c_1) + 4\sin(\bar\mu_1 c_1)\sin(\bar\mu_3 c_3) - \cos(2\bar\mu_3 c_3)\right)$$
$$+\cos^2(\bar\mu_3 c_3)\left(\sin^2(\bar\mu_2 c_2) + 4\sin(\bar\mu_1 c_1)\sin(\bar\mu_2 c_2) - \cos(2\bar\mu_1 c_1)\right)$$
$$-\left(\sin^2(\bar\mu_1 c_1)\sin^2(\bar\mu_2 c_2) + \sin^2(\bar\mu_1 c_1)\sin^2(\bar\mu_3 c_3) + \sin^2(\bar\mu_2 c_2)\sin^2(\bar\mu_3 c_3)\right)\Big]\,. \tag{24}$$

Similarly, the expression for the Kretschmann scalar turns out to be [29]

$$K = \frac{\left(\sin(\bar\mu_1 c_1 - \bar\mu_2 c_2) + \sin(\bar\mu_1 c_1 - \bar\mu_3 c_3) + \sin(\bar\mu_2 c_2 + \bar\mu_3 c_3)\right)^2}{2\gamma^4\lambda^4}\{(\cos(\bar\mu_1 c_1 - \bar\mu_2 c_2 - 2\bar\mu_3 c_3)$$
$$+2 + \cos(\bar\mu_1 c_1 - \bar\mu_2 c_2 + 2\bar\mu_3 c_3))\sin^2\left(\tfrac{\bar\mu_1 c_1 + \bar\mu_2 c_2}{2}\right)$$
$$+\sin^2(\bar\mu_1 c_1 - \bar\mu_2 c_2) + 2(\cos(\bar\mu_2 c_2) - \cos(\bar\mu_1 c_1))\sin(\bar\mu_1 c_1 - \bar\mu_3 c_3)\sin(\bar\mu_3 c_3)\}$$
$$+\frac{1}{4\gamma^2\lambda^2 V^2}\Big[\frac{\dot p_1}{p_1}(\kappa_{12} + \kappa_{13}) + \kappa_{23}\left(\frac{\dot p_2}{p_2} - \frac{\dot p_3}{p_3}\right) + 6H(\kappa_{21}\cos(\bar\mu_1 c_1 - \bar\mu_2 c_2)$$
$$+\kappa_{31}\cos(\bar\mu_1 c_1 - \bar\mu_3 c_3)) - 16\pi G(\rho + P)V\Big]\Big[\gamma\lambda\{6H\gamma\lambda(\kappa_{21}\cos(\bar\mu_1 c_1 - \bar\mu_2 c_2)$$
$$+\kappa_{31}\cos(\bar\mu_1 c_1 - \bar\mu_3 c_3)) + \kappa_{23}(\cos(\bar\mu_2 c_2)(\sin(\bar\mu_1 c_1) + \sin(\bar\mu_3 c_3))$$
$$-\cos(\bar\mu_3 c_3)(\sin(\bar\mu_1 c_1) + \sin(\bar\mu_2 c_2))) + (\kappa_{12} + \kappa_{13})\cos(\bar\mu_1 c_1)(\sin(\bar\mu_2 c_2) + \sin(\bar\mu_3 c_3))\}$$
$$+2V\left(-8\pi G\gamma^2\lambda^2(\rho + P) + \left(\sin(\bar\mu_1 c_1 - \bar\mu_2 c_2) + \sin(\bar\mu_1 c_1 - \bar\mu_3 c_3) + \sin(\bar\mu_2 c_2 + \bar\mu_3 c_3)\right)^2\right)\Big]$$
$$+\text{cyclic terms}\,. \tag{25}$$

Journal of Physics: Conference Series **360** (2012) 012008 doi:10.1088/1742-6596/360/1/012008

Unlike the case of H_i, ρ, θ and σ^2, these expressions are not bounded functions. There are only two situations in which the curvature invariants can diverge in the effective dynamics of the loop quantization of the Bianchi-I model. The first case arises if the physical volume vanishes in the physical evolution, and the second arises when the pressure of the matter content becomes infinite. Both of these scenarios must arise at a finite value of ρ, θ and σ^2, to be physically realized in the evolution. It is to be noted that all known types of singularities in the Bianchi-I model in GR, fail to satisfy the requirements for the first scenario to occur. In all known cases of singular evolution, curvature invariants diverge at a vanishing volume along with a divergence of ρ, θ and σ^2. All such events will have curvature invariants bounded in LQC, and the singularity will be resolved by the effective dynamics [29]. On the other hand, the second case has an analog in the loop quantization of the isotropic model, where the curvature scalars can diverge when pressure $P \to \pm\infty$ with ρ remaining finite. This implies a divergence in the equation of state, which can indeed occur with an appropriate choice of matter. Unlike the second scenario, there is no analog of the first possibility in the isotropic case [27].

The key question is whether the cases where curvature invariants can diverge in LQC, imply a singular evolution. To answer this question, one has to analyze the behavior of geodesics and examine the strength of singularities. Conditions to determine the strength of the singularities have been formulated in the seminal works of Tipler [5] and Królak [6], which involve integrals of the curvature components over the geodesics. For the anisotropic spacetimes, a singularity occurring at the affine parameter $\tau = \tau_e$ is strong according to Królak, if

$$\int_0^\tau \mathrm{d}\tau' R_{ab}\, u^a u^b \tag{26}$$

or

$$\int_0^\tau \mathrm{d}\tau' \left(\int_0^{\tau''} \mathrm{d}\tau'' |C_{abcd}\, u^b u^d| \right)^2 \tag{27}$$

diverges as $\tau \to \tau_e$. Here $u^a = \mathrm{d}x^a/\mathrm{d}\tau$. If the above integrals are finite, then the singularity is weak. To determine whether a singularity is strong according to Tipler's criteria one considers an additional integral in the above equations. Due to this difference in conditions, it is possible that a singularity which is strong à la Królak may turn out to be weak à la Tipler. In terms of these conditions, a detector is inevitably destroyed by a strong singularity but survives a weak singularity. Indeed it is quite possible, that an event where curvature invariants diverge may not lead to a break down of geodesics and the singularity may be weak. An example of such an event in GR, occurs for the the case of a sudden singularity which occurs due to a divergence in pressure at a finite volume and energy density [38]. Even in LQC, such an event is curvature divergent but the singularity turns out to be weak, in spatially flat [27] as well as spatially curved models [34].

For simplicity, we analyze here the details of the condition (26), and refer the reader to Ref. [29] for analysis of the integral over Weyl curvature components. The integrand in this case turns out to be,

$$\begin{aligned} R_{ab}\, u^a u^b \;=\; & \frac{k_x^2}{a_1^2}\left(H_1 H_2 + H_1 H_3 - \frac{\ddot{a}_2}{a_2} - \frac{\ddot{a}_3}{a_3}\right) + \frac{k_y^2}{a_2^2}\left(H_1 H_2 + H_2 H_3 - \frac{\ddot{a}_1}{a_1} - \frac{\ddot{a}_3}{a_3}\right) \\ & + \frac{k_z^2}{a_3^2}\left(H_3 H_1 + H_2 H_3 - \frac{\ddot{a}_1}{a_1} - \frac{\ddot{a}_2}{a_2}\right) \end{aligned} \tag{28}$$

where k_x, k_y and k_z are constants. Substituting above expression in Królak's condition for Ricci tensor, we find that due to an integral over the affine parameter, any divergence in the terms

$\ddot{a}_i/a_i$ does not translate to a divergence of the integral (26). However, if in a physical evolution, any of the scale factors vanishes at a finite value of the affine parameter, then the integral (26) can diverge. Whether or not the singularity in this case would be strong, will depend on the detailed behavior of the dynamical solutions which allow this singularity.

We are now ready to answer the questions we posed: Do curvature invariant diverging events in Bianchi-I LQC imply a strong singularity? The answer turns out to be the following (see Ref. [29] for details):

(i) If the divergence of curvature invariants is caused by the divergence in pressure of the matter component at non-vanishing scale factors and finite value of ρ, θ and σ^2, then the singularity is weak. Analysis of the geodesic equations show that they remain well defined and geodesics can be extended beyond such curvature invariant diverging events. Thus, a pressure singularity in the Bianchi-I model in LQC has very similar characteristics as in the isotropic model. In both cases, such curvature invariant diverging events turns out to be a weak singularity and is harmless.

(ii) If effective dynamics in LQC allows a vanishing of the scale factor, and hence physical volume in a finite time, at a *finite* value of ρ, θ and σ^2, then the singularity can be strong. Geodesic evolution in this case breaks down. However, it is not known, whether such a curvature divergent event is allowed by the effective dynamics. To our knowledge, an event with divergent curvature invariants and vanishing volume with a finite ρ, θ and σ^2 is not known even in GR. It will be an interesting exercise to discover such a physical solution and understand whether the allowed singularity is strong or weak.

4. Summary

In the present analysis, we investigated the effective dynamics of LQC for the Bianchi-I spacetime for matter with a vanishing anisotropic stress. A quantization of this spacetime with a massless scalar field has been recently performed in Ref. [19] (see also Ref. [20]), which demonstrated that the quantum Hamiltonian constraint is non-singular. Using effective dynamics, physics of this spacetime has been analyzed for different forms of matter [35, 32, 33, 28, 39, 40] which show the occurrence of bounces of the directional scale factors which lead to avoidance of classical singularities. Here we summarized some of the main points of the results obtained in Ref. [29], where we show that boundedness of ρ, θ and σ^2 does not imply a bound on the curvature scalars. The latter can diverge in the effective dynamics of loop quantization of Bianchi-I model. Such curvature invariant diverging events exist even for isotropic LQC. These were earlier proved by the author to be harmless as they correspond to weak singularities and geodesics can be continued beyond them [27]. The situation in Bianchi-I model turns out to be similar. We find that divergence in spacetime curvature, when occurring due to a divergence in pressure of the matter content at non-zero values of the scale factors lead neither to strong singularity nor the breakdown of geodesics. Further, all known types of strong singularities of the classical theory in the Bianchi-I spacetime are shown to be avoided.

Acknowledgments

This work is supported by NSF grant PHY1068743.

5. References

[1] G. F. R. Ellis, Bianchi models: Then and now, Gen. Rel. Grav. **38** 1003 (2006)
[2] A. G. Doroshkevich, Model of a universe with a uniform magnetic field, Astrophysics **1** 138 (1965)
[3] K. S. Thorne, Primordial Element Formation, Primordial Magnetic Fields, and the Isotropy of the Universe, Ap. J. **148** 51 (1967)

[4] S. W. Hawking, G. F. R. Ellis, *The large scale structue of Space-Time*, Cambridge (1973)
[5] F. J. Tipler, Singularities in conformally flat spacetimes, Phys. Lett. **64 A** 8 (1977)
[6] A. Królak, Towards the proof of the cosmic censorship hypothesis, Class. Quant. Grav. **3** 267 (1986)
[7] A. Ashtekar, P. Singh, Loop quantum cosmology: a status report, Class. Quant. Grav. **28** 213001 (2011)
[8] A. Ashtekar, T. Pawlowski and P. Singh, Quantum nature of the big bang, Phys. Rev. Lett. **96** 141301 (2006)
[9] A. Ashtekar, T. Pawlowski and P. Singh, Quantum nature of the big bang: An analytical and numerical investigation I, Phys. Rev. **D73** 124038 (2006)
[10] A. Ashtekar, T. Pawlowski and P. Singh, Quantum nature of the big bang: Improved dynamics, Phys. Rev. D**74** 084003 (2006)
[11] E. Bentivegna and T. Pawlowski, Anti-deSitter universe dynamics in LQC, Phys. Rev. D**77** 124025 (2008)
[12] W. Kaminski and T. Pawlowski, The LQC evolution operator of FRW universe with positive cosmological constant, Phys. Rev. D**81** 024014 (2010)
[13] T. Pawlowski, A. Ashtekar, Positive cosmological constant in loop quantum cosmology, arXiv:1112.0360.
[14] A. Ashtekar, T. Pawlowski and P. Singh, Loop quantum cosmology in the pre-inflationary epoch (in preparation).
[15] L. Szulc, W. Kaminski, J. Lewandowski, Closed FRW model in loop quantum cosmology, Class. Quant. Grav. **24** 2621 (2007)
[16] A. Ashtekar, T. Pawlowski, P. Singh and K. Vandersloot, Loop quantum cosmology of k=1 FRW models, Phys. Rev. D**75** 0240035 (2006)
[17] K. Vandersloot, Loop quantum cosmology and the $k = -1$ FRW model, Phys. Rev. D **75** 023523 (2007)
[18] L. Szulc, Open FRW model in loop quantum cosmology, Class. Quant. Grav. **24** 6191 (2007)
[19] A. Ashtekar and E. Wilson-Ewing, Loop quantum cosmology of Bianchi type I models, Phys. Rev. D**79** 083535 (2009)
[20] M. Martin-Benito, L. J. Garay, G. A. Mena Marugan, Loop quantum cosmology of the Bianchi I model: complete quantization, arXiv:1110.1941.
[21] A. Ashtekar and E. Wilson-Ewing, Loop quantum cosmology of Bianchi type II models, Phys. Rev. D**80** 123532 (2009)
[22] E. Wilson-Ewing, Loop quantum cosmology of Bianchi type IX models, Phys. Rev. D**82** 043508 (2010)
[23] J. Willis, On the low energy ramifications and a mathematical extension of loop quantum gravity. Ph.D. Dissertation, The Pennsylvaina State University (2004)
[24] V. Taveras, LQC corrections to the Friedmann equations for a universe with a free scalar field, Phys. Rev. **D78** 064072 (2008)
[25] P. Singh, V. Taveras, Effective equations for arbitrary matter in loop quantum cosmology, (In preparation).
[26] A. Ashtekar, A. Corichi and P. Singh, Robustness of predictions of loop quantum cosmology, Phys. Rev. D**77** 024046 (2008)
[27] P. Singh, Are loop quantum cosmologies never singular?, Class. Quant. Grav. **26** 125005 (2009)
[28] B. Gupt, P. Singh, Contrasting features of anisotropic loop quantum cosmologies: the role of spatial curvature, arXiv:1109.6636.
[29] P. Singh, Towards a resolution of strong singularities in Bianchi-I loop quantum cosmology (To be submitted).
[30] P. Singh, Loop cosmological dynamics and dualities with Randall-Sundrum braneworlds, Phys. Rev. D **73** 063508 (2006)
[31] P. Singh, K. Vandersloot, and G. V. Vereshchagin, Non-singular bouncing universes in loop quantum cosmology, Phys. Rev. D**74** 043510 (2006)
[32] T. Cailleteau, P. Singh and K. Vandersloot, Non-singular ekpyrotic/cyclic model in loop quantum cosmology, Phys. Rev. D**80** 124013 (2009)
[33] A. Corichi, P. Singh, A geometric perspective on singularity resolution and uniqueness in loop quantum cosmology, Phys. Rev. D **80** 124013 (2009)
[34] P. Singh, F. Vidotto, Exotic singularities and spatially curved loop quantum cosmology, Phys. Rev. D **83** 064027 (2011)
[35] D. W. Chiou and K. Vandersloot, The behavior of non-linear anisotropies in bouncing Bianchi I models of loop quantum cosmology, Phys. Rev. D **76** 084015 (2007)
[36] T. Cailleteau, A. Cardoso, K. Vandersloot, and D. Wands, Singularities in loop quantum cosmology, Phys. Rev. Lett. **101** 251302 (2008)
[37] A. Ashtekar and E. Wilson-Ewing, The covariant entropy bound and loop quantum cosmology, Phys. Rev. D **78** 06407 (2008)
[38] L. Fernandez-Jambrina, R. Lazkoz, Geodesic behavior around cosmological milestones, arXiv:0903.5527.
[39] R. Maartens, K. Vandersloot, Magnetic Bianchi-I universe in loop quantum cosmology, arXiv:0812.1889[gr-qc].
[40] M. Artymowski, Z. Lalak, Vector fields and Loop Quantum Cosmology, JCAP **09** 017 (2009)

Journal of Physics: Conference Series **360** (2012) 012009 doi:10.1088/1742-6596/360/1/012009

The diffeomorphism constraint operator in loop quantum gravity

M Varadarajan

Raman Research Institute, Bangalore, India

E-mail: madhavan@rri.res.in

Abstract. We construct the diffeomorphism constraint operator at finite triangulation from the basic holonomy- flux operators of Loop Quantum Gravity, and show that the action of its continuum limit provides an anomaly free representation of the Lie algebra of diffeomorphisms of the 3- manifold. Key features of our analysis include: (i) finite triangulation approximants to the curvature, F^i_{ab} of the Ashtekar- Barbero connection which involve not only small loop holonomies but also small surface fluxes as well as an explicit dependence on the edge labels of the spin network being acted on (ii) the dependence of the small loop underlying the holonomy on both the direction and magnitude of the shift vector field (iii) continuum constraint operators which do *not* have finite action on the kinematic Hilbert space, thus implementing a key lesson from recent studies of parameterised field theory by the authors. Features (i) and (ii) provide the first hints in LQG of a conceptual similarity with the so called "mu- bar" scheme of Loop Quantum Cosmology. We expect our work to be of use in the construction of an anomaly free quantum dynamics for LQG.

We highlight the main steps and results of our construction while suppressing most of the technical details. This work was done jointly with Alok Laddha.

1. Introductory Remarks:

One of the key open problems in LQG is that of a satisfactory definition of its quantum dynamics. The current definition of the Hamiltonian constraint operator is far from unique due to the many ad- hoc choices which go into its construction. The reason this happens is that the Hamiltonian constraint is a local expression constructed out of local fields whereas some of the basic operators of LQG are non- local and have discontinuous action. Specifically, the main source of ambiguities is that there is no known way to obtain the operator corresponding to the curvature of the Ashtekar- Barbero connection, $F^i_{ab}(x)$, from holonomy operators. While in the classical theory, $F^i_{ab}(x)$ can be obtained from a limit of holonomies around loops which shrink to the point x, the corresponding limit of operators in the quantum theory does not exist primarily because the Hilbert space structure of LQG is unable to adequately distinguish between a 'small' loop and a 'smaller' loop by virtue of its background independence.

Hence one proceeds as follows [1]. Fix a triangulation T of the spatial manifold. Choose finite triangulation approximants to the various local fields which make up the Hamiltonian constraint These approximants are constructed from the the basic holonomy- electric flux functions of the theory. Using these approximants define the finite triangulation Hamiltonian constraint $C_{ham,T}$. By construction $C_{ham,T}$ approaches the classical Hamiltonian constraint in the continuum limit of infinitely fine triangulation. Next, replace the holonomy- flux functions in $C_{ham,T}$ by their

Journal of Physics: Conference Series **360** (2012) 012009 doi:10.1088/1742-6596/360/1/012009

operator correspondents to obtain $\hat{C}_{ham,T}$. Finally, take the continuum limit of this operator, the idea being that while the individual (holonomy) operators do not have a continuum limit, the composite operator $\hat{C}_{ham,T}$ may have one.

Remarkably, a continuum limit exists [1, 2, 3] but, unfortunately, it depends on the choice of finite triangulation holonomy approximants to $F^i_{ab}(x)$. There are two aspects of this choice. One is the representation of $SU(2)$ chosen to evaluate the holonomy [4] and the second is the choice of little loops which are shrunk away to the point x. Typically the representation is chosen to be some fixed j (usually $j = \frac{1}{2}$) representation and the little loop is chosen by some subjective criterion of "simplicity". Each choice of representation and each choice of the diffeomorphism invariant way that the loop interacts with the state being acted upon yields a potentially inequivalent continuum Hamiltonian constraint operator.

Moreoever, simpler contexts suggest these choices are incorrect. In Parameterised Field Theory [9, 10], the representation of the 'holonomy' operator depends on the edge labels of state. In Loop Quantum Cosmology [11], the small loop depends on the triad operator. Given this state of affairs, notwithstanding the impressive pioneering nature of earlier works [5, 6, 7, 8, 16], it is imperative to improve upon the choice of approximants to F^i_{ab}. We seek insight into the correct choice by constructing the *diffeomorphism* constraint operator in such a way that it kills the standard diffeomorphism invariant distributions obtained by group averaging [12]. Note that the standard treatment does *not* construct this operator. The diffeomorphism constraint is imposed indirectly by the demand that states be invariant under the unitary action of *finite* diffeos. Indeed the generator of diffeomorphisms is not even defined on the kinematic Hilbert Space.

Here, we treat the diffeomorphism constraint in a manner similar to the Hamiltonian constraint through the following steps:
- Fix a triangulation T of the spatial manifold. Construct finite T approximants to relevant local fields from the holonomy- flux variables.
- From these, construct the diffeomorphism constraint $D_T(\vec{N})$ at finite triangulation, $\vec{N}$ denoting the shift vector.
- Replace the holonomy- flux functions by the corresponding operators, obtain $\hat{D}_T(\vec{N})$ and construct its continuum limit.

The above procedure necessitates a choice of finite T approximant to $F^i_{ab}(x)$ by virtue of the dependence of the diffeomorphism constraint on $F^i_{ab}(x)$. Our choice of finite T approximant to $F^i_{ab}(x)$ is guided by the requirement that at finite T:

$$\hat{D}_T(\vec{N}) = -i\hbar \frac{\hat{U}_{\phi(\vec{N},\delta)} - \mathbf{1}}{\delta}. \tag{1}$$

Here δ parameterizes the fineness of the triangulation (the continuum limit is defined as the limit $\delta \to 0$) and $\phi(\vec{N}, \delta)$ is the diffeomorphism which translates points along integral curves of the shift vector $\vec{N}$ by affine parameter length δ. This requirement has a fourfold justification:
- The expression (1) kills the standard diffeomorphism invariant states obtained by group averaging.
- If there was a generator, this form is natural.
- This form manifests in Parameterised Field Theory which is the only known field theoretic model in which all steps of the LQG program can be carried through without ambiguities.
- The continuum limit operator action obtained from (1) yields a faithful representation of its algebra on Lewandowski- Marolf habitat.

The plan of this article is as follows. In section 2 we sketch the main steps of our construction which yield the operator of equation (1). In section 3 we display our choice of curvature approximants for the simple case of the state being an edge holonomy in the $j = \frac{1}{2}$ representation. In section 4 we comment briefly on the continuum limit of the finite triangulation diffeomorphism

Journal of Physics: Conference Series **360** (2012) 012009 doi:10.1088/1742-6596/360/1/012009

constraint operator. Section 5 is devoted to a discussion of our results. Our presentation will be schematic and meant to convey the broad picture. A comprehensive account is already available [13] for the reader interested in the details.

2. Brief Sketch of Main Steps

The purpose of this section is to sketch the main steps in the construction so as to give the reader a rough global view of the logic; In this section we shall set $G = \hbar = c = 1$. We shall further simplify our presentation by choosing the Barbero- Immirizi parameter γ to be unity in this section.

The diffeomorphism constraint $D(\vec{N})$ is

$$D(\vec{N}) = \int_\Sigma \mathcal{L}_{\vec{N}} A^i_a \tilde{E}^a_i \tag{2}$$

$$= V(\vec{N}) - \mathcal{G}(N^c A^i_c) \tag{3}$$

where

$$V(\vec{N}) = \int_\Sigma N^a F^i_{ab} \tilde{E}^b_i \tag{4}$$

$$\mathcal{G}(N^c A^i_c) = \int_\Sigma N^c A^i_c \mathcal{D}_a \tilde{E}^a_i. \tag{5}$$

Here Σ is the 3- manifold, A^i_a is the Ashtekar Barbero connection, F^i_{ab} is its curvature, $\tilde{E}^a_i$ is the densitized triad and $\vec{N} \equiv N^a$ is the shift vector field.

Let $T(\delta)$ be a 1 parameter family of triangulations of Σ with the continuum limit being $\delta \to 0$ and let $D_{T(\delta)}, V_{T(\delta)}, \mathcal{G}_{T(\delta)}$ be finite triangulation approximants to the quantities $D, V, \mathcal{G}$ of the above equations. Thus $D_{T(\delta)}, V_{T(\delta)}, \mathcal{G}_{T(\delta)}$ are expressions which yield $D, V, \mathcal{G}$ in the continuum limit.

For simplicity consider a (non- gauge invariant) spin network state consisting of a single edge e with spin label j so that the state is just the m, n component of an edge holonomy $h^{(j)}_{e\,m}{}^{n}$ of the (generalized) connection along the edge e in the representation j, the indices m, n taking values in the set $1, .., 2j + 1$. In what follows we shall supress some of these labels and denote the state simply by h_e.

From (1) of section 1, our desired result is:

$$(1 + i\delta \hat{D}_{T(\delta)}) h_e = h_{\phi(\delta,\vec{N})\circ e}. \tag{6}$$

where, $\phi(\delta, \vec{N}) \circ e$ is the image of e by the diffeomorphism $\phi(\delta, \vec{N})$ which translates e by an amount δ along the integral curves of the shift vector field N^a (see Fig 1a).

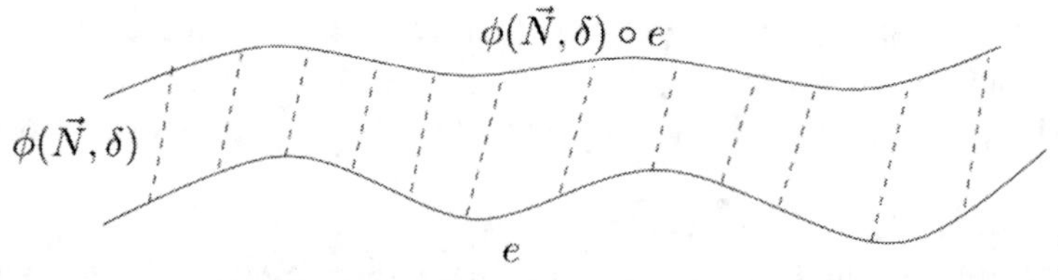

Fig 1a

We obtain the desired result through the following steps:

(i) First we set

$$(1 + i\delta \hat{D}_{T(\delta)}) := (1 + i\delta \hat{\mathcal{G}}_{T(\delta)})(1 + i\delta \hat{V}_{T(\delta)}). \tag{7}$$

(ii) Next, we show that

$$(1 + i\delta \hat{V}_{T(\delta)}) h_e = h_{\bar{e}(\vec{N},\delta)}. \tag{8}$$

Journal of Physics: Conference Series **360** (2012) 012009 doi:10.1088/1742-6596/360/1/012009

Here $\bar{e}(\vec{N},\delta)$ has the same end points as e (as it must by virtue of the gauge invariance of V) and is obtained by joining the end points of $\phi(\delta,\vec{N})\circ e$ to those of e by a pair of segments which are aligned with integral curves of N^a as shown in Fig 1b.

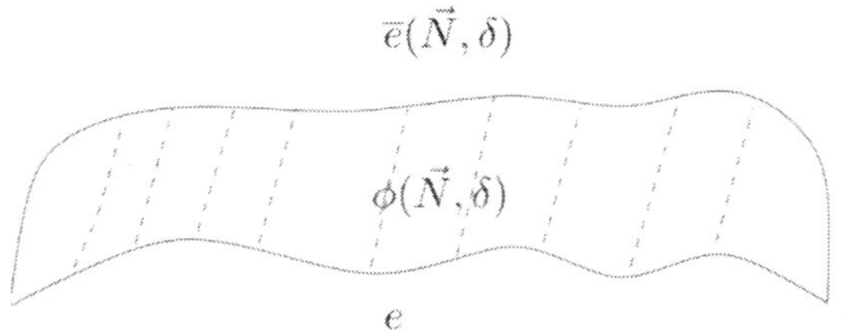

Fig-1b

(iii) Finally, we show that the Gauss Law term, $(1+i\delta\hat{\mathcal{G}}_{T(\delta)})$ removes these two extra segments (see Fig 1c).

Fig-1c

The major part of the analysis concerns the derivation of the identity (8) in step (ii) above. We proceed as follows.

$V_{T(\delta)}$ is written as a sum over contributions $V_\triangle$ where $\triangle$ denotes a 3- cell of the triangulation *dual* to $T(\delta)$, and $V_\triangle$ is a finite triangulation approximant to the integral $\int_\triangle N^a F^i_{ab}\tilde{E}^b_i$. We order the triad operator to the right in $\hat{V}_\triangle$ so that only those 3- cells contribute which intersect e.

The triangulation $T(\delta)$ is adapted to the edge e so that its restriction to e defines a triangulation of e. Thus, there is a triangulation of e by 1- cells and vertices of $T(\delta)$ so that each of these vertices $v_I, I=1,..,N$ is located at the centre of some 3- cell $\triangle=\triangle_I$. We define a finite triangulation approximant to F^i_{ab} in a such a way that the following identity holds:

$$(1+i\delta\hat{V}_{\triangle_I})h_e = h_{\bar{e}(\triangle_I)}. \tag{9}$$

$\forall\, I\, \in\, \{1,...,N-1\}$.
Here $\bar{e}(\triangle_I)$ is obtained by moving the segment of e between v_I and v_{I+1} along the integral curves of N^a by an amount δ and joining this segment to the rest of e at the points v_I, v_{I+1} by a pair of segments which run along the integral curves of N^a as shown in Fig 1d.

Fig-1d

Next, we show that the contributions from all the $\hat{V}_{\triangle_I}$ yield the edge $\bar{e}(\vec{N},\delta)$ of Fig 1b. Recall that $V_{T(\delta)}$ is obtained by summing over all the cell contributions $V_\triangle$. However, summing over the action of all the $\hat{V}_{\triangle_I}$ on h_e only yields a sum over states of the type $h_{\bar{e}(\triangle_I)}$. In order to obtain the desired result, $h_{\bar{e}(\vec{N},\delta)}$, the sum over $\triangle$ is first converted to a *product* over $\triangle$ i.e. to leading order in δ, we show that

$$1+i\delta\sum_\triangle V_\triangle \sim \prod_\triangle(1+i\delta V_\triangle). \tag{10}$$

Hence, replacing the sum over the corresponding operators by the product provides an equally legitimate definition of $\hat{V}_{T(\delta)}$. The replacement then leads, modulo some details, to the following identity

$$\prod_{\triangle}(1+i\delta\hat{V}_{\triangle})h_e = \prod_{I=1}^{N-1}(1+i\delta V_{\triangle_I})h_e. \tag{11}$$

Our definition of $\hat{V}_{\triangle_I}$ is such that each factor in the product acts independently, the Ith factor acting only on the part of e between v_I and v_{I+1}. We are then able to show that the result (8) follows essentially through the mechanism which is illlustrated schematically in Fig 1e.[1]

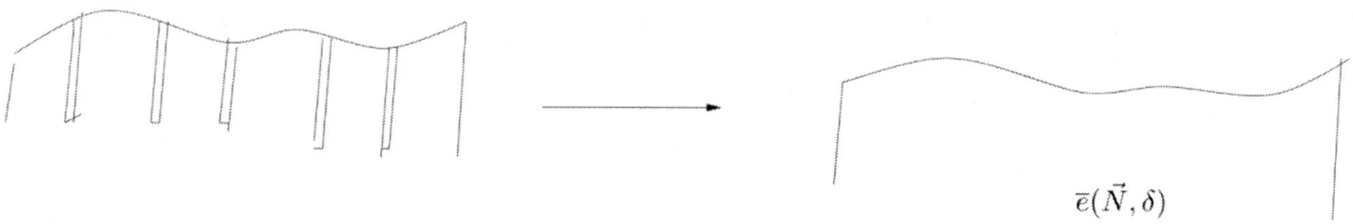

Fig-1e

To summarise, we obtain the desired result (6) through the sequence:
Fig 1d→Fig 1e→Fig 1c.

3. Curvature Approximants

For simplicity let the state h_e be in the $j=\frac{1}{2}$ representation. We seek a finite triangulation approximant to F^i_{ab} such that equation (9) holds. Hence, restrict attention to the part e_I of the edge e between v_I and v_{I+1}. Let e_I be small enough that e_I is in an open patch with coordinates (x,y,z), such that x runs along e. Let γ_I be the 'rectangular' loop obtained by running along (but opposite to) e_I from v_{I+1} to v_I, then climbing up the integral curve of N^a by an amount δ, then running along the segment $\phi(\vec{N},\delta)\circ e_I$ and finally coming down the integral curve of N^a to v_{I+1} as shown in Figure 2. Let the part of γ_I excluding e_I be $\bar{e}_I$.

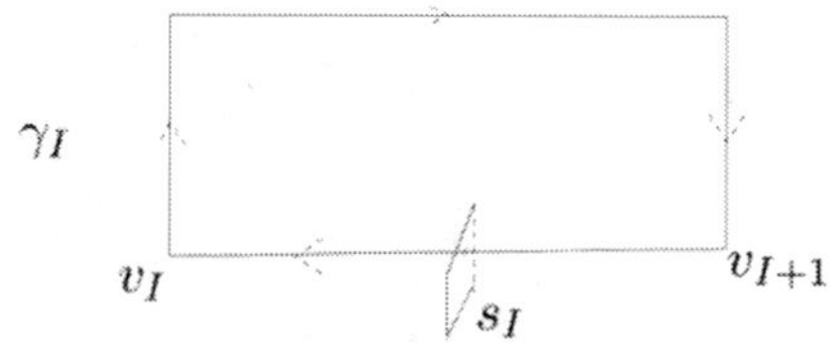

Figure 2

From the discussion in section 1, we want to approximate $V_{\triangle_I}=\int_{\triangle_I}N^aF^i_{ab}\tilde{E}^b_i$ in a way that leads to:

$$(1+i\delta V_{\triangle_I})h_{e_I}=h_{\bar{e}_I}. \tag{12}$$

$\tilde{E}^b_i$ is approximated by the electric flux through the surface S^b_I in the coordinate plane normal to the b- direction. We choose the location of these surfaces such that they are centred at the midpoint v of v_I, v_{I+1} and choose their size to be small enough that they intersect the edge γ_I only along e_I.

Clearly only the flux operator $\hat{E}_i(S_I)$ through the surface $S_I:=S^x_I$ contributes to equation (12) so that we may focus only on the contribution V^x_I to $V_{\triangle_I}$ where $V^x_I:=\int_{\triangle_I}N^aF^i_{ax}\tilde{E}^x_i$.

[1] The double lines in the figure indicate retraced paths.

Journal of Physics: Conference Series **360** (2012) 012009 doi:10.1088/1742-6596/360/1/012009

Accounting for the various small parameters coming from the coordinate volume of $\triangle_I$ and the coordinate area of S_I it turns out that we obtain the approximant:

$$V_I^x \sim (N^a F^i_{ax})_I E_i(S_I)\delta \tag{13}$$

where $(N^a F^i_{ax})_I$ is an approximant to $N^a F^i_{ax}(v_I)$.

The usual way in which curvature approximants are handled in LQG corresponds to approximating $F^i_{ax}(v_I)$ by a small loop holonomy with $N^a(v_I)$ being an overall factor. Note that this is *not* feasible here due to the product structure $\prod_I(1 + i\delta\hat{V}^x_{\triangle_I})$. If we did set $(N^a F^i_{ax})_I = N^a(v_I)F^i_{ax}(v_I)$, we would have terms containing products of shifts at vertices along e, which, in the continuum limit of the triangulation would not be well defined. This leads us to conclude that *the shift must be part of the specification of the small loop.*

Hence we focus on $(N^a F^i_{ax})_I$ as opposed to $(F^i_{ax})_I$. The most straightforward approximant is then

$$N^a F^i_{ax} \approx \frac{-\mathrm{Tr}(h_{\gamma_I}\tau^i)}{\delta^2} \Rightarrow \quad i\delta V^x_{\triangle_I} \sim -i\mathrm{Tr}(h_{\gamma_I}\tau^i)\ \ E_i(S_I) \tag{14}$$

where the insertion τ_i is made at v. Using this choice we obtain

$$(1 + i\delta\hat{V}^x_{\triangle_I})h_e \sim h_{\tilde{e}_I} - [\mathrm{Tr}(h_{\gamma_I}) - 1]h_e. \tag{15}$$

Hence we want to modify our choice of approximant so that the second, unwanted term is removed. To this end note that $\hat{E}^i(S_I)\hat{E}_i(S_I)$ is proprtional to the square of the angular momentum operator for the group element h_e so that

$$\hat{E}^i(S_I)\hat{E}_i(S_I)h_e = \lambda h_e. \tag{16}$$

This implies that the addition of a term of the form '$[\mathrm{Tr}(h_{\gamma_I}) - 1]E^i$' to $(N^a F^i_{ax})_I$ would yield the desired result. The additional term is of higher order in the small parameters δ and the coordinate area of the surface S_I and hence the modified expression is a valid curvature approximant.

Putting all the factors in, we obtain the final result:

$$(N^a F^i_{ax})_I \ := \ \frac{1}{\delta^2}\{\ -Tr(h_{\gamma_I}\cdot\tau_i)\ \ -\ \frac{2i}{3}Tr(h_{\gamma_I} - 1)\frac{E_i(S_I)}{\gamma l_p^2}\} \tag{17}$$

The above expression has two remarkable properties:
- the size and shape of the small loop γ_I depends on the shift
- the 2nd term has l_p^{-2} terms, leading one to speculate if (perhaps in contexts other that the diffeomorphism constraint) they could provide a seed for non-perturbative quantum corrections.

This concludes are our discussion of the curvature approximant for the $j = \frac{1}{2}$ case. It turns out that our considerations can be generalised, first, to the case of a single edge holonomy for any j (wherein one obtains higher order contributions in the small parameters δ and the area of S_I which are polynomial in the holonomy- flux operators) and thence to the case of an arbitrary spin network state. The latter generalization is particularly easy because of the dovetailing of the *product* form of the constraint with the fact that a general spin network is obtained as the *product* of edge holonomies (sewn together with intertwiners).

In summary, we have found a choice of curvature approximants such that equation (1) holds.

4. Continuum Limit

As sketched in previous sections we have found a choice of curvature approximants such that the diffeomorphism constraint at finite triangulation acts as $\hat{D}_T(\vec{N}) = -i\hbar\frac{\hat{U}_{\phi(\vec{N},\delta)}-1}{\delta}$. Clearly, as expected, the $\delta \to 0$ continuum limit of $\hat{D}_T(\vec{N})$ is not a well defined operator on the kinematic Hilbert space due to the discontinuous action thereon of the finite diffeomorphism unitaries $\hat{U}_{\phi(\vec{N},\delta)}$ coupled with the factor of δ in the denominator. However it turns out that the continuum limit *is* well defined on a different representation space, namely, the Lewandowski- Marolf (LM) habitat [2].

We briefly describe how this happens. Recall that the LM habitat is a subspace of the algebraic dual to the space of finite linear combinations of spin network states. A basis element of the LM habitat is labelled by a diffeomorphism equivalence class of spin nets and a 'vertex smooth function'. Specifically, let $[s]$ denote the set of spin nets diffeomorphic to the spin net s. Let n be the number of vertices of s. Clearly n is the same for every element of $[s]$. Let $f(x_1, .., x_n)$ be a smooth,complex valued function on n copies of the spatial manifold. Then we denote by $\Psi_{[s],f}$ the basis element of the LM habitat associated with $[s], f$. From [2] and equation (1) it follows that

$$\hat{D}_T(\vec{N})\Psi_{[s],f} = \Psi_{[s],g_T} \tag{18}$$

where the vertex smooth function g_T is (modulo overall factors) just the difference between the evaluation of f at the points $(x_1, ..x_n)$ and their diffeomorphic images i.e. modulo a multiplicative constant g_T is just $\frac{f(\phi^{-1}(\vec{N},\delta)x_1,..,\phi^{-1}(\vec{N},\delta)x_n)-f(x_1,..,x_n)}{\delta}$.

The continuum limit then just gives the new function

$$g = i\hbar\sum_{i=1}^{n} N^a(x_i)\frac{\partial f}{\partial x_i^a}. \tag{19}$$

It is then straightforward to show that the operator action $\hat{D}(\vec{N})\Psi_{[s],f} = \Psi_{[s],g}$ provides a representation of the constraint algebra as the Lie algebra of the group of spatial diffeomorphisms.

5. Conclusions

The diffeomorphism constraint $D(\vec{N})$ generates diffeomorphisms along the integral curves of the shift vector field $\vec{N}$. Hence, one expects the quantum constraint operator, $\hat{D}(\vec{N})$, to have a non- trivial action at *all* the (infintely many) points lying on those edges of a spin network state which are transverse to $\vec{N}$. In contrast, almost all operators of significance in LQG have a non-trivial action only at a *finite* number of points namely the vertices of the graph underlying the spin network state. Indeed, the necessity of an action at infinitely many points was thought to be an obstacle to the construction of the operator $\hat{D}(\vec{N})$ [14]. Our construction gets around this obstruction through the reformulation of the classical constraint at finite triangulation as a *product* over 3- cells of the triangulation described in section 2. This leads, in the quantum theory, to a *product* of bounded operators at finite triangulation rather than a *sum*. The product admits a satisfactory continuum limit whereas the sum does not. Thus, it is the passage to the product form which enables us to deal with the contributions from infinitely many points in the continuum limit.

A sensible product reformulation also seems to require that the shift vector $N^a(x)$ cannot appear as an overall factor multiplying the diffeomorphism constraint $D_a(x)$ at the point x because a product over all x of shift vectors at each point x is not an object which makes sense in the continuum limit. Hence it seems inevitable that the shift vector dependence in $D(\vec{N})$ at finite triangulation is taken care of by the incorporation of both its direction and magnitude in the specification of the small loop which underlies the holonomy approximant to the Ashtekar-Barbero curvature, F^i_{ab}. Indeed, what we are able to construct is the quantity $N^a F^i_{ab}$ at finite triangulation rather than F^i_{ab} itself. As a consequence, our construction of curvature approximant bears a great conceptual similarity to that of Loop Quantum Cosmology (LQC) when viewed in the following manner.

In isotropic LQC, the diffeomorphism constraint is satisfied identically and the Hamiltonian constraint reduces to its Euclidean part $H = \frac{\epsilon^{ijk} F_{abi} \tilde{E}^a_j \tilde{E}^b_k}{\sqrt{q}}$ where we have used standard notation for the densitized triad and the determinant of the 3- metric. Our work here suggests that rather than F^i_{ab} it is $\frac{\tilde{E}^a_j}{\sqrt{q}} F^i_{ab}$ which needs to be approximated at finite triangulation, and, that one should attempt to incorporate $\frac{\tilde{E}^a_j}{\sqrt{q}}$ as part of the specification of the small loop underlying the holonomy approximant. In the quantum theory such an attempt, if successful, would lead to the consideration of a loop whose size depends on the triad operator, thus exhibiting a close conceptual similarity to the "$\bar{\mu}$" scheme [11] for the Hamiltonian constraint in LQC.

Setting aside considerations of the Hamiltonian constraint, this work in itself reveals the necessity of a triad operator dependence in the construction of curvature approximants. This dependence is both explicit (as seen in the occurence of the electric flux terms in equations (17) as well as implicit in that, as mentioned in section 3, the expressions for the curvature approximants depend on the spin label j of the edge on which the curvature operator acts. Recall that the j label specifies the eigen values of the area operator which is built from the triads [15]. A similar dependence of "connection" type operators on conjugate "electric fluxes" was also seen to be crucial in recent work on Polymer Parameterised Field Theory (PPFT) [9, 10].

Apart from this "electric flux dependence", one of the key lessons of our work in PPFT [9] is the necessity of considering kinematically *singular* constraint operators in order to obtain a non- trivial representation of the constraint algebra. Here, too, the existence of a non- trivial representation of the quantum constraint algebra can be traced to the kinematically singular nature of the diffeomorphism constraint operator. That this operator is singular on $\mathcal{H}_{kin}$ is an obvious consequence of the factor of δ^{-1} in equation (1). It is this factor which leads to a non-trivial representation of the constraint algebra on the LM habitat alluded to in section 4. Had this factor been absent the action of the constraint operator would have yielded the difference of the evaluations of a vertex smooth function at points seperated by δ. This difference vanishes in the $\delta \to 0$ limit by virtue of the smoothness of the function. Instead, just as for PPFT [9], the factor of δ^{-1} converts this difference into a derivative in the continuum limit, thus yielding a non- trivial action of the diffeomorphism constraint operator on the habitat as well as a non-trivial representation of the algebra of diffeomorphism constraints thereon.

Our final goal is the construction of the Hamiltonian constraint operator in such a way as to obtain a non- trivial anomaly free representation of its algebra. Let us refer to the quantum commutator between a pair of Hamiltonian constraints as the Left Hand Side (LHS) and the quantum correspondent of the classical Poisson bracket between this pair as the Right Hand Side (RHS). The RHS is closely related to the diffeomorphism constraint operators studied here; the only difference being that the shift vector field in the RHS is operator valued. Earlier work by Thiemann [16], Lewandowski and Marolf [2] and Gambini, Lewandowski, Pullin and Marolf [17] showed that for density weight one Hamiltonian constraints, the algebra consistently trivialises i.e. the LHS and the RHS can be independently defined either with respect to the Uniform Rovelli- Smolin- Thiemann Topology on $\mathcal{H}_{kin}$ [16, 14] or on the LM habitat [2, 17] and, in both cases, both the RHS and the LHS vanish. Our work on PPFT [9] (as well as the 'rescaling by hand' in Reference [17]) suggests the use of higher density weight constraints to probe the existence of a *non- trivial* representation of the constraint algebra. Both these works also suggest

Journal of Physics: Conference Series **360** (2012) 012009 doi:10.1088/1742-6596/360/1/012009

that the current set of choices for curvature approximants are inappropriate. As emphasized in Reference [17] the current set of choices used in the LHS do not result in an RHS which can move vertices by diffeomorphisms. Since the choice of curvature approximants used in this work does result in the diffeomorphism constraint moving vertices around by diffeomorphisms, the considerations of this work should be of use for a better understanding of both the LHS as well as the RHS.

References

[1] T. Thiemann, *Class.Quant.Grav.***15** 839 (1998)
[2] J. Lewandowski and D. Marolf, *Phys.Rev.***D7** 299(1998)
[3] A. Ashtekar and J.Lewandowski, *Class.Quant.Grav.***21** R53 (2004)
[4] A. Perez, *Phys.Rev.***D73**, 044007 (2006)
[5] C. Rovelli and L. Smolin, *Nucl.Phys***B331**, 80 (1990)
[6] T. Jacobson and L. Smolin, *Nucl. Phys*B 299, 295 (1988)
[7] B. Bruegmann, J. Pullin, *Nucl. Phys.***B 90**, 399 (1993)
[8] M. Blencowe, *Nucl. Phys.* **B341**, 213 (1990)
[9] A.Laddha and M.Varadarajan, *Phys. Rev.***D83** 025019 (2011)
[10] T. Thiemann, *Lessons for Loop Quantum Gravity from Parameterized Field Theory*, e-Print: arXiv:1010.2426 [gr-qc].
[11] A. Ashtekar, T. Pawlowski and P. Singh, *Phys. Rev.***D74**, 084003 (2006)
[12] A. Ashtekar, J. Lewandowski, D. Marolf, J. Mourao and T. Thiemann, *J.Math.Phys.***36**, 6456 (1995).
[13] A. Laddha, M. Varadarajan *Class. Quant. Grav.***28** (2011)
[14] *Modern Canonical Quantum General Relativity* T. Thiemann, (Cambridge Monographs on Mathematical Physics)
[15] A. Ashtekar and J. Lewandowski, *Class. Quant. Grav.***14** A55-A82,(1997)
[16] T. Thiemann *Class. Quant. Grav.***15** 1207 (1998)
[17] R.Gambini, J.Lewandowski, D.Marolf, J. Pullin *Int.J.Mod.Phys.***D7** 97 (1999)

Journal of Physics: Conference Series **360** (2012) 012010 doi:10.1088/1742-6596/360/1/012010

3d gravity and quantum deformations: a Drinfel'd double approach

A Ballesteros[1], **F J Herranz**[1] **and C. Meusburger**[2]

[1] Departamento de Física, Universidad de Burgos, 09001 Burgos, Spain

[2] Department Mathematik, FAU Erlangen-Nürnberg, Cauerstr. 11, 91058 Erlangen, Germany

E-mail: angelb@ubu.es, fjherranz@ubu.es, catherine.meusburger@math.uni-erlangen.de

Abstract. The constant curvature spacetimes of 3d gravity and their associated symmetry algebras are shown to arise from the 6d Drinfel'd double that underlies the two-parametric 'hybrid' quantum deformation of the $\mathfrak{sl}(2,\mathbb{R})$ algebra. Moreover, the quantum deformation supplies the additional structures (star structure and pairing) that enter in the Chern–Simons formulation of the theory, thus establishing a direct link between quantum $\mathfrak{sl}(2,\mathbb{R})$ algebras and 3d gravity models. In this approach the flat spacetimes and Newtonian models arise as Lie algebra contractions that are governed by two dimensionful $\mathfrak{sl}(2,\mathbb{R})$ deformation parameters, which are directly related to the cosmological constant and to the speed of light.

1. Introduction

Several arguments suggest that the low energy limit of a quantum theory of gravity could be invariant under a certain quantum deformation of the Poincaré group (see [1] and references therein). In the particular case of 3d gravity this viewpoint is strongly supported by the fact that Poisson–Lie groups (which are just the semiclassical limit of quantum groups) arise as relevant symmetries of the classical theory [2, 3, 4]. There is evidence [6, 4] suggesting that the relevant symmetries in 3d quantum gravity are certain quantum doubles associated with the isometry groups of Lorentzian and Euclidean constant curvature spacetimes (see [5, 6, 7, 8, 9]).

Nevertheless, so far only quantum deformations based on a single deformation parameter have been explored in this context. However, multiparametric quantum deformations do exist (see [10] for a generic discussion and [11] for the specific quantum $gl(2)$ classification) and they are canonically associated to multiparametric classical Drinfel'd doubles.

In this contribution we show that the two-parametric (η, z) quantum $\mathfrak{sl}(2,\mathbb{R})$ approach presented in [12] provides a common framework for the Chern–Simons formulation of 3d gravity, since all the possible constant curvature spacetimes and isometry groups relevant in the theory arise naturally from the classical Drinfel'd double that characterizes the semiclassical limit of this 'hybrid' deformation. In fact, the two deformation parameters have a direct physical meaning: η corresponds to the cosmological constant, while z is related to the speed of light. Moreover, both the nonrelativistic [13] and the flat limit can be obtained as Lie algebra contractions.

2. Spacetimes and symmetries in 3d gravity

Due to the absence of local gravitational degrees of freedom, any solution of the 3d vacuum Einstein equations is locally isometric to one of six standard spacetimes (see Table 1), where the

Journal of Physics: Conference Series **360** (2012) 012010 doi:10.1088/1742-6596/360/1/012010

Table 1. Constant curvature spacetimes and isometry groups in 3d gravity.

	$\Lambda > 0$	$\Lambda = 0$	$\Lambda < 0$
Lorentzian	$\mathbf{dS}^{2+1} = SO(3,1)/SO(2,1)$ $\mathrm{Isom}(\mathbf{dS}^{2+1}) = SO(3,1)$	$\mathbf{M}^{2+1} = ISO(2,1)/SO(2,1)$ $\mathrm{Isom}(\mathbf{M}^{2+1}) = ISO(2,1)$	$\mathbf{AdS}^{2+1} = SO(2,2)/SO(2,1)$ $\mathrm{Isom}(\mathbf{AdS}^{2+1}) = SO(2,2)$
Euclidean	$\mathbf{S}^3 = SO(4)/SO(3)$ $\mathrm{Isom}(\mathbf{S}^3) = SO(4)$	$\mathbf{E}^3 = ISO(3)/SO(3)$ $\mathrm{Isom}(\mathbf{E}^3) = ISO(3)$	$\mathbf{H}^3 = SO(3,1)/SO(3)$ $\mathrm{Isom}(\mathbf{H}^3) = SO(3,1)$

cosmological constant is Λ. For Euclidean signature these are the three-sphere $\mathbf{S}^3$ ($\Lambda > 0$), 3d hyperbolic space $\mathbf{H}^3$ ($\Lambda < 0$) and 3d Euclidean space $\mathbf{E}^3$ ($\Lambda = 0$). The Lorentzian cases are the 3d dS space $\mathbf{dS}^{2+1}$ ($\Lambda > 0$), AdS space $\mathbf{AdS}^{2+1}$ ($\Lambda < 0$) and Minkowski space $\mathbf{M}^{2+1}$ ($\Lambda = 0$).

Moreover, 3d gravity can be formulated as a Chern–Simons (CS) gauge theory [14, 15] with gauge group given by the isometry group of the associated standard spacetime. This requires the choice of a symmetric, non-degenerate, Ad-invariant bilinear form $\langle\, , \rangle$ on the corresponding Lie algebra. The Lie algebras of the six isometry groups of 3d gravity can be written as [15]

$$[J_a, J_b] = \epsilon_{abc} J^c, \qquad [J_a, P_b] = \epsilon_{abc} P^c, \qquad [P_a, P_b] = \lambda \epsilon_{abc} J^c, \tag{1}$$

where $a = 0, 1, 2$ and indices are raised with either the 3d Minkowskian or the 3d Euclidean metric. Note that both the signature and the cosmological constant arise as structure constants ($\lambda = \Lambda$ for Euclidean signature and $\lambda = -\Lambda$ for the Lorentzian one). As shown in [15] (see also [8, 16]), the relevant bilinear form for the Chern-Simons formulation of 3d gravity is

$$\langle J_a, P_b \rangle = g_{ab}, \qquad \langle J_a, J_b \rangle = \langle P_a, P_b \rangle = 0, \tag{2}$$

where g_{ab} is either the Euclidean or the Minkowskian metric. With this bilinear form, the Einstein-Hilbert action for 3d gravity can be written as a Chern-Simons action by combining the triad e and the spin connection ω into a CS gauge field $A = e^a P_a + \omega^a J_a$. This is a connection with values in the Lie algebra (1), which has to be equipped with the star structure

$$J_a^* = -J_a, \qquad P_a^* = -P_a. \tag{3}$$

As we shall see in the sequel (see [12] for details), all the spacetimes and the symmetry algebras (1) of 3d gravity, together with the specific bilinear form (2) can be straightforwardly related to a two-parametric quantum deformation of the real Lie algebra $\mathfrak{sl}(2,\mathbb{R}) \simeq \mathfrak{so}(2,1)$.

3. The 'hybrid' quantum deformation of $\mathfrak{sl}(2,\mathbb{R})$

The $\mathfrak{sl}(2,\mathbb{R})$ Lie bracket and undeformed coproduct $\Delta_{(0)} : \mathfrak{sl}(2,\mathbb{R}) \to \mathfrak{sl}(2,\mathbb{R}) \otimes \mathfrak{sl}(2,\mathbb{R})$ are

$$[J_3, J_\pm] = \pm 2 J_\pm, \qquad [J_+, J_-] = J_3, \tag{4}$$
$$\Delta_{(0)}(J_i) = J_i \otimes 1 + 1 \otimes J_i, \qquad i = 3, \pm, \tag{5}$$

and we assume the star structure $J_3^* = -J_3, J_\pm^* = -J_\pm$. A quantum $\mathfrak{sl}(2,\mathbb{R})$ algebra is a power series deformation of the previous commutation rules and coproduct map (see [10]). If we expand the deformed coproduct map Δ_φ in the quantum deformation parameter(s) φ we get $\Delta_\varphi = \sum_{k=0}^\infty \varphi^k \, \delta_{(k)} = \Delta_{(0)} + \varphi \delta_{(1)} + o[\varphi^2]$ and each quantum deformation is uniquely characterized (see [10]) by the skew-symmetric part δ of the first-order deformation $\delta_{(1)}$. As this cocommutator δ always generates a Lie bialgebra structure, the classification of quantum deformations of a given Lie algebra reduces to the classification of its Lie bialgebra structures. In the case of $\mathfrak{sl}(2,\mathbb{R})$ such classification is well-known (see [10, 11, 17]).

In particular, the so-called 'hybrid' quantum algebra $\mathfrak{sl}_{\eta,z}(2,\mathbb{R})$ is a two-parametric quantum deformation generated by the two-parametric classical r-matrix

$$r = r_\eta + r_z = \eta\, J_+ \wedge J_- + \frac{z}{2} J_3 \wedge J_+, \tag{6}$$

which is a solution of the modified classical Yang–Baxter equation. The associated δ is

$$\delta(J_+) = \eta J_+ \wedge J_3, \qquad \delta(J_3) = z J_3 \wedge J_+, \qquad \delta(J_-) = \eta J_- \wedge J_3 + z J_- \wedge J_+. \tag{7}$$

As we will show in the following, in the context of 3d gravity both parameters η and z play essential roles and have a clear physical interpretation.

4. 3d AdS gravity from the 'hybrid' Drinfel'd double

Each coboundary Lie bialgebra gives rise to a Drinfel'd double Lie algebra [17, 18]. In our case the classical Drinfel'd double $\mathcal{D}_{\eta,z}(\mathfrak{sl}(2,\mathbb{R}),\delta)$ is the 6d Lie algebra spanned by $\{J_i\}_{i=3,\pm}$ and its dual basis $\{j^i\}_{i=3,\pm}$, with Lie brackets given by (4) and dual brackets induced from δ (7)

$$[j^3, j^+] = -\eta j^+ + z j^3, \qquad [j^3, j^-] = -\eta j^-, \qquad [j^+, j^-] = -z j^-. \tag{8}$$

The remaining brackets are the "crossed" or "mixed" Lie brackets, which read

$$\begin{array}{lll} [J_3, j^3] = z J_+, & [J_3, j^+] = -z J_3 - 2j^+, & [J_3, j^-] = 2j^-, \\ [J_+, j^3] = -\eta J_+ - j^-, & [J_+, j^+] = \eta J_3 + 2j^3, & [J_+, j^-] = 0, \\ [J_-, j^3] = -\eta J_- + j^+, & [J_-, j^+] = -z J_-, & [J_-, j^-] = \eta J_3 + z J_+ - 2j^3. \end{array} \tag{9}$$

The pairing between the basis $\{J_i\}_{i=3,\pm}$ and the dual basis $\{j^i\}_{i=3,\pm}$ is

$$\langle J_i, j^k \rangle = \langle j^k, J_i \rangle = \delta_i^k, \qquad \langle J_i, J_k \rangle = \langle j^i, j^k \rangle = 0, \qquad i,k = 3,\pm. \tag{10}$$

If both deformation parameters are real, $\mathcal{D}_{\eta,z}(\mathfrak{sl}(2,\mathbb{R}),\delta)$ inherits the star structure

$$J_3^* = -J_3, \qquad J_\pm^* = -J_\pm, \qquad j^{3*} = -j^3, \qquad j^{\pm*} = -j^\pm. \tag{11}$$

Now we introduce in $\mathcal{D}_{\eta,z}(\mathfrak{sl}(2,\mathbb{R}),\delta)$ the 'Chern–Simons basis' [12] J_a, P_a $(a = 0,1,2)$ through

$$\begin{array}{lll} J_0 = \frac{1}{2}(J_+ - J_-), & J_1 = \frac{z}{2} J_3, & J_2 = \frac{z}{2}(J_+ + J_-), \\ P_0 = \eta(J_+ + J_-) - \frac{z}{2} J_3 + j^- - j^+, & P_1 = -z^2 J_+ + 2z j^3, & P_2 = \eta z (J_+ - J_-) + \frac{z^2}{2} J_3 + z(j^+ + j^-), \end{array} \tag{12}$$

and we find that the Lie brackets of $\mathcal{D}_{\eta,z}(\mathfrak{sl}(2,\mathbb{R}),\delta)$ in the CS basis are just

$$\begin{array}{lll} [J_0, J_1] = -J_2, & [J_0, J_2] = J_1, & [J_1, J_2] = z^2 J_0, \\ [J_0, P_0] = 0, & [J_0, P_1] = -P_2, & [J_0, P_2] = P_1, \\ [J_1, P_0] = P_2, & [J_1, P_1] = 0, & [J_1, P_2] = z^2 P_0, \\ [J_2, P_0] = -P_1, & [J_2, P_1] = -z^2 P_0, & [J_2, P_2] = 0, \\ [P_0, P_1] = -4\eta^2 J_2, & [P_0, P_2] = 4\eta^2 J_1, & [P_1, P_2] = 4\eta^2 z^2 J_0. \end{array} \tag{13}$$

This means that $\mathcal{D}_{\eta,z}(\mathfrak{sl}(2,\mathbb{R}),\delta) \simeq \mathfrak{so}(2,2)$, provided that η, z are non-zero real numbers. If J_0, J_b, P_0, P_b, $(b = 1,2)$ are interpreted as the generators of rotations, boosts, time translations and spatial translations, then $\mathfrak{so}(2,2)$ is the symmetry algebra of the 3d AdS space, in which J_0, J_1, J_2 span the $\mathfrak{so}(2,1)$ subalgebra and the AdS spacetime is $\mathbf{AdS}^{2+1} = SO(2,2)/SO(2,1)$.

Thus, the deformation parameters of the hybrid deformation $\mathfrak{sl}_{\eta,z}(2,\mathbb{R})$ are directly related to the cosmological constant $\Lambda = -\lambda$ and the speed of light c through $\lambda = 4\eta^2$ and $c^2 = 1/z^2$, which means that the signature of the metric is $g = \mathrm{diag}(-1, z^2, z^2)$. Moreover, by using the relations (12), we find the star structure (3) and the pairing (10) (z^2 can be rescaled to 1)

$$\langle J_0, P_0 \rangle = -1, \qquad \langle J_1, P_1 \rangle = z^2, \qquad \langle J_2, P_2 \rangle = z^2. \tag{14}$$

Journal of Physics: Conference Series **360** (2012) 012010 doi:10.1088/1742-6596/360/1/012010

5. Analytic continuation and contractions

The previous results can be straightforwardly generalized to other signatures and values of the cosmological constant by considering imaginary values of η, z as well as the limits $\eta, z \to 0$ (which can be understood, respectively, as the "flat" and "non-relativistic" Inönü–Wigner contractions). In this way we obtain nine 3d homogenous spaces $\mathbf{X}_{\eta,z}$, which are a subfamily of the Cayley–Klein spaces [19] defined through $\mathbf{X}_{\eta,z} = \langle \mathcal{D}_{\eta,z}(\mathfrak{sl}(2,\mathbb{R}),\delta)\rangle / \langle J_0, J_1, J_2\rangle$. In this way, the quantum algebra $\mathfrak{sl}_{\eta,z}(2,\mathbb{R})$ gives rise in a unified way to the following nine homogeneous spaces:

• The Riemannian spaces for $z \in \mathrm{i}\mathbb{R}^*$: the three-sphere, the 3d hyperbolic space and the 3d Euclidean space, with $\Lambda = \lambda = 4\eta^2 = \pm 1/R^2$, where R is the radius ($R \to \infty$ for $\mathbf{E}^3$).

• The Lorentzian spaces for $z \in \mathbb{R}^*$: 3d AdS, dS and Minkowski space. Now $\Lambda = -\lambda$ with $\lambda = 4\eta^2 = \pm 1/\tau^2$, where τ is the (time) universe radius ($\tau \to \infty$ for $\mathbf{M}^{2+1}$).

• The non-relativistic or Galilean limits of 3d gravity [13] when $z = 0$ ($c \to \infty$): the two Newton–Hooke spacetimes and the Galilean one, both with degenerate pairing and metrics.

6. Concluding remarks

We emphasize that our approach is obtained from the first order (i.e. the one containing the semiclassical information) of the 'hybrid' quantum deformation. Therefore, the construction and analysis of the complete quantum counterpart of $\mathcal{D}_{\eta,z}(\mathfrak{sl}(2,\mathbb{R}),\delta)$ is worth further investigation. The explicit construction of the quantum algebra $\mathfrak{sl}_{\eta,z}(2,\mathbb{R})$ can be traced back to [11]. A nonlinear change of basis connecting this deformation with the Drinfel'd–Jimbo one [20, 21] was found in [22]. Note also that the dual viewpoint is equally tractable: the 'hybrid' Poisson–Lie group structure on $SL(2)$ can be obtained through the Sklyanin bracket for the classical r-matrix (6), and its quantization gives rise to the 'hybrid' quantum $SL(2)$ group [11, 23].

Acknowledgments

This work was partially supported by the Spanish MICINN under grant MTM2010-18556 and by the DFG Emmy-Noether fellowship ME 3425/1-1.

References

[1] Amelino-Camelia G, Smolin L and Starodubtsev A 2004 *Class. Quantum Grav.* **21** 3095
[2] Fock V V and Rosly A A 1992 Preprint ITEP-72-92; *ibid* 1999 *Am. Math. Soc. Transl.* **191** 67
[3] Alekseev A Y and Malkin A Z 1995 *Commun. Math. Phys.* **169** 99
[4] Meusburger C and Schroers B J 2008 *J. Math. Phys.* **49** 083510
[5] Podleś P and Woronowicz S L 1990 *Commun. Math. Phys.* **130** 381
[6] Buffenoir E, Noui K and Roche P 2002 *Class. Quant. Grav.* **19** 4953
[7] Meusburger C and Schroers B J 2003 *Adv. Theor. Math. Phys.* **7** 1003
[8] Meusburger C and Schroers B J 2009 *Nucl. Phys. B* **806** 462
[9] Majid S and Schroers B J 2009 *J. Phys. A: Math. Theor.* **42** 425402
[10] Chari V and Pressley A 1994 *A Guide to Quantum Groups* (Cambridge: Cambridge University Press)
[11] Ballesteros A, Herranz F J and Parashar P 1999 *J. Phys. A: Math. Gen.* **32** 2369
[12] Ballesteros A, Herranz F J and Meusburger C 2010 *Phys. Lett. B* **687** 375
[13] Schroers B J and Papageorgiou G 2009 *JHEP* **11** 009
[14] Achucarro and Townsend P K 1986 *Phys. Lett. B* **180** 89
[15] Witten E 1988 *Nucl. Phys. B* **311** 46
[16] Bonzom V and Livine E R 2008 *Class. Quantum Grav.* **25** 195024
[17] Gomez X 2000 *J. Math. Phys.* **41** 4939
[18] Ballesteros A, Celeghini E and del Olmo M A 2005 *J. Phys. A: Math. Gen.* **38** 3909
[19] Ballesteros A, Herranz F J, Ragnisco O and Santander M 2008 *Int. J. Theor. Phys.* **47** 649
[20] Drinfel'd V G 1987 *Quantum Groups*, Proc. Int. Cong. Math. Berkeley 1986, ed A V Gleason (Providence: AMS) p 798
[21] Jimbo M 1985 *Lett. Math. Phys.* **10** 63; *ibid* 1986 **11** 247
[22] Aneva B L, Arnaudon D, Chakrabarti A, Dobrev V K and Mihov S G 2001 *J. Math. Phys.* **42** 1236
[23] Kupershmidt B A 1994 *J. Phys. A: Math. Gen.* **27** L47

Journal of Physics: Conference Series **360** (2012) 012011 doi:10.1088/1742-6596/360/1/012011

Perturbative running of the Immirzi parameter

Dario Benedetti[1] **and Simone Speziale**[2]
[1]Max Planck Institute for Gravitational Physics (Albert Einstein Institute), Am Mühlenberg 1, D-14476 Potsdam, Germany
[2]Centre de Physique Théorique, UMR6207 CNRS-Luminy Case 907, 13288 Marseille, France

E-mail: dario.benedetti@aei.mpg.de, simone.speziale@cpt.univ-mrs.fr

Abstract. We report on the renormalization of the Immirzi parameter found through perturbative one-loop calculations.

If one takes the tetrad e^I_μ and the Lorentz connection ω^{IJ}_μ as independent variables, the lowest order action for gravity (with zero cosmological constant, and $\kappa^2 = 16\pi G$) has the following form,

$$S[e,\omega] = -\frac{1}{\kappa^2}\int\left\{\frac{1}{2}\epsilon_{IJKL}e^I\wedge e^J\wedge F^{KL}[\omega] + \frac{1}{\gamma}e_I\wedge e_J\wedge F^{IJ}[\omega]\right\}. \tag{1}$$

The first term is just the Einstein-Cartan action for general relativity. The second term, henceforth Holst term, is, by itself, topological, and does not change the equations of motion in the absence of torsion sources. Its coupling constant γ is known as the Immirzi, or Barbero-Immirzi, parameter in the loop quantum gravity community, where it plays an important role at the non-perturbative level [1]. This motivates the question of whether any non-trivial quantum mechanical role of γ shows up already in perturbation theory, where the gravitational theory (1) is non-renormalizable, but can be regarded as an effective field theory. To address the question, in [2] we studied the perturbative one-loop effective action of (1). Here we report on the aspects of our findings most relevant to the renormalization of γ and their implications, and further include explicit plots of the runnings.

To quantize the theory, we used the background-field method and a one-loop perturbative expansion. We took generic background fields e and ω, in particular off-shell, but we assumed invertibility of the tetrad. This allowed us to decompose $\omega = \omega(e) + K$, where $\omega(e)$ is the unique spin connection solving $d_\omega e = 0$, and K the contorsion. Within this framework, the quantization generates an infinite number of terms, that can be packaged in invariants of the background Riemann and contorsion tensors. We regularized the theory using the heat-kernel expansion with an UV cut-off Λ_{UV}. For the divergent part of the one-loop effective action, we found (here and in the following μ is the renormalization scale)

$$\Gamma^{\rm div}[e,K] = -\frac{1}{32\pi^2}\left\{\Lambda^4_{UV}\int e - \Lambda^2_{UV}\int e\mathcal{L}_1 - \ln(\Lambda^2_{UV}/\mu^2)\int e\mathcal{L}_2\right\}, \tag{2}$$

with

$$\begin{aligned}\mathcal{L}_1[e,K] &= \frac{17}{3}R(e) + \frac{3\gamma^2+5}{4\gamma^2}K_{\mu\nu\rho}K^{\mu\nu\rho} - \frac{7\gamma^2-3}{4\gamma^2}K^\mu{}_{\mu\rho}K_\nu{}^{\nu\rho} + \frac{9\gamma^2-13}{4\gamma^2}K_{\mu\nu\rho}K^{\nu\mu\rho} \\ &\quad + \frac{3}{\gamma e}\epsilon^{\mu\nu\rho\sigma}K_{\mu\nu}{}^\lambda K_{\rho\sigma\lambda} + \frac{4}{\gamma e}\epsilon^{\mu\nu\rho\sigma}K_{\mu\nu\rho}K^\lambda{}_{\lambda\sigma},\end{aligned} \tag{3}$$

and $\mathcal{L}_2$ a very long expression including all possible dimension-four invariants.

In (3), the first ϵKK invariant coincides with the Holst term once $\omega = \omega(e) + K$ is used. It is thus the divergent term candidate to renormalize the Immirzi parameter. However, on-shell $R_{\mu\nu} = K_{\mu\nu\rho} = 0$, and all the quadratic and logarithmic divergences in (2) vanish. Therefore, they can be simply renormalized through field redefinitions. Only the quartic divergence survives, but it can be reabsorbed, for instance, including also a bare cosmological constant in the action, as in [4]. Hence, like in the metric formulations [3, 4, 5], we find that pure quantum gravity is on-shell finite at one-loop, and we expect the appearance of non-renormalizable divergences at two loops [6].

The situation can change in the presence of curvature or torsion sources. As a first step in that direction, we considered a different renormalization scheme, in which the divergences are absorbed whenever possible into a redefinition of the couplings. Such a scheme, used for example in [5], comes closer to the spirit of the calculations done in the context of the asymptotic safety scenario [7], where the running of the traditionally inessential Newton's constant can be motivated by the special role it has in the theory [8]. With these considerations in mind, we considered an off-shell renormalization condition for κ^2 and γ obtained from the quadratic divergences (3). To define non-trivial beta functions from quadratic divergences, we used a non-minimal subtraction ansatz as in [9, 10]. Defining the dimensionless Newton's constant as $g \equiv \frac{1}{16\pi}\mu^2\kappa^2$, we found the following beta functions,

$$\beta_g = g(2 - \frac{17}{3\pi}g), \qquad \beta_\gamma = \frac{4}{3\pi}\gamma g. \tag{4}$$

The beta function for Newton's constant is independent of γ, and it shows an anti-screenig effect and a non-Gaussian fixed point. Although the latter lies outside the realm of perturbation theory, the result is in nice agreement with the asymptotic safety conjecture [7]. The Immirzi parameter has a non-trivial running as well. By inspection, $\gamma = 0$ and $\gamma = \infty$ are fixed points, consistently with the results of [11].[1] An explicit solution of the flow shows that $\gamma = \infty$ is UV-attractive. On the other hand, $\gamma = 0$ is not the IR asymptote of generic trajectories, because γ stops running as soon as g goes to zero. In other words, $\{g = 0, \gamma \lessgtr 0\}$ is a marginal direction, and the IR limit of γ depends on the initial conditions. The situation is depicted and further explained in Fig. 1.

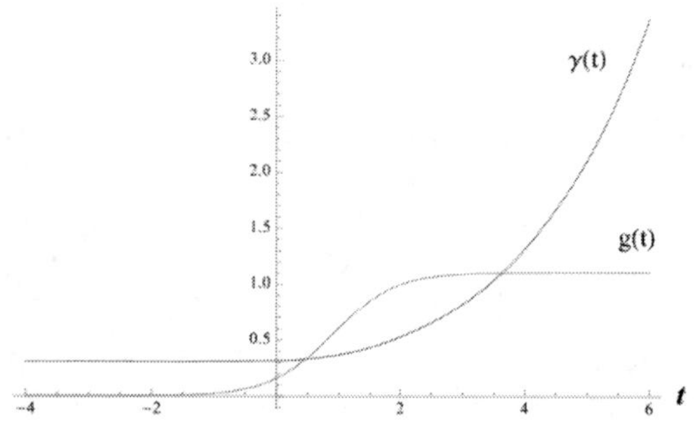

Figure 1. The running in the pure gravity case, with $t = \ln\mu$ and μ the energy scale. The running of g is characterized by two plateaus, respectively at the IR Gaussian fixed point, and at the UV non-Gaussian fixed point, with a short transient in between. The running of γ is clearly dictated by the plateaus of g, with the first corresponding to a frozen γ and the second to an exponential running.

The result is intriguing, as a UV flow from $\gamma = 0$ to $\gamma = \infty$ would be nicely consistent with the idea that the metric captures the degrees of freedom of general relativity at low energies, while the connection field becomes more important at high energies, as suggested by loop quantum gravity. The reason for this is that the $\gamma \to 0$ limit of (1) gives the second-order

[1] On the other hand, nothing happens at $\gamma^2 = 1$. These are the special values for which one is dealing with a formulation of gravity in terms of self-dual variables only.

metric formalism,[2] whereas for $\gamma \to \infty$ we obtain Einstein-Cartan theory and the connection is totally independent. The running (4) we found gives qualitative support to this picture, however, it is off-shell, thus scheme and gauge dependent.

An on-shell running of the Immirzi parameter can be obtained adding a source of torsion. To do so, we considered the following coupled gravity-fermion system,

$$S_{\text{coupled}}[e,\psi] = -\int d^4x\, e\left\{\frac{1}{\kappa^2}R(e) + \frac{i}{2}\bar\psi\gamma^I e^\mu_I \nabla_\mu(e)\psi + \frac{3\kappa^2}{128}\left(\frac{\gamma^2}{\gamma^2-1}\right)A_I A^I\right\}, \tag{5}$$

where ψ is a Majorana spinor, and $A^I = \bar\psi\gamma^I\psi$ the axial current. This action is obtained from (1) minimally coupled to the Dirac action, once the field equations for ω are used. The one-loop quantization of an action like (5) had been previously considered in [12]. Merging their results with ours, we found for the quadratic divergences

$$\mathcal{L}_1 = \frac{3}{512}\left(28\frac{\gamma^2}{\gamma^2-1} - 5\right)\kappa^4 A^2. \tag{6}$$

Unlike in the vacuum case, the logarithmic divergences do not vanish on-shell, thus establishing the non-renormalizability of the theory already at one-loop. On the other hand, the quadratic divergence (6) leads to a renormalization of the Immirzi parameter, this time on-shell, with resulting beta function

$$\beta_{\gamma^2} = -(\gamma^2-1)\frac{\mu^2\kappa^2}{(8\pi)^2}(23\gamma^2+5). \tag{7}$$

We see that neither $\gamma = 0$ nor $\gamma = \infty$ are stable under renormalization, in contrast with the pure gravity case. There is a fixed point, $\gamma^2 = 1$, and it is UV-attractive. However, it corresponds to a divergent coupling for the four-fermion interaction, hence it is out of the range of validity of perturbation theory. In the IR, the coupling flows to a finite value determined by the initial condition, and the flow depends qualitatively on whether γ^2 is larger or smaller than one, with the same UV limit and opposite IR limits. See Fig 2 for details.

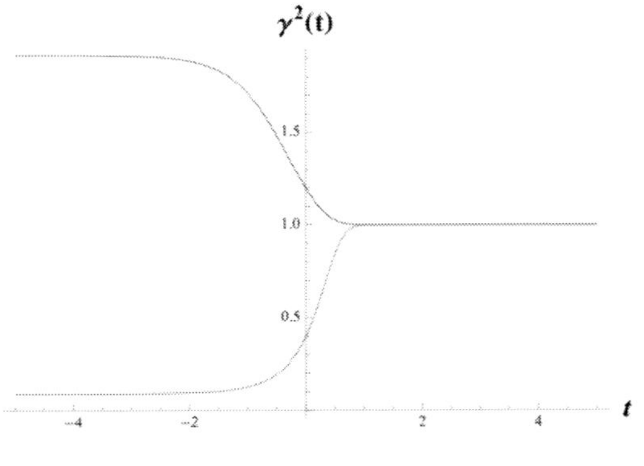

Figure 2. The running of γ^2 in the coupled case, with the two sectors $|\gamma| \lessgtr 1$. These IR asymptotes depend on the initial condition γ_0. Notice that these have to be restricted in order to guarantee that $\gamma(\mu)$ remain real for all values of μ. With these restrictions, in the IR limit $\mu \to 0$ the Immirzi parameter flows towards a value between γ_0 and $|\gamma| = +\infty$, for $\gamma_0^2 > 1$, or between γ_0 and $\gamma = 0$, for $\gamma_0^2 < 1$.

Finally, notice that the flow depends explicitly on the external parameter $\mu^2\kappa^2$. This can be interpreted as the renormalization scale measured in Planck units, as κ^2 is inessential and we are not letting it run in the present scheme. One can alternatively renormalize κ^2 as well, as we did above for pure gravity. In this off-shell scheme, one finds again a γ-independent running of

[2] To see this, notice first that the Holst term equals, up to the topological Nieh-Yan invariant, the torsion-squared $T \wedge T$. In the path integral, $\frac{1}{\gamma}T \wedge T$ is singular as $\gamma \to 0$, but it can be rewritten as $2B \wedge T + \gamma B \wedge B$ using an auxiliary 2-form field B. The limit $\gamma = 0$ now yields a Lagrange multiplier enforcing $T = 0$, i.e. the second-order theory.

Newton's constant, with beta function $\beta_g = g(2 - \frac{11}{2\pi}g)$, and in (7), $\kappa^2\mu^2/16\pi$ is replaced by a non-trivial running $g(\mu)$, bounded between $g(\mu = 0) = 0$ and $g(\mu = \infty) = 4\pi/11$. This modifies things like the velocity along the flow and the reality bounds on γ_0, but not the conclusions about the special points $\gamma^2 = 0, 1, \infty$.

The fact that $\gamma = 0$ is not stable is particularly interesting. If one starts with vanishing bare Immirzi parameter, $\gamma = 0$, the action (5) reduces to the second-order Einstein-Hilbert action coupled to fermions, with no four-fermion interaction. However, the latter is nonetheless generated by radiative corrections, see (6) with $\gamma = 0$. Namely, the radiative corrections introduce quadratic divergences which are non-renormalizable in the second-order formalism. In order to renormalize these divergences one is forced to introduce the four-fermion term in the classical action, that is, one is forced to have a non-vanishing Immirzi parameter. In this sense, the first order formulation is more suitable to quantize the coupled gravity-fermion system. Similarly, fine-tuned couplings such as the one proposed in [13] do not appear to be stable under radiative corrections.

We also looked at more general actions than (5), which include non-minimal couplings of the fermions. It can then happen that the dependence on the Immirzi parameter becomes undistinguishable from the non-minimal couplings [13, 14]. In these cases, we found that quadratic divergences non-renormalizable within the Holst action are expected, therefore one should include all dimension-two invariants, as in (3), in the bare action.

In conclusion, we showed how the Immirzi parameter gets renormalized by radiative corrections in the perturbative context. In the cases considered, the renormalization is driven by quadratic divergences, however a logarithmic contribution to the running likely arises if one includes a cosmological constant term. The results obtained could serve as a guidance in studies of semiclassical limit of loop quantum gravity, as well as a bridge to the asymptotic safety scenario.

Acknowledgments

We thank Daniel Litim for useful discussions, and we gratefully acknowledge support from the European Science Foundation (ESF) through the activity "Quantum Geometry and Quantum Gravity".

References

[1] Rovelli C 2004 *Quantum gravity* (Cambridge, UK: University Press)
[2] Benedetti D and Speziale S 2011 *J. High Energy Phys.* JHEP06(2011)107 (*Preprint* arXiv:1104.4028)
[3] 't Hooft G and Veltman M J G 1974 *Annales Poincare Phys. Theor.* A **20** 69-94
[4] Christensen S M and Duff M J 1980 *Nucl. Phys.* B **170** 480-506
[5] Buchbinder I L and Shapiro I L 1985 *Acta Physica Polonica* **B16** 103-107
[6] Goroff M H and Sagnotti A 1986 *Nucl. Phys.* **B266** 709
[7] Niedermaier M and Reuter M 2006 *Living Rev. Rel.* **9** 5
[8] Percacci R and Perini D 2004 *Class. Quant. Grav.* **21** 5035-5041 (*Preprint* hep-th/0401071)
[9] Robinson S P and Wilczek F 2006 *Phys. Rev. Lett.* **96** 231601 (*Preprint* hep-th/0509050)
[10] Niedermaier M R 2009 *Phys. Rev. Lett.* **103** 101303
[11] Daum J E and Reuter M 2010 Renormalization Group Flow of the Holst Action *Preprint* arXiv:1012.4280
[12] Barvinsky A O and Vilkovisky G A 1981 *Nucl. Phys.* **B191** 237-259
[13] Mercuri S 2006 *Phys. Rev.* **D73** 084016 (*Preprint* gr-qc/0601013)
[14] Alexandrov S 2008 *Class. Quant. Grav.* **25** 145012 (*Preprint* arXiv:0802.1221)

Journal of Physics: Conference Series **360** (2012) 012012 doi:10.1088/1742-6596/360/1/012012

S-Matrix for AdS from General Boundary QFT

Daniele Colosi, Max Dohse and Robert Oeckl

Centro de Ciencias Matemáticas (CCM), UNAM Campus Morelia, 58190 Morelia, Mexico

E-mail: colosi/max/robert@matmor.unam.mx

Abstract. The General Boundary Formulation (GBF) is a new framework for studying quantum theories. After concise overviews of the GBF and Schrödinger-Feynman quantization we apply the GBF to resolve a well known problem on Anti-deSitter spacetime where due to the lack of temporally asymptotic free states the usual S-matrix cannot be defined. We construct a different type of S-matrix plus propagators for free and interacting real Klein-Gordon theory.

1. Introduction

The General Boundary Formulation (GBF) is a new framework for studying quantum theories [1]. The respective theory is considered in some spacetime region, and the quantum states live in a Hilbert space associated to the region's boundary. Amplitudes and a probability interpretation can be assigned to the states. Not needing a fixed metric, the GBF is a promising framework for quantum theories treating the metric as a dynamical object. Using Schödinger-Feynman quantization, in [2] the GBF reproduces the usual S-matrix for real Klein-Gordon theory on Minkowski spacetime, plus an equivalent S-matrix of different type. In [3] the same has been achieved for deSitter spacetime. The techniques used for both types of S-matrices are generalized for a wide class of spacetimes in [4] and now applied here for Anti-deSitter spacetime (AdS), where due to the lack of (temporal) asymptotically free states the usual S-matrix cannot be defined. The paper is structured as follows: After short reviews of the standard S-matrix (Section 2), AdS (3), the GBF (4) and Schrödinger-Feynman quantization (5), we apply in (6) the GBF for constructing the different type of S-matrix for real Klein-Gordon (KG) theory.

2. Standard construction of the S-matrix

In standard QFT there is *one* Hilbert space $\mathcal{H}$ of free states. The S-matrix $\mathcal{S} : \mathcal{H} \to \mathcal{H}$ is a unitary operator. Matrix elements are obtained as large-time limit $\mathcal{S}_{\eta,\zeta} \sim \lim_{t\to\infty} {}_t\langle \zeta | \mathcal{U}_{[-t,t]} | \eta \rangle_{-t}$, with $\mathcal{U}_{[-t,t]}$ being the time-evolution operator.

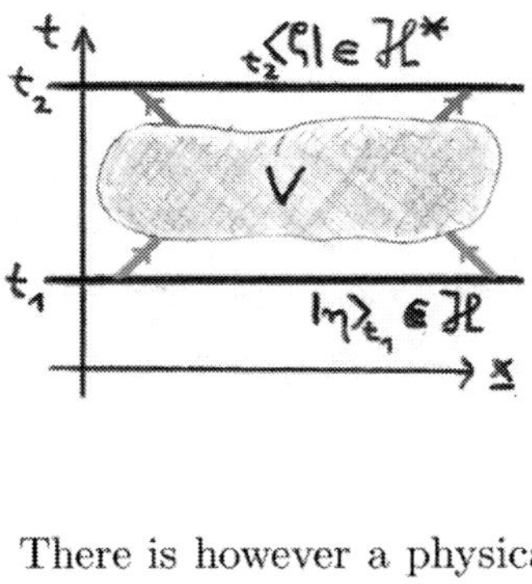

Figure 2.1. Slice of spacetime between the two equal-time hypersurfaces at t_1 and t_2. Usually it is assumed that the interaction is "switched on" only between times t_1 and t_2, then $|\eta\rangle_{t_1}$ and ${}_{t_2}\langle\zeta|$ are free states. Setting $t_1 = -t$ and $t_2 = +t$ and taking the limit $t \to \infty$ the states become asymptotically free while the interaction can now be "switched on" on all of spacetime.

There is however a physically more convincing way to describe this. Physical interactions are not switched on and off by *time*, but decrease with the *distance* between interacting particles.

Consider interacting massive particles leaving the interaction region V in 2.1. When their separation becomes large, the state can be considered free and the particles as moving along timelike geodesics. For later times the states remain free iff spacetime geometry is such that the separation remains large, which in curved spacetime cannot be taken for granted, see figure 3.1.

3. Anti-deSitter spacetime (AdS)

We coordinatize (the universal covering version of) Lorentzian AdS of dimension $(d+1)$ by $x = (t, \rho, \Omega)$, with time $t \in (-\infty, +\infty)$, radial coordinate $\rho \in [0, \frac{\pi}{2})$, and angles $\Omega = (\theta_1, ..., \theta_{d-1})$ such that $\mathbb{S}^{d-1}$ is covered. The metric is $ds^2_{\text{AdS}} = R^2_{\text{AdS}} \cos^{-2}\rho \left(dt^2 - d\rho^2 - \sin^2\rho \; ds^2_{\mathbb{S}^{d-1}}\right)$, with the curvature radius R_{AdS}. For large R_{AdS} the AdS metric approximates the Minkowski metric. The free Lagrangian of real KG theory $\mathcal{L} = \frac{1}{2}\sqrt{|g|}\left[g^{\mu\nu}(\partial_\mu\phi)(\partial_\nu\phi) - m^2\phi^2\right]$ yields as equation of motion the KG equation $(\Box + m^2)\,\phi = 0$. Its solutions are the modes $\phi_{\omega l \underline{m}}(x)$, with frequency $\omega \in (-\infty, +\infty)$ and angular quantum numbers $(l, \underline{m})$ such that the spherical harmonics $Y_l^{\underline{m}}(\Omega)$ form an orthonormal base (ONB) on $\mathbb{S}^{d-1}$. A separation ansatz [5] gives $\phi_{\omega l \underline{m}}(x) = e^{-i\omega t}\, Y_l^{\underline{m}}(\Omega)\, S_{\omega,l}(\rho)$. We consider only the case of odd d, where $S_{\omega l}$ can be any linear combination of the two linear independent solutions of the radial part of the KG equation given by ${}^1S_{\omega,l}(\rho) = \sin^l\rho \, \cos^{\tilde{m}_+}\rho \;\; F(\alpha_+, \beta_+; \gamma^{\text{S}}; \sin^2\rho)$ and ${}^2S^{\text{odd}}_{\omega,l}(\rho) = (\sin\rho)^{2-l-d} \cos^{\tilde{m}_+}\rho \; F(\alpha_+ - \gamma^{\text{S}} + 1, \beta_+ - \gamma^{\text{S}} + 1; 2 - \gamma^{\text{S}}; \sin^2\rho)$.

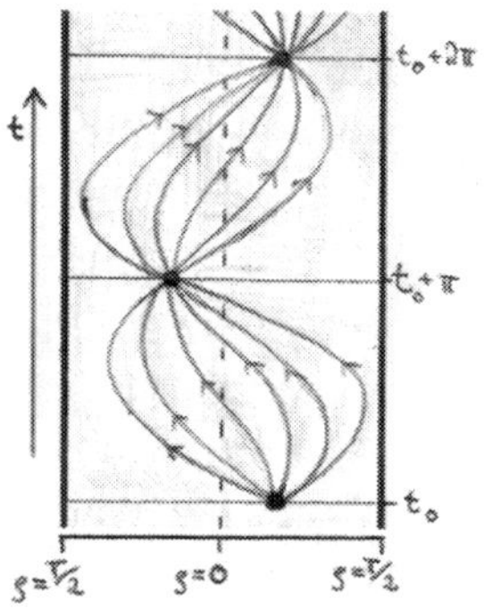

F is the hypergeometric function with $\alpha_+, \beta_+, \gamma^{\text{S}}$ depending on ω, l, mass m and spatial dimension d.

Figure 3.1. Penrose diagram of AdS. Only temporal and radial directions are drawn. The negative curvature makes the timelike geodesics emanating from an arbitrary point at t_0 reconverge in periods of π, see [6]. Thus there exists no time after which two particles leaving a scattering process are forever separated by large distances, i.e., after which interactions can be neglected forever. Therefore on AdS temporally asymptotic states do not exist and the standard S-matrix cannot be defined.

4. General Boundary Formulation (GBF)

To see how another type of S-matrix can be constructed for AdS now we outline the GBF [1]. To each hypersurface Σ (of codimension one) in a spacetime is associated its quantum state space $\mathcal{H}_\Sigma$. Each region M (of top dimension) has associated its boundary state space $\mathcal{H}_{\partial\text{M}}$, see fig. 4.1. Amplitudes for the quantum states are calculated with the amplitude map $\rho_{\text{M}} : \mathcal{H}_{\partial\text{M}} \to \mathbb{C}$.

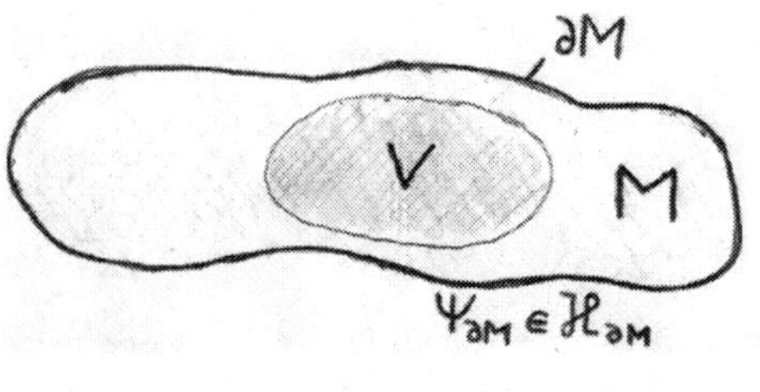

Figure 4.1.

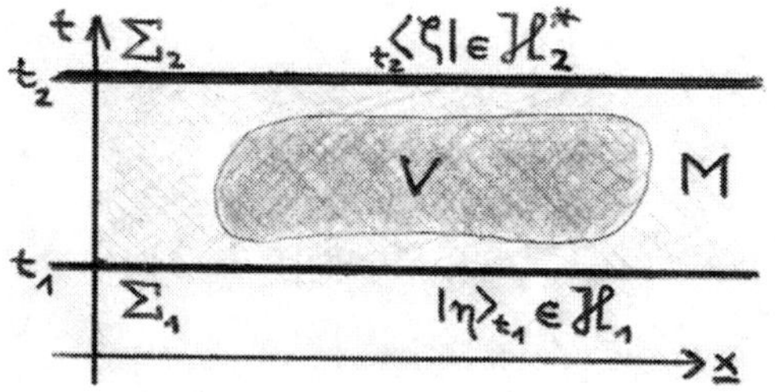

Figure 4.2.

Figure 4.2 shows how this general description includes the standard setting [1, 7]. There, M is the spacetime slice $[t_1, t_2] \times \mathbb{R}^3$ with boundary $\partial\text{M} = \Sigma_1 \sqcup \Sigma_2$, a disjoint union of two hypersurfaces. The boundary state space is the tensor product $\mathcal{H}_{\partial\text{M}} = \mathcal{H}_1 \otimes \mathcal{H}_2^*$. The amplitude map $\rho_{\text{M}} : \mathcal{H}_1 \otimes \mathcal{H}_2^* \to \mathbb{C}$ is now such that it is related to time evolution by $\rho_{\text{M}}\big(|\eta\rangle_{t_1} \otimes_{t_2}\langle\zeta|\big) = {}_2\langle\zeta\,|\,\mathcal{U}_{[t_1,t_2]}\,|\,\eta\rangle_1$. The tensor product $|\eta\rangle_{t_1} \otimes_{t_2}\langle\zeta|$ also writes as *one* boundary state $\psi_{\partial\text{M}} \in \mathcal{H}_{\partial\text{M}}$ (dropping bra-ket notation).

Journal of Physics: Conference Series **360** (2012) 012012 doi:10.1088/1742-6596/360/1/012012

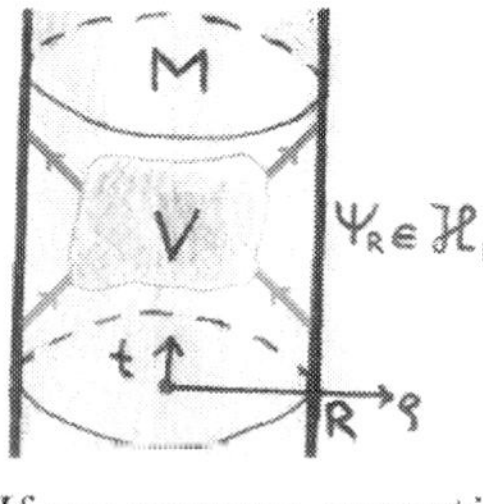

Figure 4.3. Hypercylinder region $\mathrm{M} = \mathbb{R}\times\mathbb{B}^d$: a solid ball of radius R in space times all of time [2]. The boundary $\partial\mathrm{M}$ is the *single* hypersurface $\Sigma_R := \mathbb{R}\times\mathbb{S}^{d-1}$. An incoming (outgoing) particle is now one that enters (leaves) M at any time. The boundary state space decomposes $\mathcal{H}_{\partial\mathrm{M}} = \mathcal{H}_{\mathrm{in}}\otimes\mathcal{H}_{\mathrm{out}}$ into spaces of purely incoming and outgoing states. The amplitude ρ_M of *one* boundary state $\psi_{\partial\mathrm{M}} = \psi_{\mathrm{in}}\otimes\psi_{\mathrm{out}}$ is thus related to observing its outcoming part ψ_{out} given ψ_{in} was prepared.

If we assume spacetime geometry being such that large radius R implies large distance for two points on Σ_R, then for $R \to \infty$ interactions can be neglected and the boundary state is (radially) asymptotically free. This is the case for AdS and so the problem of no (temporally) asymptotic states can be avoided. The large-R limit of $\rho_\mathrm{M}(\psi_{\partial\mathrm{M}})$ can thus be interpreted as a new type of S-matrix.

5. Schrödinger-Feynman quantization

In the Schrödinger representation [8] the states are complex function(al)s on field configurations on a hypersurface. Let K_Σ the space of field configurations on Σ, then $\psi_\Sigma : \mathrm{K}_\Sigma \to \mathbb{C}$. Feynman quantization means that amplitudes are calculated as a path-integral with Z_M called field propagator:

$$\rho_\mathrm{M}(\psi_{\partial\mathrm{M}}) = \int\limits_{\mathrm{K}_{\partial\mathrm{M}}} \mathcal{D}\varphi\ \psi_{\partial\mathrm{M}}(\varphi)\, Z_\mathrm{M}(\varphi) \qquad\qquad Z_\mathrm{M}(\varphi) = \int\limits_{\mathrm{K}_\mathrm{M},\phi|_{\partial\mathrm{M}}=\varphi} \mathcal{D}\phi\ \mathrm{e}^{\mathrm{i}S_\mathrm{M}(\phi)}\,.$$

The vacuum $\psi_{R,0}$ for a hypercylinder Σ_R is a Gaussian [9, 10], with $\hat{A}_R$ called vacuum operator:

$$\psi_{R,0}(\varphi) \sim \exp\Bigg(-\frac{1}{2}\int\limits_{\Sigma_R} \mathrm{d}t\mathrm{d}\Omega\ \varphi(\underline{x})(\hat{\mathrm{A}}_{\Sigma_R}\varphi)(\underline{x})\Bigg) \qquad\qquad \hat{A}_R = -\mathrm{i}\sqrt{|g_{\Sigma_R}g^{\rho\rho}|}\ \frac{\partial_\rho\hat{\Upsilon}(\rho)}{\hat{\Upsilon}(\rho)}\Bigg|_{\rho=R}\,.$$

An ONS on Σ_R is given by the functions $u_{\omega l\underline{m}}(t,\Omega) \sim \mathrm{e}^{-\mathrm{i}\omega t}\,Y_l^{\underline{m}}(\Omega)$, and $\hat{\Upsilon}$ is defined by its eigenvalues: $\hat{\Upsilon}(\rho)\,u_{\omega l\underline{m}}(t,\Omega) = \left[c^a_{\omega l\underline{m}}\,{}^1S_{\omega,l}(\rho) + c^b_{\omega l\underline{m}}\,{}^2S^{\mathrm{odd}}_{\omega,l}(\rho)\right] u_{\omega l\underline{m}}(t,\Omega)$. The vacuum is thus determined by fixing the coefficients $c^{a,b}$ (a,b are labels, not indices). We shall choose $c^a_{\omega l\underline{m}} = (pR_{\mathrm{Ads}})^l/(2l+1)!!$ and $c^b_{\omega l\underline{m}} = -\mathrm{i}\,(2l-1)!!/(pR_{\mathrm{Ads}})^{l+1}$ with $p := \sqrt{|\omega^2 - m^2|}$, because in the limit $R_{\mathrm{Ads}} \to \infty$ this choice yields the well known Minkowski vacuum.

Next we can define coherent states. A coherent state is determined by its complex characterising function η and usually written with creators $\hat{a}^\dagger_{\omega l\underline{m}}$ as $\psi_{R,\eta}(\varphi) \sim \left[\exp\int \mathrm{d}\omega \sum_{l\underline{m}}\ \eta_{\omega l\underline{m}}\,\hat{a}^\dagger_{\omega l\underline{m}}\right]\ \psi_{R,0}(\varphi)$. In the interaction picture it can equivalently be written as $\psi_{R,\xi}(\varphi) \sim \left[\exp\int \mathrm{d}t\mathrm{d}\Omega\ \xi(t,\Omega)\,\hat{\Upsilon}^{-1}(\rho)\,\varphi(t,\Omega)\right]\ \psi_{R,0}(\varphi)$, see [4].

6. Radial S-matrix for AdS

Labeling all quantities of the free theory by a zero, calculating the free amplitude results in

$$\rho_{R,0}(\psi_{R,\xi}) = \int \mathcal{D}\varphi\ \psi_{R,\xi}(\varphi)\, Z_{R,0}(\varphi) = \exp\left[\frac{1}{2}\int_{\Sigma_R} \mathrm{d}t\,\mathrm{d}\Omega\ \xi(t,\Omega)\,\widehat{\mathcal{B}}\left(\frac{\overline{c_b}}{c_b}\,\xi(t,\Omega) + \overline{\xi(t,\Omega)}\right)\right],$$

with $\widehat{\mathcal{B}}$ defined by the eigenvalues $p/(2R^{d-2}_{\mathrm{AdS}})\cdot(2l+1)/(2l+d-2)$ when acting on the $u_{\omega l\underline{m}}(t,\Omega)$. Since the free amplitude is independent of R (not R_{Ads}), taking the large-R limit is easy and the amplitude can then be interpreted as radial S-matrix. Now we add a source field μ to include interactions: $S_{R,\mu}(\phi) = S_{R,0}(\phi) + \int_{\rho<R} \mathrm{d}^{d+1}x\ \sqrt{|g|}\ \mu(x)\,\phi(x)$. At this point we assume the source field to vanish for $\rho \geq R$. The resulting amplitude for a coherent state is

$$\rho_{R,\mu}(\psi_{R,\xi}) = \rho_{R,0}(\psi_{R,\xi})\left[\exp\int\limits_{\rho<R}\mathrm{d}^{d+1}x\sqrt{|g|}\ \mu\hat{C}\hat{X}_a\xi\right]\left[\exp\frac{\mathrm{i}}{2}\int\limits_{\rho<R}\mathrm{d}^{d+1}x\,\mathrm{d}^{d+1}x'\,\sqrt{|g(x)g(x')|}\ \mu(x)\,G_\mathrm{F}(x,x')\,\mu(x')\right].$$

The operators are again defined by their eigenvalues: for $\hat{X}_a(\rho)$ by ${}^1S_{\omega,l}(\rho)$ and for $\hat{C}$ by $\mathrm{i}(pR_{\mathrm{AdS}})^{l+1}[R_{\mathrm{AdS}}^{d-1}(2l+d-2)(2l-1)!!]^{-1}$. The Feynman propagator G_{F} fulfills the inhomogeneous KG equation $(\square_{\mathrm{AdS}}+m^2)\,G_{\mathrm{F}}(x,x') = \delta^{(d)}(x-x')$ and is given by

$$G_{\mathrm{F}}(x,x') = \int \mathrm{d}\omega \sum_{l,\underline{m}} \left\{ \theta(\rho-\rho')\left(\hat{C}\hat{X}_a(\rho')\,\overline{u_{\omega l\underline{m}}}\right)(t',\Omega')\left(\hat{\Upsilon}(\rho)u_{\omega l\underline{m}}\right)(t,\Omega) \;+\; (x \leftrightarrow x') \right\}.$$

The amplitude is thus independent of R, so we can take the limit $R \to \infty$ (such that the source field can have support on the whole spacetime) and consider the amplitude as an S-matrix. For a field interaction with potential V and action $S_{R,V}(\phi) = S_{R,0}(\phi) + \int_{\rho<R} \mathrm{d}^{d+1}x\,\sqrt{|g|}\,V(x,\phi)$ we can use the functional derivative technique to obtain

$$\exp \mathrm{i}S_{R,V}(\phi) = \left[\exp \mathrm{i}\int_{\rho<R} \mathrm{d}^{d+1}x\,\sqrt{|g(x)|}\,\hat{V}\left(x, -\mathrm{i}\frac{\delta}{\delta\mu(x)}\right)\right] \exp \mathrm{i}S_{R,\mu}(\phi)\bigg|_{\mu=0}$$

leading to the R-independent amplitude below whose large-R limit is an S-matrix:

$$\rho_{R,V}(\psi) = \left[\exp \mathrm{i}\int_{\rho<R} \mathrm{d}^4x\,\sqrt{|g(x)|}\,\hat{V}\left(x, -\mathrm{i}\frac{\delta}{\delta\mu(x)}\right)\right] \rho_{R,\mu}(\psi)\bigg|_{\mu=0}.$$

7. Conclusions

Using the GBF, an S-matrix for AdS can be constructed by calculating (generalized) amplitudes for the hypercylinder region. This circumvents the problem of not having available (temporally) asymptotic states on AdS, needed for the S-matrix of standard QFT. Future projects consist in applying holomorphic quantization [11, 12, 13] within the GBF framework, studying properties of this S-matrix and connections with the AdS/CFT correspondence [14], and relating our S-matrix to the one found by Giddings [15].

Acknowledgments

M.D. wishes to thank the organizers of the LOOPS 11 conference for the possibility of presenting this work, which was supported in part by CONACyT grant 49093 and scholarship 213531.

References

[1] Oeckl R 2008 *Adv.Theor.Math.Phys.* **12** 319–352 [arxiv:hep-th/0509122]
[2] Colosi D and Oeckl R 2008 *Phys. Rev.* **D 78:025020** [arxiv.org/abs/0802.2274]
[3] Colosi D 2010 *OnlinePrePrint* [arxiv.org/abs/1010.1209]
[4] Colosi D and Dohse M 2010 *OnlinePrePrint* [arxiv.org/abs/1011.2243]
[5] Breitenlohner P and Freedman D 1982 *Annals of Physics* **144** 249–281
[6] Avis S, Isham C and Storey D 1978 *Phys. Rev. D* **Vol. 18, Number 10** 3565–3576
[7] Oeckl R 2010 *Conference Proceedings ''QFT and Gravity'' Regensburg 2010* [arxiv.org/abs/1101.0367]
[8] Jackiw R 1990 *Analysis on infinite dimensional manyfolds - Schrödinger representation for quantized fields* (World Scientific, River Edge) in: Field theory and particle physics (pp. 78-143)
[9] Colosi D 2009 *OnlinePrePrint* [arxiv.org/abs/0903.2476]
[10] Colosi D and Oeckl R 2009 *OnlinePrePrint* [arxiv.org/abs/0912.0556]
[11] Oeckl R 2010 *OnlinePrePrint* [arxiv.org/abs/1009.5615]
[12] Oeckl R 2011 *OnlinePrePrint* [arxiv.org/abs/1104.5527]
[13] Oeckl R 2011 *OnlinePrePrint* [arxiv.org/abs/1109.5215]
[14] Witten E 1998 *Adv.Theor.Math.phys.* **2** 253–291 [arxiv.org/abs/hep-th/9802150]
[15] Giddings S 1999 *Phys.Rev.Lett.* **83:2707-2010** [arxiv.org/abs/hep-th/9903048]

Journal of Physics: Conference Series **360** (2012) 012013 doi:10.1088/1742-6596/360/1/012013

A systematic construction of curved phase space: A gravitational gauge theory with symplectic form

Jeffrey S. Hazboun and James T. Wheeler

Department of Physics, Utah State University, 4415 Old Main Hill, Logan UT 84322-4415

E-mail: jeffrey.hazboun@gmail.com, jim.wheeler@usu.edu

Abstract. General relativity can be constructed as a gauge theory using the quotient manifold strategy of [1, 2]. We consider a conformal gauging where the geometry is far richer than normal spacetime, including a symplectic form and the necessary emergence of Lorentzian signature. The resulting 2n-dim manifold constitutes a relativistic phase space, and general relativity is recovered when we demand that the momentum space is flat. However, the full geometry allows for curved phase space.

Introduction. We construct a generalization of space-time geometry to a fully curved phase space with general relativity on the configuration submanifold. A variant of such a geometry was first considered by Max Born, attempting to introduce complementarity into general relativity [3]. More recent variants include work on regularization of 2+1 gravity [4], where curvature on the momentum space is a consequence of the theory, and in the principle of relative locality [5], where the structure of momentum space is the starting point. Abraham and Marsden define a relativistic phase space as the spacetime co-tangent bundle with a symplectic form [6]. This implies a metric of Lorentz signature[1], and a symplectic form which defines canonical conjugacy. References [3, 4, 5] add curvature of the momentum space.

In this article we examine the space endowed *naturally* with all of these characteristics from the structure of Lie groups and techniques of gauge theory. We then take the picture further by developing an action functional over the entire phase space and showing how the resulting 8-dim structure reduces to general relativity on the configuration submanifold.

In the next section we explain the quotient manifold method of gauge theory. The result is *biconformal space*, which possesses all of the necessary characteristics described above. Next, we describe how a full phase space action, together with the structure equations, leads directly to general relativity in the special case of a momentum space without curvature. Concluding, we show that a gauge change of this space curves momentum space instead while leaving spacetime flat, thereby achieving properties that we expect underlie gravitational complimentarity.

Gauging using the quotient manifold method. Standard approaches to gauge theory begin with a matter action globally invariant under some symmetry group $\mathcal{H}$. This action generally fails to be locally symmetric due to the derivatives of the fields, but can be made locally invariant by introducing an $\mathcal{H}$-covariant derivative. The connection fields used for this derivative are called

[1] This often goes unstated, since it is usually imposed from the beginning. We will discuss how its appearance is non-trivial and physically interesting.

gauge fields. The final step is to make the gauge fields dynamical by constructing their field strengths, which may be thought of as curvatures of the connection.

This procedure has led to a fiber bundle: in a neighborhood of every point of space-time, the bundle is a direct product of that neighborhood with the Lie symmetry group $\mathcal{H}$. This gives a larger manifold with dimension equal to the dimension of space-time plus the dimension of the Lie group. There exists a projection operator which projects these copies of $\mathcal{H}$ back down to space-time. Choosing a gauge amounts to picking a cross-section of this bundle, i.e., one point from each of these copies of $\mathcal{H}$. Local symmetry amounts to dynamical laws which are independent of the choice of cross-section.

Attempts to write general relativity as a gauge theory led to an inversion of the standard construction. In addition to allowing the standard model and gravity to be placed on identical mathematical footings, this inverted approach has the advantage of allowing us to specify the symmetry in advance. This newer construction uses techniques developed by Cartan [7], giving the quotient group method of Ne'emann and Regge[1, 2]. We begin by taking the quotient of a Lie group, $\mathcal{G}$, by a Lie subgroup, $\mathcal{H}$. This quotient is necessarily a manifold, $\mathcal{M}$, which we identify with space-time or phase space. The quotient turns the group manifold $\mathcal{G}$ into a fiber bundle with symmetry $\mathcal{H}$ and base manifold $\mathcal{M}$. The Maurer-Cartan equations of $\mathcal{G}$ (equivalent to the Lie algebra) immediately supply a natural connection on this bundle, giving us our gauge fields. Generalizing the Maurer-Cartan connection to allow horizontal curvature is precisely the introduction of the field strengths of the gauge fields. The horizontality property guarantees that calculations on any cross-section give identical results, making the entire structure locally gauge invariant. Finally, the curvatures and any desired matter fields are composed into an action. This action will have $\mathcal{H}$-symmetry, with the remaining, broken, group transformations of $\mathcal{G}$ being replaced by diffeomorphisms on $\mathcal{M}$ [8].

Ne'emann and Regge used the quotient manifold method to get general relativity as a gauge theory by taking the quotient of the Poincaré group by the Lorentz group, but in principle, any Lie group containing the Poincaré group is interesting in terms of gravity. MacDowell and Mansouri [9] obtained general relativity by gauging the de Sitter or anti-de Sitter groups, and using a Wigner-Inonu contraction to recover Poincaré symmetry. In their 1982 papers [10, 11], Ivanov and Niederle exhaustively considered quotients of the groups $ISO(3,1)$, $SO(4,1)$, $SO(3,2)$ and $SO(4,2)$ by various subgroups containing the Lorentz group.

Biconformal Gauging. We concern ourselves with a specific gauge theory of general relativity leading to a curved phase space, found by taking the quotient of the conformal group by its Weyl subgroup (Lorentz $\otimes$ Dilatations). This quotient, first considered in [10, 11] results in an 8-dim (n-dim) manifold called biconformal space. It has been shown to be a phase space [12], to have the co-tangent bundle of general relativity as torsion-free solutions, and to provide a natural background for quantum mechanics, [13].

Throughout the following, Latin indices run $a, b, \ldots = 1, \ldots, n$. Though we describe the 8-dim ($n = 4$) case, all results hold in any dim $2n$, with $n > 2$. Differential forms are written in boldface. Corresponding to the standard representation for the generators of the conformal group, we have connection 1-forms: the Lorentz or Euclidean spin connection, $\boldsymbol{\omega}^a_{\ b}$ and Weyl vector, $\boldsymbol{\omega}$. The phase space is spanned by basis forms $\mathbf{e}^a$ and $\mathbf{f}_a$, deriving from the translations and special conformal transformations, respectively. Because raised and lowered indices carry different conformal weight, we generally write the metric explicitly where required.

Of all the quotients looked at by Ivanov and Niederle there are two properties unique to this gauging. First, only in this gauging can the dilatational structure equation, $\mathbf{d}\boldsymbol{\omega} = \mathbf{e}^a \wedge \mathbf{f}_a + \boldsymbol{\Omega}$, give a symplectic form. This is the case if the dilatational curvature is closed, $\mathbf{d}\boldsymbol{\Omega} = 0$. Then, since the connection forms from special conformal transformations ($\mathbf{f}_a$) and translations ($\mathbf{e}^a$) span the entire quotient manifold, $\mathbf{e}^a \wedge \mathbf{f}_a$ is a manifestly non-degenerate 2-form. With $\mathbf{d}\boldsymbol{\Omega} = 0$, $\mathbf{e}^a \wedge \mathbf{f}_a$ is closed, and therefore symplectic [6].

Journal of Physics: Conference Series **360** (2012) 012013 doi:10.1088/1742-6596/360/1/012013

The second unique characteristic of biconformal space is that only in this gauging is the Killing metric non-degenerate when restricted to the base manifold. Therefore, we get a metric from the group structure of the problem, rather than declaring a metric by fiat, as is the case with all the other gaugings above, and indeed standard treatments of general relativity.

We have already spoken about the physical motivation for having a canonically conjugate basis on a metric space, but in order to connect this geometry with relativistic phase space we require separate configuration and momentum metric submanifolds. Therefore, the basis forms $\mathbf{e}^a$ and $\mathbf{f}_a$ are separately required to be in involution. Orthogonality of these sets of basis forms is required in order for the submanifolds to be metric.

When the basis is both orthogonal and canonical [12], the signature of the submanifolds is severely limited, leading to a Lorenztian configuration space, and hence the origin of time. In the case of flat biconformal space Spencer and Wheeler proved the following theorem: *Flat 8-dim biconformal space is a phase space with canonically conjugate, orthogonal, metric submanifolds if and only if the initial 4-dim space we gauge is Euclidean or signature zero. In either of these cases the resulting configuration sub-manifold is necessarily Lorentzian* [12] Thus, when we impose the conditions necessary to make biconformal space a metric phase space, only in a restricted subclass of cases can we get a metric configuration space, and that metric *must* be Lorentzian. With a suitable choice of gauge the metric can be written as $h_{ab} = \frac{1}{(y^2)^2}\left(2y_a y_b - y^2\delta_{ab}\right)$ where δ_{ab} is the Euclidean metric and the signature changing character of the metric can be easily seen if the coordinates are chosen so that $y_a/\sqrt{y^2} = (1,0,0,0)$. It is important to note that in the metric above, $y_a = W_a$, is the Weyl vector of this space. This points to another unique characteristic of flat biconformal space. The group structures of the conformal group, projected down to this quotient manifold and written as canonically conjugate, orthogonal, metric submanifolds, gives rise to a natural notion of time, given by the gauge field of dilatations.

As an illustration of the portent of this natural notion of time one can calculate the Riemannian curvature of the momentum space metric, even though its conformal curvature is zero. The Riemann curvature of this metric can be written in the form $R_{abcd} = k_{bd}k_{ca} - k_{bc}k_{da}$, where $k_{ab} = \frac{y_a y_b}{(y^2)^2} - h_{ab}$ is a projection operator in the direction of time (the Weyl vector). This is exactly the form of the metric obtained in the ADM formalism when one chooses a time direction and constructs a projection onto the spatial dimensions. In our formulation, however, these expressions are covariant.

General Relativistic Phase Space. The above discussion reveals that flat biconformal space is a $2n$-dimensional space with symplectic form. When the connection of biconformal space is generalized to include curvature, the goal of a fully curved phase space is reached. In order to have a dynamical geometry we write down an action functional on biconformal space. Such an action was first investigated by Wehner and Wheeler [14]. Since the volume element of biconformal space is dimensionless, an action functional linear in the curvatures can be conformally invariant, unlike in Weyl gravity, where curvature squared terms lead to complications, such as higher order field equations and possibly negative normed states.

Wehner and Wheeler showed when the torsion is taken to vanish general relativity on the space-time tangent bundle is obtained. However, the basis they use is not orthogonal and the metric on configuration space is not the Killing metric. As in [14], we begin with the curvature-linear action, but we use the canonical, orthogonal basis of [12]. Then, by reducing the full structure equations and field equations, we derive general relativity. All structures used follow directly from the geometry of the conformal group.

The action, in terms of curvature 2-forms, is

$$S = \int \left(\alpha \boldsymbol{\Omega}^a_b + \beta\delta^a_b \boldsymbol{\Omega} + \gamma \mathbf{e}^a \mathbf{f}_b\right) \varepsilon_{ac\ldots d}\varepsilon^{be\ldots f}\mathbf{e}^c \ldots \mathbf{e}^d \mathbf{f}_e \ldots \mathbf{f}_f$$

where $\boldsymbol{\Omega}^a_b$ is the curvature associated with Euclidean rotations and $\boldsymbol{\Omega}$ is the dilatational curvature. The torsion, $\mathbf{T}^a$, and co-torsion, $\mathbf{S}_a$, cannot appear in a linear, conformally invariant action, but do appear in the field equations.

In order to make contact with general relativity we require flatness of the momentum sector to guarantee that we have the cotangent bundle of spacetime. These flatness conditions are looked at in detail in [12]. Briefly, the flatness condition sets the maximal number of momentum curvature pieces to zero, but not the entire $\Omega^{a\ cd}_{\ b}$. If $\Omega^{a\ cd}_{\ b} = 0$ is imposed then $h_{ab} = \delta_{ab}$, which is too restrictive, and does not give us involute submanifolds (see [12]). Although we impose this condition by hand, it may be automatic since it was found in [14] that the curvature on their momentum subspace is zero as a result of the field equations.

Our approach for solving these equations is similar to Spencer and Wheeler, except that the curvatures are no longer set to zero; instead they satisfy field equations. The field equations set conditions on the curvatures, including setting parts of the co-torsion to zero. Since $\mathbf{e}^a$ is in involution, we may study the $\mathbf{e}^a = 0$ momentum submanifolds. On these submanifolds, we solve for the momentum part of the connection by making judicious gauge choices. We then extend this form of the connection back to the full space. Using the field equations, structure equations and Bianchi identities we can write down the solution in terms of the solder form coefficients, and the Weyl vector. Of particular note, combining the field equations with the Bianchi identities on the basis form structure equations shows that the configuration and momentum space dilatational curvatures are zero generically. Together, these conditions lead to the Ricci flat version of Einstein's field equation on the configuration submanifold, $R^a_{\ bac} = 0$ Since the field equations imply metric compatibility of the connection, these restrictions lead to general relativity on the spacetime submanifold. The momentum sector contains the conformally Ricci flat equivalent.

$$R^{b\ ac}_{\ a} - \frac{1}{2(n-1)}\delta^{bc}h_{de}R^{d\ ae}_{\ a} - (n-2)\left(\partial^c W^b - W^b W^c + \frac{1}{2}\delta^{bc}W^2\right) = 0$$

The gauge choice of the Weyl vector is particularly interesting, because it enables us to place the Weyl vector entirely on either the configuration or the momentum submanifold. In fact if we change this gauge, now requiring the configuration space to have zero curvature, then we have a space that describes curved momentum space. This ability to switch between a normal general relativity picture and a geometry where the roles of momentum and position are switched is exciting from the view point of complimentarity and fully achieves the goal of Born to have reciprocity of the variables of gravity.

[1] Ne'eman Y and Regge T 1978 *La Rivista del Nuovo Cimento (1978-1999)* **1**(5) 1–43 ISSN 0393-697X 10.1007/BF02724472 URL `http://dx.doi.org/10.1007/BF02724472`

[2] Ne'eman Y and Regge T 1978 *Physics Letters B* **74** 54 – 56 ISSN 0370-2693 URL `http://www.sciencedirect.com/science/article/pii/0370269378900588`

[3] Born M 1938 *Proc. Royal Soc. Series A. Math & Phys. Sci.* **165** 291 URL `http://rspa.royalsocietypublishing.org/content/165/921.toc`

[4] Freidel L and Louapre D 2004 *Classical and Quantum Gravity* **21** 5685 URL `http://stacks.iop.org/0264-9381/21/i=24/a=002`

[5] Amelino-Camelia G and et al 2011 *arXiv* URL `http://arxiv.org/abs/1101.0931v2`

[6] Abraham R H and Marsden J E 1987 *Foundations of mechanics* (Perseus Books)

[7] Kobayashi S 1969 *Foundations of Differential Geometry* (Wiley)

[8] Kibble T 1961 *J.Math.Phys.* **2** 212–221 URL `http://jmp.aip.org/resource/1/jmapaq/v2/i2/p212_s1`

[9] MacDowell S W and Mansouri F 1977 *Phys. Rev. Lett.* **38**(14) 739–742 URL `http://link.aps.org/doi/10.1103/PhysRevLett.38.739`

[10] Ivanov E A and Niederle J 1982 *Phys. Rev. D* **25**(4) 976–987 URL `http://link.aps.org/doi/10.1103/PhysRevD.25.976`

[11] Ivanov E A and Niederle J 1982 *Phys. Rev. D* **25**(4) 988–994 URL `http://link.aps.org/doi/10.1103/PhysRevD.25.988`

[12] Spencer J A and Wheeler J T 2008 *arXiv* URL `http://arxiv.org/abs/0811.0112v3`

[13] Anderson L B and Wheeler J T 2004 *arXiv* URL `http://arxiv.org/abs/hep-th/0406159v2`

[14] Wehner A and Wheeler J T 1999 *Nuclear Physics B* **557** 380 – 406 ISSN 0550-3213 URL `http://www.sciencedirect.com/science/article/pii/S0550321399003673`

Loops 11: Non-Perturbative / Background Independent Quantum Gravity IOP Publishing
Journal of Physics: Conference Series **360** (2012) 012014 doi:10.1088/1742-6596/360/1/012014

Effective relational dynamics[1]

Philipp A Höhn
Institute for Theoretical Physics, Universiteit Utrecht, 3508 TD Utrecht, The Netherlands
E-mail: p.a.hohn@uu.nl

Abstract. We provide a synopsis of an effective approach to the problem of time in the semiclassical regime. The essential features of this new approach to evaluating relational quantum dynamics in constrained systems are illustrated by means of a simple toy model.

1. The problem of time

The application of the Dirac quantization programme to generally covariant systems triggers a 'conceptual problem' since the quantization of the Hamiltonian constraint yields a seemingly 'timeless' evolution equation $\hat{H}|\psi\rangle = i\hbar\frac{\partial}{\partial t}|\psi\rangle = 0$ and thereby leads to the issue of 'frozen dynamics' in the quantum theory. While classically the presence of a Hamiltonian constraint does not lead to an impeding obstacle because we can always resort to a time coordinate along the flow of the Hamiltonian with respect to which one can order physical relations, such a time coordinate is absent in the physical Hilbert space. The standard 'conceptual solution' to this problem is *relational dynamics* [2], according to which time evolution arises by relating dynamical quantities to other internal variables which we use as clocks to keep track of internal time; such relational statements are invariant under the gauge flow of the Hamiltonian constraint and could thus, in principle, be promoted into well-defined operators on the physical Hilbert space. Nevertheless, even when adopting this 'conceptual solution' one encounters a whole plethora of *technical* problems in the quantum theory [2, 3]. Here we want to address four such problems:

(i) The *Hilbert space problem.* Which Hilbert space representation is one to use and, in particular, how is one to construct a positive-definite inner product on the space of solutions?

(ii) The *multiple choice problem.* Which relational clock variable should one employ? (Different relational clocks may yield *a priori* different quantum theories.)

(iii) The *global time problem.* Good global clocks which parametrize the gauge orbits such that every trajectory intersects every constant clock time slice once and only once may not exist.

(iv) The *observable problem.* It is a notoriously difficult challenge to construct explicit observables in generally covariant theories, especially in the quantum theory.

In the sequel we summarise the effective approach to the problem of time [4, 5] which—at least in the semiclassical regime—deals with some of these problems and circumvents others. Rather than employing special clock choices and attempting to justify these, we follow the basic premise that there are to be no distinguished clocks and that, instead, we ought to treat all clocks on an equal footing. Our goal consists in employing local internal times, possibly translating between different clocks and thereby evolving (relational) data along complete semiclassical trajectories.

[1] Based on a talk at the conference *Loops '11* in Madrid on May 24th 2011 [1].

2. Effective description of constrained systems

The central idea of the effective approach is to sidestep the *Hilbert space problem* altogether; instead of employing wave functions or density matrices to describe states in a fixed Hilbert space, we regard states as (*a priori* complex) linear functionals on an algebra of kinematical variables, say polynomials in a canonial pair (q, p), and use expectation values $\langle\hat{q}\rangle$ and $\langle\hat{p}\rangle$, and moments[2]

$$\Delta(q^a p^b) := \langle(\hat{q} - \langle\hat{q}\rangle)^a(\hat{p} - \langle\hat{p}\rangle)^b\rangle_{\text{Weyl}}, \qquad a + b \geq 2$$

to parametrize states [6]. This construction immediately carries over to an arbitrary number of canonical pairs.

This space of states can be given a (quantum) phase space structure via a Poisson bracket defined as follows for any operators $\hat{A}, \hat{B}$ polynomial in the canonical variables[3]

$$\{\langle\hat{A}\rangle, \langle\hat{B}\rangle\} := \frac{\langle[\hat{A}, \hat{B}]\rangle}{i\hbar}. \tag{1}$$

This definition can be extended to the moments by linearity and the Leibnitz rule and yields 'classical Poisson brackets' for the expectation values of basic canonical variables, i.e. $\{\langle\hat{q}\rangle, \langle\hat{p}\rangle\} = 1$, vanishing Poisson brackets between expectation values and moments $\{\langle\hat{q}\rangle, \Delta(q^a p^b)\} = 0$ and more complicated brackets between the moments [6, 7].

In this summary we focus on finite dimensional systems with a single constraint $\hat{C}$ playing the role of the Hamiltonian constraint of general relativity. The Dirac programme requires physical states to satisfy $\hat{C}|\psi\rangle = 0$. The spectrum of $\hat{C}$ is essential for the construction of the physical Hilbert space: if zero lies in the discrete part of the spectrum, the physical Hilbert space turns out to be a subspace of the kinematical Hilbert space, while a new Hilbert space with a new (positive–definite) inner product must be constructed for solutions if zero lies in the continuous part of the spectrum. There exist techniques for constructing such physical Hilbert spaces in the latter case [8], however, finding physical Hilbert spaces in practice remains a difficult task.

The effective techniques, on the other hand, work for both zero in the discrete or continuous part of the spectrum. At the effective level, physical states must clearly satisfy

$$\langle\hat{C}\rangle(\langle\hat{q}\rangle, \langle\hat{p}\rangle, \Delta(qp), \cdots) = 0 \quad , \tag{2}$$

however, this is not sufficient since the expectation value of $\hat{C}$ may be zero even if $\hat{C}|\psi\rangle \neq 0$. Furthermore, when solving one (first class) constraint classically we can eliminate an entire canonical pair, while on the quantum phase space after solving (2) and factoring out its flow we would be left with an infinite tower of (unconstrained) moments of the eliminated canonical pair. Clearly, we must impose a further set of constraints to account for this. It turns out that

$$C_{\text{pol}} := \langle\widehat{\text{pol}}\hat{C}\rangle = 0 \quad , \tag{3}$$

where $\widehat{\text{pol}}$ stands for all polynomials in the basic canonical operators, provides a correct independent set of constraint functions on quantum phase space which together with (2) is then equivalent to the Dirac condition [7, 9].[4]

[2] The subscript 'Weyl' stands for Weyl–ordering, i.e. total symmetrization of the operators inside $\langle\ldots\rangle$.
[3] This definition of a Poisson bracket on quantum phase space can be motivated by the geometrical formulation of quantum mechanics [6].
[4] Note that no Weyl–ordering is imposed on the constraints, for otherwise the constraints fail to be first class [7]. This entails complex–valued quantum constraint functions at the kinematical level, however, physical reality and positivity conditions can be imposed on Dirac observables to describe proper quantum states. In the examples studied thus far, these conditions are preserved by the dynamics [5].

Journal of Physics: Conference Series **360** (2012) 012014 doi:10.1088/1742-6596/360/1/012014

As a result, we have infinitely many constraints for infinitely many variables. In order to reduce the system to a tractable finite size, we impose a very general semiclassical approximation: assume $\Delta(q^a p^b) = O(\hbar^{(a+b)/2})$ and truncate the system at $\hbar$ by neglecting all terms of higher order. This is consistent with the quantum Poisson bracket structure (1) which preserves orders in $\hbar$ and generates the flows and dynamics on quantum phase space, once suitable initial data for the relevant expectation values and moments has been chosen.

Rather than defining a quantum theory, the effective approach in its present form provides an effective tool for evaluating the quantum dynamics of given (finite dimensional) systems.

3. A simple example without a global clock function

Let us outline the effective approach to semiclassical relational dynamics with an explicit toy model (for a more general discussion and details see [4, 5]). The model we are considering is the isotropic 2D harmonic oscillator with prescribed total energy, whose classical solutions are closed orbits in phase space such that no global clock exists [5]. It is subject to the constraint

$$\hat{C}|\psi\rangle_{\text{phys}} = (\hat{p}_1^2 + \hat{p}_2^2 + \hat{q}_1^2 + \hat{q}_2^2 - M)|\psi\rangle_{\text{phys}} = 0 \quad . \tag{4}$$

The only variables from which we could construct clock functions are the $\hat{q}_i, \hat{p}_i$. In order to obtain a notion of evolution from this *a priori* timeless system, we may choose a variable as a local clock, say q_1, and deparametrize locally at the classical level by factorizing the constraint

$$C = (p_1 + H(q_1))\,(p_1 - H(q_1))\,, \quad \text{where} \quad H(q_1) = \sqrt{M - q_1^2 - p_2^2 - q_2^2} \quad .$$

Quantization of the first factor (in the region where it vanishes) yields a local Schrödinger regime

$$i\hbar\frac{\partial}{\partial q_1}\psi(q_1, q_2) = \hat{H}(\hat{q}_2, \hat{p}_2; q_1)\psi(q_1, q_2) = \widehat{\sqrt{M - q_1^2 - p_2^2 - q_2^2}}\,\psi(q_1, q_2) \tag{5}$$

with a non-self-adjoint $\hat{H}$ (defined by spectral decomposition) due to non-unitarity in q_1 evolution (q_1 is not globally valid). Given this construction, we can calculate expectation values and moments in q_1 time and compare their evolution to the effective treatment.

At the effective level, on the other hand, we retain 14 kinematical degrees of freedom at order $\hbar$, namely four expectation values $a := \langle\hat{a}\rangle$, four spreads $\langle(\hat{a} - a)^2\rangle_{\text{Weyl}}$, and six covariances $\langle(\hat{a} - a)(\hat{b} - b)\rangle_{\text{Weyl}}$, where the $\hat{a}, \hat{b}$ are any of the $\hat{q}_i, \hat{p}_i$. At order $\hbar$ the constraint (4) translates—via (2) and (3)—into the following five first class quantum constraint functions [5]

$$\begin{aligned}
C &= p_1^2 + p_2^2 + q_1^2 + q_2^2 + (\Delta p_1)^2 + (\Delta p_2)^2 + (\Delta q_1)^2 + (\Delta q_2)^2 - M = 0 \\
C_{q_1} &= 2p_1\Delta(q_1p_1) + 2p_2\Delta(q_1p_2) + 2q_1(\Delta q_1)^2 + 2q_2\Delta(q_1q_2) + i\hbar p_1 = 0 \\
C_{p_1} &= 2p_1(\Delta p_1)^2 + 2p_2\Delta(p_1p_2) + 2q_1\Delta(p_1q_1) + 2q_2\Delta(p_1q_2) - i\hbar q_1 = 0 \\
C_{q_2} &= 2p_1\Delta(p_1q_2) + 2p_2\Delta(q_2p_2) + 2q_1\Delta(q_1q_2) + 2q_2(\Delta q_2)^2 + i\hbar p_2 = 0 \\
C_{p_2} &= 2p_1\Delta(p_1p_2) + 2p_2(\Delta p_2)^2 + 2q_1\Delta(q_1p_2) + 2q_2\Delta(q_2p_2) - i\hbar q_2 = 0.
\end{aligned}$$

These five constraints generate only four independent flows since at order $\hbar$ we have to deal with a degenerate Poisson structure [5]. It is convenient to fix three of these independent flows. Selecting q_1 as a relational clock it should not be represented as an operator; we therefore choose a gauge which 'projects this clock to a parameter' by setting its fluctuations to zero (fixing three $O(\hbar)$ flows and leaving us with one 'Hamiltonian flow')

$$(\Delta q_1)^2 = \Delta(q_1q_2) = \Delta(q_1p_2) = 0 \quad . \tag{6}$$

Indeed, the choice of an internal time variable in the effective framework is best described and interpreted in a corresponding gauge [4, 5]: we refer to such a choice and gauge (e.g. (6)) as a *Zeitgeist*. After choosing a local relational clock and corresponding Zeitgeist, local relational observables at the effective level are given by correlations of expectation values and moments with the expectation value of the chosen clock (here q_1) evaluated in its Zeitgeist. We call such state dependent local observables *fashionables* because they comprise the complete physical information about the system (at order $\hbar$) as long as the Zeitgeist is valid, but may fall out of fashion when a Zeitgeist necessarily changes at turning points of local clocks.

It turns out that the fashionables in q_1 time of the effective framework agree perfectly with those computed in the q_1 Schrödinger regime (5) [1, 5]. In its turning region, q_1 becomes complex–valued and its Zeitgeist incompatible with the semiclassical expansion, such that evolution in q_1 breaks down *before* the turning point—a signature of non–unitarity. Thus, a new local internal time (here q_2) is needed and for a full evolution we must switch between q_1 and q_2 time. Furthermore, since a given internal time is best described in a corresponding choice of gauge, we must switch also between the q_1 and q_2 Zeitgeister. Explicit gauge transformations between these Zeitgeister can, indeed, be constructed and initial data can be evolved consistently around the entire closed semiclassical orbit through the turning points of various clock variables [5]. It should be emphasized that in each gauge we evolve a *different* set of fashionables which highlights the local nature of the relational concept in the absence of global clock functions.

4. Conclusions

The effective approach summarised here sidesteps the technical problems mentioned in the introduction. In particular, the *Hilbert space problem* is avoided altogether since no use of any Hilbert space representation has been made. At the effective level one can make sense of local time evolution and switching between different local clocks can essentially be achieved by an additional gauge transformation which enables one to handle the *multiple choice* and *global time problem* and implies the equivalence of different clock choices at semiclassical order. Note, however, that if the relational clock is non–global, it assumes complex values in its turning region [4, 5]. State dependent fashionables arise naturally in this framework which simplifies the *observable problem* due to the classical treatment of the effective system. In the example outlined here and other models, the effective evolution agrees with local deparametrizations at a Hilbert space level [5]. Finally, the notion of relational evolution breaks down in highly quantum states of systems without global clocks [4, 5]. In the non–integrable closed FRW model universe filled with a massive scalar field this happens even in initially semiclassical states [10].

Acknowledgments The author would like to thank Martin Bojowald and Artur Tsobanjan for collaboration on this subject.

References

[1] slides at `http://loops11.iem.csic.es/loops11/index.php?option=com_content&view=article&id=101`
[2] C. Rovelli, *Quantum Gravity* (CUP, Cambridge, 2004); C. Rovelli, Phys. Rev. **D 42** 2638 (1990), Phys. Rev. **D 43** 442 (1991); B. Dittrich, Gen. Rel. Grav. **39** 1891 (2007), Class. Quant. Grav. **23** 6155 (2006)
[3] K. V. Kuchař, in *Proc. 4th Canadian Conference on General Relativity and Relativistic Astrophysics*, edited by G. Kunstatter, D. Vincent and J. Williams (World Scientific, Singapore, 1992); C. J. Isham, in *Integrable Systems, Quantum Groups, and Quantum Field Theories* (Kluwer Academic Publishers, London, 1993); E. Anderson (*Preprint* arXiv:1009.2157 [gr-qc]); J. Tambornino (*Preprint* arXiv:1109.0740 [gr-qc])
[4] M. Bojowald, P. A. Höhn and A. Tsobanjan, Class. Quantum Grav. **28** 035006 (2011)
[5] M. Bojowald, P. A. Höhn and A. Tsobanjan, Phys. Rev. **D 83** 125023 (2011)
[6] M. Bojowald and A. Skirzewski, Rev. Math. Phys. **18** 713–745 (2006)
[7] M. Bojowald, B. Sandhöfer, A. Skirzewski and A. Tsobanjan, Rev. Math. Phys. **21** 111 (2009)
[8] D. Marolf (*Preprint* gr-qc/9508015); T. Thiemann, Class. Quant. Grav. **23** 2211 (2006)
[9] M. Bojowald and A. Tsobanjan, Phys. Rev. **D 80** 125008 (2009)
[10] P. A. Höhn, E. Kubalová and A. Tsobanjan (*Preprint* arXiv:1111.5193 [gr-qc])

Journal of Physics: Conference Series **360** (2012) 012015 doi:10.1088/1742-6596/360/1/012015

A view on the problems of Quantum Gravity

T P Shestakova

Department of Theoretical and Computational Physics, Southern Federal University, Sorge St. 5, Rostov-on-Don 344090, Russia

E-mail: shestakova@sfedu.ru

Abstract. The existing approaches to quantization of gravity aim at giving quantum description of 3-geometry following to the ideas of the Wheeler – DeWitt geometrodynamics. In this description the role of gauge gravitational degrees of freedom is missed. A probable alternative is to consider gravitational dynamics in extended phase space, taking into account the distinctions between General Relativity and other field theories. The formulation in extended phase space leads to some consequences at classical and quantum levels. At the classical level, it ensures that Hamiltonian dynamics is fully equivalent to Lagrangian dynamics, and the algebra of Poisson brackets is invariant under reparametrizations in a wide enough class including reparametrizations of gauge variables, meantime in the canonical Dirac approach the constraints' algebra is not invariant that creates problems with quantization. At the quantum level, the approach come to the description in which the observer can see various but complementary quantum gravitational phenomena in different reference frames that answers the spirit of General Relativity and Quantum Theory. Though until now the approach was applied to General Relativity in its original formulations, its implementation in different trends, including Loop Quantum Gravity or some other representations of gravitational variables, would also be of interest.

I am grateful to the Organizers of the Conference for the opportunity to present my point of view on the problems of Quantum Gravity. In our attempts to quantize gravity we inevitably rely on approaches which work satisfactory for ordinary field theories, often without a careful analysis of applicability one or another approach to gravitation. An attractive point of Loop Quantum Gravity is that it aims at searching for its own way of description of Quantum Geometry and the Hilbert space of quantum states.

At the same time, Loop Quantum Gravity inherits the ideas of the Wheeler – DeWitt Quantum Geometrodynamics which deals with 3-geometry being described by 3-metric g_{ij}, while the other components of metric tensor, $g_{0\mu}$, are missed in this consideration. They are traditionally believed to be redundant variables, whose role in General Relativity is just to fix a reference frame, and the choice of a reference frame should not affect physical phenomena. The purpose of this my talk is to demonstrate that, on the opposite, gauge variables $g_{0\mu}$ play an important role both at classical and quantum levels and must be taken into account in any quantization scheme.

The Wheeler – DeWitt Quantum Geometrodynamics [1] is based on three cornerstones: Dirac approach to quantization of systems with constraints [2, 3], the Arnowitt – Deser – Misner (ADM) parametrization [4] and the ideas of Wheeler that the wave functional must depend only on 3-geometry [5, 6]. In the Dirac approach the central part is given to a postulate, according to which each constraint $\varphi_m(q, p) = 0$ after quantization becomes a condition on a state vector, or wave functional, Ψ: $\varphi_m\Psi = 0$. Let us emphasize that it is indeed a postulate, since it cannot be justified by any reference to the correspondence principle. As a consequence, in Quantum Geometrodynamics all the theory is reduced to the only Wheeler – DeWitt equation, that, in its turn, leads to the problem of time and other related problems. It is believed by many researchers that these problems can find a resolution without drastically altering the quantization procedure. However, the difficulties of the Wheeler – DeWitt theory have made a way for its strong criticism. In the review devoted to the problem of time Kuchař [7] wrote: "In my opinion, none of us has so far succeeded in proposing an interpretation of quantum gravity that would either solve or circumvent the problems of time". In [8] Isham made even stronger statement: "...although it may be heretical to suggest it, the Wheeler – DeWitt equation – elegant though it be – may be completely the wrong way of formulating a quantum theory of gravity". My own belief is that the mentioned problems cannot be resolved in the limits of Wheeler – DeWitt Quantum Geometrodynamics.

The ADM parametrization enables one to write gravitational constraints in the form independent of gauge variables – the lapse and shift functions N, N_i. It gave rise to an illusion that the theory in which the main equations are those of constraints does not depend on a choice of gauge conditions. However, it can be demonstrated that any choice of parametrization implies implicitly a certain choice of a reference frame. It is important to realize that, while at classical level the existing of constraints reflects gauge invariance of the theory, at quantum level imposing operator form of constraints on the wave functional does not ensure gauge invariance. Strictly speaking, we do not have any proof that the Wheeler – DeWitt theory is gauge-invariant. In this situation the quantum Hamiltonian constraint loses its sense and the whole procedure of quantization becomes questionable.

In our attempts to reconcile General Relativity with quantum principles we must be sure that we are quantizing General Relativity, and not any other theory which resembles General Relativity by its structure but is not equivalent to it. Several years ago a statement appeared in literature [9], that the ADM formulation [4] is not equivalent to the Dirac formulation of Hamiltonian dynamics for gravity [10], because the both formulations are not related by a canonical transformation. The raised question is of fundamental nature, since, if the statement were correct, we would have to agree with the declaration that "the ADM formulation might be considered as a model (geometrodynamics or ADM gravity) without any reference to Einstein GR" ([9], p. 62). In fact, this statement is a result of misunderstanding. The transformation from components of metric tensor to the ADM variables,

$$g_{00} = \gamma_{ij}N^iN^j - N^2, \qquad g_{0i} = \gamma_{ij}N^j, \qquad g_{ij} = \gamma_{ij}, \tag{1}$$

concerns gauge degrees of freedom which, from the viewpoint of Dirac, are not canonical variables at all. To pose the question, if the transformation (1) is canonical, one should extend the original phase space including into it gauge degrees of freedom and their momenta and also introduce into the Lagrangian missing velocities corresponding to gauge variables. It can be done by means of special (differential) gauge conditions that *actually extends* the phase space. For general enough gauge conditions, $f^\mu(g_{\nu\lambda}) = 0$, their differential form reads

$$\frac{d}{dt}f^\mu(g_{\nu\lambda}) = 0, \qquad \frac{\partial f^\mu}{\partial g_{00}}\dot{g}_{00} + 2\frac{\partial f^\mu}{\partial g_{0i}}\dot{g}_{0i} + \frac{\partial f^\mu}{\partial g_{ij}}\dot{g}_{ij} = 0. \tag{2}$$

Then, we can write the effective action including gauge and ghost sectors as it appears in the path integral approach to gauge field theories,

$$S = \int d^4x \left(\mathcal{L}_{(grav)} + \mathcal{L}_{(gauge)} + \mathcal{L}_{(ghost)}\right). \tag{3}$$

Journal of Physics: Conference Series **360** (2012) 012015 doi:10.1088/1742-6596/360/1/012015

It has been shown [11] for the full gravitational theory that the transformation

$$g_{0\mu} = v_\mu\left(N_\nu, g_{ij}\right), \qquad g_{ij} = \gamma_{ij}, \tag{4}$$

where $v_\mu\left(N_\nu, g_{ij}\right)$ is invertible function, is canonical in extended phase space embracing spatial components of metric tensor g_{ij} which are canonical in Dirac's sense, as well as gauge variables $g_{0\mu}$, ghosts and their momenta. The transformation to the ADM variables (1) is a particular case of (4). The canonicity of the transformation (4) not just restores the status of the ADM formulation as completely equivalent to the original Einstein formulation, but also demonstrates the role of gauge degrees of freedom which are responsible for consistency of the theory.

The extension of phase space by means of introducing a differential form of gauge conditions was suggested in the works [12, 13, 14, 15]. It is well known that the idea of extended phase space was put forward by Batalin, Fradkin and Vilkovisky (BFV) [16, 17, 18]. However, in their approach gauge variables were still considered as non-physical, secondary degrees of freedom playing just an auxiliary role in the theory. While the BFV approach aimed at reproducing the results of Dirac's canonical quantization on a path integral level, our construction of extended phase space guarantees equivalence of Lagrangian and Hamiltonian dynamics of a constrained system for a wide enough class of parametrizations, and constraints and gauge condition get a status of Hamiltonian equations in extended phase space [11]. The algebra of Poisson brackets turns out to be invariant under reparametrizations from this class.

Based upon this formulation of Hamiltonian dynamics, we have proposed Quantum Geometrodynamics in extended phase space [14, 15] making use of the path integration approach which is more powerful. We paid a special attention to the peculiarities of General Relativity. In ordinary quantum theory its gauge invariance is ensured by asymptotic boundary conditions in the path integral. The boundary conditions imply the existing of asymptotic states in which gauge and physical degrees of freedom could be separated from each other. In the case of gravity asymptotic states exists only in asymptotically flat spacetime, but for a non-trivial spacetime topology imposing asymptotic boundary conditions is not justified. Thus, path integral quantization as well as canonical quantization do not enable us to prove gauge invariance of the theory. In this situation the requirement for the wave function to satisfy the Wheeler – DeWitt equation seems to be too strong and not well-grounded. However, independently on our notion about gauge invariance or noninvariance of the theory, we expect that the wave function has to obey some Schrödinger equation. One can derive the Schrödinger equation from the path integral by the well-known standard procedure and only after that one can investigate the question if there exist any conditions which would ensure gauge invariance.

Since asymptotic boundary conditions are not justified in gravitational theory, one should define the wave function on extended configurational space depending on gauge and physical gravitational variables and ghosts (one may also add matter fields). So, the wave function is assumed to be a solution to the equation

$$H\Psi\left[g_{0\mu}, g_{ij}, \theta^\nu, \bar{\theta}_\nu, \phi\right] = i\frac{\partial\Psi}{\partial t}. \tag{5}$$

H is the Hamiltonian in extended phase space. The problem of time does not appear is this approach though the physical sense of time in Quantum Gravity require additional clarification. In any case, the existing of time is related with a reference frame in which we study spacetime geometry.

The structure of the wave function is determined by the structure of the path integral with the effective action (3). It has been demonstrated for models with finite numbers degrees of freedom that the part of the wave function depending on gauge and ghost variables can be separated out. For example, for the Bianchi IX model one gets:

$$\Psi(\tilde{N}, Q, \theta, \bar{\theta}; t) = \int \Psi_{(ph)}(Q, t)\,\delta(\tilde{N} - f(Q) - k)\,(\bar{\theta} + i\theta)\,dk. \tag{6}$$

$\tilde{N}$ stands for an arbitrary gauge variable, and Q stands for a set of physical variables. The physical part of the wave function satisfies the equation

$$H_{(ph)}\Psi_{(ph)}(Q,\,t) = i\,\frac{\partial\Psi_{(ph)}(Q,\,t)}{\partial t} \tag{7}$$

The physical Hamiltonian H_{ph} depends, in general, on a chosen parametrization and gauge. In particular, for the ADM parametrization and the condition $N = 1$ the left-hand side of (7) coincides with the left-hand side of the Wheeler – DeWitt equation [19]. It gives another evidence that the Wheeler – DeWitt Geometrodynamics is not a gauge invariant theory.

In the proposed version of Quantum Geometrodynamics spacetime geometry is described from the viewpoint of the observer who can see various but complementary quantum gravitational phenomena in different reference frames. It contradicts to the original goal to construct gauge-invariant Quantum Gravity, however, to a greater extent it answers the spirit of General Relativity and Quantum Theory where the observer plays the central role in description of physical phenomena.

I believe that it would be rather instructive to apply some ideas of extended phase space approach to Loop Quantum Gravity, or, at least, loop quantum cosmological models. The first step of the program would be the formulation of dynamics of General Relativity in new variables, in such a way that this formulation would be equivalent to the original one. It would ensure that the theory have the correct classical limits. And this is the point where the extended phase space approach could play its role, since it allows us to work with different parametrizations. On the next, quantum stage one could explore without prejudice the question if Quantum Gravity can be constructed in a gauge-invariant way. I hope that this study would enrich our understanding of quantization of Gravity.

Acknowledgments
My participation in the Loops 11 conference was partially supported by the RFBR grant 11-02-08088.

References

[1] DeWitt B S 1967 *Phys. Rev.* **160** 1113–48
[2] Dirac P A M 1950 *Can. J. Math.* **2** 129–148
[3] Dirac P A M 1958 *Proc. Roy. Soc.* A **246** 326–332
[4] Arnowitt R, Deser S and Misner C W 1962 *Gravitation, an Introduction to Current Research* ed L Witten (New York: John Wiley & Sons) pp 227–265
[5] Wheeler J A 1964 *Relativity, Groups and Topology* ed C DeWitt and B S DeWitt (New York: Gordon & Breach)
[6] Wheeler J A 1968 *Einstein's Vision* (Berlin–Heidelberg–New York: Springer Verlag)
[7] Kuchar K V 1992 *Proceedings of the 4th Canadian Conference on General Relativity and Relativistic Astrophysics* ed G Kunstatter, D Vincent and J Williams (Singapore: World Scientific) pp 211–314
[8] Isham C 1992 Canonical quantum gravity and the problem of time *Preprint* gr-qc/9210011
[9] Kiriushcheva N and Kuzmin S V 2011 *Central Eur. J. Phys.* **9** 576–615
[10] Dirac P A M 1958 *Proc. Roy. Soc.* A **246** 333–343
[11] Shestakova T P 2011 *Class. Quantum Grav.* **28** 055009.
[12] Savchenko V A, Shestakova T P and Vereshkov G M 1999 *Int. J. Mod. Phys.* A **14** 4473–90
[13] Savchenko V A, Shestakova T P and Vereshkov G M 2000 *Int. J. Mod. Phys.* A **15** 3207–20
[14] Savchenko V A, Shestakova T P and Vereshkov G M 2001 *Gravitation & Cosmology* **7** 18–28
[15] Savchenko V A, Shestakova T P and Vereshkov G M 2001 *Gravitation & Cosmology* **7** 102–116
[16] Fradkin E S and Vilkovisky G A 1975 *Phys. Lett.* B **55** 224–226
[17] Batalin I A and Vilkovisky G A 1977 *Phys. Lett.* B **69** 309–312
[18] Fradkin E S and Fradkina T E 1978 *Phys. Lett* B **72** 343–348
[19] Shestakova T P 2006 *Gravitation & Cosmology* **12** 223–226

Loop quantum effect and the fate of tachyon field collapse

Y. Tavakoli [1,*], **P. Vargas Moniz** [1,2,†], **J. Marto** [1,‡], **A. H. Ziaie** [3,§]

[1] *Departamento de Física, Universidade da Beira Interior, 6200 Covilhã, Portugal*
[2] *CENTRA, IST, Av. Rovisco Pais, 1, Lisboa, Portugal*
[3] *Department of Physics, Shahid Beheshti University, Evin, Tehran 19839, Iran*
E-mail: * tavakoli@ubi.pt, † pmoniz@ubi.pt, ‡ jmarto@ubi.pt, § ah.ziaie@gmail.com

Abstract. We study the fate of gravitational collapse of a tachyon field matter. In presence of an inverse square potential a black hole forms. Loop quantum corrections lead to the avoidance of classical singularities, which is followed by an outward flux of energy.

1. Gravitational collapse of a tachyon matter

We discuss the spacetime region inside the collapsing sphere which will contain the chosen matter content. In this region, an homogeneous and isotropic Friedmann- Robertson-Walker (FRW) metric will be considered as the setting for the gravitational collapse in comoving coordinates. However, in order to study the whole spacetime, we must match this interior spacetime to a suitable exterior region. In the model herein, it is convenient to consider a spherically symmetric and inhomogeneous spacetime such as the Schwarzschild or the generalized Vaidya geometries to model the spacetime outside the collapsing sphere (cf. see [1] for a discussion on the whole spacetime structure including the exterior geometry). Let us nevertheless be more precise for the benefit of the reader: There is a particular setting which has been of recent use, namely in investigations of loop effects in gravitational collapse involving a scalar field; it is the marginally bound case ($k = 0$), where the interior spacetime is dynamical and is parameterized by a line element as follows,

$$ds^2 = -dt^2 + a^2(t)[dr^2 + r^2 d\Omega^2], \tag{1}$$

where t is the proper time for a falling observer whose geodesic trajectories are labeled by the comoving radial coordinate r, and $d\Omega^2$ being the standard line element on the unit two- sphere; we have set the units $8\pi G = c = 1$. The dynamical evolution from Einstein's field equations for this marginal bound case regarding the interior region, can be presented as [2]

$$\rho = \frac{F_{,r}}{R^2 R_{,r}}, \quad p = -\frac{\dot{F}}{R^2 \dot{R}}, \quad \dot{R}^2 = \frac{F}{R}, \tag{2}$$

where t and r are the proper time and the comoving radial coordinate for a falling observer, respectively; we have set the units $8\pi G = c = 1$. The quantity $F(R)$ is the mass function with $R(t, r) = ra(t)$ being the area radius of the collapsing shell. Furthermore, ρ and p are energy density and pressure of the collapsing matter, respectively. The notation "$,r$" and "$\cdot$" denote a derivative with respect to r and t, respectively.

Integration of Eq. (2) gives the following relation for the mass function as $F = \frac{1}{3}\rho(t)R^3$. A spherically symmetric spacetime containing a trapped region is characterized by the corresponding boundary (trapped) surface, extracted from the equation $F = R$. In the region where the mass function satisfies the relation $F > R$, it describes a trapped region and the regions in which the mass function is less than the area radius i.e., $F < R$, are not trapped.

To investigate the gravitational collapse, we will consider herein a homogeneous tachyon field $\phi(t)$, as matter content of the collapsing cloud. Employing A. Sen's effective action, the corresponding energy density and pressure for tachyon field are given by [3]

$$\rho(\phi) = 3H^2 = \frac{V(\phi)}{\sqrt{1-\dot{\phi}^2}}, \qquad p(\phi) = -V(\phi)\sqrt{1-\dot{\phi}^2}, \tag{3}$$

where $V(\phi)$ is the tachyon potential and is assumed to be of the form $V(\phi) = V_0\phi^{-2}$. Using the equation for the energy conservation, we get the field equation for ϕ as

$$\ddot{\phi} = -(1-\dot{\phi}^2)\left[3H\dot{\phi} + \frac{V_{,\phi}}{V}\right], \tag{4}$$

where "$,\phi$" denotes the derivative with respect to the tachyon field and $H = \dot{a}/a$ is the Hubble rate which for collapsing scenario is $H < 0$. Using Eq. (3) the equation of state becomes $w = p(t)/\rho(t) = \dot{\phi}^2 - 1$.

Let us study the collapse of a tachyon field from a dynamical system perspective. We define a new time variable τ (instead of the proper time t) for our collapsing system as $\tau \equiv -\log(a/a_0)^3$, which is defined in the interval $0 < \tau < \infty$. For any time dependent function $f = f(t)$ we can write $df/d\tau \equiv -\dot{f}/3H$, where a dot over a variable denotes its derivative with respect to time t.

Introducing the new set of dynamical variables $x \equiv \dot{\phi}$ and $y \equiv V/3H^2$, the Friedmann constraint equation in (3) becomes $y/\sqrt{1-x^2} = 1$. From Eqs. (3) and (4) the corresponding autonomous system then reads

$$\frac{dx}{d\tau} = \left(\frac{\lambda}{\sqrt{3}}\sqrt{y} + x\right)(1-x^2), \qquad \frac{dy}{d\tau} = -xy\left(\frac{\lambda}{\sqrt{3}}\sqrt{y} + x\right), \tag{5}$$

where $\lambda \equiv -V_{,\phi}/V^{3/2} = 2/\sqrt{V_0} = constant$. The critical points (x_c, y_c) of the autonomous system can be obtained easily as: (a) at $(1,0)$, (b) at $(-1,0)$, and (c) at $(-\lambda\sqrt{y_0/3}, y_0)$, where $y_0 = -\lambda^2/6 + \sqrt{(\lambda^2/6)^2+1} > 0$. From the eigenvalues of the Jacobi matrix, the stability of the system can be determined;

- Fixed points (a) and (b) are asymptotically *stable* nodes (attractor) allowing to select a class of solutions in which, asymptotic time dependence of the tachyon field can be estimated as $\phi(t) \simeq \pm t + \phi_0$, where '+' and '−' denotes fixed points (a) and (b), respectively. From Eq. (3) it is seen that, the effective pressure asymptotically vanishes. This suggests a dust-like behavior where $p \simeq 0$ [4]. Then, the energy density of collapse can be obtained as $\rho = \rho_\phi \propto 1/a^3$; this expression indicates, therefore that, ρ_ϕ diverges as $a \to 0$.
- The critical point (c) is asymptotically *unstable* (source); it can not be considered as a physically relevant solution for the collapse, and hence we ignore its studying here.

The occurrence of spacetime singularities is proved by the existence of incomplete future or past directed non-spacelike geodesics. Then, it is required that the Kretschmann invariant $K = R_{abcd}R^{abcd} = 12[(\ddot{a}/a)^2 + (\dot{a}/a)^4]$ gets unboundedly large values as the singularity is approached along the non-spacelike geodesics terminating there. If such a condition is satisfied,

Journal of Physics: Conference Series **360** (2012) 012016 doi:10.1088/1742-6596/360/1/012016

then the physical system should be considered to exhibit a curvature singularity [2]. Therein one can observe that, this quantity diverges as the physical area radius vanishes, consequently signaling the occurrence of a curvature singularity; the collapse ends up with a black hole formation [1].

2. Semiclassical tachyon field collapse

In this section we will study the loop quantum effects (of the inverse triad type) on tachyon field collapse. To investigate the fate of the collapse we will consider the modified dynamical equation of the system in the semiclassical limit. The modified Friedmann and the Raychaudhury equations for the tachyon field in the semiclassical regime $a_i < a \ll a_*$ are given, respectively by [3]

$$\rho_\phi^{loop} = 3H^2 = \frac{V(\phi)}{\sqrt{1 - A^{-1}q^{-15}\dot{\phi}^2}}, \qquad p_\phi^{loop} = 2\dot{H} + 3H^2 = -\left(1 + 4\frac{\dot{\phi}^2}{Aq^{15}}\right)\rho_\phi^{loop}. \tag{6}$$

Then, using the energy conservation equation, the modified equation of motion in the semiclassical limit becomes

$$\ddot{\phi} - 12H\dot{\phi}\left(\frac{7}{2} - \frac{\dot{\phi}^2}{Aq^{15}}\right) + \left(Aq^{15} - \dot{\phi}^2\right)\frac{V_{,\phi}}{V} = 0, \tag{7}$$

where $A \equiv (12/7)^{12}$. Also $a_* = \sqrt{j/3}a_i$ is a critical scale at which the eigenvalue of the inverse scale factor has a power-law dependance on the scale factor; j is the half-integer free quantization parameter, $a_i \equiv \sqrt{\gamma}\ell_{\rm P}$ is the scale above which a classical continuous spacetime can be defined and below which the spacetime is discrete, and $\ell_{\rm P}$ is the Planck length [5]. The modified mass function within this regime ($a \ll a_*$) takes the form $F^{loop} = 1/3\rho_\phi^{loop}R^3$. The behavior of this effective mass function determines whether trapped surfaces will form during the collapse procedure, within the semiclassical regime.

To study the collapsing model using a dynamical system description we introduce a new time variable σ given by $\sigma \equiv -\log q^{3/2}$, where $q = (a/a_*)^2$, being defined in the interval $0 < \sigma < \infty$; the limit $\sigma \to 0$ corresponds to the initial condition of the collapsing system ($a \to a_*$) and the limit $\sigma \to \infty$ corresponds to $a \to 0$. For any time dependent function f, we write $df/d\sigma \equiv -\dot{f}/3H$, where dot denotes the derivative with respect to t.

We herein make use of a *new* set of convenient dynamical variables: $\tilde{x} \equiv A^{-1/2}q^{-15/2}\dot{\phi}$, $\tilde{y} \equiv V/3H^2$, $\tilde{z} \equiv A^{1/2}q^{15/2}$. Then, Eqs. (6)-(7) in terms of $\tilde{x}$, $\tilde{y}$, and $\tilde{z}$ are presented as

$$\frac{d\tilde{x}}{d\sigma} = -\frac{\xi}{\sqrt{3}}\tilde{z}\sqrt{\tilde{y}}\left(1 - \tilde{x}^2\right) + \tilde{x}\left(4\tilde{x}^2 - 9\right),$$
$$\frac{d\tilde{y}}{d\sigma} = -\frac{\xi}{\sqrt{3}}\tilde{z}\sqrt{\tilde{y}}\left(1 - \tilde{x}^2\right) + \tilde{x}\left(4\tilde{x}^2 - 9\right), \qquad \frac{d\tilde{z}}{d\sigma} = -5\tilde{z}, \tag{8}$$

where $V(\phi) = \beta\phi^{-2}$, which brings $\xi \equiv -V_{,\phi}/V^{3/2} = 2/\sqrt{\beta}$ as a constant. The Hamiltonian constraint equation (6) in terms of new variables takes the form $\tilde{y}/\sqrt{1-\tilde{x}^2} = 1$. The properties of each critical point will be determined by the eigenvalues of the (3×3)-Jacobi matrix. Finally, for the dynamical system (8), there is, under the constraints, just one physical fixed point for the physical system: (p_o) as $(0, 1, 0)$. In order to discuss the stability of (p_o), we have to determine the eigenvalues (and eigenvectors) of then Jacobi matrix. This analysis shows that, all eigenvalues are real but one is zero, the rest being negative implying that, the autonomous system is nonlinear with a non-hyperbolic point. The asymptotic properties cannot be simply determined by linearization and we need to resort to other method: the center manifold theorem. Using the center manifold analysis one can show that, (p_o) is an asymptotically *stable* critical point (for a more detailed and complete discussion on the above elements, the reader may consult ref. [1]).

The solution for the effective energy density, the effective pressure in the neighborhood of the critical point (p_o) can be easily obtained that

$$\rho_\phi^{loop} \simeq V(\phi)\left(1+\frac{1}{2}\frac{\dot{\phi}^2}{Aq^{15}}\right), \quad p_\phi^{loop} \simeq -V(\phi)\left(1+\frac{9}{2}\frac{\dot{\phi}^2}{Aq^{15}}\right). \tag{9}$$

For a sufficient choice of initial condition, this fixed point leads to a solution for the scale factor of the collapsing system as $a^{42}(\phi) = 42\sqrt{\beta/3}\ln|\phi_s/\phi|$, where ϕ_s is a constant of integration. Then, it is seen that, when the tachyon field approaches the finite value as $\phi \to \phi_s$, the scale factor vanishes as $a \to 0$. On the other hand, the potential of tachyon field decreases from its initial value, and approaches a finite value at $a = 0$. Furthermore, the effective energy density which is given by $\rho_\phi^{loop} \approx V(\phi) = \beta/\phi^2$ does not blow up, becoming instead small and remaining finite. On the other hand, the effective mass function is $F^{loop}/R \approx (\beta/3\phi^2)r^2a^2$ and vanishes as $a \to 0$; that is, no trapped surfaces form for collapsing shells. Moreover, the effective pressure in Eq. (9) evolves such that $p_\phi^{loop} \approx -\rho_\phi^{loop} = -\beta/\phi^2$, remaining finite and negative, giving rise to an outward flux of energy in semiclassical regime.

Semiclassical analysis for the Kretschmann scalar shows that, it reaches a local maximum as the singularity is approached and then converges to finite values. Consequently this behavior can be interpreted as an effective singularity avoidance.

3. Conclusion

In this paper we considered the marginally bound FRW as a spherically symmetric model for an homogeneous interior spacetime of a gravitational collapsing system constituted by a tachyon matter field with the potential of an inverse square form. We investigated the classical and semiclassical regimes which the latter characterized by inverse triad corrections in order to establish the corresponding final state. Within the classical setting, we found analytically a situation where the tachyon evolved towards a dust-like asymptotic behavior; this suggested a black hole forming. In semiclassical regime, loop quantum corrections (an inverse triad type), pointed to the regularity of the spacetime geometry and the energy density of the system, where classical trapped surfaces were absent therein. These conditions indicated that, the spacetime in semiclassical collapse is singularity-free, hence there is neither a naked singularity nor a black hole forming as final state of collapse. In other words, loop quantum corrections of the inverse triad type seem to induce an outward flux of energy at the final state of the collapse.

4. Acknowledgments

YT is supported by the Portuguese Agency Fundação para a Ciência e Tecnologia (FCT) through the fellowship SFRH/BD/43709/2008. He also would like to thank M. Bojowald for the useful discussion. This research work was also supported by the grant CERN/FP/109351/2009 and CERN/FP/116373/2010.

5. References

[1] Y. Tavakoli, P. Vargas Moniz, J. Marto, and A. H. Ziaie, Submitted to JCAP [arXiv: gr-qc/0911.1428].
[2] P. Joshi, *Gravitational Collapse and Space-Time Singularities*, (Cambridge University Press, 2007).
[3] A. A. Sen, Phys. Rev. D **74**, 043501 (2006).
[4] J. R. Oppenheimer, and H. Snyder, Phys. Rev. **56**, 455 (1939).
[5] M. Bojowald, Class. Quant. Grav. **19**, 5113 (2002).

Journal of Physics: Conference Series **360** (2012) 012017 doi:10.1088/1742-6596/360/1/012017

The geometric role of symmetry breaking in gravity

Derek K. Wise

Institute for Theoretical Physics III, University of Erlangen–Nürnberg, Staudtstr. 7/B2, 91054 Erlangen, Germany

E-mail: `wise@theorie3.physik.uni-erlangen.de`

Abstract. In gravity, breaking symmetry from a group G to a group H plays the role of describing geometry in relation to the geometry of the homogeneous space G/H. The deep reason for this is Cartan's 'method of equivalence,' giving, in particular, an exact correspondence between metrics and Cartan connections. I argue that broken symmetry is thus implicit in any gravity theory, for purely geometric reasons. As an application, I explain how this kind of thinking gives a new approach to Hamiltonian gravity in which an observer field spontaneously breaks Lorentz symmetry and gives a Cartan connection on space.

The success of spontaneous symmetry breaking in condensed matter and particle physics is famous. It explains second order phase transitions, superconductivity, the origin of mass via the Higgs mechanism, why there must be at least three generations of quarks, and so on. These applications are by now standard material for modern textbooks.

Much less famous is this: broken symmetry *links the geometry of gauge fields to the geometry of spacetime.* This, in my view, is the main role of symmetry breaking in gravity.

One clue came in 1977, when MacDowell and Mansouri wrote down an action for general relativity using a connection for the (anti-) de Sitter group, but invariant only under the Lorentz group [1]. Though their work was surely inspired by spontaneous symmetry breaking, it was Stelle and West [2] who first made the action fully gauge invariant, breaking the symmetry dynamically using a field y locally valued in (anti-) de Sitter space.

Whether one breaks the symmetry dynamically or 'by hand,' the broken symmetry of the MacDowell–Mansouri connection plays the geometric role of relating spacetime geometry to the geometry of de Sitter space. This is best understood using *Cartan geometry*, a generalization of Riemannian geometry originating in the work of Élie Cartan, in which the geometry of tangent spaces is generalized—in this case, they become copies of de Sitter space [3]. But to explain how this works, and how symmetry breaking is involved, it helps to back up further.

In geometry, inklings of spontaneous symmetry breaking date from at least 1872, in the work of Felix Klein. Ironically, to study a homogeneous space Y, with symmetry group G, one first *breaks* its perfect symmetry, artificially giving special significance to some point $y \in Y$. This gives an isomorphism $Y \cong G/G_y$ as G-spaces, where G_y is the stabilizer of y, allowing algebraic study of the geometry. While Y itself has G symmetry, this *description* of it is only invariant under the subgroup G_y. Different algebraic descriptions of Y are however related in a G-equivariant way, since $G_{gy} = gG_yg^{-1}$.

This is strikingly similar to spontaneous symmetry breaking in physics. There, one really has a family of minimum-energy states, related in a G-equivariant way under the original gauge group G. Singling out any particular state $|0\rangle$ as 'the' vacuum breaks symmetry to $G_{|0\rangle}$.

Journal of Physics: Conference Series **360** (2012) 012017 doi:10.1088/1742-6596/360/1/012017

Cartan took Klein's ideas a dramatic step further, getting an algebraic description of the geometry of a *non*homogeneous manifold M, by relating it 'infinitesimally' to one of Klein's geometries Y. Just as Klein geometry uses broken symmetry to get an isomorphism $Y \to G/G_y$, in Cartan geometry, the broken symmetry in a G connection induces an isomorphism $e\colon T_xM \to \mathfrak{g}/\mathfrak{g}_y$ for each tangent space. This is just the coframe field, also called the **soldering form** since identifying T_xM with $\mathfrak{g}/\mathfrak{g}_y \cong T_yY$ effectively solders a copy of Y to M, at each point x. These copies of Y are then related via holonomy of the Cartan connection, which can be viewed as describing 'rolling Y along M without slipping' [3] (see also [5, Appendix B]).

Physics history unfortunately skips over Cartan geometry. The Levi-Civita connection is adequate for the standard metric formulation of general relativity, and more general kinds of connections played no conspicuous role in physics until some time later. When these eventually were introduced in Yang–Mills theory, they served a purpose far removed from spacetime geometry. Yang–Mills gauge fields are really just the principal connections of Ehresmann, who, building on Cartan's ideas, liberated connections from their bondage to classical geometry. Ehresmann's definition, which lacks the crucial 'broken symmetry' in Cartan's original version, has just the flexibility needed for gauge fields in particle physics, which are concerned only with the geometry of an abstract 'internal space'—a bundle over spacetime, rather than spacetime itself. On the other hand, Cartan's original version is better when it comes to studying gravity.

Concretely, a Cartan geometry may be thought of as a connection on a principal bundle (with Ehresmann's now standard definition) *together with* a section y of the associated Y bundle. As an example, let us write a version of the MacDowell–Mansouri action, using de Sitter space $Y \cong G/H = \mathrm{SO}(4,1)/\mathrm{SO}(3,1)$ as the corresponding Klein geometry:

$$I[A,y] = \int \mathrm{tr}(F_y \wedge \star_y F_y)$$

The Cartan connection (A, y) consists of an $\mathrm{SO}(4,1)$ connection A and a locally de Sitter-valued field y. F is the curvature of A, calculated by the usual formula, and F_y is its $\mathfrak{g}_y$-valued part, where $\mathfrak{g}_y \cong \mathfrak{so}(3,1) \cong \Lambda^2\mathbb{R}^{3,1}$ has Hodge star operator $\star_y$. The easiest way to check that this recovers general relativity is by gauge fixing y to some constant value: F_y then takes the form $R + e \wedge e$, where the connection A splits into an $\mathrm{SO}(3,1)$ connection with curvature R and a complementary piece e that can be interpreted as a coframe field. The action reduces to the Palatini action with positive cosmological constant. [2, 3]

I have described additional examples of Cartan-geometric formulations of various gravity theories elsewhere [4], and there are many more. But besides the diversity of specific examples, there are deep reasons that gravity, or any related 'gauge theory of geometry,' *should* be framed in the language of Cartan geometry. This is the subject of geometric 'equivalence theorems.'

In fact, if one believes semi-Riemannian metrics are fundamental in classical gravity, one is forced to accept Cartan connections as equally fundamental. The reason for this is Cartan's **method of equivalence**, a process for proving that specified kinds of 'raw geometric data' are equivalent to corresponding types of Cartan geometry [6]. In the case of Riemannian geometry, solving the 'equivalence problem' leads to the following theorem:

Theorem. *A Riemannian metric determines a unique torsion-free Cartan geometry modeled on Euclidean space; conversely, a torsion-free Cartan geometry modeled on Euclidean space determines a Riemannian metric up to overall scale (on each connected component).*

Proof. See Sharpe's book [5]. ■

Physically, the 'overall scale' in the converse just represents a choice of length unit. One can also show that deformed versions (or 'mutations') of Euclidean geometry, namely hyperbolic and spherical geometry, lead to Cartan geometries that carry the same information [5]. The

Lorentzian analogs of these results are the real reason de Sitter and anti de Sitter geometries work in MacDowell–Mansouri gravity.

Riemannian geometry is but one application of the equivalence method. There are analogous theorems, for example, in conformal geometry or Weyl geometry, relating various types of conformal structures to Cartan geometries that take the model Y to be an appropriate kind of homogeneous conformal model. Sharpe's book [5] contains some such theorems, and some significant work has been done on applications of conformal Cartan geometry—which often goes by the name 'tractor calculus'—in physics. (See, e.g. [7] and references therein.)

For now, I just want to describe one more application of Cartan-geometric thinking in gravitational theory. Besides spacetime geometry, one can also use Cartan's ideas to describe the geometry of *space*.

Wheeler's term 'geometrodynamics' originally referred to the of evolution of spatial geometries in the metric sense. This has sometimes been contrasted with 'connection dynamics' [9]. In light of the above equivalence theorem, however, there seems little point in establishing any technical distinction between geometrodynamics and connection dynamics, at least if we mean connections in Cartan's sense. The metric and connection pictures have their own advantages, but the equivalence theorem suggests we *should* be able to translate exactly between the two.

In recent work with Steffen Gielen [8], we take an explicitly Cartan-geometric approach to evolving spatial geometries. In this case, the symmetry breaking field y lives in 3d hyperbolic space $\mathrm{SO}(3,1)/\mathrm{SO}(3)$, and can be interpretated as a **field of observers**, since the spacetime coframe field converts it into a unit timelike vector field. This can be dualized via the metric to a unit *covector* field, which we might call a field of **co-observers**. Just as observers determine a local time direction, co-observers determine local space directions, by taking their kernel. Our strategy in the Hamiltonian formulation is to fix a field of co-observers—the infinitesimal analog of picking a spacetime folitation—but let the field of observers be determined dynamically, as part of determining the metric.

The result is a model in which the observer field plays a two-part symmetry breaking role: first splitting spacetime fields into spatial and temporal parts, but then also acting as the symmetry breaking field in Cartan geometry of *space*. This gives a Cartan-geometric Hamiltonian framework in which the spatial fields fit neatly and transparently into their spacetime counterparts and transform in an equivariant way under local Lorentz symmetry.

Thanks to the equivalence theorem, this may be viewed as a concrete link between connection dynamics and geometrodynamics in the original sense.

It is conceivable that gravity descends from a more fundamental theory with larger gauge group, and so fits into the tradition of symmetry breaking in gauge theories. Such ideas are clearly worth pursuing (see, e.g. [10, 11]). At the same time, we should not ignore the lesson of Cartan geometry: broken symmetry is the means to establishing exact correspondence between geometric structures living on tangent spaces on one hand and connections on the other.

I refer the interested reader to the bibliography of [4] for additional Cartan geometry references. I would like to thank John Baez, Julian Barbour, James Dolan and Andy Randono for helpful discussions, and especially Steffen Gielen for collaboration on [8].

References

[1] S. W. MacDowell and F. Mansouri, Unified geometric theory of gravity and supergravity, *Phys. Rev. Lett.* **38**, 739 (1977). Erratum, *ibid.* **38**, 1376 (1977).

[2] K. S. Stelle and P. C. West, De Sitter gauge invariance and the geometry of the Einstein–Cartan theory, *J. Phys. A: Math. Gen.* **12** L205 (1979); Spontaneously broken de Sitter symmetry and the gravitational holonomy group, *Phys. Rev. D* **21**, 1466 (1980).

[3] D. K. Wise, MacDowell–Mansouri gravity and Cartan geometry, *Class. Quant. Grav.* **27**, 155010 (2010). arXiv:gr-qc/0611154.
[4] D. K. Wise, Symmetric space Cartan connections and gravity in three and four dimensions, *SIGMA* **5**, 080 (2009). arXiv:0904.1738.
[5] R. W. Sharpe, *Differential geometry: Cartan's Generalization of Klein's Erlangen Program* (Springer-Verlag, New York, 1997).
[6] R. B. Gardner, *The Method of Equivalence and its Applications* (Capital City Press, Montpelier, Vermont, 1989).
[7] A. Shaukat and A. Waldron, Weyl's gauge Invariance: conformal geometry, spinors, supersymmetry, and interactions, arXiv:0911.2477.
[8] S. Gielen and D. K. Wise, Spontaneous symmetry breaking for Hamiltonian gravity, arXiv:1111.7195.
[9] J. D. Romano, Geometrodynamics vs. connection dynamics, *Gen. Rel. Grav.* **25** (1993) 759-854. arXiv:gr-qc/9303032; K. Kuchař, Canonical quantum gravity, arXiv:gr-qc/9304012
[10] R. Percacci, Gravity from a particle physicist's perspective, *PoS ISFTG* **011** (2009). arXiv:0910.5167
[11] A. Randono, Gauge gravity: a forward-looking introduction, arXiv:1010.5822

Loops 11: Non-Perturbative / Background Independent Quantum Gravity IOP Publishing
Journal of Physics: Conference Series **360** (2012) 012018 doi:10.1088/1742-6596/360/1/012018

Complete loop quantization of a dimension 1+2 Lorentzian gravity theory

Rodrigo M S Barbosa, Clisthenis P Constantinidis, Zui Oporto and Olivier Piguet

Departamento de Física, Universidade Federal do Espírito Santo, Vitória, ES, Brazil

E-mail: rodrigo_martins@email.com, cpconstantinidis@gmail.com, azurnasirpal@gmail.com, opiguet@yahoo.com

Abstract. De Sitter Chern-Simons gravity in $D = 1 + 2$ spacetime is known to possess an extension with a Barbero-Immirzi like parameter. We find a partial gauge fixing which leaves a compact residual gauge group, namely SU(2). The compacticity of the residual gauge group opens the way to the usual LQG quantization techniques. We recall the exemple of the LQG quantization of SU(2) CS theory with cylindrical space topology, which thus provides a complete LQG of a Lorentzian gravity model in 3-dimensional space-time.

1. Introduction

The gauge group of D-dimensional Lorentzian spacetime, the Lorentz group SO(1,D-1), is *noncompact*. In the loop quantization framework[1], this is known to make difficult a proper definition of the internal product in the vector space of quantum states.

For space-time dimension 1 + 3, the usual way out is, starting from the Palatini-Holst action [2], to partially fix the gauge (the *"time gauge"*) in such a way that the residual group be the *compact* group SO(3) or SU(2). The presence of the *Barbero-Immirzi parameter* [3] γ appears to be crucial. However, for other dimensions, with the Lorentz group being SO(D,1), the time gauge does not lead to the compact gauge group SO(D). But a recent proposal has been given by the authors of Ref. [4], who have shown that it exists a Hamiltonian framework where the gauge group may be chosen to be compact, e.g., SO(D), even in the case of a Lorentzian theory with (1, $D - 1$) signature. In the latter case, however, as these authors have pointed out, their construction is not possible in a Lagrangian framework. The aim of the present talk is to show that a compact gauge group quantization does exist in a Lorentzian space-time of dimension 1+2 – where also the time gauge dos not help – provided there is a positive cosmological constant.

Our starting point is a formulation, due to Bonzom and Livine [5], of (1 + 2)-gravity with cosmological constant, in the presence of a Barbero-Immirzi like parameter. The gauge group of this theory is the de Sitter group SO(1,3), and we shall show the existence of a partial gauge-fixing which reduces the theory to a usual Chern-Simons theory with the *compact* SU(2) gauge group. One can then apply known results in order to quantize the theory. One may recall, e.g., the LQG results of [6], obtained by explicitly solving all the constraints, in a particular topology of 2-dimensional space.

[1] See [1] for general references on the loop quantization of General Relativity.

Journal of Physics: Conference Series **360** (2012) 012018 doi:10.1088/1742-6596/360/1/012018

Our case is somewhat different from that of the authors of [4] (second paper). On the one hand, their simplicity constraints are not needed in $1+2$ dimensions; on the other hand, we rely strongly on the existence of the Barbero-Immirzi like parameter of Bonzom and Livine, a feature very peculiar to that dimension. Moreover, here we have – and must have – a non-zero cosmological constant.

2. (2+1)-Gravity with a cosmological constant as a Chern-Simons theory with a Barbero-Immirzi parameter

The variables of 1+2 gravity in the first order formalism[2] are the triad forms $e^I = e^I_\mu dx^\mu$ and the Lorentz conections forms $\omega_I = \frac{1}{2}\varepsilon_{IJK}\,\omega^{JK} = \omega_{I\mu}dx^\mu$.

We suppose that we have a positive cosmological constant $\Lambda > 0$. The gauge invariance group of the theory is then the de Sitter gauge group SO(1,3). Note that if Λ were negative, the gauge group would be SO(2,2) and our procedure for obtaining a compact residual gauge group would fail. The basis for the Lie algebra so(1,3) is given by the three Lorentz generators $J^I \equiv \frac{1}{2}\varepsilon^{IJK}J_{JK}$ and the three "translation" generators P_I, obeying the commutation rules[3]

$$[J^I, J^J] = \varepsilon^{IJ}{}_K J^K\,, \quad [J^I, P_J] = \varepsilon^I{}_J{}^K P_K\,, \quad [P_I, P_J] = \Lambda\varepsilon_{IJK}J^K\,.$$

We shall denote by $A = \omega_I J^I + e^I P_I$ the SO(1,3) connection, obeying the gauge transformation rules $A'(x) = g^{-1}(x)dg(x) + g^{-1}(x)A(x)g(x)\,, g \in \mathrm{SO}(1,3)$ which, when written in terms of the component fields, reproduce the well-known de Sitter transformation rules.

Following [5], we shall start from the fact that the most general background independent and gauge invariant action depending only on A is of the Chern-Simons form and has two independent terms:

$$S = -\frac{\kappa}{2}\int_{\mathcal{M}}\langle A,\, dA + \frac{2}{3}A\,A\rangle_1 - \frac{\kappa}{\gamma}\int_{\mathcal{M}}\langle A,\, dA + \frac{2}{3}A\,A\rangle_2\,, \qquad (1)$$

($\wedge$ symbol omitted) where $\langle\cdot,\cdot\rangle_\alpha\,, \alpha = 1,2$ are the two invariant quadratic forms corresponding to the two quadratic Casimir operators of SO(1,3) [7]: $C_{(1)} = P_I J^I\,, C_{(2)} = \eta_{IJ}(\frac{1}{\Lambda}P^I P^J - J^I J^J)$. The parameter κ is proportional to the inverse of the gravitation constant, whereas γ is a parameter which shares with the usual Barbero-Immirzi parameter the property of not appearing in the classical field equations, which read, indeed:

$$F(A) = dA + A^2 = 0\,, \quad \text{or, in components:}$$

$$R^I \equiv d\omega^I + \varepsilon^I{}_{JK}\omega^J\omega^K = \frac{\Lambda}{2}\varepsilon^I{}_{JK}e^J e^K\,, \quad T^I \equiv de^I + \varepsilon^I{}_{JK}\omega^J e^K = 0\,.$$

One recognizes the Einstein equation with cosmological constant and the null torsion condition.

3. Decomposition of SO(1,3) in "rotations" and "boosts"

We shall now introduce new variables corresponding to a decomposition of SO(1,3) – which is isomorphic to the 4-dimensional Lorentz group – in "rotations" and "boosts". We call L_i the "rotation" generators and K_i the "boost" generators ($i = 1,2,3$).

$$[L_i, L_j] = \varepsilon_{ijk}L_k\,, \quad [L_i, K_j] = \varepsilon_{ijk}K_k\,, \quad [K_i, K_j] = -\varepsilon_{ijk}L_k\,.$$

[2] The tangent space indices $I, J, \cdots = 0,1,2$ are lowered and raised with the Minkowski metric $\eta_{IJ} = \mathrm{diag}(-1,1,1)$ and with its inverse η^{IJ}. $\mu, \nu, \cdots = t, x, y$ are world coordinates indices, and the space-time metric reads $g_{\mu\nu} = \eta_{IJ}e^I_\mu e^J_\nu$.

[3] The Levi Civita tensor ε_{IJK} is normalized as $\varepsilon_{012} = 1$. Moreover, $\varepsilon^{IJ}{}_K = \eta^{IM}\eta^{JN}\varepsilon_{MNK}$, etc. Beware of the signs!

Journal of Physics: Conference Series **360** (2012) 012018 doi:10.1088/1742-6596/360/1/012018

The new variables are the 1-forms A^i and B^i appearing in the expression of the SO(1,3) connection: $A = A^iL_i + B^iK_i$. One recognizes in A^i an SO(3) (or SU(2)) connection.

The relations between the old and new generators and variables read:

$$(L_1, L_2, L_3) = (P_2/\sqrt{\Lambda}, -P_1/\sqrt{\Lambda}, -J^0), \qquad (K_1, K_2, K_3) = (J^2, -J^1, P_0/\sqrt{\Lambda})$$
$$(A^1, A^2, A^3) = (\sqrt{\Lambda}e^2, -\sqrt{\Lambda}e^1, -\omega_0), \qquad (B^1, B^2, B^3) = (\omega_2, -\omega_1, \sqrt{\Lambda}e^0)$$

Our partial gauge fixing will consist in freezing the "boost" gauge degrees of freedom, keeping SU(2) as a residual gauge invariance.

Let us first write the action in the new variables[4]:

$$S = -\frac{\kappa}{2}\int_{\mathbb{R}} dt \int_{\Sigma} \left(\dot{\mathbf{B}}^i(\mathbf{A}^i + \mathbf{B}^i/\gamma) + \dot{\mathbf{A}}^i(\mathbf{B}^i - \mathbf{A}^i/\gamma)\right) - \int_{\mathbb{R}} dt\, H ,$$

where H is the Hamiltonian $H = \int d^2x\,(A_t\mathcal{G}_A(\mathbf{x}) + B_t\mathcal{G}_B(\mathbf{x}))$. The kinetic terms of the action determine the symplectic structure of the theory, or equivalently its Poisson bracket algebra, the non-vanishing brackets being:

$$\{A^i_a(\mathbf{x}), A^j_b(\mathbf{y})\} = \frac{1}{\kappa}\varepsilon_{ab}\delta^{ij}\frac{\gamma}{\gamma^2+1}\delta^2(\mathbf{x}-\mathbf{y}), \quad \{B^i_a(\mathbf{x}), B^j_b(\mathbf{y})\} = -\frac{1}{\kappa}\varepsilon_{ab}\delta^{ij}\frac{\gamma}{\gamma^2+1}\delta^2(\mathbf{x}-\mathbf{y}),$$
$$\{B^i_a(\mathbf{x}), A^j_b(\mathbf{y})\} = -\frac{1}{\kappa}\varepsilon_{ab}\delta^{ij}\frac{\gamma^2}{\gamma^2+1}\delta^2(\mathbf{x}-\mathbf{y}).$$

The Hamiltonian is purely constraints, the fields A_t and B_t playing the role of Lagrangian multipliers for the Gauss and curvature constraints

$$\mathcal{G}_A(\varepsilon) = \kappa\int_{\Sigma} \varepsilon^i \left(\mathbf{DB} - \frac{1}{\gamma}\left(\mathbf{F}(\mathbf{A}) - \frac{1}{2}\mathbf{B}\times\mathbf{B}\right)\right)^i \approx 0, \qquad \text{with } \mathbf{D} = \mathbf{d} + \mathbf{A}\times$$
$$\mathcal{G}_B(\eta) = \kappa\int_{\Sigma} \eta^i \left(\mathbf{F}(\mathbf{A}) - \frac{1}{2}\mathbf{B}\times\mathbf{B} + \frac{1}{\gamma}\mathbf{DB}\right)^i \approx 0, \qquad \text{and } \mathbf{F}(\mathbf{A}) = \mathbf{dA} + \tfrac{1}{2}\mathbf{A}\times\mathbf{A}.$$

These constraints are first class:

$$\{\mathcal{G}_A(\varepsilon)\ \mathcal{G}_A(\varepsilon')\} = \mathcal{G}_A(\varepsilon\times\varepsilon'), \ \{\mathcal{G}_A(\varepsilon)\ \mathcal{G}_B(\eta)\} = \mathcal{G}_B(\varepsilon\times\eta), \ \{\mathcal{G}_B(\eta)\ \mathcal{G}_B(\eta')\} = -\mathcal{G}_A(\eta\times\eta'),$$

and generate the gauge transformations:

$$\{\mathcal{G}_A(\varepsilon), \mathbf{A}\} = \mathbf{D}\varepsilon, \quad \{\mathcal{G}_B(\eta), \mathbf{A}\} = -\eta\times\mathbf{B}, \quad \{\mathcal{G}_A(\varepsilon), \mathbf{B}\} = \varepsilon\times\mathbf{B}, \quad \{\mathcal{G}_B(\eta), \mathbf{B}\} = \mathbf{D}\eta.$$

4. Partial gauge fixing

We choose to impose the axial-like gauge condition $B^i_y = 0$, i.e., $\omega_{1y} = \omega_{2y} = e^0_y = 0$, adding to the Hamiltonian a term $H \to H + \int_\Sigma d^2x\,\mu^i B^i_y$, the field μ being a Lagrange multiplier. It turns out that the constraints $B^i_y \approx 0$ and $\mathcal{G}_B \approx 0$ are second class, whereas $\mathcal{G}_A \approx 0$ remains first class. Following Dirac [8], we introduce the corresponding Dirac brackets, and treat the second class constraints as strong equalities. In particular, $\mathcal{G}_B = 0$ yields $D_yB^i_x - \gamma F^i_{xy} = 0$, which implies that B^i_x is not an independent field.

The set o independent dynamical variables reduces to the pairs of conjugate variables A^i_x and A^i_y, whose Dirac brackets read

$$\{A^i_x(\mathbf{x}), A^j_y(\mathbf{x}')\}_{\mathrm{D}} = \frac{1}{\kappa}\delta^{ij}\frac{\gamma}{\gamma^2+1}\delta^2(\mathbf{x}-\mathbf{x}').$$

[4] As usual we assume that space-time $\mathcal{M}$ may be split as $\mathbb{R}\times\Sigma$, where Σ is the 2-dimensional space manifold. Boldface letters mean forms, etc. in Σ.

With the field redefinition $\mathcal{A}_x^i = A_x^i - \gamma B_x^i$, $\mathcal{A}_y^i = A_y^i$, the brackets and the hamiltonian read

$$\{\mathcal{A}_x^i(\mathbf{x}), \mathcal{A}_y^j(\mathbf{x}')\}_{\rm D} = \frac{\gamma}{\kappa}\delta^{ij}\delta^2(\mathbf{x}-\mathbf{x}') \quad \text{and} \quad H = -\frac{\kappa}{\gamma}\int_\Sigma d^2x\, A_t^i F_{xy}^i(\mathcal{A})\,. \tag{2}$$

These are the Hamiltonian and brackets of a Chern-Simons theory for the *compact* gauge group SU(2).

5. Quantization

The problem of the quantization of the present theory is thus reduced to that of the Chern-Simons theory with gauge group SU(2). We may refer for this to the literature [7, 9, 10]. The residual gauge symmetry group SU(2) being compact, LQG methods apply. The case of 2-dimensional space being a cylinder was treated in [6], with a complete solution of the constraints.

6. Conclusions

We have thus succeeded to reduce the gauge symmetry to that of a compact group, namely SU(2), through a suitable gauge fixing. We note that the presence of a Holst-like term in the action (second term in (1)), together with the Barbero-Immirzi-like parameter γ, is crucial, much in the same way as in the $D = 3+1$ case, where the gauge symmetry of the Palatini-Holst action [2] is reduced to the same SU(2) through a time gauge fixing – which is only available in that dimension!

A complete quantization of 3 dimensional gravity with a positive cosmological constant can thus be performed in the usual loop quantization scheme, thanks to the compactness of SU(2).

A more detailed account will appear in [11].

References

[1] C. Rovelli, "Quantum Gravity", Cambridge Monography on Math. Physics (2004); A. Ashtekar and J. Lewandowski, "Background independent quantum gravity: A status report", *Class. Quantum Grav.* 21 (2004) R53) [arXiv:gr-qc/0404018]; T. Thiemann, "Modern Canonical Quantum General Relativity", Cambridge Monographs on Mathematical Physics (2008); M. Han, W. Huang and Y. Ma "Fundamental structure of loop quantum gravity", *Int. J. Mod. Phys.* D16 (2007) 1397, [arXiv:gr-qc/0509064]; H. Nicolai, K. Peeters and M. Zamaklar, "Loop quantum gravity: An outside view", *Class. Quantum Grav.* 22 (2005) R193, [arXiv:hep-th/0501114]; H. Nicolai and K. Peeters, "Loop and spin foam quantum gravity: A brief guide for beginners", [arXiv:gr-qc/0601129]; T. Thiemann, "Loop quantum gravity: An inside view", [arXiv:hep-th/0608210].

[2] S. Holst, "Barbero's Hamiltonian derived from a generalized Hilbert-Palatini action", Phys.Rev. D53 (1996) 5966-5969, [arXiv: gr-qc/9511026].

[3] J.F. Barbero, "Reality conditions and Ashtekar variables: A Different perspective", Phys. Rev. D51 (1995) 5507, [arXiv: gr-qc/9410013]; Giorgio Immirzi, "Real and complex connections for canonical gravity", Class.Quant.Grav. 14 (1997) L177-L181, [arXiv: gr-qc/9612030].

[4] N. Bodendorfer, T. Thiemann and A. Thurn "New Variables for Classical and Quantum Gravity in all Dimensions I,II,III and IV", [arXiv:1105.3703 - 1105.3706 [gr-qc]]; N. Bodendorfer, in these proceeedings; T. Thiemann, in these proceeedings; A, Thurn, in these proceeedings.

[5] V. Bonzom and E.R. Livine, "A Immirzi-like parameter for 3d quantum gravity" *Class. Quantum Grav.* 25 (2008) 195024, [arXiv:0801.4241[gr-qc]]

[6] C.P. Constantinidis, G. Luchini and O. Piguet, "The Hilbert space of Chern-Simons theory on the cylinder. A Loop Quantum Gravity approach", *Class. Quantum Grav.* 27 (2010) 065009, [arXiv:0907.3240[gr-qc]].

[7] E. Witten; "2 + 1 dimensional Gravity as an Exactly Soluble system", *Nucl. Phys.* B311 (1988) 46.

[8] P.A.M. Dirac, "Lectures on Quantum Mechanics", Dover, 2001; M. Henneaux, C. Teitelboim, "Quantization of Gauge Systems", Princeton University Press, 1994.

[9] E.Witten "Quantum Field Theory and the Jones Polynomial, *Commun. Math. Phys.* 121 (1989) 351.

[10] G.V. Dunne, R. Jackiw, C.A. Trugenberger, "Chern-Simons Theory in the Schrödinger Representation", *Ann. Phys. (N.Y.)* 194 (1989) 197; E.Guadagnini, M.Martellini, M.Mintchev, "Braids and Quantum Group Symmetry in Chern-Simons Theory, *Nucl. Phys.* B336 (1990) 581; Steven Carlip, "Quantum Gravity in 2+1 Dimensions", Cambridge Monographs on Mathematical Physics (2003).

[11] R.M.S. Barbosa, C.P. Constandinis, Z. Oporto and O. Piguet, in preparation.

Journal of Physics: Conference Series **360** (2012) 012019 doi:10.1088/1742-6596/360/1/012019

Dynamics for a simple graph using the U(N) framework for loop quantum gravity

Enrique F. Borja[1,2], Jacobo Díaz-Polo[3], Laurent Freidel[4], Iñaki Garay[1], Etera R. Livine[4,5]

[1] Institute for Theoretical Physics III, University of Erlangen-Nürnberg, Staudtstraße 7, D-91058 Erlangen (Germany).
[2]Departamento de Física Teórica and IFIC, Centro Mixto Universidad de Valencia-CSIC. Facultad de Física, Universidad de Valencia, Burjassot-46100, Valencia (Spain).
[3]Department of Physics and Astronomy, Louisiana State University. Baton Rouge, LA, 70803-4001.
[4]Perimeter Institute for Theoretical Physics, 31 Caroline St N, Waterloo ON, Canada N2L 2Y5.
[5]Laboratoire de Physique, ENS Lyon, CNRS-UMR 5672, 46 Allée d'Italie, Lyon 69007, France.

E-mail: Enrique.Fernandez@uv.es, jacobo@phys.lsu.edu, lfreidel@perimeterinstitute.ca, igael@theorie3.physik.uni-erlangen.de, etera.livine@ens-lyon.fr

Abstract. The implementation of the dynamics in loop quantum gravity (LQG) is still an open problem. Here, we discuss a tentative dynamics for the simplest class of graphs in LQG: Two vertices linked with an arbitrary number of edges. We find an interesting global U(N) symmetry in this model that selects the homogeneous/isotropic sector. Then, we propose a quantum Hamiltonian operator for this reduced sector. Finally, we introduce the spinor representation for LQG in order to propose a classical effective dynamics for this model.

1. Introduction

Loop quantum gravity (LQG) proposes a non-perturbative mathematical formulation of the kinematics of quantum gravity. The Hilbert space is generated by states defined over oriented graphs whose edges are labeled by irreducible representations of the SU(2) group and whose vertices are decorated with intertwiners (SU(2) invariant tensors). These are the so-called spin networks. Despite the several advances that have taken place in this field, one of the main challenges faced by the theory is the systematic implementation of the dynamics. Our goal is to focus on a specific model in order to propose a suitable Hamiltonian for it.

We use the U(N) framework for SU(2)-intertwiners [1, 2, 3] to study the spin network Hilbert space of the 2-vertex graph (2 nodes joined by an arbitrary number N of links) from a new point of view. We identify a global symmetry that selects a homogeneous and isotropic sector of this system [4] and we construct the operators that leave this sector invariant. They will be the building blocks to construct the Hamiltonian operator.

On the other hand, the recent spinor representation for LQG [5, 6, 7] opens a new way to study several aspects of LQG. We apply this new formalism to the 2-vertex graph and we propose a classical action with an interaction term which encodes the effective dynamics of this system. This interaction term is, indeed, the classical counterpart of the quantum Hamiltonian obtained within the U(N) framework.

2. The U(N) framework

The U(N) framework introduced in [1, 2] is very useful to study the Hilbert space of intertwiners with N legs and to build appropriate semi-classical states [3]. The basic tool is the Schwinger representation of the $\mathfrak{su}(2)$ Lie algebra in terms of a pair of harmonic oscillators a and b:

$$J_z = \frac{1}{2}(a^\dagger a - b^\dagger b), \quad J_+ = a^\dagger b, \quad J_- = ab^\dagger .$$

Labeling the N legs with the index i, we identify SU(2) invariant operators acting on pairs of (possibly equal) legs i, j [1, 3]:

$$E_{ij} = a_i^\dagger a_j + b_i^\dagger b_j, \quad (E_{ij}^\dagger = E_{ji}), \qquad F_{ij} = a_i b_j - a_j b_i, \quad (F_{ji} = -F_{ij}).$$

The operators E form a $\mathfrak{u}(N)$-algebra and they also form a closed algebra together with the operators $F, F^\dagger$. Notice that the diagonal operators give the energy on each leg, $E_{ii} = E_i$, which gives twice the spin j_i of the $\mathfrak{su}(2)$ representation carried by that leg. This spin j_i is identified geometrically as the area associated to the leg i and the total energy $E = \sum_i E_i$ gives twice the total area $J = \sum_i j_i$ associated to the intertwiner. The E_{ij}-operators change the energy/area carried by each leg, while still conserving the total energy, while the operators F_{ij} (resp. $F_{ij}^\dagger$) will decrease (resp. increase) the total area E by 2:

$$[E, E_{ij}] = 0, \qquad [E, F_{ij}] = -2F_{ij}, \quad [E, F_{ij}^\dagger] = +2F_{ij}^\dagger.$$

The operators E_{ij} allow then to navigate from state to state within each subspace $\mathcal{H}_N^{(J)}$ of N-valent intertwiners with fixed total area J; and the operators $F_{ij}^\dagger$ and F_{ij} allow to go from one subspace $\mathcal{H}_N^{(J)}$ to the next $\mathcal{H}_N^{(J\pm1)}$, thus endowing the full space of N-valent intertwiners with a Fock space structure with creation operators $F_{ij}^\dagger$ and annihilation operators F_{ij}. Besides, it was proven [2] that each subspace $\mathcal{H}_N^{(J)}$ carries an irreducible representation of U(N) generated by the E_{ij} operators. Finally, it is worth pointing out that the operators $E_{ij}, F_{ij}, F_{ij}^\dagger$ satisfy certain quadratic constraints, which correspond to a matrix algebra [4].

3. The 2 vertex model and the quantum Hamiltonian

We consider the simplest class of non-trivial graphs for spin network states in LQG: a graph with two vertices linked by N edges, as shown in fig.1.

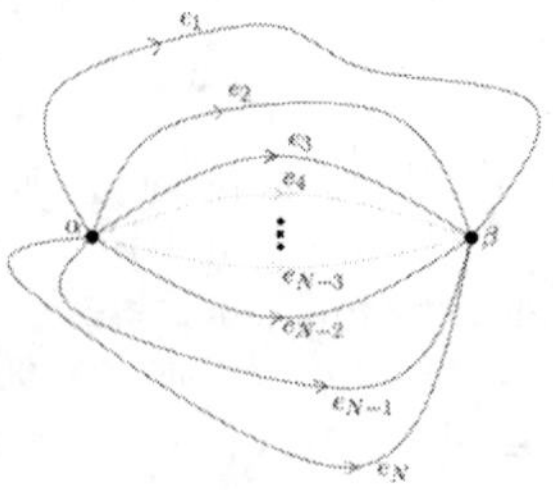

Figure 1. The 2-vertex graph with vertices α and β and the N edges linking them.

Journal of Physics: Conference Series **360** (2012) 012019 doi:10.1088/1742-6596/360/1/012019

There are matching conditions [2] imposing that each edge carries a unique SU(2) representation (same spin seen from α and from β). This translates into an equal energy condition:

$$\mathcal{E}_i \equiv E_i^{(\alpha)} - E_i^{(\beta)} = 0.$$

The only quadratic operators commuting with these conditions are: $e_{ij} \equiv E_{ij}^{(\alpha)} E_{ij}^{(\beta)}, f_{ij} \equiv F_{ij}^{(\alpha)} F_{ij}^{(\beta)}$ and $f_{ij}^{\dagger}$. Their algebra generates all the operators that, on the one hand, are compatible with the matching conditions and, on the other hand, define deformations of the geometry of the 2-vertex graph (which can be interpreted as the discrete equivalent of diffeomorphisms on this model).

The matching constraints $\mathcal{E}_k$ turn out to be part of a larger U(N) symmetry algebra. Indeed, we introduce the more general operators:

$$\mathcal{E}_{ij} \equiv E_{ij}^{(\alpha)} - E_{ji}^{(\beta)} = E_{ij}^{(\alpha)} - (E_{ij}^{(\beta)})^{\dagger},$$

that form a U(N) algebra and that reduce to the matching conditions in the case $i = j$. Now, one can show [4] that by imposing the global U(N)-invariance generated by the $\mathcal{E}_{ij}$'s on our 2-vertex system, one obtains a single state $|J\rangle$ for each total boundary area J. Thus, the U(N) invariance is restricting our system to states that are homogeneous and isotropic (the quantum state is the same at every point of space, i.e. at α and β, and all directions or edges are equivalent).

The dynamics for the 2-vertex model or, more generally, on a fixed graph[1], can be constructed in various ways: building it out of the deformation operators generating the discrete equivalent of diffeomorphisms, use the discretized and regularized Hamiltonian of loop gravity based on holonomy operators, derive an effective dynamics from spinfoam transition amplitudes, take the Hamiltonian from mini-superspace models of general relativity... In our simple case of the 2-vertex graph, all these approaches have been understood to lead to the same ansatz, made of the only three U(N)-invariant observables (up to renormalization by area factors which correspond to a choice of densitization, i.e., choice of time/lapse):

$$H \equiv \lambda \sum_{ij} E_{ij}^{(\alpha)} E_{ij}^{(\beta)} + \left(\sigma \sum_{ij} F_{ij}^{(\alpha)} F_{ij}^{(\beta)} + \bar{\sigma} \sum_{ij} F_{ij}^{\alpha\dagger} F_{ij}^{\beta\dagger} \right), \tag{1}$$

where the coupling λ is real while σ can be complex a priori, so that the operator H is Hermitian.

We studied the properties of this Hamiltonian on the U(N) invariant Hilbert space. Its action on states $|J\rangle$ is known and its spectral properties have been analyzed [4]. It turns out that it shares several mathematical analogies with the evolution operator used in loop quantum cosmology (LQC). In fact, these analogies are exact as soon as we replace the role of the area λ by the volume in LQC. At the physical level, interpreted as a cosmological model, this simple dynamical 2-vertex model describes a regularized dynamics (with big bounce) of homogeneous and isotropic FRW cosmology with curvature (with a massless scalar field used to deparameterize the system). It is a very interesting open issue to couple non-trivial matter fields to gravity in our approach.

4. Spinors and effective dynamics

Based on the Schwinger representation of SU(2) in terms of harmonic oscillators, it is possible to give a representation of the classical phase of LQG in terms of spinor variables [5, 7]. The

[1] The methods presented here can be generalized to more complicated graphs. The study of the graph with $3+N$ vertices presented in [4] could allow us to implement rotations in this kind of systems or study radiation processes in a black hole context. Other interesting generalization could be to study the continuum limit (infinite number of nodes) in a graph constructed with 4-valent vertices joined by a central loop.

Journal of Physics: Conference Series **360** (2012) 012019 doi:10.1088/1742-6596/360/1/012019

quantization of this classical system will lead back to the U(N) framework for intertwiners. Focusing on this classical system, we write an action principle with an effective dynamics of the spinors reflecting the quantum dynamics defined above.

Let us start by introducing the usual spinor notation. Let us define the spinor z and its dual:

$$|z\rangle = \begin{pmatrix} z^0 \\ z^1 \end{pmatrix} \in \mathbb{C}^2, \qquad \langle z| = \begin{pmatrix} \bar{z}^0 & \bar{z}^1 \end{pmatrix}, \qquad |z] \equiv \begin{pmatrix} -\bar{z}^1 \\ \bar{z}^0 \end{pmatrix}.$$

In order now to describe N-valent intertwiners, we consider N spinors z_i satisfying a closure condition[2] that, in terms of their components, is given by:

$$\sum_i |z_i\rangle\langle z_i| \propto \mathbb{I} \Leftrightarrow \sum_i z_i^0\,\bar{z}_i^1 = 0, \quad \sum_i \left|z_i^0\right|^2 = \sum_i \left|z_i^1\right|^2 = \frac{1}{2}\sum_i \langle z_i|z_i\rangle. \tag{2}$$

Solutions are parameterized in terms of a positive number $\lambda \in \mathbb{R}_+$ and a unitary matrix $u \in \mathrm{U}(N)$ up to $\mathrm{U}(N-2) \times \mathrm{SU}(2)$ right-transformations with $z_i^0 = \sqrt{\lambda}\, u_{i1}$ and $z_i^1 = \sqrt{\lambda}\, u_{i2}$.

The phase space is defined by the canonical Poisson bracket $\{z_i^a, \bar{z}_j^b\} \equiv i\,\delta^{ab}\delta_{ij}$. The quantization will be promoting z_i and $\bar{z}_i$ as the annihilation and creation operators of harmonic oscillators. Then the classical matrices $M_{ij} = \langle z_i|z_j\rangle$ and $Q_{ij} = [z_j|z_i\rangle$ are the classical counterparts of the operators E and F.

The U(N)-action on spinors is the simple $N\times N$ matrix action $(Uz)_i = \sum_j U_{ij} z_j$. Defining the "homogeneous cosmological" sector as the U(N)-invariant sector, satisfying $\langle z_i^\alpha|z_j^\alpha\rangle = \langle z_i^\beta|z_j^\beta\rangle$ and invariant under $z^\alpha, z^\beta \to Uz^\alpha, \bar{U}z^\beta$, imposes that all the α-spinors are equal to the β-spinors up to a global phase, $\bar{z}_i^{(\alpha)} = e^{i\phi}\, z_i^{(\beta)}$. And we get a reduced phase space with two parameters, the total area λ and its conjugate angle ϕ encoding the curvature. Our ansatz for the dynamics of this "cosmological" sector is:

$$S_{inv}[\lambda, \phi] = -2\int dt\, \left(\lambda\partial_t\phi - \lambda^2\left(\gamma^0 - \gamma^+ e^{2i\phi} - \gamma^- e^{-2i\phi}\right)\right), \tag{3}$$

which corresponds to the quantum Hamiltonian defined above. In this classical case, the equations of motion can be solved exactly [5] with certain interesting analogies with (the effective dynamics of) loop quantum cosmology, showing that the dynamics of the U(N)-invariant sector of the 2-vertex graph model can be interpreted as describing homogeneous and isotropic cosmology.

5. Conclusions

The U(N) framework and the spinor representation introduced and studied in [1, 2, 3, 4, 5, 6, 7] represents a new and refreshing way to tackle several important issues in loop quantum gravity.

In this work we have discussed these new frameworks and we have reviewed a proposal for the dynamics of the homogeneous and isotropic sector of the model, both at the quantum (using the U(N) framework) and the classical (using spinors) level. In this process, we have described the main features of the U(N) framework, like the Fock space structure of the Hilbert space of intertwiners with N legs. We further used this U(N) structure on the 2-vertex graph to define a symmetry reduction to the homogeneous and isotropic sector. We have then introduced a Hamiltonian consistent with this symmetry reduction, which can be solved exactly and shown to be analogous with the dynamics of loop quantum cosmology.

[2] We associate to each spinor z_i a 3-vector $\vec{X}(z_i) = \langle z_i|\vec{\sigma}|z_i\rangle$ by projecting it onto the Pauli matrices. Then the closure constraint is $\sum_i \vec{X}(z_i) = 0$ and we identify $\vec{X}(z_i)$ as the normal vector to the dual surface to the leg i.

Acknowledgments
This work was in part supported by the Spanish MICINN research grants FIS2008-01980, FIS2009-11893 and ESP2007-66542-C04-01 and by the grant NSF-PHY-0968871. IG is supported by the Department of Education of the Basque Government under the "Formación de Investigadores" program.

References
[1] F. Girelli and E.R. Livine. *Reconstructing Quantum Geometry from Quantum Information: Spin Networks as Harmonic Oscillators*, Class.Quant.Grav. 22 (2005) 3295 [arXiv:gr-qc/0501075].
[2] L. Freidel and E.R. Livine. *The Fine Structure of* SU(2) *Intertwiners from* U(N) *Representations*, J.Math.Phys. 51 (2010) 082502 [arXiv:0911.3553].
[3] L. Freidel and E.R. Livine. U(N) *coherent states for Loop Quantum Gravity*, J.Math.Phys. 52 (2011) 052502 [arXiv:1005.2090].
[4] E.F. Borja, J. Diaz-Polo, I. Garay and E.R. Livine. *Dynamics for a 2-vertex Quantum Gravity Model*, Class.Quant.Grav.27 (2010) 235010 [arXiv:1006.2451v2].
[5] E.F. Borja, L. Freidel, I. Garay and E.R. Livine. *U(N) tools for Loop Quantum Gravity: The Return of the Spinor*, Class.Quant.Grav. 28 (2011) 055005 [arXiv:1010.5451].
[6] Laurent Freidel and Simone Speziale. *From twistors to twisted geometries*, Phys.Rev. D82 (2010) 084041 [arXiv:1006.0199v1].
[7] Etera R. Livine and Johannes Tambornino. *Spinor Representation for Loop Quantum Gravity*. J.Math.Phys. 53 (2012) 012503 [arXiv:1105.3385v1].

Journal of Physics: Conference Series **360** (2012) 012020 doi:10.1088/1742-6596/360/1/012020

SU(2) gauge symmetry in gravity phase space

Francesco Cianfrani.
Università di Roma "Sapienza", Piazzale Aldo Moro 5, 00185, Roma.
E-mail: `francesco.cianfrani@icra.it`

Abstract. The Hamiltonian formulation of the Holst action in vacuum and in the presence of matter fields is analyzed in a generic local Lorentz frame. It is elucidated how the SU(2) gauge symmetry is inferred by reducing the set of constraints to a first-class one. The consequences of the proposed approach for Loop Quantum Gravity and Spin Foam models are discussed.

1. Introduction

The reduction of local Lorentz symmetry to SU(2) gauge invariance is the starting point for the whole Loop Quantum Gravity (LQG) framework and it constitutes a key point in order to describe the quantum behavior of the gravitational field via BF theories in Spin-Foam models. In this work we are going to review the procedure by which the Gauss constraints of the SU(2) group arise without fixing the local Lorentz frame in a Hamiltonian framework and, then, we will discuss the implications for BF theories.

2. Hamiltonian formulation in a generic local Lorentz frame

Let us consider a space-time manifold endowed with a metric tensor $g_{\mu\nu}$ and fix the local Lorentz frame by 4-bein vectors e^α_I and spin connections ω^{IJ}_α. The Holst action reads as follows

$$S = \int \sqrt{-g} \left[e^\alpha_I e^\beta_J R^{IJ}_{\alpha\beta} - \frac{1}{2\gamma} \epsilon^{IJ}{}_{KL} e^\alpha_I e^\beta_J R^{KL}_{\alpha\beta} \right] d^4x, \tag{1}$$

and conjugate momenta to spin connections are given by

$${}^\gamma\pi^a_{IJ} = \pi^a_{IJ} - \frac{1}{2\gamma}\epsilon^{KL}{}_{IJ}\pi^a_{KL}, \qquad \pi^a_{IJ} = 2\sqrt{-g} e^t_{[I} e^a_{J]}. \tag{2}$$

The Hamiltonian density is a linear combination of the following constraints

$$\begin{cases} H = \pi^a_{IM}\pi^{bM}{}_J \left(R^{IJ}_{ab} - \frac{1}{2\gamma}\epsilon^{IJ}{}_{KL} R^{KL}_{ab} \right) = 0 \\ H_a = \pi^b_{IJ} \left(R^{IJ}_{ab} - \frac{1}{2\gamma}\epsilon^{IJ}{}_{KL} R^{KL}_{ab} \right) = 0 \\ G_{IJ} = \partial_a \pi^a_{IJ} - 2\omega_{[I}{}^K{}_a \pi^a_{|K|J]} = 0 \\ C^{ab} = \epsilon^{IJKL}\pi^{(a}_{IJ}\pi^{b)}_{KL} = 0 \\ D^{ab} = \epsilon^{IJKL}\pi^c_{IM}\pi^{(aM}{}_J D_c \pi^{b)}_{KL} = 0 \end{cases} . \tag{3}$$

The set of constraints is second-class, which means that some variables are redundant and a non-trivial symplectic structure is inferred on the constraint hypersurfaces.

Journal of Physics: Conference Series **360** (2012) 012020 doi:10.1088/1742-6596/360/1/012020

The standard procedure to deal with such a second-class system is based on fixing the so-called time gauge condition, which means that the boost parameters χ_i vanish. However, in order to disentangle the fixing of the local Lorentz frame from the solution of second-class constraints, it has been given in [1] a generalized solution of second-class constraint, *i.e.*

$$\omega_i{}^j{}_a = {}^{\pi}\omega_i{}^j{}_a + \chi_i \omega^{0j}{}_a + \chi^j(\omega_i{}^0{}_a - {}^{\pi}D_a\chi_i), \qquad \pi^a_{ij} = 2\chi_{[i}\pi^a_{0j]} \tag{4}$$

where ${}^{\pi}D_a\chi_i = \partial_a\chi_i - {}^{\pi}\omega_i{}^j{}_a\chi_j$ and $\chi^i = \eta^{ij}\chi_j$. On the hypersurfaces (4) χ_i are promoted to configuration variables, such that no gauge fixing of the local Lorentz frame occurs, while second-class constraints are solved. The induced symplectic form is highly nontrivial, but as soon as phase-space is parametrized by densitized triads E^a_i and generalized Ashtekar-Barbero connections A^i_a, together with boost paremetrs and conjugate momenta π^i, the constraints of the Lorentz group are replaced by

$$G_i = \partial_a E^a_i + \gamma\epsilon_{ij}^{\ \ k} A^j_a E^a_k = 0, \qquad \pi^i = 0. \tag{5}$$

Hence, the SU(2) gauge structure arises also when the time-gauge condition is relaxed, while boost parameters are non-dynamical variables and they behave as the lapse function and the shift vector. Henceforth, one can avoid any dependence from χ_i into the wave functional and perform the quantization as in the standard treatment within the time gauge. The implications of these results on the canonical quantization procedure are that there is no need to perform any gauge fixing of the local Lorentz frame in LQG, while the spectra of geometrical operators is invariant under local boosts. This analysis has been extended to non minimally coupled matter fields [2], Immirzi field [3] and spinor fields [4].

A different parametrization of second-class constraints hypersurfaces is given in [5], where a Lorentz connection is defined. Such a connection becomes a SU(2) one only when the time-gauge condition holds.

3. Rotations and boosts on a quantum level

The generators of local Lorentz transformations take the following expression after having solved second-class constraints

$$R_i = G_i + \epsilon_i{}^j{}_k\chi_j\pi^k, \qquad K_i = (\eta_{ij} + \chi_i\chi_j)\pi^j - \beta\epsilon_i{}^{jk}\chi_j G_k. \tag{6}$$

If one want to analyze the effect of Lorentz transformation on a quantum level, variables χ_i must be retained into wave functional, such that a generic state on a path α can be written as [6]

$$\psi_\alpha = \otimes_e {}^L h_e \otimes_v I_v(\chi^i), \tag{7}$$

where e and v denote the edges and vertices of α, respectively. ${}^L h_e$ belongs to the Hilbert space of SU(2) distributional connections, while the Hilbert space associated with boost parameters can be defined as the space of square-integrable functions $I_v(\chi)$ on the hyperbolic space parametrized by χ_i. Symmetric momenta are defined as follows

$$\pi^i = -i\left(\frac{\partial}{\partial\chi_i} + \frac{\chi_i}{2(1-\chi^2)}\right), \tag{8}$$

and the scalar product in the full Hilbert space reads as

$$<\psi^1_\alpha, \psi^2_\alpha> = \Pi_{e,v}\int {}^L h^{1\dagger}_e {}^L h^2_e d\mu_{SU(2)} \int \frac{1}{\sqrt{1-\chi^2}} I^{1*}_v I^2_v d^3\chi, \tag{9}$$

where $d\mu_{SU(2)}$ denotes the Haar measure associated with the SU(2) group. Within this scheme, the wave functionals of LQG can be inferred by imposing Hamiltonian constraints (5).

In particular, the vanishing of π_i gives

$$\pi_i I_v^{LQG} = 0 \rightarrow I_v^{LQG}(\chi) \propto (1-\chi^2)^{1/4}, \tag{10}$$

and once I^{LQG} is inserted into the expression (9), the scalar product turns out to be LQG one. The condition $G_i = 0$ coincides with the Gauss constraint proper of SU(2) gauge theory, thus it can be implemented in the space of distributional connections by inserting invariant intertwiners at vertices. Therefore, it is possible to define the functionals associated with LQG by replacing $I_v(\chi)$ with SU(2) invariant intertwiners.

The expression (7) allow to investigate the effect of boost and rotation generators (6). The rotation generators are the sum of the SU(2) generators associated with Ashtekar-Barbero connections and the orbital angular momenta of χ^i. In the case of LQG functionals, I^{LQG} is the representation with vanishing angular momentum, such that the presence of SU(2) invariant intertwiners ensures rotation invariance.

As soon as boosts are concerned, let us consider the operator ordering with all momenta on the right (indeed this choice gives a non-symmetric boost generator). Actually, K_i is made of two terms, each one acting at verticies only:

- the first piece contains the conjugate momenta to χ_i and it acts on the χ-dependent part only as follows (we choose $i = 3$)

$$(\pi_3 + \chi_3\chi_j\pi^j)d_{ln}(\chi^2)Y_l^n = i(d'_{ln} - l d_{ln})\sqrt{\frac{(l+1)^2-n^2}{(2l+1)(2l+3)}}Y_{l+1}^n + i(d'_{ln} + (l+1)d_{ln})\sqrt{\frac{l^2-n^2}{(2l-1)(2l+1)}}Y_{l-1}^n,$$

  where $d'_{ln} = 2(1+\chi^2)\partial_\chi^2 d_{ln} - \frac{1}{2}d_{ln}$.
- the second term is made of the SU(2) generators times χ_i and its action reads ($i = 3$)

$$\begin{aligned}&\beta\epsilon_3{}^{ij}\chi_i G_j|j,m> \otimes d_{ln}Y_l^n = \\ &= -\tfrac{i}{2}\beta\chi\tfrac{d_{ln}}{\sqrt{2l+1}}\Big(\sqrt{(j+m)(j-m+1)}\sqrt{(l+n+2)(l+n+1)}|j,m-1> \otimes Y_{l+1}^{n+1} - \\ &-\sqrt{(j+m)(j-m+1)}\sqrt{(l-n-1)(l-n)}|j,m-1> \otimes Y_{l-1}^{n+1} - \\ &-\sqrt{(j-m)(j+m+1)}\sqrt{(l-n+2)(l-n+1)}|j,m+1> \otimes Y_{l+1}^{n-1} + \\ &+\sqrt{(j-m)(j+m+1)}\sqrt{(l+n-1)(l+n)}|j,m+1> \otimes Y_{l-1}^{n-1}\Big).\end{aligned}$$

As soon as LQG functionals are concerned, the action of the first term vanishes in the adopted operator ordering, while the second term does not provide any contribution because of SU(2) invariance. Therefore, LQG states are invariant under boost transformations.

4. From BF theory to LQG

In Spin-Foam models gravity is written as a BF theory, whose action reads

$$S = \frac{1}{2}\int \epsilon^{\alpha\beta\gamma\delta} tr(B_{\alpha\beta}^{IJ}F_{\gamma\delta}^{KL})d^4x, \tag{11}$$

with the conditions $B^{IJ} = \left[\delta^{IJ}_{KL} - \frac{1}{2\gamma}\epsilon^{IJ}{}_{KL}\right] * (e^K \wedge e^L)$. We have seen how such constraints induce on a Hamiltonian level a non trivial symplectic structure, which signals that SU(2) connections A^i_a contain all dynamical information.

In Spin-foam model, holonomies of the full Lorentz group are considered, whose associated irreducible representations of the principal series are made of an infinite tower of SU(2) representations, *i.e.*

$$H^{(k,\rho)} = \oplus_{j=k}^{+\infty} H^{(j)}, \qquad k \in \frac{\mathbf{N}}{2}, \qquad \rho \in \mathbf{R}. \tag{12}$$

In order to solve consistently second-class constraints one must implement: i) the restriction to the constraint hypersurfaces, ii) the reduction to functionals depending on A^i_a only. This task can be accomplished by taking fully projected spin-network [7], *i.e.*

$$h^{(\lambda,j)}_\alpha = \lim_{N\to+\infty} P\left\{\prod_{n=1}^{N} \pi^{(j)} h^{(\lambda)}_{e(n)} \pi^{(j)}\right\}, \qquad \alpha = \bigcup_{n=1}^{N} e(n), \tag{13}$$

where $\pi^{(j)}$ denote the projectors to the representations associated with the Ashtekar-Barbero SU(2) connections, *i.e.*

$$\pi^{(j)} : h^{(k,\rho)} \to {}^L h^j = h^{(j-r,\gamma\frac{j(j+1)}{j-r})}, \qquad \gamma = \beta_j = \frac{k\rho}{j(j+1)}.$$

In fact, the action of momenta gives

$$\pi_{ij}(S)^L h^j_e = 0, \qquad \pi_{0i}(S)^L h^j_e = o(S,e)^L h^j_{e_2} \tau_i {}^L h^j_{e_1}. \tag{14}$$

The first relation above signals that the time-gauge condition holds, while the second tells that $\omega^{ij}_a = ^\pi \omega^{ij}_a$, thus ${}^L h^j_e$ depend on A^i_a only. Henceforth, the restriction to ${}^L h^j_e$ implements properly the features of second-class constraints.

The condition $\beta_j = \gamma$ fixes the spin number of the selected SU(2) representation inside each Lorentz one.

In a path integral formulation, by adopting fully projected spin-networks, one avoids the nontrivial factor arising from the presence of second class constraints. In fact one finds

$$\int \prod_{ab} \prod_{IJ,KL} d\omega^{IJ}_a d\Pi^b_{KL} \sqrt{det[\chi_\alpha, \chi_\beta]} \Pi_\alpha \delta(\chi_\alpha) f(\omega, \Pi) = \int \prod_{ab} \prod_{ij} dA^i_a dE^b_j f(A, E), \tag{15}$$

where $\chi_\alpha = \{C^{ab}, D^{ab}\}$.

Therefore, in order to reduce the phase-space of BF theories, parametrized by holonomies of the Lorentz group, to the one of LQG, it must be addressed inside each Lorentz representation of the principal series the restriction to the SU(2) irreps for which

$$\rho = \gamma \frac{j(j+1)}{j-r}, \qquad r \in \mathtt{N}. \tag{16}$$

[1] Cianfrani F and Montani G 2009 *Phys. Rev. Lett.* 102 091301
[2] Cianfrani F and Montani G 2009 *Phys. Rev.* D 80 084045
[3] Cianfrani F and Montani G 2009 *Phys. Rev.* D 80 084040
[4] Cianfrani F and Montani G 2010 *Phys. Rev.* D 81 044015
[5] Geiller M., Lachieze-Rey M., Noui K. and Sardelli F. 2011 *SIGMA* 7 083
[6] Cianfrani F 2011 *Class. Quant. Grav.* 28 175014
[7] Alexandrov S 2010 *Phys. Rev.* D 82 024024.

Journal of Physics: Conference Series **360** (2012) 012021 doi:10.1088/1742-6596/360/1/012021

Boundary Terms and the $3+1$ Decomposition of the Holst Action

A Corichi[1,2] **and J D Reyes**[1,3,4]

[1] Centro de Ciencias Matemáticas, Unidad Morelia, Universidad Nacional Autónoma de México, UNAM-Campus Morelia, A. Postal 61-3, Morelia, Michoacán 58090, Mexico
[2] Center for Fundamental Theory, Institute for Gravitation and the Cosmos, Pennsylvania State University, University Park PA 16802, USA
[3] Instituto de Física y Matemáticas, Universidad Michoacana de San Nicolás de Hidalgo, Morelia, Michoacán, Mexico

E-mail: `corichi@matmor.unam.mx`, `jdreyes@matmor.unam.mx`

Abstract. Starting from the Holst action with surface terms and the fall-off conditions that make the variational principle and the covariant phase space formulation well-defined for asymptotically flat spacetimes, we rewrite the surface terms in the 3+1 decomposition of the action. We explore their relation with the surface terms in the Hamiltonian formulation in terms of Ashtekar variables. Just as for the Einstein-Hilbert action, if variations respecting asymptotic flatness are allowed, the energy and momentum in the Hamiltonian framework are not directly recovered from the 3+1 decomposition and gauge fixing of these terms.

1. Introduction

The Holst action [1] is the classical starting point in Loop Quantum Gravity. It is a first-order covariant action based on orthonormal tetrads and Lorentz connections, whose 3+1 decomposition plus partial gauge fixing gives a Hamiltonian formulation for General Relativity in terms of Ashtekar variables.

In [2], the form of the surface terms of the action, necessary for a proper treatment of asymptotically flat space-times, was given. These surface terms give a manifestly finite action even off-shell[5], and a well-defined variational principle, reproducing Einstein's equations under all asymptotically flat variations. Furthermore, the amended action leads naturally to a well-defined covariant phase space in which the Hamiltonians generating asymptotic symmetries provide the total energy-momentum and angular momentum of the space-time.

Here we wish to analize the form of these surface terms in the canonical or 3+1 formulation of the theory and explore its relation to the canonical treatment in [5], where surface terms and Poincaré charges are derived directly from the ADM framework.

[4] Present address: Departamento de Física, Universidad Autónoma Metropolitana-Iztapalapa, San Rafael Atlixco 186, México D.F. 09340, Mexico

[5] This is true for so called cylindrical temporal cut-offs or asymptotically time-translated spatial boundaries [3, 4]

Journal of Physics: Conference Series **360** (2012) 012021 doi:10.1088/1742-6596/360/1/012021

2. Covariant and 3+1 actions for General Relativity

In the conventional treatment of the Einstein-Hilbert action for a space-time with a boundary, two additional terms are necessary in order to have a well-defined variational principle that yields Einstein's equations from the stationary points of the action:

$$S_{EHGH} = \frac{1}{2\kappa}\left(\int_{\mathcal{M}} d^4x\sqrt{-g}\,R + 2\int_{\partial\mathcal{M}} d^3y\sqrt{|h|}(K - K_0)\right). \tag{1}$$

The first so called Gibbons-Hawking surface term [7] is inserted in the action so that its variation exactly cancels variations of the first derivatives of the metric, so only the metric is required to be fixed at the boundary. Here $\kappa = 8\pi G$, $\mathcal{M}$ is an appropriate portion of space-time and $\partial\mathcal{M}$ its boundary, R the Ricci scalar of the 4-metric g, h the induced metric on $\partial\mathcal{M}$, and K the trace of its extrinsic curvature. Since this term is generally divergent for asymptotically flat solutions the last 'non-dynamical' counter term is required to make the action finite on-shell.

In a more careful treatment though, for asymptotically flat space-times, the action should be such that asymptotically flat solutions are stationary points under all variations preserving asymptotic flatness, not just under variations of compact support. The Einstein-Hilbert action with Gibbons-Hawking term does not satisfy this requirement. Under all asymptotically flat variations, its variation gives a non-vanishing surface term when Einstein's equations are satisfied. Furthermore, the counter term becomes dynamical and since it requires an embedding of $\partial\mathcal{M}$ in a reference background, which is not always guaranteed, its variation is not even well defined [4]. Several proposals or generalizations of this term exist in the literature which aim to correct this problem [6, 4].

In contrast, in the 3+1 decomposition of space-time $\mathcal{M} \simeq \mathbb{R}\times\Sigma$, the stationarity requirement takes a different form [8, 9]. In order to have a well-defined Hamiltonian formulation and compute Poisson brackets of the constraints with various functions on the phase space, the constraints have to be finite, i.e. the integrals have to converge, in all of phase space, and they have to be functionally differentiable. So starting from the ADM action:

$$S_{ADM} = \frac{1}{2\kappa}\int dt \int_{\Sigma} d^3x(\,\dot{q}_{ab}p^{ab} - NH - N^a D_a\,), \tag{2}$$

the Hamiltonian $H[N] = \frac{1}{2\kappa}\int d^3xNH$ and vector constraints $D[N] = \frac{1}{2\kappa}\int d^3xN^aD_a$ are supplemented with boundary terms so that they are finite on the whole of phase space and

$$\delta H[N] = \int d^3x\;\frac{\delta H[N]}{\delta q_{ab}}\delta q_{ab} + \frac{\delta H[N]}{\delta p^{ab}}\delta p^{ab}$$

for all asymptotically flat variations $(\delta q_{ab}, \delta p^{ab})$. And similarly for $D[N^a]$.

Convergence of the integrals depends of course on the fall-off conditions for the lapse N and shift vector N^a. For a 3+1 decomposition, a cylindrical representation of spatial infinity i^o is most convenient so that for cartesian coordinates x^a in a neighborhood of i^o the fall-off conditions for the canonical pairs are

$$q_{ab}(x^c) = \delta_{ab} + \frac{1}{r}f_{ab}(x^c/r) + o(1/r^2)\,, \qquad p^{ab}(x^c) = \frac{1}{r^2}h^{ab}(x^c/r) + o(1/r^3) \tag{3}$$

with $f_{ab}(-x^c/r) = f_{ab}(x^c/r)$ even, and $h^{ab}(-x^c/r) = -h^{ab}(x^c/r)$ odd functions on the sphere.

For lapse and shift of order $1/r$ or odd functions of order 1, $H[N]$ and $D[N^a]$ are already finite and differentiable and generate proper gauge transformations or so called supertranslations at infinity respectively. If one allows N and N^a to asymptote to constant or linear functions then surface terms are needed to subtract divergences and to cancel unwanted variation terms so that

$H[N]$ and $D[N^a]$ are differentiable. Furthermore, with these fall-off conditions, on the constraint surface $H[N]$ and $D[N^a]$ become the generators of asymptotic Poincaré transformations at infinity, and the non-vanishing surface terms are the corresponding charges. Indeed, if $(\bar{t}, \bar{y}^a)$ is a Lorentzian frame at infinity such that Σ_t, the leaves of the foliation, coincide with $\bar{t} = const$, then their normals $n^\alpha \to \partial x^\alpha/\partial\bar{t}$, and the Hamiltonian evolution vector field goes as

$$t^\alpha \to N\left(\frac{\partial x^\alpha}{\partial \bar{t}}\right) + N^a\left(\frac{\partial x^\alpha}{\partial \bar{y}^a}\right)$$

So fall-off conditions of N and N^a determine a (10 parameter) Killing vector field of asymptotic space time metric.

We emphasize here the different roles that $H[N]$ and $D[N^a]$ play in the Hamiltonian formulation depending on the asymptotic fall-off conditions of N and N^a: as constraints and generators of gauge transformations in the bulk or as charges and generators of symmetries at infinity.

Also we note that the surface terms for energy and momentum in the 3+1 ADM action (2) can be recovered from those in the covariant action (1) only if one fixes the variations at the boundary [10]. Finally we contrast the finiteness in all of phase space of the corresponding ADM action (2) with surface terms, at least for asymptotically time-translated spatial boundaries, as opposed to finiteness on-shell for (1).

3. Actions for Loop Quantum Gravity

The Holst action with surface term is given by

$$S_{Holst} = -\frac{1}{2\kappa}\Big(\int_{\mathcal{M}} \Sigma^{IJ} \wedge (F_{IJ} + \frac{1}{\gamma} \star F_{IJ}) + \frac{1}{2\kappa}\int_{\partial\mathcal{M}} \Sigma^{IJ} \wedge (\omega_{IJ} + \frac{1}{\gamma} \star \omega_{IJ}). \quad (4)$$

Where F_{IJ} is the curvature of the spin connection ω_{IJ}, and $\Sigma^{IJ} = \frac{1}{2}\epsilon_{IJKL}e^K \wedge e^L$ is constructed from the co-tetrads. As already mentioned the action is not only finite on-shell but even off-shell for cylindrical temporal cut-offs, and it gives a well-defined covariant Hamiltonian framework leading to asymptotic symmetries and conserved charges identical to the ADM framework. It is natural to ask then what the 3+1 decomposition and temporal gauge fixing of these surface terms is and how they relate to the canonical treatment derived from the ADM action.

The transcription of the framework from the ADM action (2) in its manifestly finite and functionally differentiable form to Ashtekar variables was done in [5]. Boundary conditions are derived from those for the metric (3):

$$E^a_i = \bar{E}^a_i + \frac{F^a_i}{r} + o(1/r^2), \quad A^i_a = \frac{G^i_a}{r^2} + o(1/r^3),$$

with F^a_i and G^i_a even and odd functions on the sphere respectively, and similar conditions for Γ^i_a and K^i_a. Expressed in these variables the constraints become divergent due to the extra gauge degrees of freedom, so additional terms proportional to the Gauss constraint $\mathcal{G}_i$ have to be added to cancel the divergences. The generating functional of the canonical transformation $(q_{ab}, p^{ab}) \to (A^i_a, E^a_i)$ gives rise to these additional terms in the generators, both in the volume and the surface part:

$$\begin{aligned} D[N^a] = \frac{1}{\kappa\gamma}\int_\Sigma d^3x\big[N^a(F^i_{ab}E^b_i - A^i_a\mathcal{G}_i) + \frac{\epsilon_{abc}}{2\det E}E^a_iE^b_j\mathcal{L}_{\vec{N}}E^c_j\mathcal{G}_i\big] \\ + \frac{2}{\kappa\gamma}\int_{\partial\Sigma} dS_a(A^i_b - \Gamma^i_b)N^{[b}E^{a]}_i) \end{aligned} \quad (5)$$

$$H[N] = \frac{1}{2\kappa}\int_\Sigma d^3x N[\frac{E^a_i E^b_j}{\sqrt{\det E}}[\epsilon^{ij}_k F^k_{ab} - 2(1+\gamma^2)K^i_{[a}K^j_{b]}] + 2D_a(E^a_i \mathcal{G}_i)]$$
$$-\frac{1}{\kappa}\int_{\partial\Sigma} dS_a \left[\frac{N}{\sqrt{\det E}}E^a_i\partial_b E^b_i - D_b(N/\sqrt{\det E})E^b_i(E^a_i - \bar{E}^a_i)\right]. \quad (6)$$

On the other hand, the $3+1$ decomposition of (4) with the temporal gauge fixing gives

$$S = \int dt \int_\Sigma d^3x \left(\frac{1}{\kappa\gamma}(E^a_i \dot{A}^i_a - \Lambda^i \mathcal{G}_i) - N^a C_a - NC\right) + S_{surface}$$

with the well known constraint functionals

$$\mathcal{G}_i = \partial_a E^a_i + \epsilon^k_{ij} A^j_a E^a_k, \quad C_a = \frac{1}{\kappa\gamma}F^i_{ab}E^b_i - \frac{1+\gamma^2}{\kappa\gamma^2}K^i_a \mathcal{G}_i,$$
$$C = \frac{1}{2\kappa\sqrt{\det E}}E^a_i E^b_j[\epsilon^{ij}_k F^k_{ab} - 2(1+\gamma^2)K^i_{[a}K^j_{b]}] + \frac{(1+\gamma^2)}{\kappa\gamma^2}\frac{E^a_i\partial_a\mathcal{G}_i}{\sqrt{\det E}} \quad (7)$$

and the surface terms [11]

$$S_{surface} = \frac{1}{\kappa\gamma}\int dt \int_{\partial\Sigma_t} dS_b(\frac{N\gamma}{\sqrt{\det E}}E^a_i E^b_j \epsilon^{ij}_k A^k_a + N^a E^b_i A^i_a) \quad (8)$$
$$+\frac{1}{\kappa\gamma}\int_{\Sigma_1} E^a_i A^i_a - \frac{1}{\kappa\gamma}\int_{\Sigma_2} E^a_i A^i_a$$

It can be verified then that these additional surface terms proportional to the lapse and shift in (8) are not sufficient to cure the divergences in (7) for the constant and linear fall-off of N and N^a, nor can they be combined with terms proportional to the Gauss constraint in (7) to recover the corresponding energy and momentum charges in (6) and (5).

Thus, for asymptotically flat space-times, finiteness and well-posedness of the variational principle of the Holst action (4) do not directly descend to the canonical action framework derived from its 3+1 decomposition and temporal gauge fixing. The surface terms in (4) lead to the same ADM generators of asymptotic symmetries on the covariant phase space, defined as the space of asymptotically flat solutions. However, for the more stringent requirement of finiteness of the generators on the whole phase space of the canonical framework, additional terms are needed.

Acknowledgments

We would like to thank Irais Rubalcava and Edward Wilson-Ewing for helpful discussions. This work was in part supported by DGAPA-UNAM IN103610 grant, by NSF PHY0854743 and by the Eberly Research Funds of Penn State.

References

[1] Holst S. 1996, *Phys. Rev.* D **53** 5966
[2] Corichi A. and Wilson-Ewing E. 2010, *Class. Quant. Grav.* **27** 205015
[3] Ashtekar A., Engle J. and Sloan D. 2008, *Class. Quant. Grav.* **25** 095020
[4] Mann R. B. and Marolf D. 2006, *Class. Quant. Grav.* **23** 2927
[5] Thiemann T. 1995, *Class. Quant. Grav.* **12** 181
[6] Kraus P., Larsen F. and Siebelink R. 1999, *Nucl. Phys.* B **563** 259
[7] Gibbons G. and Hawking S. W. 1997 *Phys. Rev.* D **15** 2752
[8] Regge T. and Teitelboim C. 1974, *Ann. Phys.* **88** 286
[9] Beig R. and O' Murchadha N. 1987, *Ann. Phys.* **174** 463
[10] Hawking S. W. and Horowitz G. 1996, *Class. Quant. Grav.* **13** 1487
[11] Corichi A. and Wilson-Ewing E. 2010, unpublished

Journal of Physics: Conference Series **360** (2012) 012022 doi:10.1088/1742-6596/360/1/012022

On Ashtekar's Formulation of General Relativity

Christian Fleischhack

Institut für Mathematik, Universität Paderborn, 33095 Paderborn, Germany

E-mail: fleischh@math.upb.de

Abstract. We review Ashtekar's variables in physics and in mathematics notation. The first one uses local indices, the second one is given by global objects within fibre bundles.

1. Introduction

The coalescence of gravity and quantum theory has remained one of the greatest challenges of modern physics. Over almost one century there have been many attempts to solve that issue. One of the most promising ones was initiated a quarter of a century ago, when Abhay Ashtekar presented his "new variables" [1] in 1986. These new variables ultimately turned gravity into an SU(2) gauge field theory with constraints and opened the door to a Dirac-like canonical quantization of gravity. This theory is called loop quantum gravity today. The original success of Ashtekar's variables consisted in a remarkable simplification of the constraints in comparison to the standard Wheeler-DeWitt approach, as they removed non-polynomialities that spoiled quantization. Nevertheless, loop quantum gravity is by far not a theory of full quantum gravity yet; in particular, many dynamical issues have remained unsolved so far.

Although Ashtekar's variables form the heart of loop quantum gravity and although the early success of loop quantum gravity is heavily based on its mathematical rigor, a geometric description of Ashtekar's variables themselves in more mathematical terms has been missing for a long time. Only in recent years, there have been some efforts [2, 3, 4, 5] to describe these variables in the language of differential geometry using fibre bundles. In this short paper, we are going to summarize these developments together with a review of the usual formulation of Ashtekar's variables in physics.

We start in Section 2 with the very basic structures in classical canonical gravity. After fixing some notation in Section 3, we will review the standard index definitions of Ashtekar's variables in Section 4. We will introduce the notion of dreibeinen which lead to the first half of Ashtekar's variables, and then give the notion of spin connection and extrinsic curvature that lead to the Ashtekar connection, i.e., the second half of Ashtekar's variables. This section is closed by listing the constraints of Ashtekar gravity. Then, in Section 5, we will represent Ashtekar's variables within the language of fibre bundles. First, we are doing this for the tangent bundle, where the expressions are particularly simple. This comprises frames (they correspond to dreibeinen, see Subsection 5.1), some vector field operation (generalizing the vector product on $\mathbb{R}^3$) and the Weingarten mapping (encoding the extrinsic curvature, for both see Subsection 5.2). This will give us a covariant derivative in the tangent bundle. Using associate bundle techniques and a spin structure, we may deduce from this the SU(2) Ashtekar connection living in a certain principal fibre bundle, the so-called spin bundle (Subsection 5.4). After a naive derivation of the constraints and the application of Ashtekar's variables to Friedmann-Robertson-Walker spacetimes, we conclude by regaining the index notations in Section 6.

2. Canonical Gravity

Since its very beginning, general relativity has been described by a metric g of Lorentzian signature on some spacetime manifold, this means a 4-dimensional, connected, time-oriented and oriented Lorentz manifold. Its behaviour is governed by two (up to certain technicalities, e.g., along boundaries) equivalent formulations. First, g is required to solve Einstein's equations[1]

$$R_{\mu\nu} - \frac{1}{2} g_{\mu\nu} R \;=\; 0\,, \tag{1}$$

where $R_{\mu\nu}$ is the Ricci tensor w.r.t. g, and R the corresponding Ricci scalar. Or, second, g is a stationary point of the Einstein-Hilbert action

$$\int_M \sqrt{\det g}\, R\,, \tag{2}$$

corrected with appropriate boundary terms. In order to rewrite general relativity as a canonical theory, we need some splitting of M into $\mathbb{R} \times \Sigma$, where the $\mathbb{R}$ part is supposed to carry the time variable and Σ to denote some embedded spacelike hypersurface. For this to be possible, one usually assumes that the spacetime is globally hyperbolic as it is the case iff (M, g) is isometric[2] to $(\mathbb{R} \times \Sigma, -f\, d\tau^2 + g_\tau)$ for some $f : \mathbb{R} \longrightarrow \mathbb{R}$ and a family g_τ of Riemannian metrics on Σ. [6, 7] It is somewhat surprising that this equivalence, although conjectured for a long time, has been established many years after the invention of Ashtekar's variables that had already heavily used such a splitting. Given such a splitting, g can be uniquely reconstructed from the reduction q of g to Σ and the second fundamental form (extrinsic curvature) K describing the shape of Σ as a submanifold of M. However, not every choice of a Riemannian metric q and of a symmetric bilinear form K originates in some spacetime metric g. Indeed, certain constraints are to be fulfilled. If they are met, classical results [8, 9, 10] guarantee the (at least local) solvability of the corresponding initial-value problem. The metric q on Σ and some expression that is (up to some weight) linear in K, form a set of canonically conjugate variables.

Based on this formulation, there have been several attempts to quantize gravity. As already mentioned, one of the main obstacles has been the complicated structure of constraints. The Ashtekar variables have provided an option to overcome at least some of these problems. For the moment, we would only like to mention that the Ashtekar variables consist of an appropriately densitized dreibein field together with an SU(2) connection. Below we are now going to describe them a bit more in detail both from the physical and from the mathematical point of view, the analysis remaining classical all over the paper.

3. Notation

Throughout the whole paper, let Σ be some oriented three-dimensional manifold, and let M be some manifold diffeomorphic to $\mathbb{R} \times \Sigma$. We will identify Σ, if necessary, with $\{0\} \times \Sigma$. Next, let q be some Riemannian metric on Σ. We write $\langle X, Y \rangle := q(X, Y)$ for all vector fields X, Y on Σ. Moreover, we let g be a metric on M inducing q on Σ. Recall, that all objects we are dealing with are in the smooth category. When working with indices, we denote spatial indices by $a, b, c, \ldots$ and "internal" SO(3) or SU(2) indices by $i, j, k, \ldots$ (for expansions w.r.t. some Lie algebra basis) or $I, J, K, \ldots$ (if they are matrix indices). Lifting as well as lowering of indices is

[1] Here, we have ignored any matter or cosmological terms as we will do throughout the paper.
[2] Unless otherwise noted, all functions, metrics, isometries etc. are assumed to be smooth.

made by means of the metric q_{ab} in the spatial, and by the Kronecker delta δ_{ij} in the internal case, respectively.

To end up with the preliminaries, we would like to mention some deviation from a nowadays predominant convention. In contrast to the original version, Thiemann [11] defines the Ashtekar field as the weight-one densitized dreibein E divided by the Barbero-Immirzi parameter β. The deeper reason for this is that E/β is the canonically conjugate variable to the Ashtekar connection. However, as we are less interested in the dynamics here, we give the old version without dividing by β.

4. From indices ...

Common to any definition of Ashtekar's variables is that first the phase space of classical general relativity gets extended. But, how this is done has changed a little bit over the time. Originally [1, 12], Ashtekar started his construction by introducing the so-called soldering form σ and its canonically conjugate momentum C, a density of weight 1. The soldering form plays a similar rôle for SU(2) as the dreibein field does for SO(3). As the soldering form is determined only up to internal SU(2) transformations, C is constrained by $C^{[ab]} = 0$. Now, given a phase space point (σ, C), Ashtekar defined two connections $\mathcal{D}^{\pm}$ by

$$\mathcal{D}^{\pm}_a a_{bI} \;:=\; \nabla_a a_{bI} \pm \mathrm{i}\frac{G}{\sqrt{2\det q}}\Big[C_{aI}{}^J - \frac{1}{2}\sigma_{aI}{}^J\sigma^c{}_{KL}C_c{}^{KL}\Big]a_{bJ}\,. \tag{3}$$

Here, G is the gravitational constant. Note that the bracket term plays the same rôle w.r.t. C, as the extrinsic curvature does for the canonically conjugate momentum to q. Now, the two connections correspond to connection one-forms $(A^{\pm})^J_{aI}$ via

$$\mathcal{D}^{\pm}_a a_I \;=\; \partial_a a_I + G A_{aI}{}^J a_J\,. \tag{4}$$

One can prove that $A^{\pm}$ and $\sqrt{\det q}\,\sigma$ are (up to some constant $\pm\mathrm{i}/\sqrt{2}$) canonically conjugate variables. If then $F^{\pm}$ is the curvature two-form corresponding to $A^{\pm}$, one gets the constraints

$$\operatorname{tr}\sigma^a F^{\pm}_{ab} \;=\; 0 \qquad\text{and}\qquad \operatorname{tr}\sigma^a\sigma^b F^{\pm}_{ab} \;=\; 0\,. \tag{5}$$

Moreover, $C^{[ab]} = 0$ can be rewritten as $\mathcal{D}^{\pm}_a\sigma^a{}_{IJ} = 0$. It was mostly the polynomial form of these constraints (they are even Yang-Mills like) that triggered the success of Ashtekar's variables. This was also despite the fact that one lets oneself in for complex structures due to the i in the definition of the connection. Indeed, in order to return to the real structure, certain reality conditions had to be imposed.

The situation got somewhat changed in the mid-90s. Barbero [13] and Immirzi [14] observed that the constant i in the definition above can be replaced by any non-zero complex number β, later called Barbero-Immirzi parameter. In particular, for real β one gets rid of the reality conditions, hence one is dealing with SU(2) instead of $\mathrm{Sl}_{\mathbb{C}}(2)$ connections. This ultimately paved the road for loop quantum gravity, although its foundations (in particular, the Ashtekar-Lewandowski measure [15] needing compact gauge groups) had been laid a few years earlier. However, this switch came with a drawback: In the scalar (Hamilton) constraint, a non-polynomial term prefactored by $1 + \beta^2$, popped up. Hence, the constraints are of simple polynomial form iff $\beta = \pm\mathrm{i}$. But, this puzzle was solved as early as Thiemann [16] observed that the new term can be written by means of certain Poisson brackets.

About that time, the derivation of Ashtekar's variables changed a bit (without modifying the final point). A standard reference for this is Thiemann's book [11] that we will follow here closely (up to the overall factor β^{-1} for the Ashtekar field mentioned in Section 3). We will start with the definition of the Ashtekar fields based on dreibeinen and then define the Ashtekar connections by means of spin connection and extrinsic curvature.

4.1. Dreibeine

Consider a local[3] (oriented) orthonormal basis (e_1, e_2, e_3) of the tangent space $T\Sigma$. Using some local coordinate system x^a, we may expand each basis vector into $e_i = e_i^a \partial_a$ defining the dreibein[4] e_i^a. At each single point in Σ, we may view e_i^a as a non-singular matrix. Its inverse is denoted by e_a^i and called co-dreibein. The metric q on Σ can be reconstructed via $q_{ab} = \delta_{ij} e_a^i e_b^j$. From this, we may define the first part E_i^a of the Ashtekar variables: $E_i^a := \sqrt{\det q}\, e_i^a$ is the densitized dreibein of weight 1. Both e_i^a and e_a^i can be used to replace spatial by internal indices and vice versa. One immediately checks that multiplying e_i^a with any SO(3) matrix keeps q_{ab} invariant. Note that, using the isomorphism between $\mathfrak{su}(2)$ and $\mathfrak{so}(3)$, one can regard e_a^i also as a local SU(2)-valued one-form.

4.2. Ashtekar Connection

The second type of the Ashtekar variables is an SU(2) connection A_a^i on Σ that is built up from the spin connection and the extrinsic curvature. The spin connection, on the other hand, comes from the Levi-Civita connection associated to our given Riemannian metric q on Σ. Indeed, one first extends the spatial covariant derivative ∇ from standard tensors on Σ to tensors on Σ carrying also internal SO(3) indices. So, e.g., we have

$$\nabla_a a_i \;:=\; \partial_a a_i + \Gamma_{ai}{}^j a_j \tag{6}$$

and extend this to arbitrary tensors by linearity, Leibniz rule and contractivity. Imposing metricity $\nabla_a e_b^i = 0$, the $\Gamma_{ai}{}^j$ terms are given by

$$\Gamma_{ai}{}^j \;=\; e_i^b \left[\partial_a e_b^j - \Gamma_{ab}^c e_c^j\right] \tag{7}$$

with Γ_{ab}^c being the Christoffel symbols for q. One checks that Γ_a takes values in SO(3). Let now $\{M_1, M_2, M_3\}$ be the basis of $\mathfrak{so}(3)$ with $(M_i)_{jk} = \varepsilon_{ikj}$. We use them to expand Γ_a into $\Gamma_a^i M_i$, giving the spin connection Γ_a^i. We may interpret the internal SO(3) indices also as SU(2) indices w.r.t. the adjoint representation as this is isomorphic to the fundamental representation at Lie algebra level.

Next, let us introduce the extrinsic curvature term. For this, we now need that Σ is considered as a hypersurface of M, whose normal w.r.t. g will be denoted by n. The extrinsic curvature is then just half the Lie derivative of g w.r.t. n. More explicitly, it is given by $K_{ab} := g(\partial_a, \nabla_b n)$. Using the dreibein as above to replace the spatial index b by the internal index i, we get K_a^i. Altogether, we define the Ashtekar SU(2) connection by

$$A_a^i \;:=\; \Gamma_a^i + \beta K_a^i\,. \tag{8}$$

The corresponding covariant derivative will be denoted by ∇^A.

4.3. Constraints

Directly from metricity of the dreibein, we get the Gauß constraint

$$G_i \;=\; \nabla_a^A E_i^a \;\equiv\; \partial_a E_i^a + \varepsilon_{ij}{}^k A_a^j E_k^a \;=\; 0 \tag{9}$$

which shows exactly the structure of a gauge field theory constraint. Indeed, this constraint is the justification for considering Ashtekar gravity as an SU(2) gauge theory. Additional constraints are the (spatial) diffeomorphism constraint (also called vector constraint)

$$V_a \;=\; F_{ab}^i E_i^b \;=\; 0 \tag{10}$$

[3] As Σ is orientable, it has a spin structure and is parallelizable. [17] Therefore we may assume w.l.o.g. that this basis is even a global basis.

[4] also called triad

Journal of Physics: Conference Series **360** (2012) 012022 doi:10.1088/1742-6596/360/1/012022

and the Hamilton constraint (also called scalar constraint)

$$H = \left[F^i_{ab} - (1+\beta^2)\varepsilon^i{}_{lm}K^l_a K^m_b\right]\frac{\varepsilon_i{}^{jk}E^a_j E^b_k}{\sqrt{\det q}} = 0 \tag{11}$$

(both provided the Gauß constraint is met).

5. ... to Differential Geometry ...

Now, we are going to overcome the index notation and discuss the mathematical structures behind the Ashtekar variables. Dreibeine will correspond to (oriented) orthonormal frames. They might be discussed in any dimension, whereas the Ashtekar connection will heavily rely on the fact that the manifold Σ is three-dimensional. The deeper reason for this is the fact that the Ashtekar connection is constructed by means of the vector product, which is defined only for dimension 3. For simplicity, we will therefore restrict ourselves to dimension 3 throughout.

The presentation in this section follows[5] closely that in [5], partially as an excerpt.

5.1. *Orthonormal Frames*

As already mentioned, dreibeine correspond to certain frames in the tangent bundle $T\Sigma$. To be specific, a frame at $x \in \Sigma$ is a vector space isomorphism

$$e : \mathbb{R}^3 \longrightarrow T_x\Sigma\,. \tag{12}$$

Given a frame e, we define the corresponding Ashtekar field E to be the densitized frame field

$$E := \frac{1}{\det e}\, e \tag{13}$$

of weight 1. Here, the determinant is taken w.r.t. some (local) basis of $T_x\Sigma$ and the canonical basis of $\mathbb{R}^n$. The frame can be reconstructed from the Ashtekar field via

$$e = \frac{1}{\sqrt{\det E}}\, E\,. \tag{14}$$

We call a frame orthonormal w.r.t. the metric q on Σ iff it is an isometry, with $\mathbb{R}^3$ given the Euclidean metric. We call it oriented iff it preserves the orientation. Whereas each frame e via

$$\widetilde{q}(X,Y) := \langle e^{-1}(X), e^{-1}(Y)\rangle_{\mathrm{Eucl}} \qquad \text{for } X, Y \in T_x\Sigma \tag{15}$$

yields the unique metric $\widetilde{q}$ on Σ, such that e is orthonormal w.r.t. $\widetilde{q}$, there exist several frames that are orthonormal w.r.t. a given metric. More precisely, two frames e and e' are isometries for the same q iff $e' = e \circ L_g$ for some $g \in \mathrm{O}(3)$. Here, L_g denotes the left translation by g.

Completely analogously to the definition of the tangent bundle as an appropriately topologized union of the tangent vectors all over the points of Σ, one defines the frame bundle $\mathrm{Gl}(\Sigma)$ as such a collection of frames. The orthonormal frames form a subbundle $\mathrm{O}_q(\Sigma)$ of $\mathrm{Gl}(\Sigma)$; it can be obtained via reducing the structure group from $\mathrm{Gl}(3)$ to $\mathrm{O}(3)$. Further reduction to $\mathrm{SO}(3)$ provides us with the bundle $\mathrm{O}^+_q(\Sigma)$ of oriented orthonormal frames.

[5] This does not refer to the discussion of constraints.

Journal of Physics: Conference Series **360** (2012) 012022 doi:10.1088/1742-6596/360/1/012022

5.2. Ashtekar Connection

Let us now turn the Ashtekar connection into an intrinsic object of the tangent bundle.

For any tangent vectors X, Y on Σ and any oriented q-orthonormal frame e, we define

$$X \bullet Y \;:=\; e\big[e^{-1}(X) \times e^{-1}(Y)\big]\,, \tag{16}$$

with $\times$ being the standard vector product on $\mathbb{R}^3$. As the vector product is invariant w.r.t. SO(3), one immediately checks that $\bullet$ depends on the metric q only rather than on the oriented q-orthonormal frame e itself. The definition can be extended immediately to a smooth operation on vector fields. This can also be seen from the relation $X \bullet Y = *(X \wedge Y)$, where $\wedge$ is the standard wedge product and $*$ the Hodge operator w.r.t. q. The product $\bullet$ inherits all relevant properties of the vector product, in particular,

$$X \bullet Y \;=\; -Y \bullet X \tag{17}$$

$$\langle X \bullet Y, Z\rangle \;=\; \langle X, Y \bullet Z\rangle \tag{18}$$

$$X \bullet (Y \bullet Z) \;=\; \langle X, Z\rangle Y - \langle X, Y\rangle Z \tag{19}$$

and the Jacobi identity

$$X \bullet (Y \bullet Z) + Y \bullet (Z \bullet X) + Z \bullet (X \bullet Y) \;=\; 0\,. \tag{20}$$

The second ingredient for the Ashtekar connection is the so-called Weingarten mapping. It corresponds to the second fundamental form which is usually called extrinsic curvature in physics. The Weingarten mapping encodes the respective shape of Σ within M; indeed, we will now need to assume that Σ is an embedded submanifold of a four-dimensional Lorentzian manifold (M, g) and that q is the reduction of g to Σ. The Weingarten mapping is defined by

$$\begin{aligned} W : T\Sigma &\longrightarrow T\Sigma\,. \\ X &\longmapsto {}^4\nabla_X n \end{aligned} \tag{21}$$

Here, n is the normal to Σ within (M, g), and ${}^4\nabla$ is the Levi-Civita connection for g on M. As ${}^4\nabla$ is metric and torsion-free, the Weingarten mapping is well defined and symmetric. Moreover, it is $C^\infty(\Sigma)$-linear. The corresponding second fundamental form K is the quadratic form $K(X,Y) = \langle W(X), Y\rangle$ defined by the Weingarten mapping. Of course, W can uniquely be re-obtained from K, as q is non-degenerate.

Mixing all these ingredients appropriately, we get the Ashtekar connection. For this, fix some non-zero complex number β. This number is the Barbero-Immirzi parameter introduced above. Recall that the imaginary part of β may be non-zero. So, for non-real β, we will tacitly assume in the following that all structures (connections etc.) will be complexified.

Now, the Ashtekar connection w.r.t. β is defined by

$$\nabla^A_X Y \;:=\; \nabla_X Y + \beta\, W(X) \bullet Y\,. \tag{22}$$

One can easily show that ∇^A is metric and obeys the Leibniz rule

$$\nabla^A_X (Y \bullet Z) \;=\; \nabla^A_X Y \bullet Z + Y \bullet \nabla^A_X Z\,. \tag{23}$$

Its torsion is given by

$$T^A(X,Y) \;=\; \beta\,[W(X) \bullet Y - W(Y) \bullet X] \tag{24}$$

and its curvature by

$$\begin{aligned} R^A(X,Y)Z \;=\; & R(X,Y)Z + \beta\;[(\nabla_X W)Y - (\nabla_Y W)X] \bullet Z \\ & + \beta^2 [W(X) \bullet W(Y)] \bullet Z\,. \end{aligned} \tag{25}$$

Journal of Physics: Conference Series **360** (2012) 012022 doi:10.1088/1742-6596/360/1/012022

5.3. Reconstruction

Note that both the metric q and the second fundamental form K can be reconstructed from E and ∇^A. In fact, first reconstruct q from E as in Section 5.1. Then, let ∇ be the Levi-Civita connection to this metric q. Using $\beta \neq 0$, we now find $W(X) \bullet Y$ for all vector fields $X, Y \in T\Sigma$. Next, with the standard orthonormal basis $(\mathfrak{e}_1, \mathfrak{e}_2, \mathfrak{e}_3)$ of $\mathbb{R}^3$, we get

$$W(X) \quad = \quad \frac{1}{2} \sum_{i=1}^{3} e(\mathfrak{e}_i) \bullet (W(X) \bullet e(\mathfrak{e}_i)) \, . \qquad (26)$$

Finally, the second fundamental form is easily derived from W as mentioned above.

5.4. Spin Bundle

In order to identify Ashtekar's variables as entities of an SU(2) gauge field theory, we will now go from the tangent bundle to principal fibre bundles. For this, we will first transfer the structures to the oriented orthonormal frame bundle. This bundle, however, is an SO(3) bundle, whence we will have to introduce a spin structure to go to the SU(2) spin bundle. Of course, the tangent bundle is associated to both principal fibre bundles. Note that in the standard index notation of Subsections 4.1 and 4.2 it is barely impossible to distinguish between the SU(2) and the SO(3) case; in fact, the internal indices are Lie algebra indices, but the Lie algebras $\mathfrak{su}(2)$ and $\mathfrak{so}(3)$ are isomorphic. Within principal fibre bundles, however, the structure *group* is encoded.

To implement this for the SO(3) case, consider the canonical action ρ of SO(3) on $\mathbb{R}^3$. Given any metric q on Σ, the usual isomorphism

$$\begin{array}{rcl} \mathrm{O}^+_q(\Sigma) \times_{(\mathrm{SO}(3),\rho)} \mathbb{R}^3 & \cong & T\Sigma \\ {[e, \mathfrak{x}]} & \longmapsto & e(\mathfrak{x}) \end{array} \qquad (27)$$

identifies the tangent bundle as associated to the bundle of oriented orthonormal frames. Moreover, it induces an isomorphism between the space of metric covariant derivatives on $T\Sigma$ and the space of metric connection 1-forms in the oriented orthonormal frame bundle. In order to apply this identification to the Ashtekar connection, take some q-orthonormal local frame e on $U \subseteq \Sigma$ and define the local connection 1-form A by

$$A \quad := \quad \langle \nabla^A e, e \rangle \quad : \quad TU \longrightarrow \mathfrak{gl}(3) \, . \qquad (28)$$

More explicitly, $A(X)$ is uniquely given by

$$\langle A(X)\mathfrak{x}, \mathfrak{y} \rangle_{\mathrm{Eucl}} \quad = \quad \langle \nabla^A_X [e(\mathfrak{x})], e(\mathfrak{y}) \rangle \qquad (29)$$

for all vector fields X and $\mathfrak{x}, \mathfrak{y} \in \mathbb{R}^3$, viewing both sides as functions on U. One immediately checks that, as e is q-orthonormal, A maps to $\mathfrak{so}(3)$. Now, the Ashtekar connection ω_A in the oriented orthonormal frame bundle is simply the connection that is patched together from the local connections, i.e., mapped to A by the different pull-backs $f^*\omega_A$ with f being a mapping from some U to $\mathrm{O}^+_q(\Sigma)$.

To finally establish the Ashtekar connection as an SU(2) connection, first recall [18] that a spin structure on (Σ, q) is a pair $(S(\Sigma), \Lambda)$ consisting of an SU(2) principal fibre bundle $\widetilde{\pi} : S(\Sigma) \longrightarrow \Sigma$ and a double covering $\Lambda : S(\Sigma) \longrightarrow \mathrm{O}^+_q(\Sigma, q)$, such that the following diagram commutes:

$$\begin{array}{ccccc} S(\Sigma) \times \mathrm{SU}(2) & \longrightarrow & S(\Sigma) & & \\ \text{both 2:1} \downarrow \Lambda \times \lambda & & \text{2:1} \downarrow \Lambda & \overset{\widetilde{\pi}}{\searrow} & \\ & & & & \Sigma \, . \\ & & & \overset{\pi}{\nearrow} & \\ \mathrm{O}^+_q(\Sigma) \times \mathrm{SO}(3) & \longrightarrow & \mathrm{O}^+_q(\Sigma) & & \end{array}$$

Journal of Physics: Conference Series **360** (2012) 012022 doi:10.1088/1742-6596/360/1/012022

In the diagram, the structure groups act horizontally, and λ denotes the double covering from SU(2) to SO(3). Using the isomorphism $\lambda_* : \mathfrak{su}(2) \longrightarrow \mathfrak{so}(3)$, we may lift any connection ω in $\mathrm{O}_q^+(\Sigma)$ to a spin bundle connection $\tilde{\omega}$ via

$$\lambda_* \circ \tilde{\omega} \quad := \quad \Lambda^* \omega \,. \tag{30}$$

This way, the Levi-Civita connection is lifted to the spin connection $\tilde{\omega}_{LC}$, as well as the SO(3) Ashtekar connection ω_A is lifted to the celebrated SU(2) Ashtekar connection $\tilde{\omega}_A$. Note that the tangent bundle is associated also to the spin bundle via

$$T\Sigma \quad \cong \quad S(\Sigma) \times_{(\mathrm{SU}(2), \rho\circ\lambda)} \mathbb{R}^3 \,. \tag{31}$$

5.5. Constraints

So far, we have not been able to really derive the constraints within a purely differential-geometric framework. For this, one needs to carefully implement techniques from symplectic reduction and canonical transformations. For this short article, we refrained from doing this. Instead, we simply translate the index expressions for the constraints (see Subsection 4.3). Therefore, please be aware that the expressions might get modified later, if the "genuine" derivation has been done.

Denote the image of the standard basis of $\mathbb{R}^3$ under the frame e by (e_1, e_2, e_3). Then we have (up to constant prefactors)

$$G \quad \sim \quad \sum_i \nabla^A_{e_i} e_i \tag{32}$$

$$V \quad \sim \quad \sum_{i,j} R^A(e_i, e_j)[e_i \bullet e_j] \tag{33}$$

$$H \quad \sim \quad R^A + (1+\beta^2)\big[(\mathrm{tr}\, W)^2 - \mathrm{tr}(W^2)\big] \,. \tag{34}$$

5.6. Friedmann-Robertson-Walker Spacetimes

In highly symmetric cosmological models, the Ashtekar connection takes a particularly simple form. Here, we will review the results for the Friedmann-Robertson-Walker spacetime, where Σ is assumed to have constant sectional curvature κ. The Weingarten mapping is simply the Hubble "constant" h times the identity. From this, we get for the Ashtekar connection

$$\nabla^A_X Y \quad = \quad \nabla_X Y + \beta h \, X \bullet Y \,, \tag{35}$$

its torsion and curvature

$$T^A(X,Y) \quad = \quad 2\beta h \, X \bullet Y \tag{36}$$

$$R^A(X,Y)Z \quad = \quad \big[(\beta h)^2 - \kappa\big](X \bullet Y) \bullet Z \,, \tag{37}$$

as well as the constraints

$$G \quad = \quad 0 \tag{38}$$

$$D \quad = \quad 0 \tag{39}$$

$$H \quad \sim \quad 6(\kappa + h^2) \,. \tag{40}$$

6. ... and Back to Indices

We have already mentioned that Σ is parallelizable, whence it has a global oriented orthonormal basis for $T\Sigma$. Therefore the internal indices can be assumed to be global indices. This need not be true for the spatial ones. In fact, it might happen that the manifold Σ cannot be covered by a single chart. Therefore, the expressions derived using indices in Section 4 will, in general, be of local nature only. To rederive them from our index-free quantities, let $\chi : U \longrightarrow \mathbb{R}^3$ be a chart for some open $U \subseteq \Sigma$; this defines a local basis $\{\partial_1, \partial_2, \partial_3\}$ for the tangent space. Moreover, let $\{M_1, M_2, M_3\}$ be a basis of $\mathfrak{so}(3)$ with $[M_i, M_j] = \varepsilon_{ij}{}^k M_k$, e.g., we may take the choice $(M_i)_{jk} = \varepsilon_{ikj}$ from Subsection 4.2.

Taking some orthonormal frame e, we get the dreibein from $e(\mathfrak{e}_i) =: e^a_i \partial_a$ by means of the canonical basis $(\mathfrak{e}_i)$ of $\mathbb{R}^3$. The transition to the Ashtekar field E^a_i is now obvious. On the other hand, one can calculate [4] that

$$e^*\omega_A(\partial_a) \quad = \quad A^i_a M_i \tag{41}$$

reproduces the Ashtekar connection A^i_a. In the same way, the extrinsic curvature K^i_a can be obtained from

$$\widetilde{K}(X) \quad = \quad \langle W(\pi_* X), e \bullet e \rangle \tag{42}$$

with π being the bundle projection, as well as the spin connection Γ^i_a can be obtained from the bundle version of the Levi-Civita connection. Finally, the SO(3) indices can be reduced to the SU(2) indices by simply using the isomorphism between $\mathfrak{so}(3)$ and $\mathfrak{su}(2)$.

Acknowledgments

The author thanks Johannes Brunnemann, Stefan Funkner, Sören Graupner, Maximilian Hanusch, Thomas Leistner, Philipp Levermann, Gerd Rudolph, and Matthias Schmidt for interesting discussions. The author has been supported by the Emmy-Noether-Programm of the Deutsche Forschungsgemeinschaft under grant FL 622/1-1.

References

[1] Ashtekar A 1986 New Variables for classical and quantum gravity *Phys. Rev. Lett.* **57** 2244–2247
[2] Funkner S 2006 Die Geometrie der Ashtekarvariablen (Diplomarbeit) Universität Leipzig
[3] Graupner S 2010 Ashtekar-Variablen: Bündeltheorie und Schwarze Löcher (Diplomarbeit) Universität Leipzig
[4] Levermann Ph 2009 Geometrie des Ashtekar-Zusammenhangs (Diplomarbeit) Universität Hamburg
[5] Fleischhack Ch and Levermann Ph 2011 Ashtekar Variables: Structures in Bundles *Preprint* 1112.1262 [math-ph]
[6] Bernal A N and Sánchez M 2003 On Smooth Cauchy Hypersurfaces and Geroch's Splitting Theorem *Commun. Math. Phys.* **243** 461–470 (*Preprint* gr-qc/0306108)
[7] Bernal A N and Sánchez M 2006 Further Results on the smoothability of Cauchy hypersurfaces and Cauchy time functions *Lett. Math. Phys.* **77** 183–197 (*Preprint* gr-qc/0512095)
[8] Fourès-Bruhat Y 1952 Théorème d'existence pour certains systèmes d'équations aux dérivées partielles non linéaires *Acta Math.* **88** 141–225
[9] Straumann N 2004 *General Relativity: With Applications to Astrophysics* (Springer)
[10] Ringström H 2009 *The Cauchy Problem in General Relativity* (European Mathematical Society)
[11] Thiemann Th 2007 *Modern Canonical Quantum General Relativity* (Cambridge University Press)
[12] Ashtekar A 1987 New Hamiltonian formulation of general relativity *Phys. Rev.* **D36** 1587–1602
[13] Barbero F 1995 Real Ashtekar Variables for Lorentzian Signature Space-Times *Phys. Rev.* **D51** 5507–5510 (*Preprint* gr-qc/9410014)
[14] Immirzi G 1997 Real and complex connections for canonical gravity *Class. Quant. Grav.* **14** L177–L181 (*Preprint* gr-qc/9612030)
[15] Ashtekar A and Lewandowski J 1994 Representation theory of analytic holonomy C^* algebras *Knots and Quantum Gravity (1993, Riverside)* ed J C Baez (Oxford University Press) (*Preprint* gr-qc/9311010)
[16] Marolf D, Mourão J M and Thiemann Th 1997 The status of diffeomorphism superselection in Euclidean (2+1) gravity *J. Math. Phys.* **38** 4730–4740 (*Preprint* gr-qc/9701068)

[17] Kirby R C 1989 *The Topology of 4-Manifolds* (Springer)
[18] Friedrich Th 1997 *Dirac-Operatoren in der Riemannschen Geometrie* (Vieweg)

Journal of Physics: Conference Series **360** (2012) 012023 doi:10.1088/1742-6596/360/1/012023

Loop gravity in terms of spinors[1]

Etera Livine, Johannes Tambornino
Laboratoire de Physique, ENS Lyon,
CNRS-UMR 5672,
46 Allée d'Italie, Lyon 69007, France

E-mail: johannes.tambornino@ens-lyon.fr, etera.livine@ens-lyon.fr

Abstract. We show that loop gravity can equally well be formulated in in terms of spinorial variables (instead of the group variables which are commonly used), which have recently been shown to provide a direct link between spin network states and discrete geometries. This results in a new, unitarily equivalent formulation of the theory on a generalized Bargmann space. Since integrals over the group are exchanged for straightforward integrals over the complex plane we expect this formalism to be useful to efficiently organize practical calculations.

1. Introduction and overview

For technical reasons calculations in loop gravity can in most cases only be performed when truncating the full continuum theory to a single fixed graph. Therefore one needs to know which classical geometrical degrees of freedom are represented by the spin network functions in $\mathcal{H}_\gamma$, the Hilbert space associated to that graph[2]. As the number of degrees of freedom is always finite, they cannot represent a classical continuum geometry [2].

However, using a parametrization in terms of spinors $|z\rangle \in \mathbb{C}^2$ it was shown in [3, 4, 5, 6] that the classical phase space of loop gravity associated to a graph has a natural interpretation in terms of discrete geometries: in fact, the spin network functions in $\mathcal{H}_\gamma$ can be seen as a quantum analog of discrete, piecewise flat polyhedral geometries [7].

This formalism triggered some interesting recent developments: a better understanding of the intertwiner spaces of SU(2)-gauge-invariant loop gravity [8, 9, 10] and a new look on the simplicity constraints in spinfoam models [11, 12, 13]. See also [14, 15] for a recent generalization of these ideas to the full Lorentz-group.

Interestingly it turns out that the use of spinorial variables does not only help to clarify the classical discrete geometry of spin network functions. Also the quantum theory itself can be reformulated exclusively in terms of spinors [16]. The relevant state space $\mathcal{H}_\gamma^{\text{spin}}$ turns out to

[1] Based on a talk given by one of the authors (JT) at the conference *Loops '11* in Madrid, Spain, on May 27th 2011 [1]

[2] In loop gravity the Hilbert space associated to a graph γ is given by $\mathcal{H}_\gamma := L^2(\mathrm{SU}(2)^E, d^E g)$ where E is the number of edges of that graph and $d^E g$ the product Haar measure. This space can be interpreted as a quantization of E copies of the cotangent bundle $T^*\mathrm{SU}(2) \simeq \mathrm{SU}(2) \times \mathfrak{su}(2)$. $g \in \mathrm{SU}(2)$ is the holonomy of the Ashtekar-connection along an edge e and $X \in \mathfrak{su}(2)$ is related to the flux of the densitized triad through a surface dual to e. The Hilbert space of the continuum theory arises from the individual graph-Hilbert spaces as an inductive limit $\mathcal{H}_{\text{LQG}} := \overline{\cup_\gamma \mathcal{H}_\gamma / \sim}$, where $\sim$ denotes an equivalence relation between states living on different graphs and the completion in an appropriate topology is taken.

Journal of Physics: Conference Series **360** (2012) 012023 doi:10.1088/1742-6596/360/1/012023

be a generalization of the Bargmann space [17] of holomorphic square-integrable functions with respect to a Gaussian measure. Thus, in this formulation quantum states of the gravitational field are composed of polynomial functions over the complex numbers instead of functions on SU(2) as in the standard treatment. We expect this to simplify some computations carried out in loop gravity. One can show that $\mathcal{H}_\gamma^{\text{spin}}$ is unitarily equivalent to $\mathcal{H}_\gamma$, thus the two formulations contain exactly the same physics. Furthermore, the whole construction is compatible with the inductive limit taken to define the continuum Hilbert space $\mathcal{H}_{\text{LQG}}$, which shows that the spinor techniques do not loose their validity when considering the continuum theory.

2. The spinorial state space

The classical phase space associated to the edge-Hilbert space $\mathcal{H}_e = L^2(\text{SU}(2), dg)$ is the cotangent bundle $T^*\text{SU}(2) \simeq \text{SU}(2) \times \mathfrak{su}(2)$, which can be parameterized in terms of a group element g, a Lie algebra element X and its 'dual' $\tilde{X} = -g^{-1}Xg$. An alternative parameterization, first used in [3], is the following: consider two spinors $|z\rangle$ and $|\tilde{z}\rangle$ living at the initial and final vertex of the edge e respectively[3]. From these one can construct two vectors by projecting them on the Pauli matrices as

$$\vec{X}(z) := \langle z \mid \vec{\sigma} \mid z\rangle, \qquad \vec{\tilde{X}}(\tilde{z}) := \langle \tilde{z} \mid \vec{\sigma} \mid \tilde{z}\rangle \in \mathbb{R}^3 . \tag{1}$$

These vectors are then interpreted as oriented areas of the faces of elementary polyhedra living at each vertex. Furthermore, the following combination turns out to be in the defining representation of SU(2):

$$g(z, \tilde{z}) := \frac{|z\rangle[\tilde{z}| - |z]\langle\tilde{z}|}{\sqrt{\langle z \mid z\rangle\langle\tilde{z} \mid \tilde{z}\rangle}} \tag{2}$$

Define a constraint that generates U(1)-transformations with opposite sign on both spinors,

$$\mathcal{M} := \langle z \mid z\rangle - \langle\tilde{z} \mid \tilde{z}\rangle . \tag{3}$$

Then it can be shown [4] that (excluding some singular points) the symplectic reduction of $\mathbb{C}^2 \times \mathbb{C}^2$ with respect to that constraint gives back $T^*\text{SU}(2)$.

Starting from the spinorial formulation of $T^*\text{SU}(2)$ the most natural Hilbert space to look for a representation of this cotangent bundle is the Bargmann space [17] of holomorphic, square-integrable functions in two complex variables,

$$\mathcal{F}_2 := L^2_{\text{hol}}(\mathbb{C}^2, d\mu(z)), \quad d\mu(z) := \frac{1}{\pi^2} e^{-\langle z|z\rangle} dz^0 dz^1 .$$

The space of interest, taking into account the U(1)-constraint (3), is then

$$\mathcal{H}_e^{\text{spin}} := \mathcal{F}_2 \otimes \mathcal{F}_2 / \text{U}(1) .$$

The spinors $|z\rangle$ and $|\tilde{z}\rangle$ are represented on $\mathcal{H}_e^{\text{spin}}$ as ladder-operators. g and X are then constructed as composite operators via (1) and (2). Restricting attention to U(1)-invariant functions singles out polynomials (labelled by $\alpha, \tilde{\alpha} \in \mathbb{C}$, $j \in \frac{1}{2}\mathbb{N}$) of the form

$$\mathcal{P}^j_{\alpha\tilde{\alpha}}(z, \tilde{z}) := \frac{1}{(2j)!} \langle\alpha \mid z\rangle^{2j} [\tilde{z}|\epsilon|\tilde{\alpha}\rangle^{2j} , \tag{4}$$

[3] Our notation is as follows: a spinor $|z\rangle \in \mathbb{C}^2$ has components $|z\rangle := \begin{pmatrix} z^0 \\ z^1 \end{pmatrix}$. It has a conjugate $\langle z| := (\bar{z}^0, \bar{z}^1)$ and a dual $|z] := \epsilon|\bar{z}\rangle$, $\epsilon := \begin{pmatrix} 0 & -1 \\ 1 & 0 \end{pmatrix}$. The inner product on $\mathbb{C}^2$ is denoted by $\langle z \mid w\rangle := \bar{z}^0 w^0 + \bar{z}^1 w^1$. As a symplectic space $\mathbb{C}^2$ is equipped with the standard symplectic structure $\{z^A, \bar{z}^B\} = -i\delta^{AB}$, $A, B = 0, 1$.

which are holomorphic in both spinor variables and further have matching degree. They form an overcomplete basis of $\mathcal{H}_e^{\rm spin}$, the completeness relations can be derived as

$$
\begin{aligned}
\int d\mu(z)\int d\mu(\tilde{z})\overline{\mathcal{P}^j_{\omega\tilde{\omega}}(z,\tilde{z})}\mathcal{P}^k_{\alpha\tilde{\alpha}}(z,\tilde{z}) &= \delta^{jk}\langle\alpha\mid\omega\rangle^{2j}\langle\tilde{\omega}\mid\alpha\rangle^{2j}\,,\\
\sum_j\int d\mu(\omega)d\mu(\tilde{\omega})\frac{d_j}{(2j)!}\overline{\mathcal{P}^j_{\omega\tilde{\omega}}(z_1,\tilde{z}_1)}\mathcal{P}^j_{\omega\tilde{\omega}}(z_2,\tilde{z}_2) &= I_0(2\langle z_1\mid z_2\rangle\langle\tilde{z}_1\mid\tilde{z}_2\rangle)\,.
\end{aligned}
$$

Here $I_0(x)$ is the zeroth modified Bessel function of first kind and plays the role of the delta-distribution on $\mathcal{H}_e^{\rm spin}$. These completeness relations are (up to a missing factor of $d_j := 2j+1$ on the right side) exactly the ones fulfilled by the Wigner matrix elements in $L^2(\mathrm{SU}(2),dg)$ when written in the coherent state basis. Thus it is immediate to see that the two spaces are unitarily equivalent. This unitary map can explicitely be written in terms of an integral kernel as

$$
\begin{aligned}
\mathcal{T}_e:\quad &\mathcal{H}_e\to\mathcal{H}_e^{\rm spin}; \qquad (5)\\
&f(g)\mapsto(\mathcal{T}f)(z,\tilde{z}) := \int dg\mathcal{K}_g(z,\tilde{z})f(g)\,,\\
&\mathcal{K}_g(z,\tilde{z}) = \sum_{k\in\mathbb{N}}\frac{\sqrt{k+1}}{k!}[\tilde{z}|\epsilon g^{-1}|z\rangle^k\,.
\end{aligned}
$$

When applied to Wigner matrix elements in the coherent state basis this map has an interesting interpretation: it essentially (up to some combinatorial factors) restricts the representation matrices of SU(2), when written in terms of spinors, to their holomorphic part

$$
D^j_{\omega\tilde{\omega}}(g) = \left(\langle\omega|\frac{|z\rangle[\tilde{z}|-|z]\langle\tilde{z}|}{\sqrt{\langle z\mid z\rangle\langle\tilde{z}\mid\tilde{z}\rangle}}|\tilde{\omega}\rangle\right)^{2j} \overset{\mathcal{T}}{\mapsto} \frac{1}{(2j)!\sqrt{d_j}}\langle\omega\mid z\rangle^{2j}[\tilde{z}|\epsilon|\tilde{\omega}\rangle^{2j}\,.
$$

The unitary map (5) directly generalizes from a single edge e to an arbitrary graph γ, showing unitary equivalence between the Hilbert spaces $\mathcal{H}_\gamma$ and $\mathcal{H}_\gamma^{\rm spin} = \otimes_e\mathcal{H}_e^{\rm spin}$:

$$
\mathcal{T}_\gamma:\mathcal{H}_\gamma\to\mathcal{H}_\gamma^{\rm spin}\,.
$$

Thus, equivalence classes of spinor functions living on different graphs γ and γ' can be defined by demanding the following diagram to commute

$$
\begin{array}{ccc}
\mathcal{H}_\gamma & \xrightarrow{\mathcal{T}_\gamma} & \mathcal{H}_\gamma^{\rm spin}\\
{}^*p_{\gamma\gamma'}\downarrow & & \downarrow {}^*p^{\rm spin}_{\gamma\gamma'}\\
\mathcal{H}_{\gamma'} & \xrightarrow{\mathcal{T}_{\gamma'}} & \mathcal{H}_{\gamma'}^{\rm spin}
\end{array}
$$

Here ${}^*p_{\gamma\gamma'}$ are the isometric embeddings that define equivalence classes on the group side. Their counterparts on the spinor side ${}^*p^{\rm spin}_{\gamma\gamma'}$ are then used to define equivalence classes of spinor states. Thus, equivalence classes on the left side are mapped to equivalence classes on the right side, no matter which $\mathcal{T}_\gamma$ is used. This assures that the construction is cylindrically consistent and allows to abstractly define the *continuum spinor Hilbert space* as

$$
\mathcal{H}^{\rm spin}_{\rm LQG} := \overline{\cup_\gamma\mathcal{H}_\gamma^{\rm spin}/\sim}\,.
$$

Although the exact properties of this space are, for the moment, not very well understood, this shows that the spinor tools can be lifted from a fixed graph to the continuum level.

One surprising feature of the spinorial formalism, which was discussed in [16], is that the Haar measure on SU(2) turns out to be just a Gaussian measure[4] on $\mathbb{C}^4$ when written in terms of spinors, in the sense that

$$\int dg f(g) = \int d\mu(z) \int d\mu(\tilde{z}) f(g(z,\tilde{z})),$$

for any $f \in L^2(\mathrm{SU}(2))$ and the group element g interpreted as function of spinors as in (2) on the right side. Using spinorial variables to characterize SU(2) can be understood as choosing a coordinate system with a lot of redundant degrees of freedom. Thus, $f(g(z,\tilde{z}))$ is constant along certain directions in $\mathbb{C}^4$ which can be used to turn the Haar measure into Gaussian form. See also [14] where a similar construction was recently performed for the Haar measure on $\mathrm{SL}(2,\mathbb{C})$. This Gaussian form of the measure, together with the simple polynomial form of the holomorphic basis (4), is expected to lead to simplification for practical computations: quantities of interest concern the moments of a simple Gaussian measure on $\mathbb{C}^4$ for which combinatorial tools, such as Wick's theorem, are available.

3. Conclusion and Outlook

We showed that, based on the recent reformulation of *classical* loop gravity in terms of spinors, one can construct a spinorial Hilbert space that is unitarily equivalent to standard one build over SU(2). This space is a generalization of the Bargmann space of holomorphic, square-integrable functions over complex numbers. The construction works for an arbitrary graph and is cylincdrically consistent, therefore the lift to the continuum theory is straightforward. Within this new picture quantum states of the gravitational field are represented as holomorphic polynomials over complex numbers, and the measure on the relevant space is of Gaussian form. Detailed calculations in that framework have not been performed yet. But we expect this reformulation, which is much closer to standard field theory than the ordinary version of loop quantum gravity (for example, Wick's theorem directly applies), to push further the development of efficient calculational tools to compute quantities of physical interest (such as scattering amplitudes) within the framework of loop gravity[5].

Acknowledgments

This work was partially supported by the ANR "Programme Blanc" grant LQG-09.

[1] `http://loops11.iem.csic.es/loops11/index.php?option=com_content&view=article&id=163`
[2] Rovelli C and Speziale S 2010 *Phys. Rev.* **D82** 044018 (*Preprint* 1005.2927)
[3] Freidel L and Speziale S 2010 *Phys. Rev.* **D82** 084040 (*Preprint* 1001.2748)
[4] Freidel L and Speziale S 2010 *Phys. Rev.* **D82** 084041 (*Preprint* 1006.0199)
[5] Freidel L and Livine E R 2010 *J. Math. Phys.* **51** 082502 (*Preprint* 0911.3553)
[6] Freidel L and Livine E R 2011 *J.Math.Phys.* **52** 052502 (*Preprint* 1005.2090)
[7] Bianchi E, Dona P and Speziale S 2011 *Phys.Rev.* **D83** 044035 (*Preprint* 1009.3402)

[4] The interesting point is that the measure can be made *exactly* Gaussian without any additional measure factors. For a discussion of alternative measures see appendix D of [16].

[5] The spinorial formulation of loop gravity might also allow a different look on the definition of dynamics in the canonical formulation of quantum gravity: starting with the classical geometric intuition from the twisted geometries framework one can reexamine the definition of the Hamiltonian constraint in terms of spinorial variables and base its quantization thereon. This interesting line of research is still under investigation and no definite results have been obtained so far.

[8] Borja E F, Freidel L, Garay I and Livine E R 2011 *Class.Quant.Grav.* **28** 055005 (*Preprint* 1010.5451)
[9] Borja E F, Diaz-Polo J, Garay I and Livine E R 2010 *Class. Quant. Grav.* **27** 235010 (*Preprint* 1006.2451)
[10] Borja E F, Diaz-Polo J, Garay I and Livine E R 2010 (*Preprint* 1012.3832)
[11] Dupuis M and Livine E R 2011 *Class.Quant.Grav.* **28** 085001 (*Preprint* 1006.5666)
[12] Dupuis M and Livine E R 2011 (*Preprint* 1104.3683)
[13] Dupuis M, Freidel L, Livine E R and Speziale S 2011 (*Preprint* 1107.5274)
[14] Livine E R, Speziale S and Tambornino J 2011 (*Preprint* 1108.0369)
[15] Wieland W M 2011 (*Preprint* 1107.5002)
[16] Livine E R and Tambornino J 2011 (*Preprint* 1105.3385)
[17] Bargmann V 1962 *Rev. Mod. Phys.* **34** 829–845

Journal of Physics: Conference Series **360** (2012) 012024 doi:10.1088/1742-6596/360/1/012024

SU(2) gauge theory of gravity with topological invariants

Sandipan Sengupta

E-mail: sandipan@rri.res.in, sandi@imsc.res.in
Raman Research Institute, C.V. Raman Avenue, Bangalore-560080, India

Abstract. The most general gravity Lagrangian in four dimensions contains three topological densities, namely Nieh-Yan, Pontryagin and Euler, in addition to the Hilbert-Palatini term. We set up a Hamiltonian formulation based on this Lagrangian. The resulting canonical theory depends on three parameters which are coefficients of these terms and is shown to admit a real $SU(2)$ gauge theoretic interpretation with a set of seven first-class constraints. Thus, in addition to the Newton's constant, the theory of gravity contains three (topological) coupling constants, which might have non-trivial imports in the quantum theory, e.g. in quantum geometry.

1. Introduction

The classical dynamics of a system is not affected by the addition of topological densities in the Lagrangian. This is so because such densities can always be locally written as total divergences. However, quantum dynamics might depend on them. The cases of the Sine-Gordon quantum mechanical model or QCD provide perfect examples of such a phenomenon where topological terms leave their imprints on the quantum theory[1].

In gravity theory in 3+1 dimensions, there are three possible topological terms, namely, Nieh-Yan, Pontryagin and Euler, which can be added to the Lagrangian. In terms of tetrads and spin-connections, these can be written as[1]:

$$\begin{aligned} I_{NY} &= e\Sigma^{\mu\nu}_{IJ}\tilde{R}_{\mu\nu}{}^{IJ}(\omega) + \epsilon^{\mu\nu\alpha\beta}D_\mu(\omega)e_{I\nu}D_\alpha(\omega)e^I_\beta = \partial_\mu\left[\epsilon^{\mu\nu\alpha\beta}\, e^I_\nu\, D_\alpha(\omega)e_{I\beta}\right] \\ I_P &= \epsilon^{\mu\nu\alpha\beta}R_{\mu\nu IJ}(\omega)R_{\alpha\beta}{}^{IJ}(\omega) = 4\partial_\mu\left[\epsilon^{\mu\nu\alpha\beta}\omega_\nu{}^{IJ}\left(\partial_\alpha\omega_{\beta IJ} + \frac{2}{3}\omega_{\alpha I}{}^K\omega_{\beta KJ}\right)\right] \\ I_E &= \epsilon^{\mu\nu\alpha\beta}R_{\mu\nu IJ}(\omega)\tilde{R}_{\alpha\beta}{}^{IJ}(\omega) = 4\partial_\mu\left[\epsilon^{\mu\nu\alpha\beta}\tilde{\omega}_\nu{}^{IJ}\left(\partial_\alpha\omega_{\beta IJ} + \frac{2}{3}\omega_{\alpha I}{}^K\omega_{\beta KJ}\right)\right] \end{aligned} \quad (1)$$

where $R_{\mu\nu}^{IJ}(\omega) = \partial_{[\mu}\omega_{\nu]}{}^{IJ} + \omega_{[\mu}{}^{IK}\omega_{\nu]K}{}^J$ and $D_\mu(\omega)e^I_\nu = \partial_\mu e^I_\nu + \omega_\mu{}^{IJ}e_{\nu J}$. Although these topological densities are functions of local geometric quantities, they encode only the global properties of the manifold. The Nieh-Yan density depends on torsion and in Euclidean theory, its integral over a compact manifold is a sum of three integers associated with the homotopy maps $\pi_3(SO(5)) = Z$ and $\pi_3(SO(4)) = Z + Z$. Among the other two which depend on the curvature, the Pontryagin-class characterises the integers corresponding to the map $\pi_3(SO(4)) = Z + Z$ and the Euler-class

[1] The quantity $\tilde{X}^{IJ}$ the dual of X_{IJ} in the internal space: $\tilde{X}^{IJ} = \frac{1}{2}\,\epsilon^{IJKL}X_{KL}$

Journal of Physics: Conference Series **360** (2012) 012024 doi:10.1088/1742-6596/360/1/012024

characterises the combination of Betti numbers. While the first two densities are P and T odd, the third is P and T even (see [2] and the references within).

In order to understand their possible import in the quantum theory, it is important to set up a classical Hamiltonian formulation of the theory containing all these terms in the action. In ref.[3], such an analysis has been presented for a theory based on Lagrangian density containing the standard Hilbert-Palatini term and the Nieh-Yan density. The resulting theory, in time gauge, has been shown to correspond to the well-known canonical gauge theoretic formulation of gravity based on Sen-Ashtekar-Barbero-Immirzi *real* $SU(2)$ gauge fields [4]. Here η^{-1}, the inverse of the coefficient of Nieh-Yan term, is identified with the Barbero-Immirzi parameter γ. The framework in [3] supersedes the earlier formulation of Holst [5] in the sense that unlike the Holst term, the Nieh-Yan density
(a) does not need any further modifications for the inclusion of matter couplings and the equations of motion continue to be independent of η for all couplings;
(b) provides a topological interpretation for Barbero-Immirzi parameter, leading to a complete analogy between η and the θ-parameter of non-abelian gauge theories (from the classical perspective).
As an elucidation of these facts, the method has been applied to spin-$\frac{1}{2}$ fermions[3] and supergravity theories[6].

Here we include all three topological terms in the Hilbert-Palatini Lagrangian[2]:

$$\mathcal{L}(e,\omega) \;=\; \frac{1}{2}\, e\, \Sigma^{\mu\nu}_{IJ}\, R_{\mu\nu}{}^{IJ}(\omega) \;+\; \frac{\eta}{2}\, I_{NY} \;+\; \frac{\theta}{4}\, I_P \;+\; \frac{\phi}{4}\, I_E \tag{2}$$

where, $\Sigma^{\mu\nu}_{IJ} = \frac{1}{2}\, e^{\mu}_{[I} e^{\nu}_{J]}$. In order to understand how the canonical theory of gravity gets affected by such additions, a Hamiltonian analysis based on this Lagrangian is presented below, demonstrating how we obtain a real SU(2) formulation of gravity with all three topological densities.

2. Hamiltonian formulation

To set up a Hamiltonian description, we assume that the spacetime is of the form $\Sigma X R$ where Σ is a compact manifold. We decompose the 16 (spacetime) tetrad fields e^I_μ into the fields V^I_a, M_I, N^a and N (16=9+3+3+1) (see [2] for further details):

$$\begin{aligned}
e^I_t &= NM^I + N^a V^I_a, \qquad e^I_a = V^I_a\;; \\
e^t_I &= -\frac{M_I}{N}\,, \qquad e^a_I = V^a_I + \frac{N^a M_I}{N}; \\
M_I V^I_a &= 0\;, \qquad M_I M^I = -1\;; \\
V^I_a V^b_I &= \delta^b_a\,, \qquad V^I_a V^a_J = \delta^I_J + M^I M_J\;.
\end{aligned} \tag{3}$$

The internal space metric is Lorentzian, i.e., $\eta_{IJ} := dia(-1,1,1,1)$. Next, instead of the variables V^a_I and M^I, we define a new set of 12 variables as:

$$E^a_i \;=\; 2e\Sigma^{ta}_{0i} \;\equiv\; e\left(e^t_0 e^a_i - e^t_i e^a_0\right) = -\,\sqrt{q}\; M_{[0}V^a_{i]}, \qquad \chi_i = -M_i/M^0 \tag{4}$$

Before writing the full Lagrangian, we note that with the help of Bianchi identities $\epsilon^{abc}D_a(\omega)R_{bcIJ} = 0$ and $\epsilon^{abc}D_a(\omega)\tilde{R}_{bcIJ} = 0$, the last two terms in (2) can be written as[2]:

$$\frac{\theta}{4}\, I_P + \frac{\phi}{4}\, I_E = e^a_{IJ}\, \partial_t \omega^{(\eta)IJ}_a \tag{5}$$

[2] The quantity $X^{(\eta)IJ}$ is defined as: $X^{(\eta)IJ} = X^{IJ} + \eta\, \tilde{X}^{IJ}$

with $(1+\eta^2)\, e^a_{IJ} = \epsilon^{abc}\left\{(\theta+\eta\phi)\, R_{bcIJ}(\omega) + (\phi-\eta\theta)\, \tilde{R}_{bcIJ}(\omega)\right\}$. Using (3), (4) and (5), the Lagrangian in (2) can be written as:

$$\mathcal{L} = \pi^a_{IJ}\partial_t\omega_a^{(\eta)IJ} + t^a_I\partial_t V^I_a - NH - N^a H_a - \frac{1}{2}\omega_t^{IJ}G_{IJ} \tag{6}$$

where $\pi^a_{IJ} = e\,\Sigma^{ta}_{IJ} + e^a_{IJ}$ and

$$\begin{aligned} G_{IJ} &= -2D_a(\omega)\pi^{a(\eta)}_{IJ} - t^a_{[I}V_{J]a} \\ H_a &= \pi^b_{IJ}R^{(\eta)IJ}_{ab}(\omega) - V^I_a D_b(\omega)t^b_I \ , \\ H &= \frac{2}{\sqrt{q}}\left(\pi^{a(\eta)}_{IK} - e^{a(\eta)}_{IK}\right)\left(\pi^{b(\eta)}_{JL} - e^{b(\eta)}_{JL}\right)\eta^{KL}R_{ab}{}^{IJ}(\omega) - M^I D_a(\omega)t^a_I \ . \end{aligned} \tag{7}$$

Since there are no velocities associated with the fields N, N^a and ω_t^{IJ}, we have the constraints $H\approx 0,\ H_a\approx 0,\ G_{IJ}\approx 0$.

Next, we split the 18 spin-connection fields ω_a^{IJ} as:

$$A^i_a \equiv \omega_a^{(\eta)0i} = \omega_a^{0i} + \eta\tilde{\omega}_a^{0i}, \qquad K^i_a \equiv \omega_a^{0i} \ . \tag{8}$$

The rationale behind such a choice is to make the SU(2) interpretation transparent, as can be understood by noting that A^i_a transforms as connection and K^i_a as adjoint representation under the SU(2) gauge transformations. Also, it is convenient (although not necessary) to work in the time gauge where the boost constraints are solved by the gauge choice $\chi_i = 0$. Thus, in this gauge, the symplectic form becomes:

$$\pi^a_{IJ}\partial_t\omega_a^{(\eta)IJ} + t^a_I\partial_t V^I_a = \hat{E}^a_i\partial_t A^i_a + \hat{F}^a_i\partial_t K^i_a + t^a_i\partial_t V^i_a \tag{9}$$

with

$$\hat{E}^i_a \equiv -\frac{2}{\eta}\,\tilde{\pi}^{a(\eta)}_{0i} \equiv -\frac{2}{\eta}\left(\tilde{\pi}^a_{0i} - \eta\pi^a_{0i}\right) = E^a_i - \frac{2}{\eta}\,\tilde{e}^{a(\eta)}_{0i}(A,K) \tag{10}$$

$$\hat{F}^a_i \equiv 2\left(\eta+\frac{1}{\eta}\right)\tilde{\pi}^a_{0i} = 2\left(\eta+\frac{1}{\eta}\right)\tilde{e}^a_{0i}(A,K) \tag{11}$$

Here, the fields V^i_a and its conjugate t^a_i are not independent; they obey the following second-class constraints:

$$V^i_a - \frac{1}{\sqrt{E}}E^i_a \equiv 0\ , \qquad t^a_i - \eta\epsilon^{abc}D_b(\omega)V^i_c = \epsilon^{abc}\left(\eta D_b(A)V^i_c - \epsilon^{ijk}K^j_b V^k_c\right) \tag{12}$$

Similarly, eq.(11) shows that the momenta $\hat{F}^a_i$ obey constraints of the form

$$\chi^a_i := \hat{F}^a_i - f(A^j_b, K^k_c) \approx 0 \tag{13}$$

These imply secondary constraints:

$$[\chi^a_i(x), H(y)] \approx 0 \Rightarrow t^a_i - \left(\frac{1+\eta^2}{\eta^2}\right)\left\{\eta\epsilon^{ijk}D_b(A)\left(\sqrt{E}E^a_jE^b_k\right) + \sqrt{E}E^{[a}_jE^{b]}_iK^j_b\right\} \approx 0 \tag{14}$$

The solution of (14) can be expressed in the form: $K^i_a - \kappa^i_a(A^j_b, E^c_k) \approx 0$. Since K^i_a and $\hat{F}^a_i$ are canonically conjugate, these constraints evidently form a second-class pair with (13). The constraints (12), (13) and (14), alongwith the constraints $G^{rot}_i\approx 0,\ H_a\approx 0\ ,\ H\approx 0$, completely characterise the canonical theory corresponding to the Lagrangian density (2).

Notice that for $\theta=0$ and $\phi=0$, the momenta $\hat{F}^a_i$ in (11) vanishes. This corresponds to the Barbero-Immirzi formulation. Thus, the effect of the addition of Pontryagin and Euler terms in the Lagrangian gets reflected through a richer symplectic structure characterised by a non-vanishing $\hat{F}^a_i$. Also, for non-vanishing θ and ϕ, the canonical conjugate of the connection A^i_a is $\hat{E}^a_i$, and not the densitized triad E^a_i as in the case for $\theta=0,\ \phi=0$.

3. SU(2) interpretation

The second-class constraints can all be implemented by using the corresponding Dirac brackets instead of the Poisson brackets. After imposing all the second-class pairs strongly, we are left with a set of seven first class constraints:

$$\begin{aligned} G_i^{rot} &\equiv \eta\, D_a(A)\hat{E}_i^a \;+\; \epsilon^{ijk} K_a^j \hat{F}_k^a \;\approx\; 0 \\ H_a &\equiv \hat{E}_i^b F_{ab}^i(A) \;+\; \hat{F}_i^b D_{[a}(A) K_{b]}^i \;-\; K_a^i D_b(A) \hat{F}_i^b - \eta^{-1}\, G_i^{rot} K_a^i \;\approx\; 0 \\ H &\equiv \frac{\sqrt{E}}{2\eta} \epsilon^{ijk} E_i^a E_j^b F_{ab}^k(A) - \left(\frac{1+\eta^2}{2\eta^2}\right) \sqrt{E} E_i^a E_j^b K_{[a}^i K_{b]}^j + \frac{1}{\eta}\, \partial_a \left(\sqrt{E} G_k^{rot} E_k^a\right) \approx 0 \end{aligned} \qquad (15)$$

Evaluating the Dirac brackets of the rotation constraints G_i^{rot} with the basic fields, we find that they are the generators of the SU(2) gauge transformations:

$$\begin{aligned} \left[G_i^{rot}(x),\; \hat{E}_j^a(y)\right]_D &= \epsilon^{ijk} \hat{E}_k^a\; \delta^{(3)}(x,y)\;, \\ \left[G_i^{rot}(x),\; A_a^j(y)\right]_D &= -\eta \left(\delta^{ij}\partial_a \;+ \eta^{-1}\; \epsilon^{ikj} A_a^k\right)\; \delta^{(3)}(x,y)\;\; . \end{aligned} \qquad (16)$$

Thus, we have a SU(2) gauge theory of gravity with all three topological parameters. The Barbero-Immirzi parameter η^{-1} acts as the coupling constant of gauge field A_a^i whereas the other two parameters θ, ϕ enter in the definition of its conjugate $\hat{E}_i^a$. Since the topological densities are all functions of the geometric fields (i.e. tetrads and spin-connections), addition of matter coupling does not affect such a gauge theoretic interpretation of gravity.

4. Concluding remarks

With the Hamiltonian theory with all three topological densities in place, it is important to investigate what are the possible imports of these terms in non-perturbative quantum gravity, i.e., whether these terms imply non-trivial topological sectors and potential instanton effects in the quantum theory[7, 8]. Also, since the Barbero-Immirzi parameter is already known to appear in the area spectrum in Loop Quantum Gravity, there is no reason not to suspect a similar role of the other two parameters in the context of quantum geometry of spacetime.

We emphasize that in the real SU(2) formulation as presented here, the Dirac bracket between the phase space variables A_a^i and $\hat{E}_i^a$ is not a canonical one, unlike the Barbero-Immirzi formulation. Although this need not be an issue as far as the classical theory is concerned, quantization based on these canonical variables is not straightforward. However, as demonstrated in [2], it is possible to find a suitable canonical pair which leads to the standard bracket, thus providing a smooth passage towards the quantum theory.

Acknowledgments

It is a pleasure to thank Fernando Barbero and Madhavan Varadarajan for their critical remarks, Prof. Romesh Kaul for collaboration on this topic and the organisers of Loops-11 for the wonderful hospitality at CSIC, Madrid where this work was presented.

[1] S. Coleman, *Aspects of Symmetry* (Cambridge University Press, 1985);
R. Rajaraman, *Solitons and Instantons* (North-Holland, The Netherlands, 1982)
[2] R. K. Kaul, S. Sengupta, Phys. Rev. **D85**, 024026 (2012)
[3] G. Date, R.K. Kaul, S. Sengupta, Phys. Rev. **D79**, 044008 (2009)
[4] A. Ashtekar and J. Lewandowski, Class. Quant. Grav. **21**, R53 (2004)
[5] S. Holst, Phys. Rev. **D53**, 5966-5969 (1996)
[6] S. Sengupta and R.K. Kaul, Phys. Rev. **D81**, 024024 (2010)
[7] S. Sengupta, Class. Quantum Grav. **27**, 145008 (2010)
[8] Simone Mercuri and Andrew Randono, Class. Quant. Grav. **28**, 025001 (2011)

Journal of Physics: Conference Series **360** (2012) 012025 doi:10.1088/1742-6596/360/1/012025

Perfect discretization of path integrals[1]

Sebastian Steinhaus

Max Planck Institute for Gravitational Physics (Albert Einstein Institute), Am Mühlenberg 1, 14476 Golm, Germany

E-mail: sebastian.steinhaus@aei.mpg.de

Abstract. In order to obtain a well-defined path integral one often employs discretizations. In the case of General Relativity these generically break diffeomorphism symmetry, which has severe consequences since these symmetries determine the dynamics of the corresponding system.

In this article we consider the path integral of reparametrization invariant systems as a toy example and present an improvement procedure for the discretized propagator. Fixed points and convergence of the procedure are discussed. Furthermore we show that a reparametrization invariant path integral implies discretization independence and acts as a projector onto physical states.

1. Introduction and Motivation

Discretizing a continuum theory in order to employ numerical treatment or to regularize path integrals has become an important tool. In addition to that, in some approaches to quantum gravity discrete structures are expected to be fundamental.

However, discretizations have several drawbacks, in particular if the continuous system contains symmetries, e.g. diffeomorphism symmetry in General Relativity, which are generically broken by the discretization [2]. Furthermore, discretizations are not unique, since many different discretizations might lead to the same continuum physics[2]. As soon as discrete structures are supposed to be fundamental, these ambiguities must be addressed.

Another important issue, in case the continuous symmetries are broken, are "broken gauge" modes, i.e. degrees of freedom which are considered to be gauge in the continuum. In the discrete setting these modes become physical, that is one has to integrate them out in a path integral approach. However, in the continuum limit, the symmetries get restored such that one sums over equivalent discretizations, which might lead to divergencies.

These issues can be overcome by implementing the symmetry of the continuous system into the discrete. In addition to restoring the continuous symmetry, the system is independent of the discretization and the discretization might be even unique [3]. However, to achieve this one has to pull back the dynamics of the continuous system onto the lattice, i.e. to solve the dynamics. In case this is not possible, one can improve the discretization by a Wilsonian Renormalization procedure and implement the dynamics step by step.

Such a discretization is called a perfect discretization, for instance discussed in [4] for asymptotically free theories. This has been extended to systems with a diffeomorphism(-like) symmetry, e.g. reparametrization invariant systems [5], Regge calculus without [6] and

[1] Based on a talk given at Loops'11, Madrid, May 27 [1]

[2] Actually extracting large scale physics from discrete quantum gravity approaches is one of the main problems.

with a cosmological constant [7]. However, these approaches only discuss classical systems. In [3] we have successfully extended the program of perfect discretizations to the quantized reparametrization invariant particle, the results will be presented here briefly.

2. Reparametrization invariant system

A classical system is described by its Lagrangian $L(q, \dot{q})$, which is a function of generalized coordinates $q(t)$ and its first time-derivative $\dot{q}(t)$, with the action given by $S = \int dt\, L(q, \dot{q})$. By introducing an auxiliary parameter s, the sytem is parametrised such that time t becomes a variable itself. One can check that

$$S(q(s), t(s)) = \int ds\, L\Big(q, \frac{q'}{t'}\Big) t' \tag{1}$$

is invariant under reparametrizations, i.e. transformations of the form $q(s), t(s) \to q(f(s)), t(f(s))$. As a consequence, the solutions to the equations of motion to (1) are not uniquely determined by the boundary data. In the canonical framework, this is manifest as the canonical momenta p_q and p_t (conjugated to q, t respectively) are not linearly independent, which results in a constraint identical to the Hamiltonian $\tilde{H} = p_t + H = 0$ of the system, where H denotes the Hamiltonian of the non-parametrized system.

To discretize the system, $q(s)$ and $t(s)$ are replaced by a finite set of variables, q_n and t_n with $n = 0, \ldots, N$. The action is discretized by approximating derivatives by difference quotients and choosing a discretization of the potential term:

$$S^{(0)}(q_n, t_n) := \sum_{n=0}^{N-1} S_n^{(0)} := \sum_{n=0}^{N-1} \left[\frac{1}{2} \left(\frac{q_{n+1} - q_n}{t_{n+1} - t_n} \right)^2 + \frac{1}{2} \left(V(q_n) + V(q_{n+1}) \right) \right] (t_{n+t} - t_n) \tag{2}$$

This is extended to a quantum system by defining the propagator:

$$K^{(0)}(q_n, t_n, q_{n+1}, t_{n+1}) := \sqrt{\frac{1}{2\pi\hbar(t_{n+1} - t_n)}} \exp\left\{ -\frac{1}{\hbar} S_n^{(0)} \right\} \tag{3}$$

In order to improve this discretization, consider a small discretization consisting of three lattice sites, where only the central lattice site with variables q_1 and t_1 is dynamical. The propagation between the lattice sites is described by $K^{(n)}(q_i, t_i, q_{i+1}, t_{i+1})$, see also Fig. 1. By integrating out q_1 and t_1, one obtains an improved propagator

$$K^{(n+1)}(q_0, t_0, q_2, t_2) := \int dq_1\, dt_1\, K^{(n)}(q_0, t_0, q_1, t_1) K^{(n)}(q_1, t_1, q_2, t_2). \tag{4}$$

Iterating this procedure gives the renormalization group flow of the system.

In order to discuss convergence of this procedure, we consider a class of systems by replacing the propagator by arbitrary coefficient functions. In case of the Harmonic oscillator, this gives:

$$K^{(n)}(q_0, q_1, T) := \eta^{(n)}(T) \exp\left\{ -\frac{1}{\hbar} \left(\alpha_1^{(n)}(T)(q_0^2 + q_1^2) + \alpha_2^{(n)} q_0 q_1 \right) \right\} \tag{5}$$

Applying the discussed improving procedure to (5), one obtains recursion relations, e.g.:

$$\alpha_1^{(n+1)}(2T) = \alpha_1^{(n)}(T) - \frac{\alpha_2^{(n)}(T)^2}{8\alpha_1(T)}, \quad \alpha_2^{(n+1)}(2T) = -\frac{\alpha_2^{(n)}(T)^2}{4\alpha_1^{(n)}(T)}. \tag{6}$$

Journal of Physics: Conference Series **360** (2012) 012025 doi:10.1088/1742-6596/360/1/012025

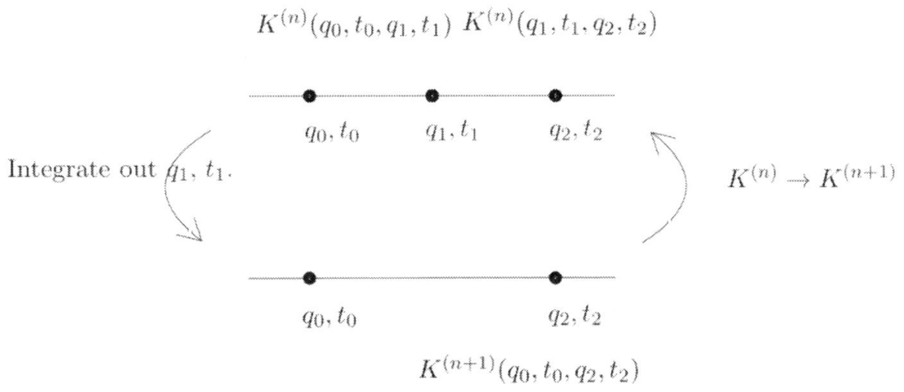

Figure 1. By integrating out the variables q_1 and t_1, one obtains an improved propagator $K^{(n+1)}$.

Fixed point equations can be obtained from (6) by dropping the superscripts. Solving these fixed point equations is equivalent to obtaining the continuum limit / solving the dynamics of the system. It turns out that the convergence of the initial system to the fixed point depends crucially on the choice of initial functions, in particular the measure. The procedure is convergent if

$$\eta^{(0)}(T) \sim \sqrt{\frac{1}{2\pi\hbar T}}. \tag{7}$$

in leading order in T.

3. Reparametrization invariance implies discretization independence

So far we have not discussed the relation between the fixed point of the improvement procedure, i.e. a discretization independent system, and reparametrization invariance.

Reparametrization invariance in the discrete is invariance of the action / path integral under translations of the (inner) vertices in time [3]. Assume the propagator K is invariant under vertex translations and consider the discretization described in Fig. 1. The path integral of this system is given by:

$$\langle q_0, t_0 | q_2, t_2 \rangle := \int dq_1\, dt_1\, K(q_0, t_0, q_1, t_1) K(q_1, t_1, q_2, t_2) \tag{8}$$

Since the K is invariant under vertex translations, i.e. translations of t_1, we can gauge fix t_1 to an arbitrary value t_1^f and drop the t_1 integration. Thus

$$\langle q_0, t_0 | q_2, t_2 \rangle = \int dq_1\, K(q_0, t_0, q_1, t_1^f) K(q_1, t_1^f, q_2, t_2) \tag{9}$$

Since t_1^f is arbitrary, we can consider the limit $t_1^f \to t_2$. Since $K(q_1, t_1^f, q_2, t_2)$ gives the amplitude for a particle to propagate from q_1 to q_2 in the time interval $(t_2 - t_1^f)$, $K(q_1, t_1^f, q_2, t_2) \to \delta(q_2 - q_1)$ in the limit $t_1^f \to t_2$. But t_1^f is arbitrary, such that we obtain:

$$K(q_0, t_0, q_2, t_2) = \int dq_1 K(q_0, t_0, q_1, t_1) K(q_1, t_1, q_2, t_2) \tag{10}$$

Hence we have shown that reparametrization invariance realized in the discrete implies discretization independence of the system. Furthermore, (10) is the usual propagator property of quantum mechanics.

4. Example: Anharmonic oscillator

The same techniques allow to find the perfect discretization of the anharmonic oscillator with anharmonic term λq^4, as a perturbation series in λ. For details, see [3].

One solution to the fixed point equations for the anharmonic oscillator to first order in the coupling constant λ is given by:

$$
\begin{aligned}
K(q_0, q_1, T) = &\sqrt{\frac{\omega}{2\pi\hbar\sinh(T\omega)}}\exp\left\{-\frac{1}{\hbar}S_{harm}(q_0,q_1,T)\right\}\times\left(1-\frac{\lambda}{\hbar}\times\right.\\
&\times\left[\frac{1}{768\omega\sinh^4(\omega T)}\Big(12T\omega-8\sinh(2T\omega)+\sinh(4T\omega)\Big)\left(q_0^4+q_1^4\right)+\right.\\
&+\frac{1}{192\omega\sinh^4(T\omega)}\Big(-12T\omega\cosh(T\omega)+9\sinh(T\omega)+\sinh(3T\omega)\Big)\left(q_0^3q_1+q_0q_1^3\right)+\\
&+\frac{1}{64\omega\sinh^4(T\omega)}\Big(2T\omega(2+\cosh(2T\omega))-3\sinh(2T\omega)\Big)q_0^2q_1^2+\\
&+\hbar\left(\frac{2+\cosh^2(T\omega)-3T\omega\coth(T\omega)}{32\omega^2\sinh^2(T\omega)}(q_0^2+q_1^2)+\frac{4T\omega+2T\omega\cosh(2T\omega)-3\sinh(2T\omega)}{32\omega^2\sinh^3(T\omega)}q_0q_1\right)-\\
&\left.\left.-\frac{\hbar^2}{64\omega^3}\left(3\coth(T\omega)-T\omega\left(2+\frac{3}{\sinh^2(T\omega)}\right)\right)\right]+O(\lambda^2)\right) \qquad (11)
\end{aligned}
$$

where S_{harm} is the perfect action of the harmonic oscillator. Only the full expression (11) satisfies the constraint equation (to first order in λ). For the interpretation and a more general solution of the fixed point equations, we refer to [3].

5. Summary and Outlook

To address the problem of broken gauge symmetries in discretized reparametrization invariant systems, we introduced the method of perfect discretizations. With this we can construct a reparametrization symmetry preserving discretization of the path integral as a fixed point of a renormalization flow. We have also proven that in these systems the issues of (broken) reparametrization invariance and discretization independence are equivalent. For the discussion of approximate solutions of the fixed point equations, uniqueness and (pseudo-)gauge modes, we refer to [3].

Applying coarse graining / to improve the discretization for quantum gravity models is left for future work. For some first steps, see [8]. Furthermore, in [9] we will discuss a measure for linearized Regge Calculus invariant under Pachner moves.

References

[1] `http://loops11.iem.csic.es/loops11/index.php?option=com_content&view=article&id=75&Itemid=73`
[2] Bahr B and Dittrich B 2009 (Broken) Gauge Symmetries and Constraints in Regge Calculus *Class. Quant. Grav.* **26** 225011. (*Preprint* arXiv:0905.1670 [gr-qc]).
[3] Bahr B, Dittrich B and Steinhaus S 2011 Perfect discretization of reparametrization invariant path integrals *Phys. Rev.* **D83** 105026. (*Preprint* arXiv:1101.4775 [gr-qc]).
[4] Hasenfratz P and Niedermayer F 1994 Perfect lattice action for asymptotically free theories *Nucl. Phys.* **B414** 785-814. (*Preprint* hep-lat/9308004).
[5] Marsden J E and West M 2001 Discrete mechanics and variational integrator *Acta Numerica* , **10** 357-514.
[6] Rocek M and Williams R M 1981 Quantum Regge Calculus *Phys. Lett.* **B104** 31.
[7] Bahr B and Dittrich B 2009 Improved and Perfect Actions in Discrete Gravity *Phys. Rev.* **D80** 124030. (*Preprint* arXiv:0907.4323 [gr-qc]).
[8] Dittrich B, Eckert F C and Martin-Benito M 2011 Coarse graining methods for spin net and spin foam models *Preprint* arXiv:1109.4927 [gr-qc].
[9] Dittrich B and Steinhaus S 2011 Path integral measure and triangulation independence in discrete gravity *Phys. Rev.* **D85** 044032 (*Preprint* arXiv:1110.6866 [gr-qc]).

Journal of Physics: Conference Series **360** (2012) 012026 doi:10.1088/1742-6596/360/1/012026

Quantum back-reaction in a universe with positive cosmological constant

David Brizuela

Institute for Gravitation and the Cosmos, The Pennsylvania State University, 104 Davey Lab, University Park, PA 16802, USA, and
Max-Planck-Institut für Gravitationsphysik, Albert-Einstein-Institut, Am Mühlenberg 1, 14476 Potsdam-Golm, Germany.

E-mail: brizuela@aei.mpg.de

Abstract. Semiclassical techniques have proven to be a very powerful method to extract physical effects from different quantum theories. Therefore, it is expected that in the near future they will play a very prominent role in the context of quantum gravity. In this work we develop systematic tools to derive semiclassical approximations for any quantum theory with one degree of freedom. In our approach, the wave function is decomposed in terms of an infinite set of moments, which encode the complete quantum information of the system. Semiclassical regimes can then be properly described by truncation of this infinite system. The use of efficient computer algebra tools allows us to compute the equations of motion up to a very high order. In this way, we can study very precisely the quantum back reaction of the system as well as the convergence of the method with the considered order. Finally, these tools are applied to the particular case of a homogeneous universe filled with a massless scalar field and positive cosmological constant, which provide interesting physical results.

1. Introduction

Nowadays, one of the main goals of the research in the context of quantum gravity is to extract physical predictions from the different theories proposed in the literature. This is a crucial step since, in the best scenario, we will be able to make contact with the experiments in order to verify or falsify these theories. As in any other quantum theory, semiclassical approximations are expected to play a very prominent role in this respect. On the one hand, they will allow us to extract new physical phenomena in their validity regime, where quantum effects are relevant but not dominant. On the other hand, even if deep quantum phases can be conceptually very interesting, their direct experimental observation is not expected in a near future due to technical difficulties. Thus, in order to obtain potentially observable effects, one should rather focus on the semiclassical region. In the particular case of quantum cosmology, these approximations should provide consistency with the large-scale behavior of the universe and, probably, some potential low energy observation, for instance through the cosmic microwave background.

In this regard, the main available tool in quantum gravity is that of effective equations [1, 2]. This approach describes the evolution of the expectation values of the quantum observables of the system as well as its fluctuations, correlations and higher-order moments. In the general case they form an infinite system of coupled differential equations. However in semiclassical regimes

Journal of Physics: Conference Series **360** (2012) 012026 doi:10.1088/1742-6596/360/1/012026

it can be truncated and provides, to a very good approximation, the evolution of the quantum system.

Therefore, the main goal of this work is to derive systematic semiclassical approximations for any quantum system with one degree of freedom. In order to do so, we will first analyze the general formalism and then construct efficient computer algebra tools. Finally, the developed machinery will be applied to the particular case of a homogeneous universe with a matter content of a massless scalar field and positive cosmological constant.

2. General formalism

Let us assume we have a classical system with one degree of freedom which is thus completely described by a couple of conjugate variables (V, P). Once the quantization of this system is performed promoting those variables to operators, the observable information must be extracted through expectation values. Hence we define the infinite set of moments,

$$G^{a,b} := \langle (\hat{P} - P)^a (\hat{V} - V)^b \rangle_{\rm Weyl}, \tag{1}$$

with $V := \langle \hat{V} \rangle$ and $P := \langle \hat{P} \rangle$, which will encode the complete quantum information of the system. In last formula the subscript Weyl stands for totally symmetric ordering. These moments are purely quantum variables that would vanish in a classical regime.

The evolution of these variables is then given by the effective Hamiltonian, defined as the expectation value of the quantum Hamiltonian $\hat{H}$. By Taylor expansion, it can be written as

$$H_{\rm eff}(V, P, G^{a,b}) := \langle \hat{H} \rangle_{\rm Weyl} = H(V, P) + \sum\sum \frac{1}{a!b!} \frac{\partial^{a+b} H}{\partial P^a \partial V^b} G^{a,b}, \tag{2}$$

where H is the classical Hamiltonian. Making use of the definition $\{\langle \hat{X} \rangle, \langle \hat{Y} \rangle\} = -i\hbar^{-1} \langle [\hat{X}, \hat{Y}] \rangle$, for any two operators $\hat{X}$ and $\hat{Y}$, the equations of motion generated by this effective Hamiltonian are guaranteed to be equivalent to the Schrödinger flow of states or the Heisenberg flow of operators.

As has been made explicit, the effective Hamiltonian is a state-dependent object and hence a function of expectation values and all the moments $G^{a,b}$. Therefore, the computation of the equations of motion for different variables implies calculating the Poisson bracket between two generic moments. This bracket is given by,

$$\{G^{a,b}, G^{c,d}\} = a\, d\, G^{a-1,b}\, G^{c,d-1} - b\, c\, G^{a,b-1}\, G^{c-1,d} + \sum_n \left(\frac{i\hbar}{2}\right)^{n-1} K^n_{abcd}\, G^{a+c-n,b+d-n}, \tag{3}$$

where the sum over n runs over all odd numbers from 1 to $\mathrm{Min}(a+c, b+d, a+b, c+d)$ and the coefficients K^n_{abcd} depend on all its indices. This formula was already presented in the literature [1], but a typo remain discovered. Due to the importance of this formula to systematize the computation of the effective equations, in Ref. [3], we have corrected it and provided a complete proof.

3. Algebraic implementation

In order to automatize the computation of Poisson brackets between two generic moments $\{G^{a,b}, G^{c,d}\}$, we have developed two different *Mathematica* codes. The first code works iteratively. That is, it begins with the definition of the moments given by (1) and then uses the generic relation $\{\langle \hat{X} \rangle, \langle \hat{Y} \rangle\} = -i\hbar \langle [\hat{X}, \hat{Y}] \rangle$, as well as properties of the Poisson brackets and of the commutator, until all terms are reduced to simple commutator between isolated operators $\hat{V}$ and $\hat{P}$. This is a long process and, in fact, the time the computer requires to compute a given

Poisson bracket $\{G^{a,b}, G^{c,d}\}$ by this method increases exponentially with the considered order $(a+b+c+d)$.

On the other hand, the second code makes use of the formula (3) and it is much more efficient. For instance, the timing to compute a 40th-order bracket like $\{G^{11,8}, G^{9,12}\}$ is less than a thousandth of a second, which clearly shows the practical importance of equation (3). This generic code is one of the main results of our research and it allows us to obtain the equations of motion to a very high-order for any quantum system with one degree of freedom. In the following section we will develop one example in the context of quantum cosmology.

4. An example: massless scalar field with positive cosmological constant

The Friedmann equation for a homogeneous universe with a massless scalar field ϕ and a positive cosmological constant Λ reads,

$$\left(\frac{a'}{a}\right)^2 = \frac{4\pi G}{3}\frac{p_\phi^2}{a^6} + \Lambda, \tag{4}$$

where a is the scale factor, p_ϕ the conjugate momentum of the scalar field and derivatives are with respect to cosmological time. We define the canonical gravitational variables $V := 2a^3/3$ and $P := a'/a$, and choose the scalar field as the internal time. In this way, at classical level, the momentum p_ϕ turns out to be the physical Hamiltonian,

$$H := p_\phi = \frac{3}{2}V\sqrt{P^2 - \Lambda}. \tag{5}$$

In order to obtain the effective Hamiltonian that will describe the dynamics of all moments, the Taylor expansion (2) is applied

$$H_{\text{eff}} := H + \sum_{n=2}^{\infty}[V\alpha_n(P,\Lambda)G^{n,0} + \beta_n(P,\Lambda)G^{n-1,1}]. \tag{6}$$

Here α_n and β_n are coefficients that only depend on the order n, the cosmological constant Λ and the momentum P. Note that, due to the linear dependence of the classical Hamiltonian on the variable V, only moments of the form $G^{a,0}$ and $G^{a,1}$ appear on the definition of the effective Hamiltonian.

In order to extract physical information from the system, we have obtained the equations of motion and solve them numerically at every order up to ten. As initial state we have chosen an unsqueezed Gaussian pulse in the volume V. The corresponding moments $G^{a,b}$ are non-vanishing only if both a and b are even numbers. The initial values of the expectation values V and P, as well as the width of the Gaussian are chosen so that initially the state is peaked, that is, the relative fluctuations $\Delta V/V$ and $\Delta P/P$ are small. Finally, three different values have been taken for the cosmological constant: small, intermediate and large.

The behavior of this system with vanishing cosmological constant ($\Lambda = 0$) is particularly simple. In this case, the effective Hamiltonian is only composed by a finite amount of terms and all equations of motion decouple. Therefore there is no back reaction and the evolution of all variables is simply exponential $V \approx e^{\phi}$, $P \approx e^{-\phi}$, and $G^{a,b} \approx e^{(b-a)\phi}$. This harmonic system is used as a reference to check that the quantum back reaction becomes stronger when considering larger values for Λ. On the other hand, the negative cosmological case has been analyzed with these effective techniques in [4] and with a complete numerical solution of the wave equation in [5]. However, for $\Lambda < 0$, the volume is bounded by recollapse and the moments do not grow for large volumes. These two features are not present for $\Lambda > 0$ case, making this analysis of high-order moments necessary.

Regarding the evolution of the expectation values V and P, we would like to point out two main features we have observed. Classically, the volume $V(\phi)$ of the universe is an increasing function and reaches infinity at a finite value of time $\phi = \phi_{\rm div}$. Our analysis shows that the quantum back reaction enhances this divergence. That is, at higher order the divergence of the volume is reached faster. The second feature is related to the convergence. When we truncate the system at a given order, we implicitly assume that the higher-order moments can not change significantly the observed behavior. In order to test this convergence, we have defined $\Delta V_n(\phi) := 1 - V_n(\phi)/V_{n+1}(\phi)$, where $V_n(\phi)$ is the trajectory of the expectation value of the volume at order n. This object is time dependent, so its behavior will change with ϕ. We have studied three different values of ϕ and in all the considered cases, even for very late times, and for all values of Λ, we have obtained an exponential convergence of the system with the order n. This is a very important result and shows the reliability of our method even near the divergence of the volume.

Concerning the behavior of the moments, we observe that they depart from their harmonic behavior ($\Lambda = 0$ case) as the value of the cosmological constant is increased. As expected, several moments grow and become dominant, providing thus strong quantum back reaction. In all cases, even for small values of Λ, as soon as the evolution is started, the initially vanishing moments are immediately excited. In this way, the initial Gaussian shape is broken. Remarkably, after a rapid initial increase of these moments, the state adapts itself to the evolution and moments change less severely. One of the open questions in this respect is whether this natural state is independent of the initial conditions. We can not give any definitive answer without performing further numerical analysis but, due to the rapid adaptation to the evolution, the consequences of assuming an initial Gaussian state might, after all, be not so special.

5. Conclusions

We have presented a systematic method to analyze the effective behavior of any quantum system with one degree of freedom. The key ingredients of this method are the generic formula for the Poisson bracket between any two moments (3) and an extensive use of computer algebra tools. As an example, these tools have been applied to the particular case of a homogeneous universe with a matter content of a massless scalar field and positive cosmological constant. The main results we have obtained is that the quantum back reaction enhances the classical divergence of the volume. In addition, we have observed that the initial Gaussian state is immediately adapted to another state better suited to the evolution. And remarkably, exponential convergence of the results with the considered order has been obtained even for late times (near the divergence of the volume), which validates the applied methods.

Acknowledgments

I would like to thank M. Bojowald, H.H. Hernández, M. J. Koop, and H. A. Morales-Técotl for joint collaboration on which the presented talk is based. This work was supported by the Spanish Ministry of Education through National Program No. I-D+i2008-2011 and by the Spanish MICINN Project No. FIS2008-06078-C03-03.

References

[1] M. Bojowald and A. Skirzewski, Rev. Math. Phys. **18**, 713 (2006).
[2] M. Bojowald, B. Sandhöfer, A. Skirzewski, and A. Tsobanjan, Rev. Math. Phys. **21**, 111 (2009).
[3] M. Bojowald, D. Brizuela, H. H. Hernández, M. J. Koop, H. A. Morales-Tecotl, Phys. Rev. D **84**, 043514 (2011).
[4] M. Bojowald and R. Tavakol, Phys. Rev. D **78**, 023515 (2008).
[5] E. Bentivegna and T. Pawlowski, Phys. Rev. D **77**, 124025 (2008).

Journal of Physics: Conference Series **360** (2012) 012027 doi:10.1088/1742-6596/360/1/012027

Inflationary spectra and observations in loop quantum cosmology

Gianluca Calcagni
Max Planck Institute for Gravitational Physics (Albert Einstein Institute)
Am Mühlenberg 1, D-14476 Golm, Germany

E-mail: calcagni@aei.mpg.de

Abstract. We review some recent progress in the extraction of inflationary observables in loop quantum cosmology. Inverse-volume quantum corrections induce a growth of power in the large-scale cosmological spectra and are constrained by observations.

Observational implications of quantum gravity present a delicate issue. Corrections to the general relativistic dynamics are expected to arise in different ways. For instance, loop corrections are always present in perturbative graviton field theory, which can be captured in effective actions with higher-curvature corrections to the Einstein–Hilbert action. The additional terms change the Newton potential as well as the cosmological dynamics. However, in currently observable regimes the curvature scale is very small, and so one expects only tiny corrections of (adimensional) comoving size at most $\ell_{\rm Pl}\mathcal{H}$, where $\ell_{\rm Pl}$ is the Planck length and $\mathcal{H}^{-1} = a/a'$ is the comoving radius of the Hubble region (a is the scale factor in the flat Friedmann–Lemaître–Robertson–Walker, FLRW, background and primes denote derivatives with respect to conformal time τ). In such cases, tests of quantum gravity are possible at best indirectly, for instance if it provides concrete and sufficiently constrained models for inflation. So far, however, models do not appear tight enough.

In background-independent frameworks such as loop quantum gravity (LQG) [1], stronger modifications of the theory are possible since the usual covariant continuum dynamics is generalized, and entirely new effects may be contemplated. In LQG, gauge transformations as well as the dynamics are generated by constraint equations. Since the latter are modified with respect to the classical constraints, gauge transformations change and new spacetime structures become apparent. Some of these modifications arise as follows. An inverse of metric operators is needed to construct the quantized constraints, but this inverse does not exist because metric operators have discrete spectra containing zero as an eigenvalue. Certain quantization procedures allow one to construct suitable densely-defined operators which give rise to quantum corrections sensitive to the discreteness scale. The modifications are controlled by the requirement that covariance not be broken but at most deformed with respect to its ordinary incarnation, leading to anomaly-free and consistent sets of equations. The resulting dynamics and gauge-invariant observables provide the basis for a cosmological analysis.

Inverse-volume corrections constitute the first example that has been consistently implemented, when they are small, in the dynamics of loop quantum cosmology (LQC [2]). Effective dynamical equations with inverse-volume corrections are known for pure FLRW and linear perturbations in all sectors (scalar, vector and tensor) [3, 4, 5, 6]. The setting is an inflationary era driven by a slowly rolling scalar field, the only difference with respect to the standard case being the presence of the quantum corrections. For instance, the background equations of motion are $\mathcal{H}^2 = (8\pi G/3)\alpha[\varphi'^2/(2\nu) + a^2 V(\varphi)]$ and $\varphi'' + 2\mathcal{H}(1 - \mathrm{d}\ln\nu/\mathrm{d}\ln a^2)\varphi' + \nu a^2 V_{,\varphi} = 0$, where V is the scalar field potential and

$$\alpha \approx 1 + \alpha_0 \delta_{\rm Pl}\,, \qquad \nu \approx 1 + \nu_0 \delta_{\rm Pl}\,, \qquad \delta_{\rm Pl} := (a_*/a)^\sigma \tag{1}$$

are the inverse-volume quantum corrections. These are parametrized as an inverse power law of the scale factor. In pure mini-superspace (no inhomogeneities), the parameters α_0 and ν_0 are calculable from the quantization ambiguities of the Hamiltonian constraint and also depend on σ, a constant determined by the way an elementary holonomy cell evolves with the Hubble expansion. A natural choice is the so-called 'improved quantization' scheme [7], where $\sigma = 6$ and $\alpha_0 \sim O(0.1) \sim \nu_0$. In a flat or open universe, the Hamiltonian formalism is well defined only in a finite comoving volume $\mathcal{V}_0$, which is chosen *ad hoc*. However, the quantum corrections are of the form $\delta_{\rm Pl} \sim (\ell_{\rm Pl}^3/\mathcal{V}_0)^{\sigma/3} a^{-\sigma}$, and they depend explicitly on the unphysical fiducial volume (or, in other words, on the constant a_* which can be rescaled arbitrarily).

Switching on inhomogeneities, the interpretation of inverse-volume corrections and the fiducial-volume problem change considerably. Once a closed constraint algebra and the full set of linearized perturbed equations are obtained, one can study the role of different parametrization schemes and a number of other issues. We mention the conservation law for the curvature perturbation, the construction of the inflationary spectra and other cosmological observables and, finally, the extraction of observational constraints from large-scale data. These aspects, analyzed in [6, 8, 9] will be presently reviewed here. Another characteristic effect of LQC quantization, holonomy corrections, should also enter the picture, but so far the closure of the constraint algebra has been obtained only in the tensor and vector sector [3, 10]. Consequently, the phenomenology of holonomy effects has been limited to the structure of the primordial bounce and to the inflationary tensor signal [11].

The first step is to implement first-order perturbation theory in the classical constraints. The fundamental canonical variables, the densitized triad and the Ashtekar–Barbero connection are given by a background contribution plus a linear inhomogeneous correction: $E^a_i = p\delta^a_i + \delta E^a_i$, $A^i_a = c\delta^i_a + (\delta\Gamma^i_a + \gamma\delta K^i_a)$, where indices $i = 1, 2, 3$ and $a = 1, 2, 3$ run over spacetime and internal (tangent) space directions, $p = |a|^2$, $c = \gamma\dot{a}$ classically, γ is the Barbero–Immirzi parameter, Γ is the spin connection and K is the extrinsic curvature. Perturbations obey the canonical Poisson relation $\{\delta K^i_a(\mathbf{x}), \delta E^b_j(\mathbf{y})\} = 8\pi G \delta^b_a \delta^i_j \delta(\mathbf{x}, \mathbf{y})$. The smeared effective Hamiltonian constraint with inverse-volume correction function reads $H[N] \sim \int \mathrm{d}^3x N[\alpha(E)\mathcal{H}_g + \nu(E)\mathcal{H}_\pi + \rho(E)\mathcal{H}_\nabla + \mathcal{H}_V]$, where N is the lapse function and different contributions $\mathcal{H}_{g,\pi,\nabla,V}$ pertain the gravitational sector and the scalar field kinetic, gradient, and potential terms, respectively. Similarly, one considers the perturbed Gauss and diffeomorphism constraints, and imposes closure of the effective constraint algebra, $\{C_\alpha, C_\beta\} = f_{\alpha\beta}{}^\gamma(A, E)C_\gamma$. The resulting perturbed equations contain counterterms which fix the functions α, ν and ρ and guarantee anomaly cancellation in the constraint algebra [3, 5]. These counterterms only depend on the three parameters α_0, ν_0 and σ, but a consistency condition further reduces the parameter space to two dimensions. Notably, inflationary and de Sitter background solutions exist for $0 \lesssim \sigma \lesssim 3$ [6], which is in tension with the mini-superspace parametrization $\sigma = 6$. This problem is bypassed in the lattice refinement framework, where the proper fiducial volume $\mathcal{V} = \mathcal{V}_0 a^3$ is replaced by the volume $L^3 = \mathcal{V}/\mathcal{N}$ of a microscopic cell of size L, determined by the number of vertices (or 'patches') of an underlying discrete state. The number of cells can be parametrized as $\mathcal{N} = \mathcal{N}_0 a^{6n}$, where $0 \leq n \leq 1/2$ (both $\mathcal{N}$ and L must not decrease with the volume in a discrete geometrical setting). Then, $\delta_{\rm Pl} := (\ell_{\rm Pl}/L)^m = (\ell_{\rm Pl}^3 \mathcal{N}_0/\mathcal{V}_0)^{m/3} a^{-(1-2n)m} =: (a_*/a)^\sigma$, where m is a positive parameter. Eventually, one can argue that the parameter range is $\sigma \geq 0$ and $\alpha_0 \geq 0$, $\nu_0 \geq 0$ [6].

Scalar fluctuations Ψ and $\delta\varphi$ in the metric and in the scalar field generate the gauge-invariant curvature perturbation on comoving hypersurfaces $\mathcal{R} = \Psi + (\mathcal{H}/\varphi')(1 - \sigma\nu_0\delta_{\rm Pl}/6)\delta\varphi$. At large scales, this quantity is conserved thanks to a delicate cancellation of counterterms [6]: $\mathcal{R}' = [1 + (\alpha_0/2 + 2\nu_0)\delta_{\rm Pl}][\mathcal{H}/(4\pi G\varphi'^2)]\Delta\Psi$, where Δ is the Laplacian. Because of this property, one can argue (and also rigorously show) that the Mukhanov scalar variable $u = z\mathcal{R}$, where $z := (a\varphi'/\mathcal{H})[1 + (\alpha_0/2 - \nu_0)\delta_{\rm Pl}]$, obeys the simple dynamical equation $u'' - (s^2\Delta + z''/z)u = 0$, where $s(\alpha_0, \nu_0, \sigma)$ is the propagation speed of the perturbation. According to the inflationary paradigm, observables are expanded in terms of small slow-roll parameters $\epsilon := 1 - \mathcal{H}'/\mathcal{H}^2$, $\eta := 1 - \varphi''/(\mathcal{H}\varphi')$, $\xi^2 := \mathcal{H}^{-2}(\varphi''/\varphi')' + \epsilon + \eta - 1$. Asymptotic solutions to the Mukhanov equation at large scales, evaluated at horizon crossing, eventually yield the scalar spectrum, the scalar index and its running:

$$\mathcal{P}_{\rm s} := \frac{k^3}{2\pi^2 z^2}\left\langle |u_{k\ll\mathcal{H}}|^2\right\rangle\Big|_{k|\tau|=1} = \frac{G}{\pi}\frac{\mathcal{H}^2}{a^2\epsilon}(1 + \gamma_{\rm s}\delta_{\rm Pl}), \tag{2}$$

$$n_{\rm s} - 1 := \frac{{\rm d}\ln\mathcal{P}_{\rm s}}{{\rm d}\ln k} = 2\eta - 4\epsilon + \sigma\gamma_{n_{\rm s}}\delta_{\rm Pl}, \tag{3}$$

$$\alpha_{\rm s} := \frac{{\rm d}n_{\rm s}}{{\rm d}\ln k} = 2(5\epsilon\eta - 4\epsilon^2 - \xi^2) + \sigma(4\tilde{\epsilon} - \sigma\gamma_{n_{\rm s}})\delta_{\rm Pl} = O(\epsilon^2) + O(\sigma\delta_{\rm Pl}), \tag{4}$$

where $\gamma_{\rm s}$ and $\gamma_{n_{\rm s}}$ depend on the parameters α_0, ν_0, σ in a precise way [6]. One can notice a large-scale enhancement of power via the term $\delta_{\rm Pl} \sim a^{-\sigma} \sim (1/|\tau|)^{-\sigma} \sim k^{-\sigma}$. If large enough, quantum corrections dominate and $\alpha_{\rm s} = \sigma f_{\rm s}(\alpha_0, \sigma)\delta_{\rm Pl}$, where f_s is a specific function of the parameters. Bounds on the scalar running, in fact, turn out to be the main constraint on the parameters. The power spectrum can be expressed in terms of the comoving wavenumber k of perturbations and of a pivot scale k_0, which must be chosen within the scale range probed by the given experiment. Letting $x := \ln(k/k_0)$, one has

$$\frac{\mathcal{P}_{\rm s}(k)}{\mathcal{P}_{\rm s}(k_0)} = \exp\left\{[n_{\rm s}(k_0) - 1]x + \frac{\alpha_{\rm s}(k_0)}{2}x^2 + f_{\rm s}\delta_{\rm Pl}(k_0)\left[x\left(1 - \frac{1}{2}\sigma x\right) + \frac{1}{\sigma}(e^{-\sigma x} - 1)\right]\right\}. \tag{5}$$

Due to cosmic variance, there is an intrinsic uncertainty in the determination of the spectrum at large scales (small multipoles ℓ), which should be compared with the strength of the typical signal from quantum corrections. Tensor observables can be calculated analogously and display the same type of corrections. From those, one can extract a consistency relation between scalar and tensor perturbations: $r = -8\{n_{\rm t} + [n_{\rm t}(\gamma_{\rm t} - \gamma_{\rm s}) + \sigma\gamma_{\rm t}]\delta_{\rm Pl}\}$.

In order to compare with observations, one can choose an inflationary potential and recast all observables as its functions. Then, for any given choice of σ, one can find an upper bound for the quantum parameter $\delta := \alpha_0\delta_{\rm Pl}$ (we recall that ν_0 is not independent). An example of likelihood contours and power-spectrum enhancement is illustrated in figure 1, for a quadratic potential. We use the 7-year WMAP data combined with large-scale structure, the Hubble constant measurement from the Hubble Space Telescope, Supernovae type Ia and Big Bang Nucleosynthesis. For a range of e-folds between 45 and 65, the probability distribution of ϵ is consistent with the theoretical range $0.008 < \epsilon < 0.011$, and the quadratic potential is compatible with observations as in the standard case. In the figure, cosmic variance is greater than the LQC signal, but this is *not* the case for $\sigma \lesssim 1$ [9]. For $\sigma > 2$, even tighter bounds are obtained, corresponding to practically unobservable inverse-volume effects. The 95%-confidence-level upper limits of δ constrained by observations for the potential $V(\varphi) = V_0\varphi^2$, with $k_0 = 0.002\ {\rm Mpc}^{-1}$ and different values of σ, are [9]

$$\delta(\sigma = 0.5) = 0.27, \quad \delta(\sigma = 1) = 3.5\times10^{-2}, \quad \delta(\sigma = 1.5) = 1.7\times10^{-3},$$
$$\delta(\sigma = 2) = 6.8\times10^{-5}, \quad \delta(\sigma = 3) = 4.3\times10^{-7}. \tag{6}$$

Figure 1. Left: Combined marginalized distribution for the quantum-gravity parameter $\delta(k_0) = \alpha_0\delta_{\rm Pl}(k_0)$ and the slow-roll parameter $\epsilon(k_0)$ with the pivot $k_0 = 0.002$ Mpc^{-1} for $V \propto \varphi^2$ and $\sigma = 2$. Internal and external solid lines correspond to the 68% and 95% confidence levels, respectively. Right: Primordial scalar power spectrum $\mathcal{P}_{\rm s}(\ell)$ for the same potential and pivot wavenumber (corresponding to $\ell_0 = 29$), with three different values of $\delta(k_0)$: 0 (dotted line), 7×10^{-5} (experimental upper bound, solid line), 4.8×10^{-4} (from the a-priori upper bound $\delta_{\rm Pl}(\ell = 2) < 0.1$, dashed line). The shaded region is affected by cosmic variance [8, 9].

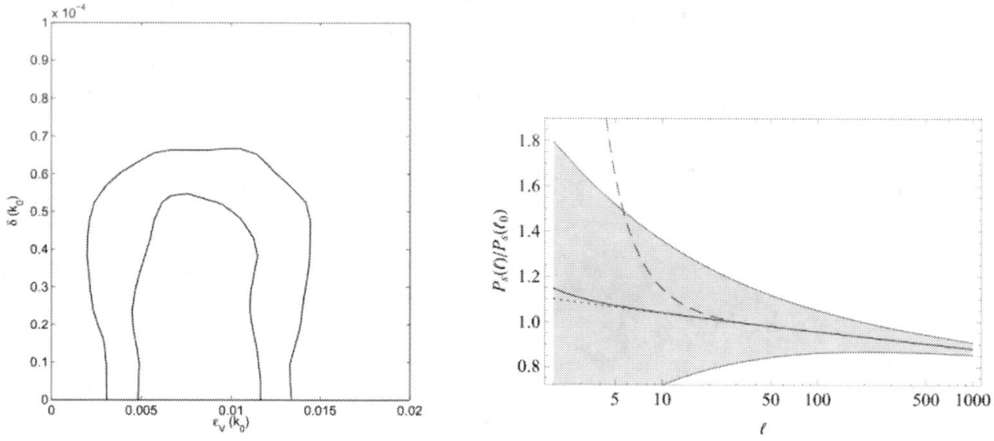

The likelihood analysis has not been performed for $\sigma = 6$ since the signal is below the cosmic variance threshold already when $\sigma = 2$. For $\sigma = 3$, the parameter $\delta = \nu_0\delta_{\rm Pl}$ has been used instead. The analysis of [9] includes also different potentials and values of the pivot scale.

To conclude, observations highlight a tension between mini-superspace parametrizations of FLRW LQC ($\sigma > 1$) and lattice refinement parametrizations. The latters are the only one compatible (at least in this model) with anomaly cancellation in inhomogeneous LQC and power-law inflationary solutions. Tight upper bounds can be obtained for inverse-volume quantum corrections; their improvement with future missions such as Planck will further constrain the parameter space of the theory and, hopefully, stimulate our understanding of the semi-classical limit of loop quantum cosmology.

References

[1] Rovelli C 2004 *Quantum Gravity* (Cambridge: Cambridge University Press); Thiemann T 2007 *Introduction to Modern Canonical Quantum General Relativity* (Cambridge: Cambridge University Press)
[2] Ashtekar A and Singh P 2011 *Class. Quant. Grav.* **28** 213001; Banerjee K, Calcagni G and Martín-Benito M 2011 *SIGMA* to appear (*Preprint* arXiv:1109.6801)
[3] Bojowald M and Hossain G M 2007 *Class. Quant. Grav.* **24** 4801; 2008 *Phys. Rev.* D **77** 023508
[4] Calcagni G and Hossain G M 2009 *Adv. Sci. Lett.* **2** 184
[5] Bojowald M, Hossain G M, Kagan M and Shankaranarayanan S 2008 *Phys. Rev.* D **78** 063547; 2009 *Phys. Rev.* D **79** 043505; 2010 *Phys. Rev.* D **82** 109903(E)
[6] Bojowald M and Calcagni C 2011 *J. Cosmol. Astropart. Phys.* JCAP03(2011)032
[7] Ashtekar A, Pawłowski T and Singh P 2006 *Phys. Rev.* D **74** 084003
[8] Bojowald M, Calcagni G and Tsujikawa S 2011 *Phys. Rev. Lett.* **107** 211302
[9] Bojowald M, Calcagni G and Tsujikawa S 2011 *J. Cosmol. Astropart. Phys.* JCAP11(2011)046
[10] Mielczarek J, Cailleteau T, Barrau A and Grain J 2011 *Preprint* arXiv:1106.3744
[11] Mielczarek J 2008 *J. Cosmol. Astropart. Phys.* JCAP11(2008)011; Copeland E J, Mulryne D J, Nunes N J and Shaeri M 2009 *Phys. Rev.* D **79** 023508; Grain J and Barrau A 2009 *Phys. Rev. Lett.* **102** 081301

Journal of Physics: Conference Series **360** (2012) 012028 doi:10.1088/1742-6596/360/1/012028

Consistent probabilities in sLQC

David A Craig[1] **and Parampreet Singh**[2]

[1] Department of Physics, Le Moyne College, Syracuse, New York, 13214, USA *and* Perimeter Institute for Theoretical Physics, Waterloo, Ontario, N2L 2Y5, Canada

[2] Department of Physics, Lousiana State University, Baton Rouge, Louisiana, 70803, USA

E-mail: `craigda@lemoyne.edu`, `psingh@phys.lsu.edu`

Abstract. We describe a complete, coherent framework within which probabilities may be consistently extracted from amplitudes for quantum histories in loop quantum cosmology. The decoherence functional is constructed for a loop quantization of a flat Friedmann-Lemaître-Robertson-Walker cosmological model, thereby permitting consistent quantum predictions to be made in a mathematically precise model of quantum cosmology. Consistent families of quantum histories are exhibited. Singularity avoidance by generic quantum states in loop-quantized models is contrasted with the failure of parallel Wheeler-DeWitt-quantized universes to avoid the big-bang singularity. It is shown that all loop quantum states approach arbitrarily large volume Wheeler-DeWitt states in both the asymptotic "past" and "future" of emergent time. The critical role played by decoherence of histories in these predictions is illustrated.

1. Introduction

The program of quantum cosmology is the application of a quantum theory of gravity[1] to arrive at quantum predictions for the behavior of cosmological models. Because the universe is homogeneous and isotropic on large scales, these models exhibit a high degree of symmetry. Quantum cosmology therefore serves in part also as a proving ground for candidate theories of quantum gravity because it allows ideas and techniques from the putative full theory to be applied to simplified models in which the classical symmetries are imposed prior to quantization. Loop quantum cosmology[2] (LQC) is a recent example of the successful application of this principle to loop quantum gravity (LQG). Perhaps the most striking result of LQC is the removal of the singularities of the corresponding classical model in a robust range of examples, in which the classical infinite-density "big bang" is replaced by a quantum "bounce" at the Planck density.

Easy to overlook in any application of a quantum theory of gravity to a cosmological model is that, in principle, the universe is a *closed system* – there is by definition nothing outside of it. In quantum mechanics, however, the orthodoxy is that probabilities may only be assigned to physical quantities which are *measured* by external observers, leading to the collapse of the wave function and the destruction of interference between alternatives. The canonical illustrative example is the two slit experiment. If the slit through which the particle passed is not measured, there is interference between the wave functions for the branches $\psi_{\mathrm{upper}}(y)$ and $\psi_{\mathrm{lower}}(y)$ passing through each slit and ending up at position y on the screen, and the probability $p(y)$ for ending up at y is not simply the sum of the probabilities for passing through each slit separately,

[1] See Ref. [1] for an overview of many of the currently viable candidates.

[2] See Ref. [2] for a thorough recent review.

$p(y) = |\psi_{\rm upper}(y) + \psi_{\rm lower}(y)|^2 \neq |\psi_{\rm upper}(y)|^2 + |\psi_{\rm lower}(y)|^2$. In this sense it is not logically consistent to assert the particle "did" pass through one slit or the other if the slit was not measured: the putative probabilities simply do not add up.

2. Consistent histories

There is, nonetheless, a framework for making consistent predictions for closed quantum systems for which no notion of external observers or measurements is available. This is known as the *consistent* or *decoherent histories* formulation of quantum theory [3, 4, 5]. The main idea is to take seriously the observation that the principle function served by external measurements, or, more generally, environmental decoherence [6] in producing well-defined probabilities is the destruction of interference between alternatives: probabilities may be assigned when, and only when, interference between alternatives vanishes. The central mathematical constructions of the theory are *class operators* C_h which specify physically distinct alternative histories h from an exclusive, exhaustive set of alternatives $\{h\}$; *branch wave functions* $|\psi_h\rangle$ constructed from the intitial states and the class operator for the history h, and which correspond to the particular quantum state which has "followed" the history h; and the *decoherence functional* $d(h, h')$ which measures the quantum interference between the histories h and h'. For theories defined on a Hilbert space with a homogeneous background time, class operators corresponding to histories $h = (\Delta a_{k_1}, \Delta a_{k_2}, \ldots)$ for which observable α_1 is in the range Δa_{k_1} at time t_1; observable α_2 is in the range Δa_{k_2} at time t_2; and so on, are defined by products of Heisenberg projections onto the ranges of eigenvalues of interest:

$$C_h = P^{\alpha_1}_{\Delta a_{k_1}}(t_1) P^{\alpha_2}_{\Delta a_{k_2}}(t_2) \cdots P^{\alpha_n}_{\Delta a_{k_n}}(t_n). \tag{1}$$

The corresponding branch wave function is given by

$$|\psi_h\rangle = C_h^\dagger |\psi\rangle. \tag{2}$$

The decoherence functional is then defined as

$$d(h, h') = \langle \psi_{h'} \,|\, \psi_h \rangle. \tag{3}$$

Probabilities $p(h)$ for the histories $\{h\}$ may then be sensibly defined – in the sense that $\sum_h p(h) = 1$ on an exclusive, exhaustive set of alternative histories – when the interference between h and h' vanishes for all pairs of histories in the set, as measured by the decoherence functional. This probability is given by the Lüders-von Neumann formula: $p(h) = \langle \psi_h \,|\, \psi_h \rangle$. When the decoherence functional is diagonal on a complete set of alternative histories, $\langle \psi_{h'} \,|\, \psi_h \rangle = 0$ for $h \neq h'$, that set is said to "decohere" or "be consistent", and the probabilities of the individual histories h are given by the diagonal elements of the decoherence functional: $d(h, h') = p(h)\, \delta_{h'h}$. The decoherence functional thus provides an objective, internally defined measure for when probabilities may be consistently defined in a closed quantum system. We now apply this construction to loop quantum cosmology.

3. Loop quantum cosmology

The model we consider is a loop quantization of a flat Friedmann-Lemaître-Robertson-Walker cosmology with a massless, minimally coupled scalar field ϕ as a matter source [7, 8, 2, 9]. We refer to these references for further definitions and technical details. The classical solutions of this model are $\phi = \pm \ln(V/V_c)/\sqrt{12\pi G} + \phi_c$, corresponding to disconnected expanding and collapsing branches. Within the framework of loop quantum cosmology, quantization in the harmonic gauge leads to an exactly solvable model dubbed "sLQC" [8]. In terms of the rescaled volume variable $\nu = \varepsilon V / 2\pi\gamma l_p^2$, where V is the volume of a fiducial spatial cell and $\varepsilon = \pm 1$, solutions $\psi(\nu, \phi)$

to the quantum Hamiltonian constraint split up into identical positive and negative frequency sectors. Restricting to positive frequencies, these states satisfy $-i\,\partial_\phi\psi(\nu,\phi) = \sqrt{\Theta}\,\psi(\nu,\phi)$, where Θ is the gravitational part of the constraint. In sLQC, however, Θ is a *difference* operator, and solutions may be chosen to have support only on the lattice $\nu = 4\lambda n$, with n an integer, where $\lambda = \sqrt{4\sqrt{3}\pi\gamma}$ is related to the "area gap" of loop quantum gravity. In this way, volume in sLQC becomes discrete, and the positive definite inner product given by group averaging is $\langle\psi\,|\,\chi\rangle = \sum_\nu \psi(\nu,\phi)^*\chi(\nu,\phi)$. It is seen that ϕ may for convenience be viewed as an emergent background "clock", and we will define evolution with respect to it via the propagator $U(\phi) = \exp(i\sqrt{\Theta}\phi)$.

General solutions to the constraint may be written

$$\Psi(\nu,\phi) = \int_{-\infty}^{\infty} dk\,\tilde{\Psi}(k)e_k^{(s)}(\nu)e^{i\omega_k\phi}, \tag{4}$$

where $\omega_k = \sqrt{12\pi G}|k|$ and the $e_k^{(s)}(\nu)$ are symmetric eigenfunctions of the evolution operator Θ with eigenvalues ω_k^2. Exact expressions for the $e_k^{(s)}(\nu)$ are available [7, 10, 11, 12], and much about them is known. Two pertinent facts are (i) the $e_k^{(s)}(\nu)$ approach at large volume a symmetric superposition of the corresponding expanding and contracting eigenfunctions of the Wheeler-DeWitt quantization of the same physical model; and (ii) the $e_k^{(s)}(\nu)$ exhibit an ultraviolet cutoff at $|k| = |\nu/2\lambda|$, decaying exponentially to zero for $|k|$ larger than that limit [13, 12].

Observables include $p_\phi = -i\hbar\partial/\partial\phi$, the momentum conjugate to the scalar field and a constant of the motion (since it commutes with $\sqrt{\Theta}$ via the constraint), and the volume at scalar field value ϕ^*, given by $\hat{\nu}_{|\phi^*}\,\psi(\nu,\phi) = U(\phi^*-\phi)^\dagger\,\hat{\nu}\,U(\phi^*-\phi)\psi(\nu,\phi)$. Defining Heisenberg-picture projection operators via the propagator $U(\phi)$, class operators, branch wave functions, and the decoherence functional in this model may be defined just as in Eqs. (1), (2) and (3). It is then straightforward to show, for example, that the probability the universe is found in a range of volume $\Delta\nu$ at ϕ is given by $p_{\Delta\nu}(\phi) = \sum_{\nu\in\Delta\nu}|\Psi(\nu,\phi)|^2$ [14].

It is known that these quantum cosmological models are non-singular, in the sense that the density is bounded above by approximately the Planck density [7, 8, 12], unlike the Wheeler-DeWitt quantization of the same model, in which states are invariably "sucked in" to the classical singularity in one of the limits $\phi \to \pm\infty$ [8, 9]. Indeed, numerical solutions [7] show that quasiclassical states peaked on a contracting classical solution "bounce" at small volume and thereafter remain peaked on an expanding classical solution.

In Ref. [9] the Wheeler-DeWitt quantization of the same model was studied from the consistent histories point of view. Taking the cue from the fact that the sLQC eigenstates are symmetric superpositions of expanding and contracting Wheeler-DeWitt eigenstates at large volume, the question of whether the quantum "bounce" in sLQC could be due to this superposition was investigated. A calculation of the probability that the volume of such a universe becomes arbitrarily small naively suggested that a bounce might indeed be possible for such a superposed state. A more careful consistent histories analysis showed, however, that this was not the case, and that Wheeler-DeWitt universes are invariably singular. This arose from the recognition that the physical statement that the universe "bounces" is an assertion about a *sequence* of values of ϕ – precisely the situation in which decoherence matters, just as in the two-slit experiment. It was then shown that the branch wave function for a bouncing cosmology is in fact zero in the limit $\phi \to \pm\infty$ in which the histories {bounce,singular} decohere.

Here, we show in a similar way that, unlike in the Wheeler-DeWitt model, *all* states in sLQC – quasiclassical or not – achieve arbitrarily large volume volume in both the "past" ($\phi \to -\infty$) and "future" ($\phi \to +\infty$). This is fairly easy to see. Indeed, if one chooses any arbitrary volume ν^* and defines $\Delta\nu^* = [-\nu^*,\nu^*]$ and $\overline{\Delta\nu^*}$ to be its complement, the probability the universe has volume less than ν^* at ϕ is $p_{\Delta\nu^*}(\phi) = \sum_{\nu\in\Delta\nu^*}|\Psi(\nu,\phi)|^2$. Because of the discreteness of

Journal of Physics: Conference Series **360** (2012) 012028 doi:10.1088/1742-6596/360/1/012028

volume in sLQC and the UV cutoff in the $e_k^{(s)}(\nu)$, rapid oscillation of the exponential factor in Eq. (4) sends this probability to zero in both limits $\phi \to \pm\infty$. Since ν^* is arbitrary, this says the probability the universe assumes arbitrarily large volume as both $\phi \to \pm\infty$ is unity.

However, the statement that a quantum universe "bounces" is the statement that it has large volume *both* in the past and the future. As in the Wheeler-DeWitt case, it must be shown that the coarse-grained histories {bounce, singular} decohere before probabilities may be assigned. The class operator for the history in which the universe bounces is

$$C_{\text{bounce}} = \lim_{\substack{\phi_1 \to -\infty \\ \phi_2 \to +\infty}} P^{\nu}_{\overline{\Delta\nu_1^*}}(\phi_1) P^{\nu}_{\overline{\Delta\nu_2^*}}(\phi_2), \tag{5}$$

and for the complementary history C_{sing} in which the universe takes on zero volume at either or both of $\phi \to \pm\infty$, $C_{\text{sing}} = 1 - C_{\text{bounce}}$. In the same way as above it may be shown that $\lim_{\phi\to\pm\infty} P^{\nu}_{\Delta\nu^*}(\phi)|\Psi\rangle = 0$ and $\lim_{\phi\to\pm\infty} P^{\nu}_{\overline{\Delta\nu^*}}(\phi)|\Psi\rangle = |\Psi\rangle$. In this way, in the limit $\phi \to \pm\infty$, $|\Psi_{\text{bounce}}\rangle = C^{\dagger}_{\text{bounce}}|\Psi\rangle = |\Psi\rangle$ and $|\Psi_{\text{sing}}\rangle = C^{\dagger}_{\text{sing}}|\Psi\rangle = 0$, so that the histories {bounce, singular} decohere, and $p_{\text{bounce}} = 1$ and $p_{\text{sing}} = 0$ for all states in sLQC, completely the opposite of the case for the Wheeler-DeWitt quantization. We have thus shown in an objective, observer independent way, that a quantum bounce is inevitable in these models.[3]

4. Summary

We have described the consistent histories formulation of loop quantum cosmology, and showed how it can be used to arrive at consistent quantum predictions for histories of alternative outcomes without any need for observers or external measurements. By way of example, we have shown that arbitrary states in the theory – not merely quasiclassical ones – "bounce" from an arbitrarily large volume contracting Wheeler-DeWitt state in the "past" ($\phi \to -\infty$) to the corresponding symmetrically related large volume expanding Wheeler-DeWitt state in the "future" ($\phi \to +\infty$).

Acknowledgments

Research at the Perimeter Institute is supported by the Government of Canada through Industry Canada and the Province of Ontario through the Ministry of Research & Innovation.

References

[1] Kiefer C 2007 *Quantum gravity* 2nd ed (Oxford University Press, Oxford)
[2] Ashtekar A and Singh P 2011 *Class. Quant. Grav.* **28** 213001 (*Preprint* 1108.0893)
[3] Omnès R 1994 *The interpretation of quantum mechanics* (Princeton University Press, Princeton)
[4] Griffiths R B 2008 *Consistent quantum theory* (Cambridge University Press, Cambridge)
[5] Hartle J B 1995 Spacetime quantum mechanics and the quantum mechanics of spacetime *Gravitation and Quantizations, Proceedings of the 1992 Les Houches Summer School* ed Julia B and Zinn-Justin J (North Holland, Amsterdam) (*Preprint* gr-qc/9304006)
[6] Schlosshauer M 2007 *Decoherence and the quantum-to-classical transition* (Springer-Verlag, Berlin)
[7] Ashtekar A, Pawlowski T and Singh P 2006 *Phys. Rev.* **D74** 084003 (*Preprint* gr-qc/0607039)
[8] Ashtekar A, Corichi A and Singh P 2008 *Phys. Rev.* **D77** 024046 (*Preprint* 0710.3565)
[9] Craig D A and Singh P 2010 *Phys. Rev.* **D82** 123526–123546 (*Preprint* 1006.3837v1)
[10] Ashtekar A, Campiglia M and Henderson A 2010 *Class. Quant. Grav.* **27** 135020 (*Preprint* 1001.5147v2)
[11] Ashtekar A, Campiglia M and Henderson A 2010 *Phys. Rev.* **D82** 124043 (*Preprint* 1011.1024)
[12] Craig D A 2012 Dynamical eigenfunctions and critical density in loop quantum cosmology (in preparation)
[13] Ashtekar A, Pawlowski T and Singh P 2006 *Phys. Rev.* **D73** 124038 (*Preprint* gr-qc/0604013v3)
[14] Craig D A and Singh P 2012 Consistent probabilities in loop quantum cosmology (in preparation)

[3] As an aside, we note that this is precisely the role of the $\phi \to \pm\infty$ limit in these calculations: to guarantee the results hold for all choices of state, and are therefore generic, certain predictions of the theory, rather than being particular to a sub-class of states such as quasi-classical ones.

Journal of Physics: Conference Series **360** (2012) 012029 doi:10.1088/1742-6596/360/1/012029

2-point functions in quantum cosmology[1]

Steffen Gielen

Max Planck Institute for Gravitational Physics (Albert Einstein Institute), Am Mühlenberg 1, 14476 Golm, Germany

E-mail: `gielen@aei.mpg.de`

Abstract. We discuss the path-integral formulation of quantum cosmology with a massless scalar field as a sum-over-histories, with particular reference to loop quantum cosmology. Exploiting the analogy with the relativistic particle, we give a complete overview of the possible two-point functions, deriving vertex expansions and composition laws they satisfy. We clarify the tie between definitions using a group averaging procedure and those in a deparametrised framework. We draw some conclusions about the physics of a single quantum universe and multiverse field theories where the role of these sectors and the inner product are reinterpreted.

1. Introduction and Motivation

Quantum cosmology is the quantisation of dimensionally reduced models where one makes symmetry assumptions about the geometry already at the classical level. Both in the traditional approaches and in loop quantum cosmology, the motivation for studying such models has been twofold: One could hope to make predictions for cosmology, where the FRW spacetimes give a good approximation to our own Universe, or one could study mathematical and conceptual issues of quantum gravity in a simplified model where explicit calculations are possible. We focus on the second point and on the definition of the *physical inner product*, which for a constrained system plays the role of a transition amplitude, as a sum over histories. Using the analogy with the relativistic particle, we study different possible choices and focus on the *composition laws* they satisfy, in situations where a massless scalar field represents an internal time variable.

We assume a homogeneous, isotropic $k = 0$ FRW universe with a massless scalar field, with Hamiltonian constraint (here Θ can be a difference operator or a differential operator)

$$\hat{\mathfrak{C}}\Psi(\nu,\phi) \equiv \left(-\Theta - \partial_\phi^2\right)\Psi(\nu,\phi) = 0\,, \tag{1}$$

where ν parametrises the gravitational field and Θ is an operator acting on the gravitational part of the kinematical Hilbert space $\mathcal{H} = \mathcal{H}^{\rm g}_{\rm kin} \otimes \mathcal{H}^{\phi}_{\rm kin}$; ϕ appears as a relational time variable.

In loop quantum cosmology (LQC), one has a difference operator Θ of the form

$$(\Theta\Psi)(\nu,\phi) \equiv A(\nu)\Psi(\nu+\nu_0,\phi) + B(\nu)\Psi(\nu,\phi) + C(\nu)\Psi(\nu-\nu_0,\phi)\,, \tag{2}$$

where the functions A, B, C depend on the quantisation scheme; there is a superselection of subspaces $\{|\nu' + n\nu_0\rangle | n \in \mathbb{Z}\}$ and one may restrict wave functions to $\nu_0\mathbb{Z}$. The solutions to (1) can be split into positive and negative frequency parts satisfying $(\hat{p}_\phi \mp \sqrt{\Theta})\Psi_\pm(\nu,\phi) = 0$.

[1] Based on a talk given at Loops '11, Madrid, on 26 May 2011 [1].

In [2], a spin foam-type expansion was done for the inner product in loop quantum cosmology. Starting from a slight modification of the group-averaging expression used in timeless systems,

$$G_{\rm NW}(\nu_{\rm f},\phi_{\rm f};\nu_{\rm i},\phi_{\rm i}) \equiv \int_{-\infty}^{\infty} d\alpha\ \langle \nu_{\rm f},\phi_{\rm f}|e^{{\rm i}\alpha\mathfrak{C}}2|\hat{p}_\phi||\nu_{\rm i},\phi_{\rm i}\rangle\,, \tag{3}$$

one splits α into N parts and characterises each history $(\nu_{\rm i},\bar{\nu}_1,\ldots,\nu_{\rm f})$ by the number of volume transitions. Then the limit $N\to\infty$ can be taken to yield the sum-over-histories form [2] [2]

$$\begin{aligned} G_{\rm NW}(\nu_{\rm f},\phi_{\rm f};\nu_{\rm i},\phi_{\rm i}) &= \sum_{M=0}^{\infty}\sum_{\substack{\nu_{M-1},\ldots,\nu_1\\ \nu_m\neq\nu_{m+1}}}\Theta_{\nu_{\rm f}\nu_{M-1}}\ldots\Theta_{\nu_2\nu_1}\Theta_{\nu_1\nu_{\rm i}} \\ &\times\prod_{k=1}^{p}\frac{1}{(n_k-1)!}\left(\frac{\partial}{\partial\Theta_{w_kw_k}}\right)^{n_k-1}\sum_{m=1}^{p}\frac{2\cos(\sqrt{\Theta_{w_mw_m}}(\phi_{\rm f}-\phi_{\rm i}))}{\prod_{\substack{j=1\\ j\neq m}}^{p}(\Theta_{w_mw_m}-\Theta_{w_jw_j})}\,. \end{aligned} \tag{4}$$

Alternatively, one can pass to a deparametrised formalism, considering transition amplitudes $G^{+}_{\rm NW}(\nu_{\rm f},\phi_{\rm f};\nu_{\rm i},\phi_{\rm i}) = \langle\nu_{\rm f}|e^{{\rm i}\sqrt{\Theta}(\phi_{\rm f}-\phi_{\rm i})}|\nu_{\rm i}\rangle$ for the "square root" of (1), $-{\rm i}\partial_\phi\Psi(\nu,\phi)=\sqrt{\Theta}\Psi(\nu,\phi)$,

$$\begin{aligned} G^{+}_{\rm NW}(\nu_{\rm f},\phi_{\rm f};\nu_{\rm i},\phi_{\rm i}) &= \sum_{M=0}^{\infty}\sum_{\substack{\nu_{M-1},\ldots,\nu_1\\ \nu_m\neq\nu_{m+1}}}(\sqrt{\Theta})_{\nu_{\rm f}\nu_{M-1}}\ldots(\sqrt{\Theta})_{\nu_2\nu_1}(\sqrt{\Theta})_{\nu_1\nu_{\rm i}}\times \\ &\prod_{k=1}^{p}\frac{1}{(n_k-1)!}\left(\frac{\partial}{\partial(\sqrt{\Theta})_{w_kw_k}}\right)^{n_k-1}\sum_{m=1}^{p}\frac{e^{{\rm i}(\sqrt{\Theta})_{w_mw_m}(\phi_{\rm f}-\phi_{\rm i})}}{\prod_{\substack{j=1\\ j\neq m}}^{p}((\sqrt{\Theta})_{w_mw_m}-(\sqrt{\Theta})_{w_jw_j})}\,. \end{aligned} \tag{5}$$

Our notation for the different definitions should become clear in the next section. (4) and (5) define two different expansions of the same function (after restricting to positive frequency in (4))[3]. Are there other possible definitions for the inner product? Do (4) and (5) have good composition properties? In quantum gravity one might want to assume a relation like $G[g,g'] = \int\mathcal{D}g''\,G[g,g'']G[g'',g']$ which can be checked in this model for all possible two-point functions. We will see that all of them satisfy the same composition laws as for the relativistic particle. For details of all calculations see [3].

2. Two-point functions

2.1. Particle analogy

First we detail the two-point functions for the relativistic particle following [4]. Writing the relativistic particle in parametrised form $S=\int d\tau\ \left(p_a\dot{x}^a + N(p_x^2-p_t^2+m^2)\right)$ and fixing $\dot{N}=0$, one can define two-point functions through a sum-over-histories, such as the *Hadamard function*

$$G_{\rm H}(x'',t'';x',t') = \int_{-\infty}^{\infty} dT\ g(x'',t'';T|x',t';0)\,, \tag{6}$$

where $g(x'',t'';T|x',t';0)$ is a non-relativistic transition amplitude for $H=p_x^2-p_t^2+m^2$ in proper time T. Restricting the range of integration of proper time one obtains the *Feynman propagator*

$${\rm i}G_{\rm F}(x'',t'';x',t') = \int_{0}^{\infty} dT\ g(x'',t'';T|x',t';0)\,, \tag{7}$$

[2] Notation: $\Theta_{\nu_i\nu_j}=\langle\nu_i|\Theta|\nu_j\rangle$ are matrix elements of Θ; in the second line of (4) the distinct values appearing in $(\nu_{\rm i},\nu_1,\ldots,\nu_{M-1},\nu_{\rm f})$ are labelled by $w_1,\ldots,w_p$, with multiplicities $n_1,\ldots,n_p$ so that $\sum n_k = M+1$.

[3] We show directly in [3] that they are the same function, the positive-frequency Newton-Wigner function $G^{+}_{\rm NW}$.

Journal of Physics: Conference Series **360** (2012) 012029 doi:10.1088/1742-6596/360/1/012029

which is a proper Green's function: $(\Box_{x'} - m^2)G_F(x'',t'';x',t') = -\delta(x''-x';t''-t')$. It satisfies a relativistic composition law involving a normal derivative,

$$G_F(x'',t'';x',t') = -\int_\Sigma ds\, G_F(x'',t'';x(s),t(s))\left(\overrightarrow{\partial_n} - \overleftarrow{\partial_n}\right) G_F(x(s),t(s);x',t')\,, \tag{8}$$

which is absent for the Hadamard function: $G_H \circ G_H = G_c$, the causal two-point function [4].

Other two-point functions which require a splitting into positive and negative frequency are the Wightman functions $G^\pm$, so that $G_H = G^+ + G^-$ and $iG_F = G^+ \cdot \theta(t_f - t_i) + G^- \cdot \theta(t_i - t_f)$, and the non-Lorentz invariant *Newton-Wigner function*

$$G^+_{NW}(x'',t'';x',t') = \langle x''|e^{-i\sqrt{m^2+k^2}(t''-t')}|x'\rangle = \int \frac{dk}{2\pi} e^{ik(x''-x')} e^{-i\sqrt{m^2+k^2}(t''-t')} \tag{9}$$

which satisfies the non-relativistic composition law $G^+_{NW}(x'',t'';x',t') = \int dx\, G^+_{NW}(x'',t'';x,t) \times G^+_{NW}(x,t;x',t')$. The similarity of these expressions to quantum cosmology is obvious. The role of Lorentz invariance in minisuperspace is however rather unclear.

2.2. (Loop) quantum cosmology: Definitions and composition properties

We can now derive vertex expansions for all two-point functions for quantum cosmology; the Hadamard function and Feynman propagator can be defined without deparametrisation. The analogue of (6), $G_H(\nu_f,\phi_f;\nu_i,\phi_i) \equiv \int_{-\infty}^{\infty} d\alpha\, \langle \nu_f,\phi_f|e^{i\alpha\hat{\mathcal{C}}}|\nu_i,\phi_i\rangle$, gives the Hadamard function

$$\begin{aligned} G_H(\nu_f,\phi_f;\nu_i,\phi_i) \;=\; & \sum_{M=0}^{\infty} \sum_{\substack{\nu_{M-1},\ldots,\nu_1 \\ \nu_m \neq \nu_{m+1}}} \Theta_{\nu_f\nu_{M-1}} \ldots \Theta_{\nu_2\nu_1}\Theta_{\nu_1\nu_i} \times \\ & \prod_{k=1}^{p} \frac{1}{(n_k-1)!}\left(\frac{\partial}{\partial\Theta_{w_kw_k}}\right)^{n_k-1} \sum_{m=1}^{p} \frac{1}{\sqrt{\Theta_{w_mw_m}}} \frac{\cos(\sqrt{\Theta_{w_mw_m}}(\phi_f-\phi_i))}{\prod_{\substack{j=1\\ j\neq m}}^{p}(\Theta_{w_mw_m}-\Theta_{w_jw_j})}\,. \end{aligned} \tag{10}$$

Analogously, one derives the Feynman propagator by restricting the range of integration to positive α, as in (7), and following the "iϵ" contour in the complex p_ϕ plane to get

$$\begin{aligned} iG_F(\nu_f,\phi_f;\nu_i,\phi_i) \;=\; & \sum_{M=0}^{\infty} \sum_{\substack{\nu_{M-1},\ldots,\nu_1 \\ \nu_m \neq \nu_{m+1}}} \Theta_{\nu_f\nu_{M-1}} \ldots \Theta_{\nu_2\nu_1}\Theta_{\nu_1\nu_i} \prod_{k=1}^{p} \frac{1}{(n_k-1)!} \times \\ & \left(\frac{\partial}{\partial\Theta_{w_kw_k}}\right)^{n_k-1} \sum_{m=1}^{p} \frac{e^{-i\sqrt{\Theta_{w_mw_m}}\Delta\phi}\theta(\Delta\phi) + e^{i\sqrt{\Theta_{w_mw_m}}\Delta\phi}\theta(-\Delta\phi)}{2\sqrt{\Theta_{w_mw_m}}\prod_{\substack{j=1\\ j\neq m}}^{p}(\Theta_{w_mw_m}-\Theta_{w_jw_j})} \end{aligned} \tag{11}$$

with $\Delta\phi \equiv \phi_f - \phi_i$. The ϕ dependence of these expressions is analogous to the two-point functions for the relativistic particle; they also satisfy the same composition laws.

First one can show that the Newton-Wigner function restricted to the positive frequency sector,

$$G^+_{NW}(\nu_f,\phi_f;\nu_i,\phi_i) \equiv \int_{-\infty}^{\infty} d\alpha \int_0^{\infty} \frac{dp_\phi}{2\pi} \langle \nu_f|e^{-i\alpha\Theta}|\nu_i\rangle (2|p_\phi|) e^{i\alpha p_\phi^2} e^{ip_\phi\Delta\phi} = \langle \nu_f|e^{i\sqrt{\Theta}\,\Delta\phi}|\nu_i\rangle\,, \tag{12}$$

satisfies the usual non-relativistic composition law

$$G^+_{NW}(\nu_f,\phi_f;\nu_i,\phi_i) = \sum_\nu G^+_{NW}(\nu_f,\phi_f;\nu,\phi) G^+_{NW}(\nu,\phi;\nu_i,\phi_i)\,. \tag{13}$$

There is no composition law if one includes both positive- and negative-frequency sectors. It is essential to be able to separate these sectors.

For the Hadamard function, one has, just as for the relativistic particle,

$$G_{\rm c}(\nu_{\rm f},\phi_{\rm f};\nu_{\rm i},\phi_{\rm i}) = \sum_{\nu} G_{\rm H}(\nu_{\rm f},\phi_{\rm f};\nu,\phi)\left(\overrightarrow{\partial_\phi} - \overleftarrow{\partial_\phi}\right) G_{\rm H}(\nu,\phi;\nu_{\rm i},\phi_{\rm i}) \tag{14}$$

because of different composition law for positive and negative frequency Wightman functions.

Finally, the Feynman propagator satisfies, again in agreement with the relativistic particle,

$$\mathrm{i}G_{\rm F}(\nu_{\rm f},\phi_{\rm f};\nu_{\rm i},\phi_{\rm i}) = \mathrm{i}\sum_{\nu} \mathrm{i}G_{\rm F}(\nu_{\rm f},\phi_{\rm f};\nu,\phi)\left(\overrightarrow{\partial_\phi} - \overleftarrow{\partial_\phi}\right) \mathrm{i}G_{\rm F}(\nu,\phi;\nu_{\rm i},\phi_{\rm i})\,. \tag{15}$$

Recall that it is not a projector on solutions to the constraint $\mathfrak{C}$, but a proper Green's function.

3. Summary and Outlook

Exploiting the analogy with the relativistic particle, we have given explicit expressions for all two-point functions (Hadamard, Feynman, Newton-Wigner, etc), for constrained dynamics of the form $\mathfrak{C} = p_\phi^2 - \Theta$. They satisfy the same composition properties as their analogues. For the definition of some of the two-point functions the existence of an explicit splitting into positive and negative frequency is essential; such a splitting is not generally available for spin foams or more general cosmological models. Without the frequency splitting there is no two-point function defining a physical inner product (i.e. a projector on solutions to the constraint) and satisfying a "nice" composition law. Possible issues with the composition properties of the usual definition of the inner product through spin foams have been discussed as the "cosine problem", which is closely related to the different composition laws for positive and negative frequency Wightman functions that we have seen above.

In our model, the splitting of solutions into two sectors is of course possible and for a single particle (=Universe) everything is consistent when one restricts to one sector. The different choices of two-point functions become relevant in a field theory (third quantisation) picture, where one might follow an argument by Kuchař [5] for geometrodynamics: If one needs a splitting analogous to positive and negative frequency to have a consistent one-Universe quantum mechanics, one would require a conserved quantity analogous to energy, corresponding to a Killing vector, on (metric) superspace, which however does not exist. If one takes this argument (which is only based on an analogy) seriously, one has to consider the equivalent of QFT on curved spacetime, formulated without the fundamental concept of a single particle.

For the cosmological model at hand, this reasoning leads naturally to consider a GFT-like model of a quantum field theory on the 2-dimensional space spanned by ν and ϕ, where one could add interactions to implement topology change. One possible physical interpretation of that would be the creation of inhomogeneities; other questions addressed in this framework include the role of the "GFT coupling constant" λ. Work on this is currently in progress.

References

[1] `http://loops11.iem.csic.es/loops11/index.php?option=com_content&view=article&id=75&Itemid=73`

[2] Ashtekar A, Campiglia M and Henderson A 2009 Loop Quantum Cosmology and Spin Foams *Physics Letters* **B681** 347–352; Ashtekar A, Campiglia M and Henderson A 2010 Casting Loop Quantum Cosmology in the Spin Foam Paradigm *Classical and Quantum Gravity* **27** 135020

[3] Calcagni G, Gielen S and Oriti D 2011 Two-point functions in (loop) quantum cosmology *Classical and Quantum Gravity* **28** 125014

[4] Halliwell JJ and Ortiz ME 1993 Sum-over-histories origin of the composition laws of relativistic quantum mechanics and quantum cosmology *Physical Review* **D48** 748–768

[5] Kuchař K 1981 General relativity: Dynamics without symmetry *Journal of Mathematical Physics* **22** 2640–2654

Journal of Physics: Conference Series **360** (2012) 012030 doi:10.1088/1742-6596/360/1/012030

Plane Gravitational Waves and Loop Quantization

F Hinterleitner[1] **and S Major**[2]

[1] Department of Theoretical Physics and Astrophysics, Faculty of Science of the Masaryk University, Kotlářská 2, 61137 Brno, Czech Republic

[2] Department of Physics, Hamilton College, Clinton NY 13323 USA

Abstract. Starting from the polarized Gowdy model in Ashtekar variables, the Killing equations characteristic for plane-fronted parallel gravitational waves are introduced in part as a set of first-class constraints, in addition to the standard ones of General Relativity. These constraints are expressed in terms of quantities that have an operator equivalent in Loop Quantum Gravity, making plane wave space-times accessible to loop quantization techniques.

1. Introduction

The motivation for studying plane gravitational waves is two fold: First, these waves are interesting objects on which to test possible quantum theories of gravity. Being homogenous in two directions and inhomogeneous in the third direction, the degree of difficulty of these models lies between the complicated theory of General Relativity (GR) and homogenous cosmological models. Second, the last decade has seen an abundance of conjectures on Lorentz invariance (LI) violation, dependence of the speed of light on energy, "Doubly Special Relativity", and so on, see e.g. [1]. Some of these conjectures are inspired by the granularity of space predicted by Loop Quantum Gravity (LQG). So far there are plausibility arguments in favor of or against these conjectures, for example [2], but, to our knowledge, no calculation from an exactly solvable model. There is important prior work: plane gravitational waves were quantized [3]; LI has been investigated in the context of spherical symmetry [4], and [5] provides a derivation of gravitational wave dispersion in LQG in a cosmological context by perturbative methods. Gravitational plane waves appear to be simple enough for deducing whether LQG techniques yield dispersion, without further simplifications.

We start from a slightly more general system, the polarized Gowdy model in a form presented by Banerjee and Date [6, 7]. Like space with plane waves, this model is homogenous in two dimensions. The essential step in the reduction is to single out waves going into one direction and so to avoid colliding plane waves, which lead to complicated interaction processes and singularities. This reduction is carried out by means of a set of first-class constraints, so that the system becomes accessible to loop quantum techniques. See [8] for more details.

2. The polarized Gowdy model in Ashtekar variables

We assume homogeneity in the x, y plane and wave propagation in the z direction. We choose adapted spatial triads with one leg in the z direction and two arbitrary orthogonal vectors in the x, y plane. Densitized inverse triad variables are denoted by $E^a{}_i$, $a = x, y, z$ and $i = 1, 2, 3$. Following [6] we write $\mathcal{E} := E^z{}_3$ and introduce "polar" coordinates for the triad vectors in the x, y plane with the "radial" components E^x and E^y. The associated "angular" variable (the

Journal of Physics: Conference Series **360** (2012) 012030 doi:10.1088/1742-6596/360/1/012030

same for E^x and E^y in the polarized Gowdy model) is pure gauge. They are fixed and disappear when the Gauß constraint is strongly imposed [6]. All variables depend only on z and the time variable t. The canonically conjugate diagonal elements of the Ashtekar connection (divided by the Barbero-Immirzi parameter γ) to E^x and E^y are the extrinsic curvature components K_x and K_y, the conjugate element to $\mathcal{E}$ is denoted by $\mathcal{A}$. The fundamental Poisson brackets are

$$\{K_a(z), E^b(z')\} = \kappa\delta_a^b\delta(z-z'), \quad \{\mathcal{A}(z), \mathcal{E}(z')\} = \kappa\delta(z-z'). \tag{1}$$

$\kappa = 8\pi G_{\rm Newton}$ is the gravitational constant.

In terms of these variables, and modulo the strongly imposed Gauß constraint, we have the following sets of first-class constraints of GR: The diffeomorphism constraint

$$C = \frac{1}{\kappa}\left[K_x' E^x + K_y' E^y - \mathcal{E}'\mathcal{A}\right], \tag{2}$$

(the prime denotes the derivative with respect to z) generating diffeomorphisms along the z-axis, and the Hamiltonian constraint

$$\begin{aligned} H = &-\frac{1}{\kappa\sqrt{\mathcal{E}E^xE^y}}\left[E^xK_xE^yK_y + (E^xK_x + E^yK_y)\mathcal{E}\mathcal{A} - \frac{1}{4}\mathcal{E}'^2 - \mathcal{E}\mathcal{E}''\right. \\ &\left. - \frac{1}{4}\mathcal{E}^2\left[\left(\ln\frac{E^y}{E^x}\right)'\right]^2 + \frac{1}{2}\mathcal{E}\mathcal{E}'(\ln E^xE^y)'\right]. \end{aligned} \tag{3}$$

3. Reduction to plane waves

Plane waves are characterized by a null Killing vector field in the direction of propagation. With a lapse function $N(t,z)$ we have the space-time metric

$$ds^2 = -N^2\,dt^2 + \mathcal{E}\,\frac{E^y}{E^x}\,dx^2 + \mathcal{E}\,\frac{E^x}{E^y}\,dy^2 + \frac{E^xE^y}{\mathcal{E}}\,dz^2. \tag{4}$$

For this metric a null vector field in the positive z-direction has the form $k^\mu = \left(\sqrt{E^xE^y/\mathcal{E}}\,k, 0, 0, \pm Nk\right)$ with $k = k(t,z)$. With the minus sign chosen, the Killing equations $k_{\mu;\nu} + k_{\nu;\mu} = 0$ give rise to the following conditions

$$U_x = E^xK_x - \frac{1}{2}\mathcal{E}' - \frac{1}{2}\mathcal{E}\left(\frac{E^{y\prime}}{E^y} - \frac{E^{x\prime}}{E^x}\right) = 0, \quad U_y = E^yK_y - \frac{1}{2}\mathcal{E}' + \frac{1}{2}\mathcal{E}\left(\frac{E^{y\prime}}{E^y} - \frac{E^{x\prime}}{E^x}\right) = 0. \tag{5}$$

These two relations render the spatial diffeomorphism and Hamiltonian constraints equivalent,

$$C \approx -\sqrt{\frac{E^xE^y}{\mathcal{E}}}\,H = -\sqrt{g_{zz}}\,H.$$

4. Constraints

With test functions the expressions U_x and U_y become

$$U_a[f] := \int dz\, f(z)U_a(z). \tag{6}$$

The Poisson bracket structure may be expressed as

$$\begin{pmatrix} \{U_x[f], U_x[g]\} & \{U_x[f], U_y[g]\} \\ \{U_y[f], U_x[g]\} & \{U_y[f], U_y[g]\} \end{pmatrix} = \frac{1}{2}\begin{pmatrix} 1 & -1 \\ -1 & 1 \end{pmatrix}\int dz\,(f'g - fg')\,\mathcal{E}. \tag{7}$$

This matrix has a zero eigenvalue and is diagonalized by the combinations $U_+ = U_x + U_y$ and $U_- = U_x - U_y$, or explicitly

$$U_+ = E^x K_x + E^y K_y - \mathcal{E}', \qquad U_- = E^x K_x - E^y K_y - \mathcal{E}\left(\ln \frac{E^y}{E^x}\right)'. \tag{8}$$

The algebra of these new constraints is

$$\{U_+[f], U_+[g]\} = \{U_+[f], U_-[g]\} = 0, \qquad \{U_-[f], U_-[g]\} = 2\int \mathrm{d}z\,(f'g - fg')\,\mathcal{E}. \tag{9}$$

U_+ weakly Poisson-commutes also with C, and H:

$$\{U_+[f], C[g]\} = -\frac{1}{\kappa} U_+[f'g] \approx 0, \qquad \{U_+[f], H[g]\} = \frac{1}{\kappa} U_+\left[\sqrt{\frac{\mathcal{E}}{E^x E^y}}\, f'g\right] - H[fg] \approx 0. \tag{10}$$

Thus U_+ can be added as another first-class constraint. Together with the Poisson bracket relations of the standard constraints,

$$\begin{aligned} \{C[f], C[g]\} &= C[fg' - f'g], \qquad \{C[f], H[g]\} = H[fg'], \\ \{H[f], H[g]\} &= C\left[(fg' - f'g)\frac{\mathcal{E}}{E^x E^y}\right], \end{aligned} \tag{11}$$

they establish the enlarged Poisson bracket algebra of the first-class constraints C, H, and U_+. The constraint U_- can be written in terms of the three first-class constraints [8].

Concerning physical degrees of freedom counting, the gauge generators C and H reduce the number from originally three canonical pairs to one, when applied together with corresponding gauge conditions. The constraints $U_+(z)$ on the other hand, are not gauge generators, there are no gauge conditions associated with them, so they fix one function. Thus, when all constraints and gauge conditions are imposed, there is one free function left, like in the standard approach to plane waves. In this way the association of classical variables in canonical pairs gets lost. In the Dirac quantization approach however, one does not impose the constraints strongly, but as conditions on physical state functions, so in quantum theory the canonical structure remains untouched.

5. Preparation for quantization

In view of quantization it is important that U_+ can be given a meaning as a well-defined operator. Both $E^x K_x + E^y K_y$ and $\mathcal{E}'$ are scalar densities that can be naturally integrated along z. The integral over some interval $\mathcal{I}$ is

$$U_+[\mathcal{I}] = \int_{\mathcal{I}} \mathrm{d}z (E^x K_x + E^y K_y) - \mathcal{E}_+ + \mathcal{E}_-, \tag{12}$$

where $\mathcal{E}_\pm$ are the values of $\mathcal{E}$ at the endpoints of $\mathcal{I}$. $\mathcal{E}$ has a meaningful operator equivalent in the adapted LQG framework [7]. In analogy to full LQG the integral in (12) can be obtained as the Poisson bracket

$$\left\{\int \mathrm{d}x \frac{E^x K_x E^y K_y}{\sqrt{\mathcal{E} E^x E^y}}, \int \mathrm{d}y \sqrt{\mathcal{E} E^x E^y}\right\} = \frac{1}{2}\int \mathrm{d}z (E^x K_x + E^y K_y). \tag{13}$$

The first expression in the bracket is the restriction of the Euclidean Hamiltonian constraint to the x, y plane, the second one is the volume of a slice of space, constructed from a fiducial (unit) area in the x, y plane as basis and some perpendicular interval in the z-direction.

The analogs of both of them have operator equivalents in standard LQG, for the present case we find the corresponding operators in [7], equations (31) and (32). Now we are in a position to express all first-class constraints in terms of loop quantum operators, acting on one-dimensional spin network states, as demonstrated in [7].

6. Conclusion and outlook

In classical GR the existence of a null Killing vector field in the direction of propagation of gravitational plane waves guarantees dispersion-free propagation of such waves at a constant speed. In the present paper it was shown that a linear combination of two Killing equations, concerning area and volume expansion, can be expressed as a set of first class constraints in the polarized Gowdy model of [6]. The complementary linear combination U_- is dependent on the set of first-class constraints, so the latter one contains the full information about plane gravitational waves. An ongoing quantum investigation in the spirit of [7] should illuminate the issues of LI and dispersion of plane waves in LQG.

Acknowledgments

The work was supported by the Ministry of Education of the Czech Republic, contract no. MSM 0021622409.

[1] G. Amelino-Camelia, Doubly-Special Relativity: Facts, Myths and Some Key Open Issues. *Symmetry* **2** (2010) 230-271 arXiv:1003.3942.

[2] C. Rovelli and S. Speziale, Lorentz invariance of loop quantum gravity, *Phys. Rev.* **D83** (2009) 104029, arXiv:1012.1739, S. Hossenfelder, The Box-Problem in Deformed Special Relativity. arXiv:0912.0090, L. Smolin, Classical paradoxes of locality and their possible quantum resolution in deformed special relativity. arXiv:1004.0664.

[3] G. A. Mena Marugan and M. Montejo, "Quantization of pure gravitational plane waves" *Phys. Rev.* **D 58** (1998) 104017 arXiv:gr-qc/9806105.

[4] R. Gambini, S. Rastgoo, J. Pullin, "Small Lorentz violations in quantum gravity: do they lead to unacceptably large effects?" *Class. Quantum Grav.* **28** (2011) 155005 1106.1417; "Quantum scalar field in quantum gravity: the propagator and Lorentz invariance in the spherically symmetric case" *Gen. Rel. Grav.* **43** 3569 (2011) arXiv:1105.0667.

[5] M. Bojowald and G. M. Hossain, Loop quantum gravity corrections to gravitational wave dispersion. *Phys. Rev.* **D77** (2008) 023508 arXiv:0709.2365.

[6] K. Banerjee and G. Date, Loop Quantization of Polarized Gowdy Model on T^3: Classical Theory. *Class. Quantum Grav.* **25** (2008) 105014 arXiv:0712.0683

[7] K. Banerjee and G. Date, Loop Quantization of Polarized Gowdy Model on T^3: Kinematical States and Constraint Operators. *Class. Quantum Grav.* **25** (2008) 145004 arXiv:0712.0687.

[8] F. Hinterleitner and S. Major, Toward loop quantization of plane gravitational waves, *Class. Quantum Grav.* **29** in print, arXiv:1106.1448.

Journal of Physics: Conference Series **360** (2012) 012031 doi:10.1088/1742-6596/360/1/012031

Loop quantum cosmology of the Bianchi I model: complete quantization

M Martín-Benito[1], L J Garay[2,3], G A Mena Marugán[2], E. Wilson-Ewing[4]

[1] MPI für Gravitational Physics, Albert Einstein Institute, Am Mühlenberg 1, D-14476 Potsdam, Germany.
[2] Instituto de Estructura de la Materia, IEM-CSIC, Serrano 121, 28006 Madrid, Spain.
[3]Departamento de Física Teórica II, Universidad Complutense de Madrid, 28040 Madrid, Spain.
[4]Centre de Physique Théorique, CNRS-Luminy Case 907, 13288 Marseille Cedex 09, France.

E-mail: mercedes@aei.mpg.de, luis.garay@fis.ucm.es, mena@iem.cfmac.csic.es, wilson-ewing@cpt.univ-mrs.fr

Abstract. We complete the canonical quantization of the vacuum Bianchi I model within the improved dynamics scheme of loop quantum cosmology, characterizing the Hilbert structure of the physical states and providing a complete set of observables acting on them. In order to achieve this task, it has been essential to determine the structure of the separable superselection sectors that arise owing to the polymeric quantization, and to prove that the initial value problem obtained when regarding the Hamiltonian constraint as an evolution equation, interpreting the volume as the evolution parameter, is well-posed.

1. Introduction

Loop quantum cosmology (LQC) [1, 2, 3] adapts the techniques of loop quantum gravity [4] in the quantization of models with high degree of symmetry, such us homogeneous models. Remarkably, the quantization of (homogeneous and isotropic) Friedman-Lemaître-Robertson-Walker models within (the improved dynamics of) LQC succeeds in solving the singularity problem: the classical big bang turns out to be replaced by a quantum bounce happening at Planck scales, and no observable diverges in the quantum theory, as shown for the first time in the seminal work of reference [5].

In order to extent this quantization to (homogeneous but anisotropic) Bianchi models, we have focused our attention in the simplest case: the vacuum Bianchi I model. Our interest in this model also comes from the necessity of including inhomogeneities in LQC. Actually, with that aim we have analyzed from the LQC perspective the simplest inhomogeneous cosmologies: the linearly polarized Gowdy model with three-torus topology [6, 7]. This model can be regarded as a homogeneous Bianchi I background with three-torus topology filled with linearly polarized gravitational waves traveling in a single direction. The above quantization of the Gowdy model employs the polymeric quantization of the Bianchi I model when representing the homogeneous background and, therefore, first we need to have under control the Bianchi I model itself.

The quantization of the Bianchi I model in LQC is subject to several ambiguities. One of them concerns the representation of the curvature tensor of the connection in the quantum theory. Different definitions of this object lead to different schemes of quantization, being the most satisfactory one the *improved dynamics* put forward in Ref. [8], where although the Hamiltonian constraint was constructed, no further analysis of the physical solutions was carried out.

In this note we further study this quantization of the Bianchi I model. It summarizes some results already presented in [6, 7]. We will start by adopting a different factor ordering of that of [8] when symmetrizing the Hamiltonian constraint operator. With our factor ordering the operator is well-defined in the octant of $\mathbb{R}^3$ given by the positive eigenvalues of the operators representing the coefficients of the densitized triad. In this way our model displays sectors of superselection that are simpler than those of [8]. We will analyze the structure of the superselection sectors (aspect not studied in [8]) and see that they are separable.

The Hamiltonian constraint provides a difference equation in the volume. Then, it seems natural to interpret it as an internal time and the constraint as an evolution equation with respect to it. We will see that this interpretation is valid inasmuch as the corresponding initial value problem is well-possed: physical solutions are completely determined by a countable set of data given in the initial section of the volume (which displays a non-vanishing minimum value). The previous result allows us to identify the physical Hilbert space with the Hilbert space of initial data, whose inner product is determined by imposing reality conditions in a (over) complete set of physical observables.

2. Quantization of the model

We consider the vacuum Bianchi I model with three-torus spatial topology. Then, we use global coordinates $\{t, \theta, \sigma, \delta\}$, with $\theta, \sigma, \delta \in S^1$. In order to prepare the model for its loop quantization, we describe it in the Ashtekar-Barbero formalism [4]. Using a diagonal gauge, the nontrivial components of the densitized triad are $p_i/4\pi^2$, with $i = \theta, \sigma, \delta$, and $c_i/2\pi$ are those of the $su(2)$ connection. They satisfy $\{c_i, p_j\} = 8\pi G\gamma\delta_{ij}$, where γ is the Immirzi parameter and G is the Newton constant. The spacetime metric in these variables reads

$$ds^2 = -dt^2 + \frac{|p_\theta p_\sigma p_\delta|}{4\pi^2}\left(\frac{d\theta^2}{p_\theta^2} + \frac{d\sigma^2}{p_\sigma^2} + \frac{d\delta^2}{p_\delta^2}\right). \quad (1)$$

The phase space is constrained by the Hamiltonian constraint

$$C_{\rm BI} = -\frac{2}{\gamma^2}\frac{c^\theta p_\theta c^\sigma p_\sigma + c^\theta p_\theta c^\delta p_\delta + c^\sigma p_\sigma c^\delta p_\delta}{V} = 0. \quad (2)$$

In this expression, $V = \sqrt{|p_\theta p_\sigma p_\delta|}$ is the volume of the compact spatial sections.

In order to represent the phase space in the quantum theory we choose the kinematical Hilbert space of the Bianchi I model constructed in LQC (see e.g. [8]), that we call $\mathcal{H}_{\rm kin}$. We recall that, on $\mathcal{H}_{\rm kin}$, the operators $\hat{p}_i$ have a discrete spectrum equal to the real line. The corresponding eigenstates, $|p_\theta, p_\sigma, p_\delta\rangle$, form an orthonormal basis (in the discrete norm) of $\mathcal{H}_{\rm kin}$. Owing to this discreteness, there is no well-defined operator representing the connection, but rather its holonomies. They are computed along straight edges in the fiducial directions. The so-called improved dynamics prescription states that, when writing the curvature tensor in terms of holonomies, we have to evaluate them along edges with a certain minimum dynamical (state dependent) length $\bar{\mu}_i$. We use the specific improved dynamics prescription put forward in [8]: the elementary operators which represent the matrix elements of the holonomies, called $\hat{\mathcal{N}}_{\bar{\mu}_i}$, produce all a constant shift in volume V. In order to simplify the analysis, it is convenient to relabel the basis states in the form $|v, \lambda_\sigma, \lambda_\delta\rangle$, where $|v|$ is proportional to V such that the operators $\widehat{\mathcal{N}}_{\pm\bar{\mu}_i}$ cause a shift on it equal to ± 1, and the parameters λ_i are all equally defined in terms of the corresponding parameters p_i, and verify that $v = 2\lambda_\theta\lambda_\sigma\lambda_\delta$.

Out of the basic operators $\hat{p}_i$, $\widehat{\mathcal{N}}_{\bar{\mu}_i}$ we represent the Hamiltonian constraint as an operator $\widehat{\mathcal{C}}$. We choose a very suitable symmetric factor ordering such that $\widehat{\mathcal{C}}$ decouples the states of $\mathcal{H}_{\text{kin}}$ with support in $v = 0$, namely, the states with vanishing volume [6, 7]. Moreover, our operator $\widehat{\mathcal{C}}$ does not relate states with different orientation of any of the eigenvalues of the operators $\hat{p}_i$. Owing to this property we can then restrict the domain of definition of $\widehat{\mathcal{C}}$ to e.g. the space spanned by the states $|v > 0, \lambda_\sigma > 0, \lambda_\delta > 0\rangle$. We call the resulting Hilbert space $\mathcal{H}^+_{\text{kin}}$. Remarkably, for this restriction we do not need to impose any particular boundary condition or appeal to any parity symmetry. Note that both $\mathcal{H}_{\text{kin}}$ and $\mathcal{H}^+_{\text{kin}}$ are non-separable Hilbert spaces, feature not desirable for a physical theory. This problem is overcome by the own action of the Hamiltonian constraint operator. Indeed, $\widehat{\mathcal{C}}$ defined on $\mathcal{H}^+_{\text{kin}}$ turns out to leave invariant some separable Hilbert subspaces that provide superselection sectors. In concrete, they are spanned by the states $|v = \varepsilon + 4n, \lambda_\sigma = \lambda^\star_\sigma\omega_\varepsilon, \lambda_\delta = \lambda^\star_\delta\omega_\varepsilon\rangle$. Here $\varepsilon \in (0, 4]$ and $n \in \mathbb{N}$, and therefore v takes support in semilattices of constant step equal to 4 starting in a minimum non-vanishing value ε. In addition, $\lambda^\star_a$ $(a = \sigma, \delta)$ is some positive real number and ω_ε runs over the subset of $\mathbb{R}^+$ given by

$$\mathcal{W}_\varepsilon = \left\{ \left(\frac{\varepsilon - 2}{\varepsilon}\right)^z \prod_k \left(\frac{\varepsilon + 2m_k}{\varepsilon + 2n_k}\right)^{p_k} \right\},$$

where $m_k, n_k, p_k \in \mathbb{N}$, and $z \in \mathbb{Z}$ when $\varepsilon > 2$, while $z = 0$ otherwise. One can check that in deed this set is dense in $\mathbb{R}^+$ and countable [6]. Therefore, any of these sectors provide separable Hilbert spaces contained in $\mathcal{H}^+_{\text{kin}}$. We denote them as $\mathcal{H}_{\varepsilon,\lambda^\star_\sigma,\lambda^\star_\delta} = \mathcal{H}_\varepsilon \otimes \mathcal{H}_{\lambda^\star_\sigma} \otimes \mathcal{H}_{\lambda^\star_\delta}$. Note that the removal of the states with support in $v = 0$ means that there is no analog of the cosmological singularity (classically located in $p_i = 0$) in our quantum theory. We thus solve the singularity already at the level of superselection in a very simple way.

We then restrict the study to any of these superselection sectors, and expand a general solution in the form $(\psi| = \sum_{v\in\mathcal{L}^+_\varepsilon} \sum_{\omega_\varepsilon\in\mathcal{W}_\varepsilon} \sum_{\bar{\omega}_\varepsilon\in\mathcal{W}_\varepsilon} \psi(v, \omega_\varepsilon\lambda^\star_\sigma, \bar{\omega}_\varepsilon\lambda^\star_\delta)\langle v, \omega_\varepsilon\lambda^\star_\sigma, \bar{\omega}_\varepsilon\lambda^\star_\delta|$. One obtain that the constraint $(\psi|\widehat{\mathcal{C}}^\dagger = 0$ leads to a recurrence equation that relates the combination of states

$$\begin{aligned}\psi_+(v+4, \lambda_\sigma, \lambda_\delta) = {}& \psi\left(v+4, \lambda_\sigma, \frac{v+4}{v+2}\lambda_\delta\right) + \psi\left(v+4, \frac{v+4}{v+2}\lambda_\sigma, \lambda_\delta\right) + \psi\left(v+4, \frac{v+2}{v}\lambda_\sigma, \lambda_\delta\right) \\ & + \psi\left(v+4, \lambda_\sigma, \frac{v+2}{v}\lambda_\delta\right) + \psi\left(v+4, \frac{v+2}{v}\lambda_\sigma, \frac{v+4}{v+2}\lambda_\delta\right) + \psi\left(v+4, \frac{v+4}{v+2}\lambda_\sigma, \frac{v+2}{v}\lambda_\delta\right)\end{aligned} \tag{3}$$

with data in the previous sections v and $v - 4$. Therefore, if we regard v as an internal time the constraint can be interpreted as an evolution equation in it. It turns out that, when particularized to the initial time $v = \varepsilon$, the constraint simply gives the combination of states $\psi_+(\varepsilon + 4, \lambda_\sigma, \lambda_\delta)$ in terms of some initial data $\psi(\varepsilon, \lambda'_\sigma, \lambda'_\delta)$. From the combinations $\psi_+(\varepsilon + 4, \lambda_\sigma, \lambda_\delta)$, it is possible to determine any of the individual terms $\psi(\varepsilon + 4, \lambda_\sigma, \lambda_\delta)$, since the system of equations that relate the former ones with the latter ones is formally invertible. This issue is very non-trivial. Actually, both the separability of the support of λ_a and the fact that it is dense in $\mathbb{R}^+$ have been essential to show it (the details can be found in [7]). Therefore, the initial value problem is well-posed and it is indeed feasible to make the above interpretation of the constraint as an evolution equation in v.

In conclusion, the physical solutions of the Hamiltonian constraint are completely determined by the set of initial data $\{\psi(\varepsilon, \lambda_\sigma, \lambda_\delta) = \psi(\varepsilon, \omega_\varepsilon\lambda^\star_\sigma, \bar{\omega}_\varepsilon\lambda^\star_\delta), \omega_\varepsilon, \bar{\omega}_\varepsilon \in \mathcal{W}_\varepsilon\}$, and we can identify solutions with this set. We can also characterize the physical Hilbert space as the Hilbert space of the initial data. In order to endow the set of initial data with a Hilbert structure, one can take a complete set of observables forming a closed algebra, and impose that the quantum counterpart of their complex conjugation relations become adjointness relations between operators. This determines a unique (up to unitary equivalence) inner product.

Before doing that, it is most convenient changing the notation. Let us introduce the variables $x_a = \ln(\lambda_a) = \ln(\lambda_a^\star) + \rho_\varepsilon$. Note that ρ_ε takes values in a dense set of the real line, given by the logarithm of the points in the set $\mathcal{W}_\varepsilon$. We will denote that set by $\mathcal{Z}_\varepsilon$. Then, a set of observables acting on the initial data $\tilde{\psi}(x_\sigma, x_\delta) := \psi(\varepsilon, x_\sigma, x_\delta)$ is that formed by the operators $\widehat{e^{ix_a}}$ and $\widehat{U}_a^{\rho_a}$, with $\rho_a \in \mathcal{Z}_\varepsilon$ and $a = \sigma, \delta$, defined as

$$\widehat{e^{ix_\sigma}}\tilde{\psi}(x_\sigma, x_\delta) = e^{ix_\sigma}\tilde{\psi}(x_\sigma, x_\delta), \qquad \widehat{U}_\sigma^{\rho_\sigma}\tilde{\psi}(x_\sigma, x_\delta) = \tilde{\psi}(x_\sigma + \rho_\sigma, x_\delta), \tag{4}$$

and similarly for $\widehat{e^{ix_\delta}}$ and $\widehat{U}_\delta^{\rho_\delta}$. These operators provide an overcomplete set of observables and are unitary in $\mathcal{H}_{\lambda_\sigma^\star} \otimes \mathcal{H}_{\lambda_\delta^\star}$, according with their reality conditions. Therefore, we conclude that this Hilbert space is precisely the physical Hilbert space of the vacuum Bianchi I model.

3. Conclusions

Adopting the improved dynamics of reference [8], we have completed the quantization of the Bianchi I model in LQC providing the physical Hilbert space and a complete set of physical observables, task so far not achieved.

With that aim, first we have needed to determine the structure of the superselection sectors that arise in the quantum model owing to the polymeric representation of the geometry. These sectors display an involved structure concerning the anisotropies, as a consequence of the complicated action of the Hamiltonian constraint on those variables, inherent to the adopted improved dynamics. Actually, the values of the anisotropies run over a dense set of the positive real line that turns out to be countable. In contrast, the volume takes values in simple semilattices of constant step. As a result, the superselection sectors are separable.

The Hamiltonian constraint provides a difference equation in the volume, and can be interpreted as an evolution equation, being the volume the variable playing the role of the internal time. The separability of the support of the anisotropies and the fact that it is dense in $\mathbb{R}^+$ are essential to show that the associated initial value problem is well-posed, and therefore the above interpretation is valid. Since the initial data determines the solution, we can characterize the physical Hilbert space as the Hilbert space of the initial data. The physical inner product can be determined by imposing reality conditions on a complete set of observables. The result is that the physical Hilbert space coincides with the tensor product of the superselection sectors of the two anisotropy variables.

Acknowledgments

This work was in part supported by the Spanish MICINN Projects No. FIS2008-06078-C03-03, No. FIS2011-30145-C03-02, and the Consolider-Ingenio Program CPAN No. CSD2007-00042.

4. References

[1] Bojowald M 2008 Loop quantum cosmology *Living Rev. Rel.* **11** 4
[2] Ashtekar A and Singh P 2011 Loop quantum cosmology: a status report *Preprint* arXiv:1108.0893
[3] Banerjee K, Calcagni G and Martín-Benito M 2011 Introduction to loop quantum cosmology *Preprint* arXiv:1109.6801
[4] Thiemann T 2007 *Modern canonical quantum general relativity* (Cambridge: Cambridge University Press)
[5] Ashtekar A, Pawlowski T and Singh P 2006 Quantum nature of the big bang: improved dynamics, *Phys. Rev.* D **74** 084003
[6] Garay L J, Martín-Benito M and Mena Marugán G A 2010 *Phys. Rev.* D **82** 044048
[7] Martín-Benito M, Mena Marugán G A and Wilson-Ewing E 2010 *Phys. Rev.* D **82** 084012
[8] Ashtekar A and Wilson-Ewing E 2009 *Phys. Rev.* D **79** 083535

Journal of Physics: Conference Series **360** (2012) 012032 doi:10.1088/1742-6596/360/1/012032

Inhomogenous loop quantum cosmology with matter

D Martín-de Blas[1], M Martín-Benito[2] and G A Mena Marugán[1]

[1] Instituto de Estructura de la Materia, IEM-CSIC, Serrano 121, 28006 Madrid, Spain.
[2] MPI für Gravitational Physics, Albert Einstein Institute, Am Mühlenberg 1, D-14476 Potsdam, Germany.

E-mail: daniel.martin@iem.cfmac.csic.es, mercedes@aei.mpg.de, mena@iem.cfmac.csic.es

Abstract. The linearly polarized Gowdy T^3 model with a massless scalar field with the same symmetries as the metric is quantized by applying a hybrid approach. The homogeneous geometry degrees of freedom are loop quantized, fact which leads to the resolution of the cosmological singularity, while a Fock quantization is employed for both matter and gravitational inhomogeneities. Owing to the inclusion of the massless scalar field this system allows us to modelize flat Friedmann-Robertson-Walker cosmologies filled with inhomogeneities propagating in one direction. It provides a perfect scenario to study the quantum back-reaction between the inhomogeneities and the polymeric homogeneous and isotropic background.

1. Introduction

Loop quantum cosmology (LQC) [1, 2, 3] is a quantization for cosmological systems based on ideas and techniques of loop quantum gravity (LQG) [4]. Several homogeneous cosmologies have been satisfactory quantized within LQC [2]. The loop quantization leads to a resolution of the cosmological singularities, that turn out to be replaced by quantum bounces [5].

In order to study inhomogeneous cosmological systems in the framework of LQC, a hybrid quantization has been developed and applied to the Gowdy T^3 model with linear polarization [6, 7, 8, 9]. This hybrid quantization applies to reduced models in which only global constraints remain to be imposed at the quantum level. It combines the polymeric quantization of LQC for the homogeneous degrees of freedom with a Fock quantization for the inhomogeneities, so that one can deal with the field complexity. Therefore, this hybrid procedure assumes that there exists a hierarchy in the loop quantum geometry effects such that the most relevant ones are those associated with the homogeneous degrees of freedom.

With the aim of studying more physically interesting models we introduce a massless scalar field in the linearly polarized Gowdy T^3 model [10]. This field is minimally coupled and has the same symmetries of the geometry. Like in the vacuum model, the application of the hybrid quantization turns out in the resolution of the initial singularity, achieved thanks to the particular action of the Hamiltonian constraint whose homogeneous terms are loop quantized, and in the recovering of the standard Fock quantization for the inhomogeneities. The interest of the current model lies in the fact that, unlike the vacuum case, it admits as subset of homogeneous an isotropic solutions the flat Friedmann-Robertson-Walker (FRW) model, so that we can regard the system as an approximated flat-FRW background filled with inhomogeneities propagating in one direction. Therefore this model is perfect to analyze the quantum effects of anisotropies and inhomogeneities on a flat FRW background that is loop quantized, and vice versa, to study how the polymeric background affects the evolution of the anisotropies and inhomogeneities. Actually, we are now developing approximations to obtain physically interesting solutions of the Hamiltonian constraint [12].

The Gowdy T^3 model with linear polarization presents a subset of classical solutions with local rotational symmetry (LRS). To simplify the calculations one can work with the reduced *LRS-Gowdy model* without loss of generality. We will see how to implement this reduction at the quantum level using the map introduced in Ref. [11].

2. Hybrid Quantization

We apply a hybrid quantization to the linearly polarized Gowdy T^3 model with a minimally coupled massless scalar field. These cosmologies are globally hyperbolic spacetimes with spatial three torus topology and with two axial and hypersurface orthogonal Killing vectors fields ∂_σ and ∂_δ. We choose coordinates $\{t, \theta, \sigma, \delta\}$ adapted to the symmetries such that all our fields only depend on time and on the spatial coordinate $\theta \in S^1$. After performing a symmetry reduction and a partial gauge fixing as in Ref. [7], we obtain a reduced phase space that can be split in two sectors by expanding the fields in Fourier modes. The zero modes form the *homogeneous sector*, which is equivalent to the phase space of the Bianchi I model with an homogeneous massless scalar field, ϕ. The *inhomogeneous sector* contains the non-zero modes of both gravitational and matter fields, ξ and φ respectively. With the aim of quantizing the homogeneous sector of the geometry by using LQC techniques, we describe it in terms of the Ashtekar-Barbero variables of the Bianchi I model with three-torus topology. In a diagonal gauge these variables are given by the three components of the densitized triad p_j and of the $su(2)$-connection c_j, with $j = \theta, \sigma, \delta$. They satisfy $\{c_i, p_j\} = 8\pi\gamma G\delta_{ij}$, where γ is the Immirzi parameter and G is the Newton constant. It has been proven that the deparametrized model admits a unique satisfactory Fock quantization [13]. Then we describe the inhomogeneities employing the annihilation and creation variables chosen in [13]. Two global constraints remain to be imposed at the quantum level, a momentum constraint $\mathcal{C}_\theta$ that generates translations in the circle and only involves the inhomogeneous sector, and the Hamiltonian constraint, $\mathcal{C} = \mathcal{C}_{\text{hom}} + \mathcal{C}_{\text{inh}}$, formed by a homogeneous term that is not other thing but the Hamiltonian constraint of the Bianchi I model with a homogeneous scalar ϕ, and by an additional term that couples homogeneous and inhomogeneous sectors.

In order to quantize the system we have to define a representation of the basic variables on a kinematical Hilbert space, that in our case takes this form: $\mathcal{H}_{\text{kin}} = \mathcal{H}_{\text{kin}}^{\text{BI}} \otimes L^2(\mathbb{R}, d\phi) \otimes \mathcal{F}^\xi \otimes \mathcal{F}^\varphi$. Here $\mathcal{H}_{\text{kin}}^{\text{BI}}$ is the kinematical Hilbert space for the Bianchi I model within the improved dynamics of LQC [11], $\mathcal{F}^\xi$ and $\mathcal{F}^\varphi$ are identical Fock spaces, for the gravitational waves and for the matter inhomogeneities respectively, and $L^2(\mathbb{R}, d\phi)$ is the Hilbert space accounting for the homogeneous scalar ϕ for which we employ a standard Schrödinger representation. The construction of the kinematical Hilbert space $\mathcal{H}_{\text{kin}}^{\text{BI}}$ in LQC mimics that of LQG in the sense that the connection is not defined in the quantum theory but only its holonomies. The inner product is discrete, so that, the operators $\hat{p}_j$ have a discrete spectrum equal to the real line, and their mutual eigenstates $|p_\theta, p_\sigma, p_\delta\rangle$ form a orthonormal basis of $\mathcal{H}_{\text{kin}}^{\text{BI}}$. Because of the improved dynamics, the action of the basic holonomy operators, $\hat{\mathcal{N}}_{\pm\bar{\mu}_j}$, on the states $|p_\theta, p_\sigma, p_\delta\rangle$ is quite involved. It is convenient to relabel them as $|\lambda_\theta, \lambda_\sigma, v\rangle$, where $\lambda_j \propto \text{sgn}(p_j)\sqrt{|p_j|}$ and $v = 2\lambda_\theta\lambda_\sigma\lambda_\delta$ is proportional to the Bianchi I volume, or *homogeneous* volume. The holonomy operators $\hat{\mathcal{N}}_{\pm\bar{\mu}_j}$ scale the label λ_j in such a way that the label v is simply shifted by one [11]. For the Fock spaces, $\mathcal{F}^\alpha(\alpha = \xi, \varphi)$, an orthonormal basis is given by the n-particle states. We call n_m^α the occupation number of the field α in the mode m. The creation-annihilation variables are promoted to creation-annihilation operators, denoted by $\hat{a}_m^{(\alpha)\dagger}, \hat{a}_m^{(\alpha)}$, respectively.

Now we can construct the operators that represent the constraints. The generator of the translations in the circle is represented by the operator

$$\widehat{C}_\theta = \sum_{m=1}^{\infty} m\left(\widehat{X}_m^{\xi} + \widehat{X}_m^{\varphi}\right), \qquad \widehat{X}_m^{\alpha} = \hat{a}_m^{(\alpha)\dagger}\hat{a}_m^{(\alpha)} - \hat{a}_{-m}^{(\alpha)\dagger}\hat{a}_{-m}^{(\alpha)}. \tag{1}$$

The imposition of this constraint on $\mathcal{F}^{\xi}\otimes\mathcal{F}^{\varphi}$ leads to the condition, $\sum_{m=1}^{\infty} m(X_m^{\xi}+X_m^{\varphi})=0$, where $X_m^{\alpha}=n_m^{\alpha}-n_{-m}^{\alpha}$. The n-particle states that satisfy this condition form a proper Fock subspace $\mathcal{F}_{\rm p}\subset\mathcal{F}^{\xi}\otimes\mathcal{F}^{\varphi}$ that is unitarily equivalent to the physical space of the Fock quantization of the deparametrized model, obtained as in vacuum case [13].

For the Hamiltonian constraint operator we follow the construction of the vacuum case [8], so that the operator decouples the states of zero homogeneous volume and does not relate states with different sign of any of the homogeneous geometry labels $\lambda_\theta, \lambda_\sigma$ and v. Consequently, we can remove from our theory the states which are the analog of the classical singularity (those with $v=0$) and restrict the study to the sector with strictly positive labels $\lambda_\theta, \lambda_\sigma$ and v, for instance. The Hamiltonian constraint operator reads,

$$\widehat{C} = -\sum_{i\neq j}\sum_{j}\frac{\widehat{\Theta}_i\widehat{\Theta}_j}{16\pi G\gamma^2} - \frac{\hbar^2}{2}\left[\frac{\partial}{\partial\phi}\right]^2 + 2\pi\hbar\widehat{|p_\theta|}\widehat{H}_0 + \hbar\left[\widehat{\frac{1}{|p_\theta|^{\frac14}}}\right]^2\frac{\left(\widehat{\Theta}_\delta+\widehat{\Theta}_\sigma\right)^2}{16\pi\gamma^2}\left[\widehat{\frac{1}{|p_\theta|^{\frac14}}}\right]^2\widehat{H}_{\rm int}, \tag{2}$$

with $i,j\in\{\theta,\sigma,\delta\}$. Here we have introduced the quantum version of c_jp_j, given by

$$\widehat{\Theta}_j = i\pi\gamma G\hbar\widehat{\sqrt{|v|}}\left[\left(\hat{\mathcal{N}}_{-2\bar{\mu}_j}-\hat{\mathcal{N}}_{2\bar{\mu}_j}\right)\widehat{\mathrm{sgn}(p_j)}+\widehat{\mathrm{sgn}(p_j)}\left(\hat{\mathcal{N}}_{-2\bar{\mu}_j}-\hat{\mathcal{N}}_{2\bar{\mu}_j}\right)\right]\widehat{\sqrt{|v|}}, \tag{3}$$

and the operator $[\widehat{1/|p_\theta|^{1/4}}]$, which is regularized as usual by taking the commutator of some suitable power of p_θ with the holonomies [8]. The operators $\widehat{H}_{\rm int}$ and $\widehat{H}_0$ act non-trivially only on the inhomogeneous sector of the kinematical Hilbert space and are given by

$$\widehat{H}_0 = \sum_{\alpha\in\{\xi,\varphi\}}\sum_{m=1}^{\infty} m\widehat{N}_m^{\alpha}, \qquad \widehat{H}_{\rm int} = \sum_{\alpha\in\{\xi,\varphi\}}\sum_{m=1}^{\infty}\frac{1}{m}\left(\widehat{N}_m^{\alpha}+\hat{a}_m^{(\alpha)\dagger}\hat{a}_{-m}^{(\alpha)\dagger}+\hat{a}_m^{(\alpha)}\hat{a}_{-m}^{(\alpha)}\right), \tag{4}$$

where $\widehat{N}_m^{\alpha}=\hat{a}_m^{(\alpha)\dagger}\hat{a}_m^{(\alpha)}+\hat{a}_{-m}^{(\alpha)\dagger}\hat{a}_{-m}^{(\alpha)}$. Note that the inhomogeneities of both fields contribute to the constraints in exactly the same way.

The Hamiltonian constraint operator does not relate all states with different values of v and λ_j ($j=\theta,\sigma$), but *superselects* different sectors. The superselection sectors in v are semilattices of step four in $\mathbb{R}^+$ from a minimum value $\epsilon\in(0,4]$. The superselection sectors in λ_j ($j=\theta,\sigma$) are more involved. Given initial data λ_j^* and ϵ, the values of λ_j in the corresponding sector are of the form $\lambda_j=\omega_\epsilon\lambda_j^*$, where ω_ϵ runs over a certain countable set, that is dense in $\mathbb{R}^+$ (see [8]).

The quantum Hamiltonian constraint leads to a difference equation in the variable v, and then we can regard it as an evolution equation in v. As in the vacuum model [8], the solutions are completely determined by a dense set of data in the initial section $v=\epsilon$. This property allows us to characterize the physical Hilbert space as the Hilbert space of these initial data, whose inner product is determined imposing reality conditions in a complete set of observables. The result is $\mathcal{H}_{\rm phys}=\mathcal{H}_{\rm phys}^{\rm BI}\otimes L^2(\mathbb{R},d\phi)\otimes\mathcal{F}_{\rm p}$, where $\mathcal{H}_{\rm phys}^{\rm BI}$ is the physical Hilbert space for Bianchi I cosmologies given in [9].

3. *Projection* to LRS-Gowdy Model

The linearly polarized Gowdy T^3 model is symmetric under the interchange $\sigma \longleftrightarrow \delta$. Therefore, there is a subset of classical solutions with local rotational symmetry and we will call LRS-Gowdy model the system that they define. The quantum LRS-Gowdy model can be obtained from the quantum Gowdy model that we have constructed above by using the map

$$|\Psi(\lambda_\theta, \lambda_\sigma, v)\rangle \qquad \longrightarrow \qquad \sum_{\lambda_\sigma} |\Psi(\lambda_\theta, \lambda_\sigma, v)\rangle \equiv |\psi(\lambda_\theta, v)\rangle, \tag{5}$$

from the Gowdy states spanned in the basis of Bianchi I states $|\Psi(\lambda_\theta, \lambda_\sigma, v)\rangle$ to the LRS-Gowdy states spanned in the basis of LRS-Bianchi I states $|\psi(\lambda_\theta, v)\rangle$. Here the sum in λ_σ is carried out over the considered superselection sector. A similar map was introduced in [11] to project from the loop quantized Bianchi I model to its isotropic sector, the flat FRW model. Note that in our case we cannot sum over λ_θ because the interchange $\theta \longleftrightarrow \sigma$ is not a symmetry.

4. Conclusions

We have obtained a complete quantization of the Gowdy T^3 model with linearly polarized gravitational waves and a minimally coupled inhomogeneous massless scalar field as matter content using hybrid techniques in the framework of LQC. This hybrid quantization follows exactly in the same way as that of the vacuum model because the matter and gravitational inhomogeneities can be treated in the very same way.

Owing to the polymeric quantization of the homogeneous gravitational sector we have been able to eliminate the analog states to the classical cosmological singularities at the level of superselection. Furthermore, the standard quantization of both matter and gravitational inhomogeneities is recovered in physical states.

The inclusion of the massless scalar field not only has allowed us to study matter inhomogeneities in the framework of LQC by means of this hybrid quantization, but also leads to a more physical interesting model. In fact, the homogeneous sector of the model admits now flat Friedmann-Robertson-Walker (FRW) solutions.

This hybrid Gowdy T^3 model, or more specifically the simpler hybrid LRS-Gowdy T^3 model, provides a perfect scenario for further study of the quantum back-reaction between anisotropies and inhomogeneities and a loop quantized flat FRW background, as well as for the study of approximated methods in LQC in order to extract physical predictions from the theory [12].

Acknowledgments

This work was supported by the Spanish MICINN Projects No. FIS2008-06078-C03-03, No. FIS2011-30145-C03-02, and the Consolider-Ingenio Program CPAN No. CSD2007-00042. D M-dB is supported by CSIC and the European Social Fund under the Grant No. JAEPre-09-01796.

References

[1] Bojowald M 2008 *Living Rev. Rel.* **11** 4
[2] Ashtekar A and Singh P 2011 *Preprint* arXiv:1108.0893
[3] Banerjee K, Calcagni G and Martín-Benito M 2011 *Preprint* arXiv:1109.6801
[4] Thiemann T 2007 *Modern canonical quantum general relativity* (Cambridge: Cambridge University Press)
[5] Ashtekar A, Pawłowski T and Singh P 2006 *Phys. Rev.* D **74** 084003
[6] Martín-Benito M, Garay L J and Mena Marugán G A 2008 *Phys. Rev.* D **78** 083516
[7] Martín-Benito M and Mena Marugán G A 2009 *Int. J. Mod. Phys.* A **24** 2820
[8] Garay L J, Martín-Benito M and Mena Marugán G A 2010 *Phys. Rev.* D **82** 044048
[9] Martín-Benito M, Mena Marugán and Wilson-Ewing E 2010 *Phys. Rev.* D **82** 084012
[10] Martín-Benito M, Martín-de Blas and Mena Marugán G A 2011 *Phys. Rev.* D **83** 084050
[11] Ashtekar A and Wilson-Ewing E 2009 *Phys. Rev.* D **79** 083535
[12] Martín-Benito M, Martín-de Blas D and Mena Marugán G A *in preparation*
[13] Cortez J, Mena Marugán G A and Velhinho J M 2007 *Phys. Rev.* D **75**, 084027.

Journal of Physics: Conference Series **360** (2012) 012033 doi:10.1088/1742-6596/360/1/012033

Inflation and inhomogeneities: a hybrid quantization

J Olmedo, M Fernández-Méndez and G A Mena Marugán

Instituto de Estructura de la Materia, IEM-CSIC, Serrano 121, 28006 Madrid, Spain

E-mail: olmedo@iem.cfmac.csic.es, m.fernandez@iem.cfmac.csic.es, mena@iem.cfmac.csic.es

Abstract. We provide a complete quantization of a homogeneous and isotropic spacetime with positive spatial curvature coupled to a massive scalar field in the framework of Loop Quantum Cosmology. The physical Hilbert space is constructed out of the space of initial data on the minimum volume section. By means of a perturbative treatment we introduce inhomogeneities and thereafter we adopt a hybrid quantum approach, in which these inhomogeneous degrees of freedom are described by a standard Fock quantization. For the considered case of compact spatial topology, the requirements of: i) invariance of the vacuum state under the spatial isometries, and ii) unitary implementation of the quantum dynamics, pick up a privileged set of canonical fields and a unique Fock representation (up to unitary equivalence).

1. Introduction

Loop Quantum Cosmology (LQC) [1] is an approach to quantize symmetry reduced systems in cosmology which has provided new insights into the early universe physics. In particular, the analysis of a homogeneous and isotropic Friedmann-Robertson-Walker (FRW) spacetime coupled to a massless scalar field has been carried out to completion, showing that the classical singularity is replaced with a quantum bounce. Besides, less symmetric models, such as anisotropic homogeneous spacetimes [2], have been successfully quantized.

To discuss the inflationary paradigm [3] that naturally explains the present state of our universe, one can add a potential for the scalar field, like in the simple case of a mass term. A careful analysis of the effective dynamics expected to correspond to this model [4] demonstrates the high probability that the inflaton produces the necessary number of e-foldings, solving the fine-tuning problem that appears in General Relativity. However, we still lack a rigorous derivation of such an effective dynamics from a complete quantization within LQC of a massive scalar field in a FRW spacetime. On the other hand, an even more realistic scenario is achieved by introducing inhomogeneities coming from the vacuum fluctuations of the inflaton. The application of perturbative methods, together with a standard Fock representation for the inhomogeneities, should satisfactorily describe the origin of the structures currently observed. Nevertheless, in general, one has available an infinite number of inequivalent Fock quantizations for fields in FRW spacetimes, resulting in a loss of prediction capability unless one can justify the selection of a preferred Fock representation.

In this work, we will consider a closed FRW spacetime coupled to a massive scalar field. We will provide a polymeric quantization of this model, in which any solution to the Hamiltonian constraint is totally determined by the data on the minimum (nonvanishing) volume section. The physical Hilbert space can be identified with such initial data space once it has been completed with a suitable inner product, which is uniquely determined by requiring the self-adjointness of a complete set of real observables. Moreover, we will introduce small inhomogeneities around homogeneous solutions. After linearization of the system and gauge fixing, the inhomogeneities are described by a (linear) scalar field theory. Finally, we will provide a unique $SO(4)$-invariant Fock quantization for them just by demanding the unitary implementation of the dynamics.

2. Homogeneous system

The homogeneous and isotropic system has two global degrees of freedom. One is the pair (ϕ, p_ϕ), consisting of a massive scalar field and its canonically conjugate momentum. As for the geometry, we consider a description in terms of a densitized triad adapted to the fiducial metric of the 3-spheres, parametrized by the variable p, and an Ashtekar-Barbero $su(2)$-connection, determined by a variable c. In order to maintain the parallelism with models where the spatial slices have open, flat topology, we parametrize the phase space so that some fundamental structures, as the Poisson brackets, are independent of the fiducial volume $l_0^3 = 2\pi^2$. Therefore, $\{c, p\} = 8\pi G\gamma/3$. Here, G and γ are the Newton constant and the Immirzi parameter, respectively.

For the quantization of the geometry, we apply a polymeric quantization. The basic variables in LQC are fluxes of densitized triads through "square" surfaces, enclosed by four geodesic edges, which basically reduce to p, and holonomies of the connection along integral curves of the fiducial triads of fiducial length $\bar{\mu}l_0$. In this respect, we adopt the improved dynamics scheme: $\bar{\mu} = \sqrt{\Delta/p}$, with the constant Δ determined by the infrared spectrum of the area operator defined in Loop Quantum Gravity. The gravitational part of the kinematical Hilbert space will be $\mathcal{H}_{\text{kin}}^{\text{grav}} = L^2(\mathbb{R}_{\text{Bohr}}, d\mu_{\text{Bohr}})$, where $\mathbb{R}_{\text{Bohr}}$ is the Bohr compactification of the real line, with Haar measure given by $d\mu_{\text{Bohr}}$. In this Hilbert space, one can find a basis of normalizable states $\{|v\rangle\}$ in which the matrix elements of the holonomies, i.e., $\hat{\mathcal{N}}_{\bar{\mu}}$, create quanta of volume, while $\hat{p}|v\rangle = \text{sign}(v)(2\pi\gamma G\hbar\sqrt{\Delta}|v|)^{2/3}|v\rangle$. For the scalar field, we will consider a standard Schrödinger representation. Now, we can promote the Hamiltonian constraint to a quantum operator (see, e.g., [5]). Adopting a suitable factor ordering, we get for the quantum Hamiltonian constraint

$$\hat{C} = \widehat{\left[\frac{1}{V}\right]}^{1/2}\left[\hat{C}_g + 8\pi G(\hat{p}_\phi^2 + m^2\hat{V}^2\hat{\phi}^2)\right]\widehat{\left[\frac{1}{V}\right]}^{1/2}, \quad \hat{C}_g = \hat{C}_g^0 - \frac{6}{\gamma^2}\left[(1+\gamma^2)l_0^2\hat{V}^{4/3} - \frac{\hat{V}^2}{\Delta}\sin^2\bar{\mu}l_0\right],$$

$$\hat{C}_g^0 = -\frac{3\pi^2G^2\hbar^2}{2}\left[\hat{\mathcal{N}}_{2\bar{\mu}}f(\hat{v})\hat{\mathcal{N}}_{2\bar{\mu}} + \hat{\mathcal{N}}_{-2\bar{\mu}}f(\hat{v})\hat{\mathcal{N}}_{-2\bar{\mu}} - f_o(\hat{v})\right], \quad f(\hat{v}) = \sqrt{|\hat{v}^2-4|}|\hat{v}|s_+(\hat{v})s_-(\hat{v}),$$

$$f_o(\hat{v}) = |\hat{v}|\left[|\hat{v}+2|s_+^2(\hat{v}) + |\hat{v}-2|s_-^2(\hat{v})\right], \quad s_\pm(\hat{v}) = \text{sgn}(\hat{v}\pm 2) + \text{sgn}(\hat{v}).$$

The constraint is a second order difference operator, with a step of 4 units in the label v. Moreover, we find that $f(v) = 0\ \forall v \in [-2, 2]$. Therefore, $\hat{C}_g$ has an invariant domain composed of states with support on semilattices of the type $\mathcal{L}_\varepsilon^\pm = \{v = \pm(\varepsilon + 4n), n \in \mathbb{N}\}$, where $\varepsilon \in (0, 4]$ is a bounded continuous parameter labeling each sector. Besides, all the relevant operators under consideration preserve such sectors, being possible to interpret them as superselection sectors. For convenience, we restrict our considerations to semilattices with $v > 0$ from now on. Moreover, any solution to the constraint is totally determined at any $v = \varepsilon + 4n$ with $n > 0$ by its value at the minimum volume, $\Psi(\phi, v = \varepsilon)$. The physical Hilbert space can be constructed out of the functional space of initial data at $v = \varepsilon$. Its inner product can be determined by imposing reality conditions on a complete set of observables as self-adjointness relations. In this way, the physical Hilbert space of this homogeneous system can be taken as $\mathcal{H}^{\text{phy}} = L^2(\mathbb{R}, d\phi)$.

3. Inhomogeneities and gauge fixing

It is natural to assume that small initial inhomogeneities may arise from vacuum fluctuations of the inflaton. In this context, a perturbative treatment of them seems justified. We will consider only scalar perturbations around homogeneous solutions. Thus, we parametrize the lapse, the shift and the 3-metric in the geometry sector in the form $N = \sigma(N_0 + \delta N)$, $N_a = \sigma^2 \delta N_a$, and $h_{ab} = \sigma^2 e^{2\alpha}(\Omega_{ab} + \epsilon_{ab})$. We also introduce a massive scalar field $\Phi = \sigma^{-1}(l_0^{-3/2}\varphi + \delta\varphi)$. We have defined $\sigma^2 = 4\pi G/3l_0^3$, and the scale factor is e^α [6]. Besides, the compact spatial sections are isomorphic to S^3, which is endowed with a natural basis of spherical harmonics. In this basis, we can write

$$\delta N = l_0^{3/2} N_0 \sum_n g_n Q_n, \quad \delta N_a = l_0^{3/2} e^\alpha \sum_n j_n P_a^n, \quad \epsilon_{ab} = 2 l_0^{3/2} \sum_n (a_n Q_n \Omega_{ab} + 3 b_n P_{ab}^n),$$

and $\delta\varphi = \sum_n f_n Q_n$. We will consider perturbations to second order in the Hamiltonian, which yields

$$H = N_0 \left[C_0 + \sum_n (C_2^n + g_n C_1^n) \right] + \sum_n j_n D_1^n, \tag{1}$$

where C_0 corresponds to the scalar constraint of the homogeneous setting $(\alpha, \pi_\alpha, \varphi, \pi_\varphi)$ (with α replacing p and π_α being its momentum), and each C_2^n is quadratic in the scalar modes (a_n, b_n, f_n) and their conjugate momenta $(\pi_{a_n}, \pi_{b_n}, \pi_{f_n})$, while C_1^n and D_1^n are the modes of the linear perturbation of the scalar and the diffeomorphism constraints, respectively. Note that, in our formalism, the perturbations g_n and j_n just play the role of Lagrange multipliers.

In order to reduce the system, we introduce a partial gauge fixing. One possibility is the longitudinal gauge, what in our canonical formalism is equivalent to the conditions $b_n = 0$ and $\Pi_n = \pi_{a_n} - \pi_\alpha a_n - 3\pi_\varphi f_n = 0$. These conditions are second class with D_1^n and C_1^n. By means of the dynamical stability conditions $\dot{b}_n = 0$ and $\dot{\Pi}_n = 0$, together with $D_1^n = 0$, one can see that $j_n = 0$ (or equivalently $\pi_{b_n} = 0$) and $g_n = -a_n$. If the linear scalar constraint is imposed, we can finally solve for a_n and π_{a_n} in terms of f_n and π_{f_n}. After gauge fixing, the reduced symplectic structure is nondiagonal. To remedy this, we perform a transformation in the phase space so as to arrive at a symplectic structure with canonical form. The whole transformation can be reconstructed from the redefinition $\tilde{\pi}_{f_n} = \pi_{f_n} - 3\pi_\varphi a_n$. Finally, in order to be able to apply the uniqueness results of [7, 8] to our system, we conveniently reformulate it by introducing a canonical transformation that replaces the mode functions $\tilde{f}_n$ and $\tilde{\pi}_{f_n}$ with a new set, $\bar{f}_n = e^\alpha \tilde{f}_n$ and $\bar{\pi}_{f_n} = e^{-\alpha}(\tilde{\pi}_{f_n} - \pi_\alpha \tilde{f}_n)$. The transformation is such that the second order equation of motion for $\bar{f}_n$ approaches (in the ultraviolet limit) sufficiently fast the equation which corresponds to (the modes of) a scalar field propagating in a static spacetime, but with a time dependent mass. Besides, one can easily extend the canonical transformation to the homogeneous sector.

4. Hybrid quantization

To carry out a complete quantization of the system, we will apply the polymeric quantization prescription explained above for the homogeneous sector of the geometry, whereas, for the matter content (background + perturbations), we will consider a standard representation. Now, the problem that we have to face concerns the choice of representation for the canonical commutation relations (CCR's) of the inhomogeneous variables. The usual procedure to pick up a unique Fock quantization is to invoke the classical symmetries (e.g., in Minkowski or pure deSitter spacetimes). In our case, the classical symmetries are not enough, and hence such a criterion is insufficient. Fortunately, significant progress on this issue has been made recently. In particular, it has been shown that any representation which is invariant under the spatial isometries and permits the unitary implementation of the dynamics belongs to a unique equivalence class of Fock quantizations [7]. Moreover, these criteria not only fix the representation of the CCR's, but also the pair of canonical fields is uniquely determined, among all the possibilities related by a time dependent scaling of the field [8].

In our situation, the variables $\bar{f}_n$ and $\bar{\pi}_{f_n}$ satisfy those requirements. Therefore, we can choose a (unique) $SO(4)$-invariant Fock representation with unitary dynamics for them, identifying the kinematical Hilbert space $\mathcal{H}^{\rm kin}_{\rm inh}$ with the corresponding Fock space $\mathcal{F}$ constructed from the one-particle Hilbert space. For the homogeneous degrees of freedom, we adopt the polymeric quantization presented above. One can see then that the total kinematical Hilbert space is $\mathcal{H}^{\rm kin} = \mathcal{H}^{\rm kin}_{\rm grav} \otimes L^2(\mathbb{R}, d\phi) \otimes \mathcal{F}$.

We are now ready for the next step: studying the solutions to the Hamiltonian constraint. The strategy is based on proving that any solution is completely determined by its value on the minimum volume section. Therefore, following the lines explained above for the homogeneous scenario, one can construct the physical Hilbert space by equipping the space of initial data on the minimum volume section with a suitable inner product, which can be determined by requirements of self-adjointness on a complete set of observables (see, e.g., [9]).

Finally, in order to avoid any dependence on the gauge choice, it is natural to work with gauge invariant quantities [10]. In particular, we are interested in finding a suitable choice of gauge invariant potentials that can be identified with our fundamental variables $\bar{f}_n$ and $\bar{\pi}_{f_n}$ by means of a canonical transformation which is unitarily implementable in the quantum theory. All those aspects will be a matter of future research.

5. Conclusions

We have started by analyzing a homogeneous massive scalar field propagating in a closed FRW spacetime. We have quantized the system to completion, combining both Schrödinger and polymeric representations. The physical Hilbert space has been constructed out of the functional space defined on the minimum volume section and which determines the solutions to the constraint. Later on, with the purpose of studying a more realistic setting, we have introduced inhomogeneities by means of perturbation theory, under the assumption that the observed inhomogeneities come from small quantum fluctuations of the inflaton. We have considered scalar perturbations to second order in the action. We have carried out a partial gauge fixing, removing two of the three fieldlike degrees of freedom. To fulfill the uniqueness criteria of invariance and unitarity of the Fock quantization, we have introduced a canonical transformation consisting in a scaling of the remaining field which depends on the homogeneous phase space variables. Finally, we have quantized the homogeneous system combining the "loop" techniques for the homogeneous degrees of freedom with the standard ones for the inhomogeneities. Besides, we have constructed the quantum Hamiltonian constraint. Finally, this constraint must be solved by means of a characterization of its solutions, and the Hilbert space constructed out of them.

Acknowledgments

This work was supported by the MICINN/MINECO Projects No. FIS2008-06078-C03-03 and No. FIS2011-30145-C03-02, and the Consolider-Ingenio Program CPAN No. CSD2007-00042.

References

[1] Bojowald M 2008 *Living Rev. Relativity* **11**, 4
[2] Ashtekar A and Wilson-Ewing E 2009 *Phys. Rev.* D **79**, 083535; Martín-Benito M, Mena Marugán G A and Pawłowski T 2008 *Phys. Rev.* D **78**, 064008
[3] Liddle A R and Lyth D H 2000 *Inflation and Large-Scale Structure* (Cambridge University Press)
[4] Ashtekar A and Sloan D 2010 *Phys. Lett.* B **694**, 108
[5] Ashtekar A, Pawłowski T, Singh P and Vandersloot K 2007 *Phys. Rev.* D **75**, 024035
[6] Halliwell J J and Hawking W 1985 *Phys. Rev.* D **31**, 8
[7] Cortez J, Mena Marugán G A and Velhinho J M 2010 *Phys. Rev.* D **81**, 044037
[8] Cortez J, Mena Marugán G A, Olmedo J and Velhinho J M 2010 *JCAP* **10**, 030
[9] Garay L J, Martín-Benito M and Mena Marugán G A 2010 *Phys. Rev.* D **82**, 044048
[10] Bardeen J M 1980 *Phys. Rev.* D **22**, 8

Journal of Physics: Conference Series **360** (2012) 012034 doi:10.1088/1742-6596/360/1/012034

Palatini approach to bouncing cosmologies and DSR-like effects

Gonzalo J. Olmo

Departamento de Física Teórica and IFIC, Centro Mixto Universidad de Valencia - CSIC. Facultad de Física, Universidad de Valencia, Burjassot-46100, Valencia, Spain.

E-mail: gonzalo.olmo@csic.es

Abstract. It is shown that a quadratic gravitational Lagrangian in the Palatini formulation is able to capture different aspects of quantum gravity phenomenology in a single framework. In particular, in this theory field excitations propagating with different energy-densities perceive different background metrics, a fundamental characteristic of the DSR and Rainbow Gravity approaches. This theory, however, avoids the so-called *soccer ball problem*. Also, the resulting isotropic and anisotropic cosmologies are free from the big bang singularity. This singularity avoidance occurs non-perturbatively and shares some similitudes with the effective dynamics of loop quantum cosmology.

Introduction. How could the Planck length, $l_P = \sqrt{\hbar G/c^3} \sim 10^{-35}$m, affect our understanding of space-time and quantum theory? To have a chance to answer this question, we should first understand how l_P could be made compatible with the principles of relativity and general covariance. The tension between the coexistence of relativity and an invariant length scale is mainly due to our view that space-time assumes a Minkowskian continuous structure up to arbitrarily short scales. In this sense, to accommodate l_P within our physical theories it may be useful to follow Riemann's advise and modify conventional geometric concepts as one applies them *both in the direction of the immeasurably large , and in the direction of the immeasurably small.* Einstein modified the space-time structure and dynamics at large scales, but kept the Minkowskian structure locally. Here it will be shown that a further revision of the classical space-time structure and dynamics at microscopic scales allows to introduce l_P in an invariant and universal way compatible with general covariance. The resulting theory has interesting phenomenological aspects. On the one hand, we find that particles with different energy-momentum densities perceive different background geometries, a property closely related to doubly or deformed special relativity (DSR) [1, 2] and its gravitational extension, known as Rainbow Gravity (RG) [3]. On the other hand, the cosmological dynamics of the theory closely matches that of GR at all times except near the big bang singularity, which is replaced by a regular bounce that occurs near the Planck density.

Palatini approach to the idea of fundamental length. Special relativity was built by requiring that the speed of light were an invariant and universal magnitude. To combine the speed of light and the Planck length in a way that preserves the invariant and universal nature of both quantities, we first note that c^2 has the dimensions of a squared velocity but that c is not a 3-velocity. Analogously, we may see l_P^2 as an invariant with dimensions of length squared but l_P not being a 3-length[2]. Dimensional compatibility with a curvature suggests that $l_P^2 \equiv 1/R_P$ could be introduced in the theory via the gravitational sector. The situation is not as simple as it may seem at first because an action like

$$S[g_{\mu\nu},\psi] = \frac{\hbar}{16\pi l_P^2}\int d^4x\sqrt{-g}\left[R + l_P^2 R^2\right] + S_m[g_{\mu\nu},\psi] \ , \tag{1}$$

where $S_m[g_{\mu\nu},\psi]$ represents the matter sector, contains l_P but not in the invariant form that we wished. In fact, the field equations that follow from (1) are equivalent to those of a Brans-Dicke theory with $\phi \equiv 1 + 2R/R_P$, a non-zero potential $V(\phi) = \frac{R_P}{4}(\phi-1)^2$, and $w = 0$ [4]. As is well-known, in Brans-Dicke theory the observed Newton's constant is promoted to the status of field: $G_{eff} \sim G/\phi$. This allows the effective Newton's constant G_{eff} to change in time and in space and, which makes l_P^{eff} also vary in space and time. This is quite different from the assumed constancy and universality of the speed of light in special relativity and is closer to the philosophy of varying speed of light theories [5]. The situation does not improve if we introduce higher curvature invariants in (1) because they generate new degrees of freedom which turn Newton's constant into a dynamical field.

A way out of this problem can be found by modifying the geometric structure of the theory. If we construct the theory *à la Palatini* [6], that is in terms of a connection not a priori constrained to be given by the Christoffel symbols, then the resulting equations do not necessarily contain new dynamical degrees of freedom (as compared to GR), and the Planck length may remain space-time independent in much the same way as the speed of light. The field equations that follow from (1) when metric and connection are varied independently are

$$f_R R_{\mu\nu}(\Gamma) - \frac{1}{2} f g_{\mu\nu} = \kappa^2 T_{\mu\nu} \ , \ \nabla_\alpha\left(\sqrt{-g} f_R g^{\beta\gamma}\right) = 0 \ , \tag{2}$$

where $f = R + l_P^2 R^2$, $f_R \equiv \partial_R f$, and $\kappa^2 \equiv 8\pi G$. The connection equation can be easily solved after noticing that the trace of (2), $Rf_R - 2f = \kappa^2 T$, represents an algebraic relation between $R \equiv g^{\mu\nu}R_{\mu\nu}(\Gamma)$ and T, which implies that $R = R(T)$ and hence $f_R = f_R(T)$. For the particular Lagrangian (1), we find that $R = -\kappa^2 T$, like in GR. The connection turns out to be the Levi-Civita connection of an auxiliary metric $h_{\mu\nu}$ conformally related with the physical metric, $h_{\mu\nu} = f_R(T) g_{\mu\nu}$. We can now rewrite (2) as follows

$$G_{\mu\nu}(h) = \frac{\kappa^2}{f_R(T)} T_{\mu\nu} + \Lambda(T) h_{\mu\nu} \ , \ \nabla_\alpha\left(\sqrt{-h} h^{\beta\gamma}\right) = 0 \ , \tag{3}$$

where $\Lambda(T) \equiv (f - Rf_R)/(2f_R^2) = -(\kappa^2 T)^2 l_P^2$. Note that $\frac{\kappa^2}{f_R(T)} \equiv 8\pi G/(1 - \kappa^2 T/R_P)$ and the function $\Lambda(T)$ can be seen as a T-dependent Newton's constant and cosmological *constant*, respectively. This structure is very similar to that proposed in [3] for RG, $G_{\mu\nu}(E) = 8\pi G(E) T_{\mu\nu}(E) + g_{\mu\nu}\Lambda(E)$, with the difference that the energy E of the particle probing the geometry is here replaced by T which is a generally covariant function of its energy-density. From the structure of (3) and the relation $g_{\mu\nu} = (1/f_R)h_{\mu\nu}$, it follows that $g_{\mu\nu}$ is affected by the matter-energy in two different ways. The first contribution corresponds to the cumulative effects of matter, and the second contribution is due to the dependence on the local density distributions of energy and momentum. In fact, the structure of (3) is similar to that of GR, which implies that $h_{\mu\nu}$ is determined by integrating over all the sources (gravity as a cumulative effect). Besides that, $g_{\mu\nu}$ is also affected by the local sources via the factor $f_R^{-1}(T)$. To illustrate

[2] Note that this idea is radically different from lattice-type approaches to the problem of a minimum length.

Journal of Physics: Conference Series **360** (2012) 012034 doi:10.1088/1742-6596/360/1/012034

this point, consider a region of the spacetime containing a total mass M and filled with sources of low energy-density as compared to the Planck scale ($|\kappa^2 T/R_P| \ll 1$). For the quadratic model $f(R) = R + R^2/R_P$, in this region (3) boils down to $G_{\mu\nu}(h) = \kappa^2 T_{\mu\nu} + O(\kappa^2 T/R_P)$, and $h_{\mu\nu} \approx (1 + O(\kappa^2 T/R_P))g_{\mu\nu}$, which implies that the GR solution is a very good approximation . This confirms that $h_{\mu\nu}$ is determined by an integration over the sources, like in GR. Now, if this region is traversed by a particle of mass $m \ll M$ but with a non-negligible ratio $\kappa^2 T/R_P$, then the contribution of this particle to $h_{\mu\nu}$ can be neglected, but its effect on $g_{\mu\nu}$ via de factor $f_R^{-1} = 1 - \kappa^2 T/R_P$ on the region that supports the particle (its classical trajectory) is important. This phenomenon is analogous to that described in RG, where particles of different energies (energy-densities in our case) perceive different metrics. Our approach, however, naturally cures the so-called *soccer ball problem* for macroscopic bodies thanks to the dependence of the metric on the local energy-momentum densities rather than on the total energy-momentum of the particle.

Beyond $f(R)$ theories. The energy-density properties of the Palatini theory (1) is maintained with a much richer phenomenology when the $f(R)$ family is extended to Palatini $f(R, Q)$ Lagrangians, where $Q \equiv R_{\mu\nu}R^{\mu\nu}$, and $R_{\mu\nu}$ is the symmetric Ricci tensor. In this case, the connection can also be expressed as the Levi-Civita connection of a metric $\tilde{h}_{\mu\nu}$ related with $g_{\mu\nu}$ by $\tilde{h}^{\mu\nu} = \frac{g^{\mu\alpha}{\Sigma_\alpha}^\nu}{\sqrt{\det \Sigma}}$, where ${\Sigma_\alpha}^\nu = f_R\delta_\alpha^\nu + 2f_Q{P_\alpha}^\nu$ and ${P_\mu}^\nu = R_{\mu\alpha}g^{\alpha\nu}$ are functions of ${T_\mu}^\nu = T_{\mu\alpha}g^{\alpha\nu}$ and, therefore, depend on the local densities of energy and momentum (see [7, 8] for details). The field equations for the metric can be written in compact form as

$$ {R_\mu}^\nu(\tilde{h}) = \frac{1}{\sqrt{\det \hat{\Sigma}}}(\frac{f}{2}{\delta_\mu}^\nu + \kappa^2 {T_\mu}^\nu) \ . \tag{4} $$

For a scalar field with kinetic energy $\chi \equiv g^{\mu\nu}\partial_\mu\phi\partial_\nu\phi$ and Lagrangian $\mathcal{L} = \chi + 2V(\phi)$, $\tilde{h}_{\mu\nu}$ and $g_{\mu\nu}$ turn out to be related by

$$ g_{\mu\nu} = \frac{1}{\Omega}\tilde{h}_{\mu\nu} + \frac{\Lambda_2}{\Lambda_1 + \chi\Lambda_2}\partial_\mu\phi\partial_\nu\phi \tag{5} $$

where $\Omega = [\Lambda_1(\Lambda_1 + \chi\Lambda_2)]^{1/2}$, $\Lambda_1 = \sqrt{2f_Q}\lambda + \frac{f_R}{2}$, $\Lambda_2 = \sqrt{2f_Q}(-\lambda \pm \sqrt{\lambda^2 + \kappa^2\chi})/\chi$, and $\lambda^2 = f/2 + f_R^2/8f_Q - \kappa^2\mathcal{L}/2$. For the particular model $f(R,Q) = R - \frac{R^2}{2R_P} + \frac{Q}{R_P}$, the low energy-density limit $|\kappa^2\mathcal{L}/R_P| \ll 1$ leads to

$$ R_{\mu\nu}(\tilde{h}) \approx \kappa^2\left(\partial_\mu\phi\partial_\nu\phi + \frac{V}{2}\tilde{h}_{\mu\nu}\right) + \frac{\kappa^2}{R_P}\left[(V - \tilde{\chi})\partial_\mu\phi\partial_\nu\phi + \left(\frac{2\kappa^2V^2 + \kappa^2\tilde{\chi}^2}{4}\right)\tilde{h}_{\mu\nu}\right] \tag{6} $$

which is in agreement with GR up to corrections of order $O(1/R_P)$. Note that to this order $\chi \approx \tilde{h}^{\mu\nu}\partial_\mu\phi\partial_\nu\phi \equiv \tilde{\chi}$. Similarly as in the $f(R)$ case, this indicates that $\tilde{h}_{\mu\nu}$ is mainly determined by integrating over the sources (cumulative effects of gravity), whereas Ω and the last term of (5) represent the local energy-density contributions to the metric. The DSR limit of this theory can be obtained by neglecting the cumulative effects of gravity, which corresponds to the limit $\tilde{h}_{\mu\nu} \approx \eta_{\mu\nu}$. In this limit, for the $f(R,Q)$ model of above the metric becomes

$$ g_{\mu\nu} \approx \eta_{\mu\nu} + \frac{2\kappa^2}{R_P}(V\eta_{\mu\nu} + \partial_\mu\phi\partial_\nu\phi) + O\left(\frac{1}{R_P^2}\right) . \tag{7} $$

From (7) we see that the leading order corrections to the Minkowski metric are strongly suppressed by inverse powers of the Planck curvature, which indicates that a perturbative study of such contributions in field theories should be feasible at low energy densities (see [7] for more details). At high densities, however, non-perturbative effects arise. To illustrate this point, consider the scalar Q in the $f(R,Q)$ model of above with a perfect fluid of density ρ and pressure P

$$Q = \frac{3R_P^2}{8}\left[1 - \frac{2\kappa^2(\rho+P)}{R_P} + \frac{2\kappa^4(\rho-3P)^2}{3R_P^2} - \sqrt{1-\frac{4\kappa^2(\rho+P)}{R_P}}\right] \approx Q_{GR} + \frac{3(P+\rho)^3}{2R_P} + \dots . \tag{8}$$

At low densities the GR solution is recovered but, as $\frac{4\kappa^2(\rho+P)}{R_P}$ approaches unity, important departures from GR must arise to avoid that the argument of the square root of (8) becomes negative. This implies that ρ and P must be bounded if the dynamics is to be consistent. And this is so regardless of the symmetries of the theory. In the case of an isotropic FRW universe with constant equation of state $w = P/\rho$, the Hubble function in the spatially flat case is [9]

$$H^2 = \frac{1}{6(\Lambda_1 - \Lambda_2)} \frac{[f + \kappa^2(\rho+3P)]}{\left[1+\frac{3}{2}\Delta_1\right]^2} , \tag{9}$$

where $\Delta_1 = -(1+w)\rho\partial_\rho\Omega$. At low densities, this expression exactly recovers the linear ρ-dependence of GR, but near the Planck scale, when $\kappa^2\rho/R_P \sim 0.1$, its behavior changes reaching a maximum and then vanishing at a higher density, which implies a cosmic bounce (see [7, 9] for illustrative plots). These isotropic bouncing solutions occur for all equations of state comprised within the interval $-1 < w < 11$ and persist in anisotropic (Bianchi I) spacetimes [9]. It should be noted that the effective dynamics of isotropic loop quantum cosmology [10] for a massless scalar was exactly reproduced by an $f(R)$ Palatini theory in [11]. The extension of that result to other matter sources and spacetimes with less symmetry could be naturally implemented in the context of $f(R,Q)$ theories.

In this talk I have shown that without imposing any a priori phenomenological structure, Palatini models predict an energy-density dependence of the geometry that closely matches the structure conjectured by Magueijo and Smolin in [3, 2] but without the problems of that approach. At the same time, it avoids the big bang singularity in quite general situations and for reasonable sources of matter and energy such as dust and radiation. This confirms that Palatini theories represent a new and powerful framework to address different aspects of quantum gravity phenomenology. An important lesson that can be extracted from this analysis is that quantum gravitational phenomena associated with a minimum length scale may arise non-perturbatively, being the perturbative corrections strongly suppressed.

Acknowledgments. Work supported by the Spanish grants FIS2008-06078-C03-02, FIS2008-06078-C03-03, and the Programme CPAN (CSD2007-00042).

References

[1] G.Amelino-Camelia, *Int.J.Mod.Phys.* D**11**, 35 (2002).
[2] J. Magueijo and L. Smolin, Phys.Rev.Lett.**88**, 190403(2002); Phys.Rev. D**67**, 044017(2003).
[3] J.Magueijo and L.Smolin, Class. Quant. Grav.**21**, 1725 (2004).
[4] G. J. Olmo, Phys.Rev.Lett. **95**, 261102(2005); Phys.Rev. D72, 083505(2005); Phys.Rev. D75, 023511(2007).
[5] J. Magueijo, *Rept. Prog. Phys.* **66**, 2025 (2003); J. Moffat, *Int. J. Mod. Phys.* D **2**, 351 (1993).
[6] G. J. Olmo, Int. J. Mod. Phys. D **20**, 413 (2011) [arXiv:1101.3864 [gr-qc]].
[7] G. J. Olmo JCAP10, 018 (2011), [arXiv:1101.2841 [gr-qc]].
[8] G. J. Olmo, H. Sanchis-Alepuz and S. Tripathi, Phys. Rev. D **80**, 024013 (2009) [arXiv:0907.2787 [gr-qc]].
[9] C. Barragan and G. J. Olmo, arXiv:1005.4136 [gr-qc].
[10] M. Bojowald, Living Rev. Rel. 11, 4 (2008); A. Ashtekar, Nuovo Cim. B 122 (2007) 135; A. Ashtekar, T. Pawlowski and P. Singh, *Phys. Rev. Lett.***96** (2006) 141301; *Phys. Rev.* D **74** (2006) 084003; A. Ashtekar, A. Corichi and P. Singh, *Phys. Rev.* D **77** (2008) 024046.
[11] G. J. Olmo and P. Singh, JCAP 0901, 030 (2009).

Journal of Physics: Conference Series **360** (2012) 012035 doi:10.1088/1742-6596/360/1/012035

Black hole entropy in loop quantum gravity

Iván Agulló[1], J. Fernando Barbero G.[2], Enrique F. Borja[3,4], Jacobo Díaz-Polo[5,1], Eduardo J. S. Villaseñor[6]

[1] Institute for Gravitation and the Cosmos & Physics Department, Penn State, University Park, PA, 16802-6300, USA.
[2] Instituto de Estructura de la Materia, CSIC, Serrano 123, 28006 Madrid, Spain.
[3] Institute for Theoretical Physics III, University of Erlangen-Nürnberg, Staudtstraße 7, D-91058 Erlangen, Germany.
[4]Departamento de Física Teórica and IFIC, Centro Mixto Universidad de Valencia-CSIC. Facultad de Física, Universidad de Valencia, Burjassot-46100, Valencia (Spain).
[5] Department of Physics & Astronomy, Louisiana State University, Baton Rouge, LA, 70803-4001, USA.
[6] Instituto Gregorio Millán, Grupo de Modelización y Simulación Numérica, Universidad Carlos III de Madrid, Avda. de la Universidad 30, 28911 Leganés, Spain

E-mail: `Ivan.Agullo@uv.es`, `fbarbero@iem.cfmac.csic.es`, `Enrique.Fernandez@uv.es`, `jacobo@phys.lsu.edu`, `ejsanche@math.uc3m.es`

Abstract. We discuss the recent progress on black hole entropy in loop quantum gravity, focusing in particular on the recently discovered discretization effect for microscopic black holes. Powerful analytical techniques have been developed to perform the exact computation of entropy. A statistical analysis of the structures responsible for this effect shows its progressive damping and eventual disappearance as one increases the considered horizon area.

1. Introduction

The description of black hole entropy [1] is one of the most successful results of loop quantum gravity (LQG). The framework is based on the quantization of a spacetime containing an isolated horizon as an internal boundary. Upon quantization, the horizon is described by a $U(1)$ Chern-Simons theory of level k on a punctured sphere. The N distinguishable punctures concentrate the horizon degrees of freedom, codified in integer (mod k) numbers a_i, $i = 1, \ldots, N$ quantifying the distributional curvature. The spacetime surrounding the horizon is described, as usual in LQG, by spin-networks. The punctures on the horizon are created by those spin-network edges that pierce it. These edges carry two labels, spins $j_i \in \mathbb{N}/2$ and magnetic numbers $m_i \in \{-j_i, -j_i+1, \ldots, j_i\}$, $i = 1, \ldots, N$, and endow the horizon with an area given by

$$A_H = 8\pi\gamma\ell_P^2 \sum_{i=1}^{N} \sqrt{j_i(j_i+1)}, \tag{1}$$

where γ is the so-called Barbero-Immirzi parameter.

The promotion of the isolated horizon boundary conditions to a quantum operator equation implies a relationship between a_i and m_i labels, namely

Journal of Physics: Conference Series **360** (2012) 012035 doi:10.1088/1742-6596/360/1/012035

$$2m_i = -a_i \mod k, \quad i = 1, \ldots, N. \tag{2}$$

Additionally, the spherical topology of (the spacial sections of) the horizon implies a constraint on the horizon labels a_i, a quantum version of the Gauss-Bonet theorem:

$$\sum_{i=1}^{N} a_i = 0. \tag{3}$$

These three constraints give rise to a precise combinatorial problem, namely to count all possible ordered lists of integer numbers a_i satisfying (3) and compatible through (2) and (1) with a given value of the horizon area $A \pm \delta A$, (the logarithm of) whose solution yields the entropy of a black hole as described by LQG.

The precise combinatorial problem was posed in [2] and its asymptotic solution (in the large area limit) was found in [3] to be given by

$$S(A) = \frac{\gamma_0}{\gamma} \frac{A}{4\ell_P^2} - \frac{1}{2} \ln \frac{A}{\ell_P^2} + O(A^0), \tag{4}$$

where γ_0 is a constant of known fixed value.

On the other hand, an exact computation in the low area regime [4] showed an effective discretization of entropy as a function of area for microscopic black holes, as shown in Figures 1 and 2.

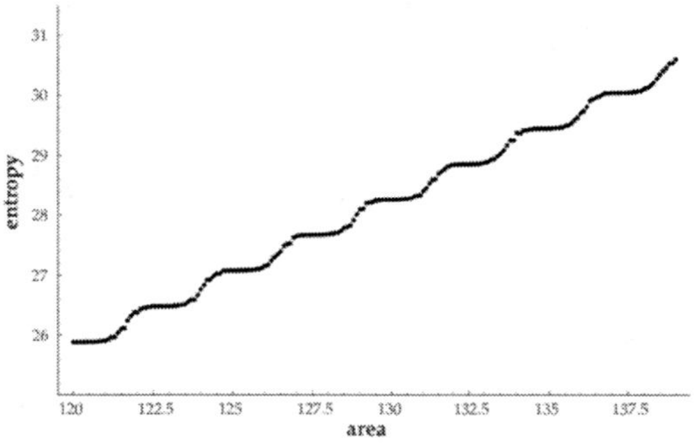

Figure 1. The entropy is plotted as a function of the horizon area in Plank units. The effective discretization effect is observed. The smearing interval δA is responsible for smoothing the band structure observed in Figure 2.

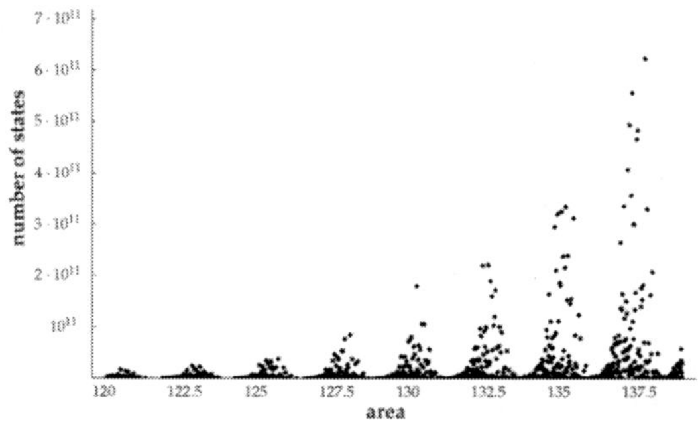

Figure 2. Degeneracy spectrum. The number of quantum states corresponding to each eigenvalue of the area operator is plotted. No smearing interval is considered. The periodic structure observed is responsible for the entropy discretization.

The question we are going to address here is whether these two results are compatible and what is the intermediate behavior in the area region between these two very different regimes.

2. Exact computational techniques

The first step toward understanding the compatibility of the results obtained by the two distinct approaches to the problem was taken in [5], where it was shown that, although the solution given in [3] correctly captures the first order asymptotic behavior, it neglects a mathematical subtlety that, when properly accounted for, leaves room for an additional oscillatory behavior of the kind responsible for the discretization effect. This solves the incompatibility problem and raises the new question of whether or not the discrete behavior of entropy does indeed extend to the large area limit.

In order to answer this, a series of technical developments were necessary, with the aim of extending the exact computations to larger values of area, and to obtain analytic expressions that capture the discrete nature of the problem, allowing an asymptotic analysis of the discrete structures.

In [6], new exact techniques to solve the combinatorial problem were introduced, based on number-theoretical strategies and the use of generating functions. The problem was divided in a four-step procedure:

(i) Characterization of the area spectrum (1) in terms of spins j, using the so-called Pell equation, that provided full control on the spin configurations compatible with every single area eigenvalue.

(ii) Implementation of the constraint (3) with techniques derived from the solution to the partition problem in number theory.

(iii) Computation of the degeneracy due to reordering of the labels, using standard combinatorics.

(iv) Summation of the results of the previous three steps for all the relevant values of area (i.e., introduction of the smearing interval δA), by means of an appropriately designed generating function.

These new techniques offer several advantages when studying the entropy computation. On the one hand, they allow to extend the exact computational results to larger values of area, as they can be efficiently implemented in a computational algorithm. On the other hand, they provide a structured understanding of the combinatorial problem, that allowed a thorough analysis [7] of many detailed features of the solution. In particular, the ability to separate the different sources of degeneracy in the spectrum and recognize the ones responsible for the discretization effect, and the possibility to find strategies to isolate partial sectors of the spectrum and study their particular properties, resulted of great interest to the analysis of the asymptotic behavior of entropy. Finally, the introduction of a solution fully based on generating functions [8] also provided analytical expressions that can be used as a departure point for asymptotic expansions.

Furthermore, these techniques have a wide range of applicability, and have been proven useful in related combinatorial problems, arising from different formulations of the black hole framework in LQG. It is particularly remarkable that in [9], by using the same techniques, it was possible to perform the entropy counting for the recently developed framework based on a $SU(2)$ invariant quantization of the isolated horizon [10].

3. The asymptotic behavior of entropy

Using the above presented techniques as a starting point, it was finally possible to answer the question about the asymptotic behavior of entropy in [11]. Following the introduction in [12] of a 'peak selector' variable $p \in \mathbb{N}$ (a particular combination of the number of punctures and the total sum of spins j on the horizon), it was possible to define a generating function that isolates each individual discrete structure (peak of degeneracy) in the black hole degeneracy spectrum (Figure 2):

$$G(\nu, z; x_1, x_2, \ldots) := \left(1 - \sum_{i=1}^{\infty}\sum_{n=1}^{\infty}(z^{k_n^i} + z^{-z_n^i})\nu^{3k_n^i+2}x_i^{y_n^i}\right)^{-1}. \tag{5}$$

The pairs of numbers $\{k_n^i, y_n^i\}$ are obtained from the solutions to the Pell equation. The coefficients in the series expansion of $G(\nu, z; x_1, x_2, \ldots)$ of terms with powers p of the variable ν, correspond to points within a single peak of degeneracy in the black hole degeneracy spectrum. Therefore, for each given value of the peak selector p, this generating function produces a fraction of the spectrum corresponding to a single peak of degeneracy.

The statistical properties of this generating function, and thus of the shape of the peaks, were studied using techniques from analytic combinatorics. In particular, it was possible to show that the distribution generated by (5) tends to a Gaussian as the peak selector p – and therefore the area of the horizon – tends to infinity. Furthermore, the asymptotic values of the mean μ_p and variance σ_p^2 corresponding to that Gaussian can be computed as a function of p. The result is given by the expressions:

$$\mu_p = (0.34959022\ldots)\cdot p, \quad \sigma_p^2 = (0.00009817\ldots)\cdot p \tag{6}$$

Therefore, though very slowly compared to the separation between consecutive peaks, the variance grows linearly with area, making the peak structures grow wider as the area tends to infinity. The consequence of this is that, progressively, consecutive peaks will start to overlap each other, smoothing out the discrete behavior of entropy and eventually washing it out completely. This prediction was tested in [11] by constructing a model based on the combination of all these consecutive Gaussians to obtain the behavior of entropy. The model showed the progressive damping of the staircase structure for increasing areas.

There is a caveat to that analysis, though. The value of the slope obtained for the asymptotic curve with the Gaussian model is slightly different (disagreement starting at the sixth significant figure) than the one obtained by the standard asymptotic computation (4). This implies that, although the agreement is extremely good for low values of area, there is an exponential divergence between the Gaussian model and the actual entropy in the asymptotic limit. The agreement is good enough for the disappearance of the discrete behavior to be observed at areas much lower than the ones for which the model starts to significantly disagree with the complete calculation, so the conclusions reached above stand. However, one cannot rule out a possible revival of the discretization effect (or some related kind of oscillatory behavior) for larger areas.

4. Conclusions

We have reviewed some of the latest contributions to the computation of black hole entropy within loop quantum gravity. The discretization effect for microscopic black holes is a purely quantum effect, and it is fully compatible with the linear asymptotic behavior predicted in the large area limit. The transition between these two distinct regimes occurs by a progressive smoothing of the staircase structure when the area grows out of the deep Planck regime.

Acknowledgments

This work was partially supported by the Spanish MICINN grants FIS2009-11893, FIS2008-06078-C03-02, FIS2008-01980, ESP2007-66542-C04-01, AYA 2009-14027-C05-01, the NSF grants PHY-0854743 and PHY-0968871, and the Eberly research funds of Penn State.

References

[1] A. Ashtekar, J. Baez, A. Corichi, K. Krasnov, Phys. Rev. Lett. **80**, 904-907 (1998);
[2] M. Domagala and J. Lewandowski, Class. Quant. Grav. **21** (2004) 5233.
[3] K. A. Meissner, Class. Quant. Grav. **21** (2004) 5245.
[4] A. Corichi, J. Diaz-Polo, E. Fernandez-Borja, Phys. Rev. Lett. **98**, 181301 (2007).
[5] J. F. Barbero G. and E. J. S. Villasenor, Class. Quant. Grav. **26** (2009) 035017.
[6] I. Agullo, J. F. Barbero G., J. Diaz-Polo, E. Fernandez-Borja and E. J. S. Villasenor, Phys. Rev. Lett. **100** (2008) 211301.
[7] I. Agullo, J. F. Barbero, E. F. Borja, J. Diaz-Polo and E. J. S. Villasenor, Phys. Rev. D **82** (2010) 084029.
[8] J. F. Barbero G. and E. J. S. Villasenor, Phys. Rev. D **77** (2008) 121502.
[9] I. Agullo, J. F. Barbero G., E. F. Borja, J. Diaz-Polo and E. J. S. Villasenor, Phys. Rev. D **80** (2009) 084006.
[10] J. Engle, A. Perez, K. Noui, Phys. Rev. Lett. **105**, 031302 (2010).
[11] J. F. Barbero G., E. J. S. Villasenor, Phys. Rev. **D83**, 104013 (2011).
[12] I. Agullo, J. Diaz-Polo and E. Fernandez-Borja, Phys. Rev. **D77** (2008) 104024.

Loops 11: Non-Perturbative / Background Independent Quantum Gravity IOP Publishing
Journal of Physics: Conference Series **360** (2012) 012036 doi:10.1088/1742-6596/360/1/012036

Hawking radiation from Loop Black Holes

Emanuele Alesci[1] **and Leonardo Modesto**[2]

[1] Universität Erlangen, Institut für Theoretische Physik III, Lehrstuhl für Quantengravitation Staudtstrasse 7, D-91058 Erlangen, EU

[2] Perimeter Institute for Theoretical Physics, Waterloo, Ontario N2L 2Y5, Canada

E-mail: alesci@theorie3.physik.uni-erlangen.de, lmodesto@perimeterinstitute.ca

Abstract. We review a recent black hole particle production analisys in a regular spacetime metric obtained in a minisuperspace approach to loop quantum gravity. Repeating the Hawking analysis, which leads to a thermal flux of particles at the future infinity, one finds an infinite evaporation time and unitarity recovered due to the regularity of the singularity free spacetime and to the characteristic behavior of the surface gravity.

1. Introduction

Since classical black holes radiate particles [1, 2], their mass slowly decrease and finally disappear, leaving a paradox. Supposing that matter which enters the black hole was in a pure quantum state at early time, then the existence of an horizon transforms such matter into the mix thermal state of Hawking radiation at late time for an external observer, and the complete evaporation of the singularity destroys the correlations between the radiation and the information "swallowed" by the black hole that would allow to reconstruct the original pure quantum state. This is the Hawking scenario [3]. The role of the singularity is crucial in this paradox. Its absence in the presence of quantum gravitational effects has consequences for the entire global structure [4], and its removal is essential for resolving the black hole information loss problem [5, 6]. It is thus promising that a resolution of the big bang as well as the black hole singularites [7, 8, 9] has been achieved in a simplified version of LQG [10], known as loop quantum cosmology [11]. A regular static black hole metric (LBH) with quantum gravity corrections inspired by LQG was recently derived in [12, 13]. Here we review, the Hawking calculation of particle creation, based on an extension of the procedure in [14] [15], in a LBH background [16] to show that in this simplified model, the whole process, collapse and complete evaporation, is unitary.

2. The LBH spacetime: Collapse, Particle Creation, Evaporation and Unitarity

The quantum gravitationally corrected Schwarzschild metric of the LBH is

$$ds^2 = -G(r)dt^2 + \frac{dr^2}{F(r)} + H(r)d\Omega^2 \,, \qquad G(r) = \frac{(r-r_+)(r-r_-)(r+r_*)^2}{r^4+a_o^2}$$
$$F(r) = \frac{(r-r_+)(r-r_-)r^4}{(r+r_*)^2(r^4+a_o^2)} \,, \qquad H(r) = r^2 + \frac{a_o^2}{r^2} \tag{1}$$

with $d\Omega^2 = d\theta^2 + \sin^2\theta d\phi^2$. Here, $r_+ = 2m$ and $r_- = 2mP^2$ are the two horizons, and $r_* = \sqrt{r_+ r_-} = 2mP$. P is the polymeric function with $P \ll 1$, such that r_- and r_* are very close to $r = 0$. $a_o = A_{\min}/8\pi$, $A_{\min}$ being the area gap of LQG. In the limit $r \to 0$ the solution does not have a singularity: but instead has another asymptotically flat Schwarzschild region. The Penrose diagram [16] shows two horizons and two pairs of asymptotically flat regions.

If we proceed by combining the static metric with a radially ingoing null-dust, we obtain a dynamical space-time for a black hole formed from such dust. In the present model this process, usually described by the Vaidya metric [17], will have corrections, negligible in the asymptotic region, but crucial to avoid the formation of a singularity in the strong-curvature region.

Inserting the LBH metric in the wave-equation for a massless scalar field in a general spherically symmetric curved space-time one obtains obtains the radial equation for the field $\left[\frac{\partial^2}{\partial r^{*2}} + \omega^2 - V(r(r^*))\right]\psi(r) = 0$ where r^* is the tortoise coordinate implicitly defined by $\frac{dr^*}{dr} := \frac{1}{\sqrt{GF}}$. The potential $V(r)$ (see [16]) is zero at $r = r_+$ and r_- as for the classical Reissner-Nordström black hole and can be approximated near the horizons via $V(r^*) \propto e^{2\kappa_+ r^*}$, for$r \to r_+$ or $r^* \to -\infty$, $V(r^*) \propto e^{-2\kappa_- r^*}$, for $r \to r_-$ or $r^* \to +\infty$, $V(r^*) \to 0$, for $r \to 0$ or $r \to +\infty$. For this metric (1) we find the following values for the surface gravity on the inner and outer horizons:

$$\kappa_- = \frac{4m^3 P^4(1-P^2)}{16m^4 P^8 + a_o^2}, \quad \kappa_+ = \frac{4m^3(1-P^2)}{16m^4 + a_o^2}, \tag{2}$$

With these tools one can evaluate the particle production looking for the Bogulibov transformations needed to connect the positive ingoing f_ω and outgoing p_ω frequency modes of the field, between the initial I^- and final I^+ stationary regions with natural time $v = t + r^*$ and $u = t - r^*$ respectively. Assuming the frequency to be discrete

$$p_\omega = \sum_{\omega'} A_{\omega\omega'} f_{\omega'} + B_{\omega\omega'} f^*_{\omega'} \quad , \qquad A_{\omega\omega'} = (p_\omega, f_{\omega'}) \;\; \& \;\; B_{\omega\omega'} = -(p_\omega, f^*_{\omega'}) \tag{3}$$

which satisfy the following matrix relations $AA^\dagger - BB^\dagger = 1$, $AB^T - BA^T = 0$ and the matrix elements are $A_{\omega\omega'} = -i\int_{I^-} dv H d\Omega (p_\omega \partial_v f^*_\omega - f^*_{\omega'}\partial_v p_\omega)$,. If any of the $B_{\omega,\omega'}$ are non zero the particle content of the vacuum state at I^- (which we indicate with the ket $|\text{in}\rangle$) respect to the Fock space at I^+ is non trivial: $\langle \text{in}|\hat{N}^{I^+}_\omega|\text{in}\rangle = \sum_{\omega'} |B_{\omega,\omega'}|^2$, where $\hat{N}^{I^+}_\omega$ is the particle number operator at frequency ω at I^+. In contrast, if all the coefficient $B_{\omega,\omega'}$ are equal to zero then the positive frequency mode basis f_ω and p_ω are related by a unitary transformation.

To evaluate A we need to know the behavior of the modes p_ω at I^-. This has been found with a *geometric optic approximation* in [16], and using it, is possible to calculate A. From the relations between A and B [17, 18] one gets, the number of particles in the ωth mode at I^+ :

$$\langle \text{in}|\hat{N}^{I^+}_\omega|\text{in}\rangle = (BB^\dagger)_{\omega,\omega} = \frac{1}{e^{2\pi\omega/\kappa_+} - 1}. \tag{4}$$

which is the Planck distribution of thermal radiation for bosons at temperature $T_{BH} = \kappa_+/2\pi$.

The evaporation proceeds through the Hawking emission at r_+, and the black hole's Bekenstein-Hawking temperature is [13] $T_{BH}(m) = \frac{(2m)^3(1-P^2)}{4\pi[(2m)^4 + a_0^2]}$. T_{BH} coincides with the Hawking temperature in the limit of large masses but goes to zero for $m \to 0$. From the luminosity one can deduce the mass loss and this implies that the black hole needs an infinite time to completely evaporate. In the complete evaporation process, we neglected the backreaction for any value of the mass because at $v \approx +\infty$, when $m \lesssim m_P$, $dm/dv << m$ and then, contrary to the classical case, such approximation is valid also in the final stages of evaporation.

Combining the black hole formation and evaporation, considering the creation of (massless) particles on the horizon such that locally energy is conserved, one obtains the complete dynamics as depicted in the resulting causal diagram Fig.1, fundamental to study the black hole paradox.

Journal of Physics: Conference Series **360** (2012) 012036 doi:10.1088/1742-6596/360/1/012036

The paradox presents essentialy two features: the loss of information, strictly linked to the existence of an horizon, and the not unitary evolution, that is essentially related to the presence of a singularity. The effective LBH spcetime, seems to be a good candidate to solve both problems. The spacetime is in fact singularity free and the Bogoulibov coefficients evolve with the mass allowing a unitary tranformation between the initial and final vacuum. For small but non zero black hole mass we have thermal radiation but when the mass goes to zero, in an infinite amount of time, the unitarity is restored. More concretely looking at the surface gravity κ_+, we see the mass goes to zero together with the temperature. This is the crucial difference with the classical explosive case where the temperature goes to infinity when the mass reduces to zero. From the Planck spectrum in (4) we deduce that in an infinite amount of time

$$\lim_{m\to 0}(BB^\dagger)_{\omega,\omega}=0 \quad \forall\, \omega, \tag{5}$$

because $\kappa \to 0$ for $m \to 0$. Since $(BB^\dagger)_{\omega,\omega}$ is positive semi-define, this vanish iff $B = 0$. Since $B = 0$ the positive frequency *mode basis* f_ω *and* p_ω *are related by the unitary transformation* A. The quantum state at I^+ is described by a thermal *density matrix* $\rho = \prod_\omega (1-e^{-2\pi\omega/\kappa_+})\sum_{N=0}^{+\infty} e^{-2\pi\omega N/\kappa_+}|N_\omega\rangle\langle N_\omega|$, where $|N_\omega\rangle$ is the state at I^+ with N particles in the mode ω: when the balck hole mass goes to zero it reduces to a pure state

$$\lim_{m\to 0}\rho = \prod_\omega |0_\omega\rangle\langle 0_\omega| \equiv |\text{in}\rangle\langle\text{in}|, \tag{6}$$

Up to now we have restriced our attention to the surface I^+. When the black hole mass is non zero the pre state $|\text{in}\rangle$ could appear mixed but it is still a pure state since we have to add the modes crossing the event horizon H^+. Because we have ignored this sector we have lost all the possible correlations with the quanta entering into the black hole. If we properly take into account those modes, the purity of the $|\text{in}\rangle$ state is restored. Following [17] we find an explicit relation between $|\text{in}\rangle$ and the N-particles states with frequency ω at I^+ and H^+, respectively $|N_\omega^{I^+}\rangle$, $|N_\omega^{H^+}\rangle$: $|\text{in}\rangle = \prod_\omega \sqrt{1-e^{-\frac{2\pi\omega}{\kappa_+}}}\;\sum_{N=0}^{+\infty} e^{-\frac{\pi N\omega}{\kappa_+}}|N_\omega^{I^+}\rangle \otimes |N_\omega^{H^+}\rangle$.This relation shows we have an independent emission of quantum entangled states of outgoing and ingoing radiation.

In Fig.1 we can see that the evolution is unitary at a fundamental level, in fact the absence of singularity implies that any Cauchy surface Σ of the spacetime is a Cauchy surface for the whole spacetime and the Hawking scenario is avoided. If we use Σ as a "time label" for the spacetime describing the whole evaporation, we don't find any discontinuity in contrast to the Hawking scenario [18]. We can describe the state $|\text{in}\rangle$ in terms of a mass dependent splitting between internal and external spacetime of the final Cauchy surface $\Sigma_f^m = \Sigma_{\text{int}}^m \bigcup \Sigma_{\text{out}}^m$: which is based on the decomposition of the final one particle Hilbert space in terms of the ones internal and external to the horizon, $H_{\text{fin}}^m = H_{\text{int}}^m \oplus H_{\text{out}}^m$. For $m = 0$, $H_{\text{fin}}^0 = H_{\text{r}=0} \oplus H_{\text{I}^+}$ and the initial vacuum reduces to $|0\rangle_{in} = |0\rangle_{\text{out}} \otimes |0\rangle_{\text{int}} = |0\rangle_{\text{fin}}$ and the density matrix to the pure state (6).

We can better understand the properties of the radiation introducing an abserver O in the classical region of the dynamical spacetime describing the collapse. (Figure 1). O_2 at $u << u_c$ will not see any remarkable radiation being too far from the event horizon. Any O_3 for $-\infty > u \geq u_c$ will see thermal radiation as an incoming flux of particles from the region where the matter collapsed. The essential point is the presence of an horizon: the pure state is perceived as a flux of particles because the observer is obliged to trace over the internal degrees of freedom, losing the information about the particles inside the black hole. O_3 will experience a thermal particle bath of increasing or decreasing temperature depending on the stage of the evaporation. Finally O_1 (r= constant) sees a gas of particles of increasing or decreasing temperature moving in time towards i^+ and at i^+ the initial vacuum state evolves in itself when $\kappa \to 0$. We wish to mention the fact that the observer in i^+ is in causal contact with both the radiation outcoming towards I^+ and the one ingoing to $r = 0$, this fact in principle allows him to detect the previously hidden correlations needed to restore purity at all the times.

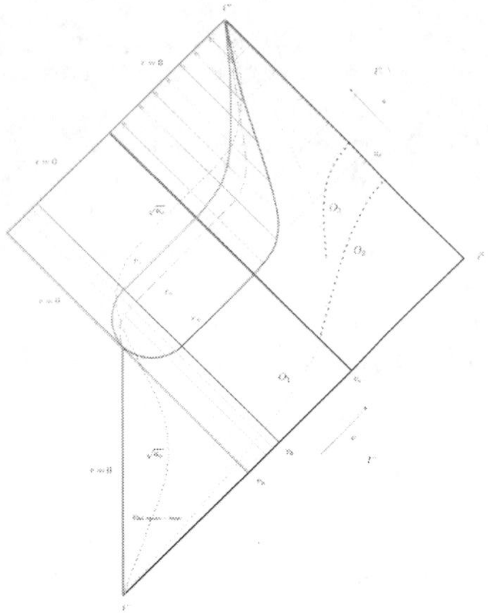

Figure 1. Penrose diagram for the formation and evaporation of the LBH. The red and dark blue solid lines depict the two trapping horizons r_- and r_+. The brown, dotted line is the curve of $r = \sqrt{a_o}$ and the brown, long dashed one is r_*. In the region $v < v_a$ we have a flat and empty region. For all times $v > v_a$, the inner and outer trapping horizons are present. These horizons join smoothly at $r = 0$ in an infinite time. A black hole begins to form at $v = v_a$ from null dust which has collapsed completely at $v = v_b$ to a static state with mass m_0. It begins to evaporate at $v = v_c$, and the complete evaporation takes an infinite time. The observer at I^+ sees particle emission set in at some retarded time u_c. The region with $v > v_c$ is then divided into a static region for $u < u_c$, and the Vaidya region for $u > u_c$, which is further subdivided into an ingoing and an outgoing part. The ingoing negative flux during evaporation (magenta arrows) has a flipped sign region.

3. Conclusions

We have calculated the particle creation by an entirely singularity-free LBH. The Schwarzschild surface gravity (proportional to the temperature) diverges when the black hole mass goes to zero. For the LBH instead it goes to zero with the mass but in an infinite time. The approximation used in [16] is that the massless scalar field obey the usual wave equations on an effective LBH metric, solution of Einstein equations with an effective energy-tensor with contributions that violate the positive energy condition and prevent the formation of a singularity. Even if the model is very simple and the analysis similar to the classical one, the result is that in an infinite time *the whole process, collapse and complete evaporation, is unitary.*

References

[1] S. W. Hawking, 1975 *Comm. Math. Phys.* **43**, 199
[2] J. D. Bekenstein, 1973 *Phys. Rev. D* **7**, 2333
[3] S. W. Hawking, 1976 *Phys. Rev. D* **14**, 2460
[4] A. Ashtekar and M. Bojowald, 2005 *Class. Quantum Grav.* **22** 3349
[5] S. Hossenfelder and L. Smolin, 2010 *Phys. Rev. D* **81** 064009
[6] A. Ashtekar, V. Taveras and M. Varadarajan, 2008 *Phys. Rev. Lett.* **100**, 211302 .
[7] A. Ashtekar and M. Bojowald, 2006 *Class. Quant. Grav.* **23** 391 [arXiv:gr-qc/0509075].
[8] L. Modesto, 2004 *Phys. Rev. D* **70** 124009 [gr-qc/0407097].
[9] L. Modesto, 2006 *Class. Quant. Grav.* **23**, 5587 [arXiv:gr-qc/0509078].
[10] C. Rovelli, 2004 *Quantum Gravity* (Cambridge: Cambridge University Press).
[11] M. Bojowald, 2005 *Living Rev. Rel.* **8**, 11
[12] L. Modesto, 2008 "Space-Time Structure of Loop Quantum Black Hole" [arXiv:0811.2196 [gr-qc]].
[13] L. Modesto and I. Premont-Schwarz, 2009 *Phys. Rev. D* **80**, 064041
[14] S. A. Hayward, 2006 *Phys. Rev. Lett.* **96**, 031103
[15] S. Hossenfelder, L. Modesto and I. Premont-Schwarz, 2010 *Phys. Rev. D* **81**, 044036
[16] E. Alesci and L. Modesto, 2011 "Particle Creation by Loop Black Holes" [arXiv:1101.5792 [gr-qc]].
[17] A. Fabbri and J. Navarro-Salas, 2005 Modeling black hole evaporation (Imperial College Press)
[18] R. M. Wald, 1994 Quantum field theory in curved space-time and black hole thermodynamics (Chicago: Univ. Pr.)

Journal of Physics: Conference Series **360** (2012) 012037 doi:10.1088/1742-6596/360/1/012037

Entropy of generic quantum isolated horizons

C Beetle and J Engle[1]
Florida Atlantic University, 777 Glades Road, Boca Raton, FL 33431, USA
E-mail: cbeetle@physics.fau.edu, jonathan.engle@fau.edu

Abstract. We review our recent proposal of a method to extend the quantization of *spherically symmetric* isolated horizons, a seminal result of loop quantum gravity, to a phase space containing horizons of *arbitrary* geometry. Although the details of the quantization remain formally unchanged, the physical interpretation of the results can be quite different. We highlight several such differences, with particular emphasis on the physical interpretation of black hole entropy in loop quantum gravity.

The confirmation of the Bekenstein–Hawking entropy formula by Ashtekar, Baez, Corichi and Krasnov [1–3] (ABCK) is one of the triumphs of loop quantum gravity. The ABCK approach begins by quantizing a classical phase space whose points correspond to spacetimes with inner boundary at a *spherically symmetric* isolated horizon [4–6]. The calculation relies on spherical symmetry (just of the intrinsic geometry of the horizon itself) to make the symplectic structure on that phase space, and thus the ensuing quantization, well-defined.

This paper is concerned with an apparent inconsistency in the roles played by spherical symmetry before and after quantization in the ABCK approach. Imposing spherical symmetry classically restricts the allowed *bulk* fields such that they induce a round metric on the horizon. After quantization, however, the bulk spin network states in ABCK are (virtually) generic at the horizon, no different from those allowed on an *arbitrary* 2-surface in loop quantum gravity. In this sense, the *only* place that spherical symmetry is used at all in deriving the *quantum* theory of ABCK is in making the symplectic structure of the *classical* theory well-defined.

Here, we review a new way [7] to justify the ABCK symplectic structure classically, including its crucial Chern–Simons surface term. Most importantly, our proposed scheme does not rely on spherical symmetry, or indeed on any restriction of the intrinsic horizon geometry, to make the symplectic structure well-defined. Rather, it allows one simply to quantize the phase space of *all* isolated horizons of a given total area a_Δ. The intrinsic geometry of the horizon, its shape, *is not fixed a priori*. This approach renders moot the question of how to impose spherical symmetry at the horizon quantum mechanically, which is the key missing ingredient that would make ABCK a consistent quantization of a spherically symmetric horizon. Rather, our approach renders the ABCK quantization entirely consistent in another way, namely, by broadening the *classical* picture to eliminate the requirement of symmetry *ab initio*.

1. ABCK Phase Space and Quantization

ABCK use a covariant phase space, each point of which corresponds a space*time* $\mathcal{M}$ of the form shown in Figure 1(a) in which the classical (Einstein) equations of motion hold. It is bounded

[1] Presenter

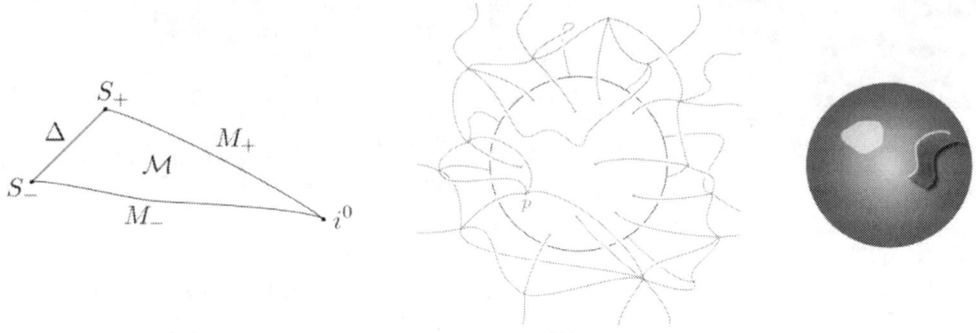

(a) (b) (c)

Figure 1. (a) The space-time arena considered both here and in ABCK. (b) Bulk spin networks fix charges for the quantum Chern–Simons theory via the quantum horizon boundary condition. (c) Classical horizon shape is probed by transverse, not pullback, fluxes.

to the future and past by partial Cauchy slices $M_\pm$, which extend to spatial infinity i^0, and has an inner boundary at a null surface Δ diffeomorphic to $S^2 \times \mathbb{R}$. Boundary conditions at Δ make it is a spherically symmetric *isolated horizon*. In a precise sense [4], this means that Δ models the surface of a quiescent black hole in perfect equilibrium with its immediate surroundings.

Under the above conditions (and one or two additional technical assumptions), the integral

$$\Omega(\delta_1, \delta_2) := {}^\beta\Omega_{\mathrm{B}}(\delta_1, \delta_2) + {}^\beta\Omega_{\mathrm{S}}(\delta_1, \delta_2) := \frac{1}{4\pi G\beta} \int_M \delta_{[1}\Sigma^i \wedge \delta_{2]}{}^\beta\! A_i + \frac{1}{2\pi} \frac{a_\Delta}{4\pi G\beta} \oint_S \delta_1 V \wedge \delta_2 V \quad (1)$$

takes the same value over *any* partial Cauchy slice M through $\mathcal{M}$ with inner boundary S. Here, $\delta_{1,2}$ represent a pair of tangent vectors to the ABCK phase space, *i.e.*, solutions of the linearized equations of motion on the given background that preserve the total area and spherical symmetry of S. The bulk term ${}^\beta\Omega_{\mathrm{B}}(\delta_1, \delta_2)$ in (1) is the standard symplectic structure of loop quantum gravity. The surface term ${}^\beta\Omega_{\mathrm{S}}(\delta_1, \delta_2)$ has a form familiar from a Chern–Simons theory for the $U(1)$ *spin connection* V_a induced on S by the bulk geometry. The curvature of V_a is

$$\mathrm{d}V = -\tfrac{1}{4}\mathcal{R}\,\epsilon \qquad \rightsquigarrow \qquad \mathrm{d}V = -\tfrac{2\pi}{a_\Delta} \underleftarrow{\Sigma}_i\, r^i \quad \text{if } \mathcal{R} \text{ is } \textit{constant}, \quad (2)$$

where $\mathcal{R}$ is the the intrinsic scalar curvature of S, ϵ is its 2-from area element, and r^i is a gauge-fixed internal radial direction. It is the latter equation here, which holds only in spherical symmetry, that makes (1) independent of M, and thus a well-defined symplectic structure.

The sum of bulk and surface terms in the symplectic structure (1) suggests that the Hilbert space of the quantum theory should be a tensor product $\mathcal{H}^{\mathrm{pre}} = \mathcal{H}_{\mathrm{B}} \otimes \mathcal{H}_{\mathrm{S}}$ of bulk and surface factors. This is indeed what happens at first in the ABCK quantization. The bulk Hilbert space $\mathcal{H}_{\mathrm{B}}$ is the standard one of loop quantum gravity, spanned by spin network states whose underlying graphs may include edges that end at one of a finite set $\mathcal{P}$ of *puncture points* on the horizon. Such a spin network is shown in Figure 1(b). The surface Hilbert space $\mathcal{H}_{\mathrm{S}}$ is a direct limit [1] of Hilbert spaces for a quantum Chern–Simons theory, where the limit runs over such sets $\mathcal{P}$ of punctures, ordered by inclusion. The two Hilbert space factors at this stage are entirely independent of one another and, in particular, the quantum Chern–Simons connection on the horizon is totally unrelated to the geometric degrees of freedom in the bulk. We denote that connection by X, rather than V, to emphasize the absence of such a relation.

The initial Hilbert space $\mathcal{H}^{\mathrm{pre}}$ of the ABCK quantization is reduced to the true, physical Hilbert space $\mathcal{H}^{\mathrm{phys}}$ of the model in a series of steps. The first reasserts the physical relationship between the Chern–Simons connection X and the bulk variables by restricting to states $|\psi\rangle$ in $\mathcal{H}^{\rm pre}$ that satisfy the *quantum horizon boundary condition*

$$\Big[\hat{I}_{\rm B} \otimes \hat{U}[X,C]\Big]|\psi\rangle = \Big[\exp\Big(-\frac{2\pi{\rm i}}{a_\Delta}\,\hat{\Sigma}[{\rm int}\,C,r]\Big) \otimes \hat{I}_{\rm S}\Big]|\psi\rangle. \tag{3}$$

Here, $\hat{U}[X,C]$ is the Chern–Simons holonomy operator around an arbitrary loop C in S, while $\hat{\Sigma}[{\rm int}\,C,r]$ is the canonical flux of loop quantum gravity, in the gauge-fixed internal direction r^i, through the interior of C within S. The quantum horizon boundary condition (3) is modeled on the classical relation (2), under the assumption that $\mathcal{R}$ is *constant*. The kinematical Hilbert space $\mathcal{H}^{\rm kin}$ of states $|\psi\rangle$ obeying (3) is then further reduced to the physical Hilbert space $\mathcal{H}^{\rm phys}$ of the ABCK model by imposing the diffeomorphism and Hamiltonian constraints.

2. Area Connection and Quantization of Generic Horizons

Our proposed application [7] of the ABCK quantization to the phase space of *all* isolated horizons with total area a_Δ hinges on the definition

$$\mathring{V}_a := V_a + \tfrac{1}{4}({*}{\rm d}\psi)_a \qquad \text{with} \qquad \triangle\psi := {*}{\rm d}{*}{\rm d}\psi := \mathcal{R} - \langle\mathcal{R}\rangle \tag{4}$$

of a new $U(1)$ connection on the horizon in the classical theory. Here, $\triangle$ is the scalar Laplacian and $*$ is the Hodge dual operation, both intrinsic to S, while $\langle\mathcal{R}\rangle := \oint_S \mathcal{R}\epsilon/a_\Delta$ denotes the *average* value of the curvature of S. We call the solution ψ of this Poisson equation the *curvature potential* of S and the $U(1)$ connection $\mathring{V}_a$, the *area connection* because

$${\rm d}\mathring{V} = -\tfrac{1}{4}\langle\mathcal{R}\rangle\,\epsilon = -\tfrac{2\pi}{a_\Delta}\textstyle\sum_i \underleftarrow{\Sigma_i}\, r^i \qquad \rightsquigarrow \qquad \exp\oint_C {\rm i}\mathring{V} = \exp\Big(-\frac{2\pi{\rm i}}{a_\Delta}\int_{{\rm int}\,C}\Sigma_i r^i\Big) \tag{5}$$

for *any* geometry. That is, in close analogy to (3), the *classical* holonomy of the area connection about any closed loop $C \subset S$ depends *solely* on the area of S interior to C.

Using the area connection $\mathring{V}_a$ from (5), we show [7] that the symplectic structure integral

$$\Omega(\delta_1,\delta_2) := {}^{\beta}\Omega_{\rm B}(\delta_1,\delta_2) + {}^{\beta}\mathring{\Omega}_{\rm S}(\delta_1,\delta_2) := {}^{\beta}\Omega_{\rm B}(\delta_1,\delta_2) + \frac{1}{2\pi}\frac{a_\Delta}{4\pi G\beta}\oint_S \delta_1\mathring{V}\wedge\delta_2\mathring{V} \tag{6}$$

is independent of M throughout the entire phase space of *all* isolated horizons of area a_Δ. Remarkably, (6) differs from (1) only in that the surface term involves the area connection rather than the spin connection, and yet is well-defined on a much larger phase space. Immediately after quantization, *i.e.*, at the level of $\mathcal{H}^{\rm pre}$, this classical distinction is irrelevant because the quantum Chern–Simons connection X on S has no physical relation to the bulk variables. When that relation is restored by the quantum horizon boundary condition (3), the meaning given to X is exactly analogous to that of the classical area connection exhibited in (5).

3. Quantum Horizon Shape and Entropy

We now consider the situation from a purely quantum mechanical perspectve. The mere existence of a classical connection $\mathring{V}_a$ with holomies (5) on a generic isolated horizon dispels the notion that just imposing the quantum horizon boundary condition (3) makes a quantum horizon spherically symmetric. Moreover, since (3) *does* determine the boundary state uniquely in terms of the bulk, we must look to the bulk to see whether symmetry is actually present.

The geometric content of bulk loop quantum gravity states is probed [8] by flux operators

$$\frac{1}{8\pi G\beta}\,\hat{\Sigma}[T,f] := \frac{\hbar}{2}\sum\nolimits_{x\in T} f^i(x)\sum\nolimits_{e\text{ at }x}\kappa(T,e)\,\hat{J}_i^{(x,e)}, \tag{7}$$

where the sums are over all analytic curves e extending from each point x of a transversely oriented 2-surface T in M, $f^i(x)$ is a Lie-algebra valued smearing function, $\kappa(T,e) = \pm 1$ (or 0) according to the orientation of e relative to T, and $\hat{J}_i^{(x,e)}$ is a generator of internal $SU(2)$ gauge along e. We distinguish between *pullback fluxes*, wherein T is a subset of S, and *transverse fluxes*, wherein T intersects S in a curve. The two cases are illustrated in Figure 1(c).

The quantum horizon boundary condition (3) involves only pullback fluxes. Classically, however, information about the intrinsic dyad induced on S inheres in the transverse fluxes. Because the ABCK quantization yields no restriction on the transverse fluxes, we argue that it is properly viewed as the *consistent* quantization of an isolated horizon of fixed total area a_Δ, but *arbitrary* shape.

In the months since Loops '11, an invitation from Abhay Ashtekar to visit him at Penn State resulted in some very helpful discussions that revealed an important subtlety in our proposed scheme. Namely, even though we argue that the final Hilbert space contains quantum states corresponding to all possible classical shapes of the horizon, and all of these enter into the statistical ensemble underlying the ABCK entropy, it is *not* these shapes that are counted in determining the entropy. Rather, what one counts are possible states of the "quantum area element" — i.e., distinct eigenvalues of the pull-back fluxes (labeled by 'm's') and (possibly) partial areas (labeled by 'j's') [9]. The classical analogue of these quantities taken together are *pure gauge*; however, due to the distributional nature of quantum geometry, they are no longer pure gauge in the quantum theory.

Shortly after this work was finished, a different strategy for allowing arbitrary shapes in the calculation of quantum entropy was published by Perez and Pranzetti [10], which, in contrast to the present work, is $SU(2)$*-covariant.* Let us remark on the relative strengths and weaknesses of their work as compared to the present one. A strength of both is that quantum black holes *with arbitrary horizon shape* are described, and an advantage of Perez-Pranzetti over the present approach is that it is $SU(2)$ covariant, something necessary to obtain the correct coefficient in the next to leading order term in the entropy [11–15]. However, a disadvantage of [10] is that one has to perform a separate quantization for each horizon shape (i.e., diffeomorphism equivalence class of horizon geometries) and then combine them, a procedure which neglects information in the Poisson algebra of observables describing the horizon shape. As in ABCK, it is also not clear that this fixing of the horizon shape is actually reflected in each elementary quantization.

Acknowledgments
We thank the organizers of the conference for the opportunity to present these results. We also are indebted to Abhay Ashtekar, who invited us to Penn State to discuss this work.

References

[1] Ashtekar A, Baez J, Corichi A and Krasnov K 1998 *Phys. Rev. Lett.* **80** 904–907
[2] Domagala M and Lewandowski J 2004 *Class. Quantum Grav.* **12** 5233–5244
[3] Meissner K 2004 *Class. Quantum Grav.* **21** 5245–5251
[4] Ashtekar A and Krishnan B 2004 *Living Rev. Rel.* **7** 10
[5] Ashtekar A, Beetle C and Fairhurst S 1999 *Class. Quantum Grav.* **16** L1–L7
[6] Ashtekar A, Corichi A and Krasnov K 1999 *Adv. Theor. Math. Phys.* **3** 419–478
[7] Beetle C and Engle J 2010 *Class. Quantum Grav.* **27** 235024
[8] Ahtekar A and Lewandowski J 2004 *Class. Quantum Grav.* **21** R53–R152
[9] Ghosh A and Mitra P 2005 *Phys.Lett.* **B616** 114–117
[10] Perez A and Pranzetti D 2011 *Entropy* **13** 744–777
[11] Engle J, Noui K and Perez A 2010 *Phys. Rev. Lett.* **105** 031302
[12] Agullo I, Barbero J F, Borja E, Diaz-Polo J and Villasenor E 2009 *Phys. Rev. D* **80** 084006
[13] Ghosh A and Mitra P 2005 *Phys. Rev. D* **71** 027502
[14] Gour G 2002 *Phys. Rev. D* **66** 104022
[15] Carlip S 2000 *Class. Quantum Grav.* **17** 4175–4186

Loops 11: Non-Perturbative / Background Independent Quantum Gravity IOP Publishing
Journal of Physics: Conference Series **360** (2012) 012038 doi:10.1088/1742-6596/360/1/012038

The effective kinetic term in CDT

T G Budd
Institute for Theoretical Physics, Utrecht University, 3508 TD, Utrecht, The Netherlands
E-mail: t.g.budd@uu.nl

Abstract. We report on recently performed simulations of Causal Dynamical Triangulations (CDT) in 2+1 dimensions aimed at studying its effective dynamics in the continuum limit. Two pieces of evidence from completely different measurements are presented suggesting that three-dimensional CDT is effectively described by an action with kinetic term given by a modified Wheeler-De Witt metric. These observations could strengthen an earlier observed connection between CDT and Hořava-Lifshitz gravity. One piece of evidence comes from measurements of the modular parameter in CDT simulations with spatial topology of a torus, the other from measurements of local metric fluctuations close to a fixed spatial boundary.

1. Introduction

Causal dynamical triangulation (CDT) provides an explicit realization of the sum over space-time geometries which appears in the path-integral approach to quantum gravity. Its simplicity allows us to feed the model into a computer and find expectation values of observables numerically. We attempt to exploit the simplicity to address one of the hardest challenges faced by non-perturbative approaches to quantum gravity: showing that at low energies a classical geometrical theory emerges.

CDT has provided several pieces of evidence suggesting that it might possess a continuum limit showing classical behaviour. However, at present it is unclear what effective theory governs this classical regime. In this paper we describe two distinct sets of observables which probe the semi-classical regime of CDT in 2+1 dimensions and allow us to obtain non-trivial information about the character of the effective theory. Before describing these, let us first briefly introduce the set-up of CDT in 2+1 dimensions.

The motivation for CDT comes from the assumption that path integrals over Lorentzian geometries can be computed with the help of statistical path integrals over Euclidean geometries that are equipped with a time-foliation. The leaves of a foliated geometry are interpreted as spatial geometries and their topology is assumed not to change in time. In CDT the path integral is replaced by a discrete sum over all such geometries that can be assembled from a finite number of equilateral tetrahedra. More precisely we have the partition function

$$Z_{\mathrm{CDT}} = \sum_{\text{triangulations } T} e^{-S_{\mathrm{CDT}}[T]} \quad \text{with} \quad S_{\mathrm{CDT}}[T] = k_3 N_3 - k_0 N_0. \tag{1}$$

The action S_{CDT} is derived from the Regge action and depends linearly on the number N_3 of tetrahedra and the number N_0 of vertices in the triangulation T. The three-dimensional triangulations are required to be foliated in the sense that they can be constructed from a sequence of two-dimensional triangulations connected by slices of tetrahedra (see figure 4).

We use Monte Carlo methods to randomly generate CDT configurations according to the Boltzmann distribution in (1). By performing measurements on individual triangulations we can approximate expectation values of observables. A particularly simple observable is the volume $V(t)$ of the spatial triangulation at time t. It has been previously observed[1] that its expectation value $\langle V(t)\rangle$ for CDT with spatial topology of a 2-sphere corresponds to that of a proper-time slicing of a round 3-sphere, which is the Euclidean counterpart of de Sitter space. Both the expectation value and its quantum fluctuations turn out to be well described by an effective action

$$S_{\text{eff}}[V] = \int dt \left(c_0 \frac{\dot{V}^2}{V} - c_1 V \right) \quad \text{with} \quad c_0, c_1 > 0, \tag{2}$$

which is very similar to the (Euclidean) Einstein-Hilbert action $\int d^3x\sqrt{g}(-R+2\Lambda)$ evaluated on a spherical cosmology $ds^2 = dt^2 + V(t)d\Omega^2$. The only difference is an overall minus sign, which ensures that for fixed 3-volume (2) is bounded below while the Einstein-Hilbert one is not.

This is a general property of the Euclidean Einstein-Hilbert action and is known as the conformal mode problem. We can illustrate this by writing the metric in proper-time gauge $ds^2 = dt^2 + g_{ab}dx^a dx^b$, in which case the Einstein-Hilbert action takes the form

$$S_{\text{EH}} = \kappa \int dt \int d^2x\sqrt{g} \left(\frac{1}{4}\dot{g}_{ab}\mathcal{G}^{abcd}\dot{g}_{cd} - R + 2\Lambda \right), \quad \mathcal{G}^{abcd} = \frac{1}{2}\left(g^{ac}g^{bd} + g^{ad}g^{bc} \right) - g^{ab}g^{cd}, \tag{3}$$

and $\mathcal{G}^{abcd}$ is referred to as the Wheeler-De Witt supermetric. This metric on the linear space of metric deformations is indefinite: it is positive definite on traceless deformations but negative definite on conformal deformations. As a consequence (3) is not a suitable ansatz for an effective action for CDT: we want such an effective action to describe, at least in a semi-classical way, the quantum fluctuations we observe in our simulations and to do that it needs to have a (local) minimum at the classical solutions.

Inspired by the Hořava-Lifshitz approach to gravity [3] and the presence of a preferred time-slicing in CDT, we propose a modification of the ansatz (3). Namely we modify the Wheeler-De Witt metric by inserting a free parameter λ,

$$\mathcal{G}^{abcd} \quad \rightarrow \quad \mathcal{G}_\lambda^{abcd} = \frac{1}{2}\left(g^{ac}g^{bd} + g^{ad}g^{bc} \right) - \lambda g^{ab}g^{cd}. \tag{4}$$

This way we obtain the most general ultralocal kinetic term compatible with foliation-preserving diffeomorphism invariance. This metric is positive definite in the regime $\lambda < 1/2$, while in general relativity we have the value $\lambda = 1$.

In the remainder of this paper we will put this ansatz to test by measuring observables that are only sensitive to the kinetic term in the effective action.

2. CDT on a torus

We would like to see what consequence our ansatz has for cosmological models. Clearly the spatial volume is an observable which only depends on the conformal degrees of freedom in the metric. In order to probe the kinetic term in a non-trivial way, we need another observable that depends also on traceless degrees of freedom, i.e. an observable that measures the shape of space. Taking the spatial topology to be that of the torus, we have a prime candidate for such an observable: the complex modular parameter τ describing the conformal structure. It is well-known that any metric on the torus is conformally related to a unique flat metric of unit volume. Up to diffeomorphisms the latter form a 2-parameter family and can be obtained by gluing sides of a parallelogram in the Euclidean plane (see figure 1).

[1] See [1] for results in 2+1 dimensions and [2] for more recent results in 3+1 dimensions.

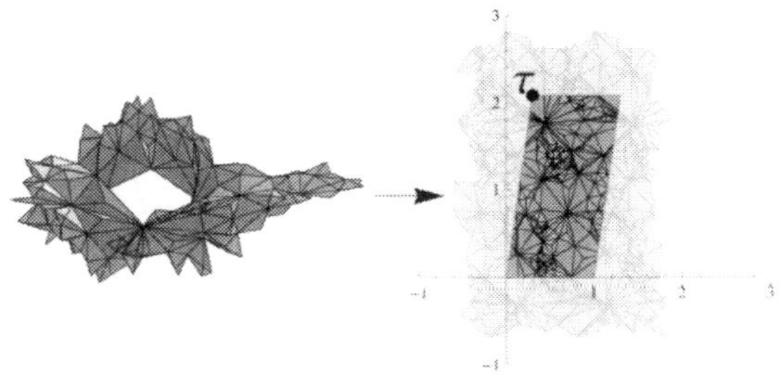

Figure 1. A harmonic embedding of a triangulation of the torus. The modular parameter $\tau = \tau_1 + i\tau_2$ is associated with the shape of the parallelogram in the manner shown.

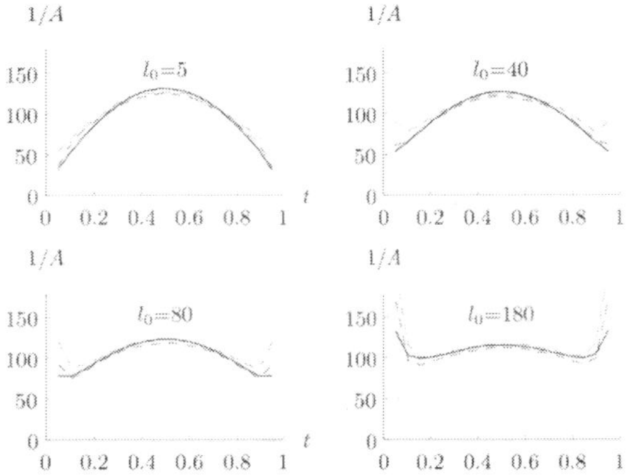

Figure 2. The quantity $1/A[g]$ deduced from direct measurement (solid curve) and from the correlations of τ_1 and τ_2 (dashed curves).

In [4] an algorithm is described to associate a modular parameter to a triangulation of the torus. Basically, this is done by constructing a periodic embedding of the torus in the Euclidean plane which we require to be harmonic, by which we mean that any vertex is located at the centre of mass of its neighbours. Such an embedding is unique up to scaling, rotations, and translations if we insist that the sum of all squared edge lengths is minimized (figure 1).

Let's see how this modulus appears in the effective action according to our ansatz. Restricting the kinetic term given by the modified Wheeler-DeWitt metric $\mathcal{G}_\lambda$ to the spatial volume V and the modulus $\tau = \tau_1 + i\tau_2$ we obtain

$$\kappa \int dt \left((\frac{1}{2} - \lambda)\frac{\dot{V}^2}{V} + \frac{1}{2A[g]}\frac{\dot{\tau}_1^2 + \dot{\tau}_2^2}{\tau_2^2} \right), \quad \text{where} \quad A[g] = \frac{\int d^2x\sqrt{g}\exp(2\Delta^{-1}R)}{\left(\int d^2x\sqrt{g}\exp(\Delta^{-1}R)\right)^2}. \tag{5}$$

In a CDT simulation we can perform direct measurements of the functional $A[g]$ on the spatial triangulation at time t, the result of which is shown in black in figure 2 for several different boundary conditions (which we will not discuss here). However, we can also deduce the ratio $(1/2 - \lambda)A[g]$ of the prefactors in the kinetic term by measuring the correlation functions $\langle V(t)V(t + \Delta t)\rangle$ and $\langle \tau_i(t)\tau_j(t + \Delta t)\rangle$ for small Δt. The results are shown in orange and red in figure 2, where we have determined (for all four graphs simultaneously) the overall factor $1/2 - \lambda$ by a best fit. We see that the overlap is reasonably good, which is in favour of our ansatz. This way we have determined λ for various values of the CDT coupling k_0 as is shown in blue in figure 3.

3. Extrinsic curvature at a fixed boundary

Another more local way of probing the kinetic term is by studying the geometry near a fixed boundary. It can be shown that the fluctuations of the spatial metric $g_{ab}(\Delta t)$ a small time Δt after a fixed initial two-dimensional surface are governed by the inverse $\mathcal{G}^\lambda_{abcd}$ of the modified Wheeler-De Witt metric in (4),

$$\langle g_{ab}(\Delta t, x)g_{cd}(\Delta t, y)\rangle - \langle g_{ab}(\Delta t, x)\rangle\langle g_{cd}(\Delta t, y)\rangle \propto \delta(x - y)\mathcal{G}^\lambda_{abcd}. \tag{6}$$

The natural way to test this ansatz would be to consider three-dimensional CDT configurations of which we fix the first spatial triangulation and observe how the geometry of the second

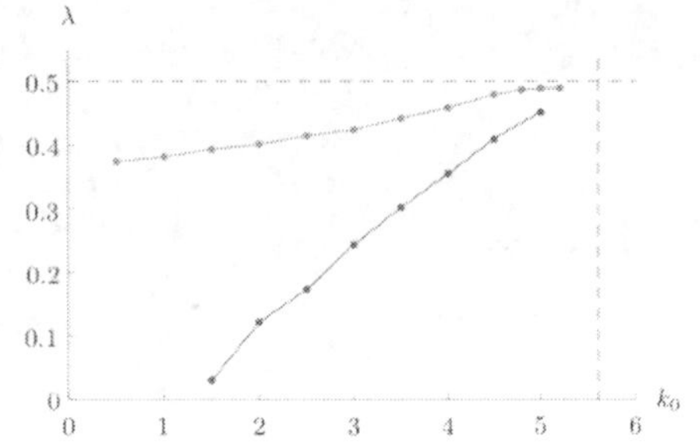

Figure 3. Determined values of λ from the minisuperspace method (bottom curve) and from extrinsic curvature (top curve).

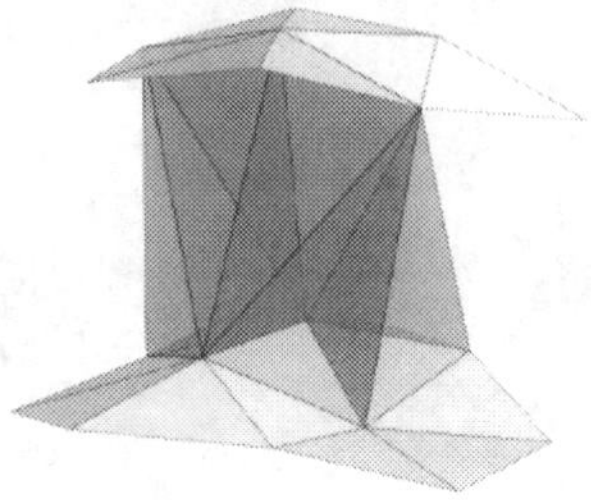

Figure 4. The extrinsic curvature at an edge in the initial boundary is proportional to the number of tetrahedra connecting to it.

spatial triangulation fluctuates. However, defining observables that capture these fluctuations and relating them to the continuum metric turns out to be difficult and highly ambiguous. Therefore, we make the assumption that the geometry of the second spatial triangulation is to large extent captured indirectly by the extrinsic curvature at the initial surface. Accordingly we test the ansatz (6) in which we replace $g_{ab}(\Delta t)$ by the extrinsic curvature K_{ab}, but have to keep in mind that this substitution might have a systematic effect on λ.

Viewing the CDT configuration as a piece-wise linear manifold, the extrinsic curvature has support on the edges and is (up to a constant) proportional to the number $N(e)$ of tetrahedra connecting to the edge e (figure 4). We have checked numerically that the fluctuations in $N(e)$ only correlate locally and the correlation does not depend on the system size. This means that if we send the system size to infinity only an ultralocal correlation will survive. We take our boundary to be a regular triangulation of the torus, because this allows us to unambiguously extract the ultralocal part of the correlations in $N(e)$. We have determined λ by comparing this to our ansatz for the correlation of the extrinsic curvature. The results are shown in red in figure 3.

4. Discussion

At an effective level comparing CDT to the Euclidean Einstein-Hilbert action is problematic due to the conformal mode problem. However, if we replace its kinetic term by a non-covariant but positive-definite one given by $\mathcal{G}_\lambda$ for $\lambda < 1/2$ we obtain a sensible ansatz and we find qualitative agreement with the experiments described. These findings seem to strengthen an earlier observed connection between CDT and Hořava-Lifshitz gravity in [5].

The discrepancy in figure 3 between the results of both methods indicates that the determination of λ is quite subtle and probably sensitive to the precise ensemble of triangulations used. However, the overall behaviour is clear: as we approach the phase transition in CDT, corresponding to a critical coupling $k_0 \approx 5.6$, λ increases to $1/2$ at which point $\mathcal{G}_\lambda$ becomes degenerate.

A more detailed exposition of the presented measurements will appear soon.

5. References

[1] Ambjørn J, Jurkiewicz J and Loll R 2001 *Phys. Rev.* D **64** 044011 (*Preprint* arXiv:hep-th/0011276)
[2] Ambjørn J, Görlich A, Jurkiewicz J and Loll R 2008 *Phys. Rev. Lett.* **100** 091304 (*Preprint* arXiv:0712.2485)
[3] Hořava P 2009 *Phys. Rev.* D **79** 084008 (*Preprint* arXiv:0901.3775)
[4] Ambjørn J, Barkley J and Budd T G 2011 *Preprint* arXiv:1110.4649
[5] Ambjørn J, Görlich A, Jordan S, Jurkiewicz J and Loll R 2010 *Phys. Lett.* B **690** 413-419 (*Preprint* arXiv:1002.3298)

Journal of Physics: Conference Series **360** (2012) 012039 doi:10.1088/1742-6596/360/1/012039

Curved geometry and Graphs[1]

Francesco Caravelli
Perimeter Institute for Theoretical Physics,
Waterloo, Ontario N2L 2Y5 Canada,
and
University of Waterloo, Waterloo, Ontario N2L 3G1, Canada,
and
Max Planck Institute for Gravitational Physics, Albert Einstein Institute,
Am Mühlenberg 1, Golm, D-14476 Golm, Germany

E-mail: fcaravelli@perimeterinstitute.ca

Abstract. Quantum Graphity is an approach to quantum gravity based on a background independent formulation of condensed matter systems on graphs. We summarize recent results obtained on the notion of emergent geometry from the point of view of a particle hopping on the graph. We discuss the role of connectivity in emergent Lorentzian perturbations in a curved background and the Bose–Hubbard (BH) model defined on graphs with particular symmetries.

1. Introduction

The gravitational interaction may be an effective description of an underlying theory which does not suffer from the well-known problems plaguing Einstein's gravity, as for instance perturbative non-renormalizability [2]. These ideas are the starting point of Quantum Graphity [1]. The target *per se* is understanding the emergence of gravity through model building of discrete quantum systems. A similar approach is Analogue Models [4], although in the continuum and considering Quantum Field Theory as the main tool. The motivation for the introduction of simplified models for Quantum Gravity is understanding the features of background independence in contexts in which not all the obstructions typical of gravity are present and, in particular, better understanding the phenomenon of emergence in background independent contexts.

The first Quantum Graphity model was introduced in [5, 6]. The basic motivation in [5] was to construct a Hamiltonian on a Hilbert space associated to the degrees of freedom of a graph (a set of nodes V with cardinality $\mathcal{N}$ and of links $\mathcal{E}$ of cardinality $\mathcal{N}(\mathcal{N}-1)/2$) such that the ground state of the Hamiltonian is a graph which has geometrical properties. The initial hope, in [5], was that by fixing the parameters of the reduced model (the one with *on/off* links), the ground state of the Hamiltonian would have been a graph of average degree d and with geometrical properties similar to simplicial complexes[3].

[1] Based on a talk given at Loops '11, Madrid, on 24 May 2011.

[2] However, there is still hope that gravity is non-perturbativly renormalizable. This is the starting point of Asymptotic Safety and Loop Quantum Gravity[2, 3].

[3] However the results obtained numerically in [6] suggested that the low-energy structure of the graph is string-like: the ground state is a one of a 1-dimensional object. This result is compatible with the mean field theory analysis performed in the reduced model[7], in which the model was mapped to an Ising-type Hamiltonian. This

A second model, which we now describe, was introduced in [9]. The main motivation for the introduction of [9] was the interpretational issue of the external bath (i.e. the temperature of the system): how can it be interpreted in a closed system (i.e. the Universe)? The same problem arises in closed quantum systems, where one could ask why thermalization is such a general phenomenon despite the unitary time evolution on the Hilbert space. In general the solution to the problem is strongly dependent on the *observability* of full Hilbert space, i.e. what part of it should be traced out in order to observe decoherence in the local observables. From this perspective, [9] is a simplified model in which these questions can be answered in a background independent context. While the graph degrees of freedom are the same as the model introduced in [5], additional degrees of freedom associated to bosonic particles with a Bose–Hubbard (BH) interaction are present. The result is a dynamical graph with local interactions.

2. Matter coupling: a graph-dynamical Bose–Hubbard model

The model introduced in [9] and recently further studied in [10], focuses on the study of interaction between matter (in the specific case, bosonic degrees of freedom on the vertices of the graph) and the graph. The energy terms for links and particles are of the form

$$\widehat{H}_0 = \sum_i \mu\, \widehat{a}_i^\dagger \widehat{a}_i + \sum_{ij} \nu\, \widehat{b}_{ij}^\dagger \widehat{b}_{ij}, \tag{1}$$

with $\widehat{a}_i$, $\widehat{b}_{ij}$ are hard core bosonic ladder operators on the space of vertices and links respectively and μ,ν coupling constants. There are two other terms, namely a BH interaction,

$$\widehat{H}_{BH} = -E \sum_{ij} (\widehat{a}_i^\dagger \widehat{a}_j + h.c.) \otimes \widehat{P}_{ij}, \tag{2}$$

where $\widehat{P}_{ij}$ is a projector on the *on* links ij, and an interaction between the bosonic particles and the links,

$$\widehat{H}_{int} = -E \sum_{ij} \widehat{P}_{ij}^L (\widehat{a}_i^\dagger \widehat{a}_j^\dagger \otimes \widehat{b}_{ij} + h.c.). \tag{3}$$

$\widehat{P}_{ij}^L$ is a nonlocal projector which annihilates graph states unless the link ij is not on a triangle. This keeps the dynamics local.
Thermalization. Simulations[9] showed that, in the long time regime, the classical model evolves into random graphs. In the quantum case, instead, the damping of oscillations of graphs observables, e.g. the degree of the graph, indicates that the model thermalizes to a metastable state. We studied the case of a 4-vertices graph with a link turned off and the others on. We observed dumping for the vertex degree observables. For the studied case, the asymptotic state turned out to be a mixed state of the graph with the link turned on and the link turned off.

3. Graphs and curved geometry

Trapped surfaces. An interesting feature of this model, first noticed in [9], is that regions of high connectivity are able to trap particles of low energy. This was studied in detail in [10] on a fixed graph. The graph chosen is the one of Fig. 1(a), which has a rotational symmetry: single particle

mapping allowed a straightforward use of the mean field theory techniques well known from the study of Ising-type models, after having identified the average degree of the graph as an order parameter. In [8] a different Hamiltonian based on graphs and close in spirit to [5] found, instead, a 2-dimensional complex in a low-energy phase. This last result gives some hope that a generic mechanism to obtain low energy d-dimensional simplicial complexes exists.

Journal of Physics: Conference Series **360** (2012) 012039 doi:10.1088/1742-6596/360/1/012039

states can be labeled by two quantum numbers, the shell position n and the internal position, θ, thus states of the form $|i, \theta\rangle$. In this case, the problem can be simplified by noticing that the Hamiltonian is diagonal in blocks of constant angular momentum. We can thus introduce, in the 1-particle sector, the *delocalized* states, defined as:

$$|n\rangle = \frac{1}{\sqrt{K_n}} \sum_{\theta=1}^{K_n} |n, \theta\rangle, \tag{4}$$

where K_n is the number of sites inside a shell n (see Fig. 1(b)) The full quantum evolution of a delocalized state can be *fully* reduced to a 1-dimensional BH model if the initial state is delocalized, where the coefficients of the Hamiltonian are opportunely chosen. This allowed to prove that the ground state, for the graph in Fig. 1(a), is $|n = 0\rangle$.

In fact, this reduction showed that particles inside the highly connected central region of Fig. 1a feel a potential which is proportional to the degree. In particular, between particle states inside the central region and the outside regions there is an energy gap proportional the relative degree. This prevents low energy particles from escaping the trap, thus confirming the argument of [9].

Emergent curved space. Another interesting finding of [10] is that, in the reduced models, expectation values of number operators on the graph exhibit an emergent Lorentz symmetry. The reduced effective 1-dimensional BH Hamiltonian, for a generic rotationally invariant graph, takes the form,

$$\widehat{H}_{BH} = -\sum_i f_{i\ i-1}(\widehat{b}_i^\dagger \widehat{b}_{i-1} + h.c.), \tag{5}$$

with $f_{ij} = d_{ij}E$ and d_{ij} is the relative degree between the shell i and the shell j, $\widehat{b}_i^\dagger$,$\widehat{b}_i$ is a symmetry reduced ladder operators. The idea then is to study the following expectation values (e.v.) $\langle \widehat{b}_i^\dagger \widehat{b}_i \rangle = \Psi_i(t)$. The interesting result is that these e.v. satisfy the following (closed) relation on the manifold of classical states:

$$\begin{aligned} \frac{\hbar^2}{2}\partial_t^2 \Psi_n(t) =& f_{n-1,n}^2 \left(\Psi_{n+1}(t) + \Psi_{n-1}(t) - 2\Psi_n(t)\right) \\ &+ \left(f_{n+1,n}^2 - f_{n-1,n}^2\right)\left(\Psi_{n+1}(t) - \Psi_n(t)\right). \end{aligned} \tag{6}$$

This fully describes the time evolution of the probability density. In the continuum limit, eqn. (6) becomes

$$\left[\partial_t^2 - \partial_x\Big(n(x)\partial_x\Big)\right]\Psi(x,t) = 0,$$

so it has a site-dependent speed of propagation for the density. This equation is everywhere locally Lorentz-invariant, with a local speed of propagation given by $c_x = 1/\sqrt{n(x)}$. The higher-dimensional version of this equation is known to be related to the Gordon background[4]. In fact, there is no obstruction to repeating the same procedure for graphs which can be foliated in more than one dimension. Thus, curved space is encoded in the connectivity of the graph (i.e. the local degree).

4. Conclusions

The emergence of gravity in a background independent contexts is a complete new research direction within the field of Analogue Models. In the case of graph-based models as ours it is not always obvious what are the right questions to ask. Indeed, we have shown in [10] that by focusing on a particular set of graphs and by asking the right question, "How does the particle probability density evolve?", the phenomenon of emergence (in our case of a Lorentz symmetry)

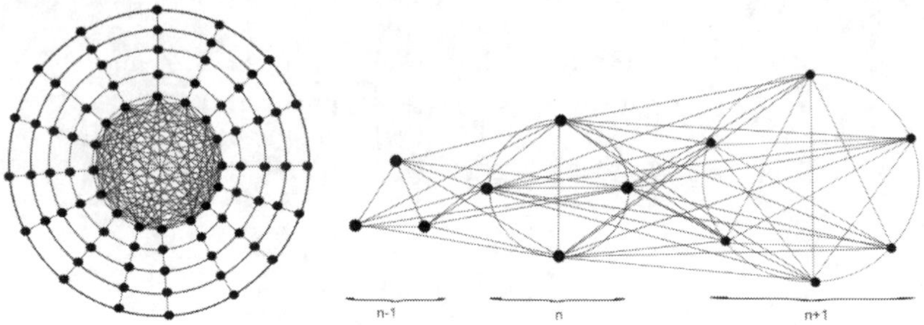

(a) The $\mathcal{K}_N$ configuration. (b) A non-trivial graph in which the coupling constants of the 1d-reduced model are site-dependent.

Figure 1: Two graphs on which the wavefunction can be symmetry-reduced.

can be identified. This happens because the graph we considered can be foliated, and then the states classified, according to this foliation. More recently, the same technique has been used to study the effects of disordered locality on the Lorentz symmetry, showing that a mass term appears [11]. We have identified graph configurations which, in a appropriate limit, can be considered as trapping surfaces. By considering delocalized states (i.e. symmetry-reduced wavefunctions), we showed that high connectivity implies a high energy gap between particles being inside the region or outside.

It is not clear, at this point, how general these models are and how many of the emergent phenomena of Analogue Models can be reproduced. We believe that many interesting questions related to background independence (and emergent gravity [12]) can be addressed in such simplified models. The full connection between a graph and the low energy manifold associated to it is an open problem to address. It is tempting, given the large literature on Regge calculus, to associate a simplicial complex to the graph when possible. On the other hand, it may be more physical to require that only the scaling properties - for instance looking at the Heat Kernel of the Laplacian of the graph[13]- match those of manifolds of a specific dimension.

Aknowledgements

We would like to thank A. Hamma, F. Markopoulou and A. Riera for several stimulating discussions and for the collaboration in the main work that we summarized here. Also, we thank L. Sindoni for many fruitful conversations about emergent gravity. Research at Perimeter Institute is supported by the Government of Canada through Industry Canada and by the Province of Ontario through the Ministry of Research & Innovation.

[1] F. Markopoulou, A. Hamma, New J. Phys. 13:095006 (2011) [arXiv:1011.5754]
[2] C. Rovelli, (Lectures given at Zakopane 2011) [arXiv:1102.3660]
[3] M. Reuter, F. Saueressig, (Lectures given at Zakopane 2007) [arXiv:0708.1317v1]
[4] C. Barcelo, S. Liberati, M. Visser, Living Rev. Rel. 8 12 (2005) [arXiv:gr-qc/0505065]
[5] T. Konopka, F. Markopoulou, L. Smolin, [arXiv:hep-th/0611197]; T. Konopka, F. Markopoulou, S. Severini, Phys. Rev. D 77, 104029 (2008) [arXiv:0801.0861]
[6] T. Konopka, Phys. Rev. D 78,044032 (2008) [arXiv:0805.2283]
[7] F. Caravelli, F. Markopoulou, Phys. Rev. D 84, 024002 (2011), [arXiv:1008.1340]
[8] F. Conrady, J. Statist. Phys. 142:898 (2011) [arXiv:1009.3195]
[9] A. Hamma, F. Markopoulou, S. Lloyd, F. Caravelli, S. Severini, K. Markstrom, Phys. Rev. D 81, 104032 (2010) [arXiv:0911.5075]
[10] F. Caravelli, A. Hamma, F. Markopoulou, A. Riera, Phys. Rev. D 85, 044046 (2012) [arXiv:1108.2013]
[11] F. Caravelli, F. Markopoulou, [arXiv:1201.3206]
[12] L. Sindoni, Contribution to SIGMA Special Issue "Loop Quantum Gravity and Cosmology" (to appear) [arXiv:1110.0686]
[13] T. Filk, Class.Quant.Grav.17:4841-4854 (2000) [arXiv:hep-th/0010126]

Journal of Physics: Conference Series **360** (2012) 012040 doi:10.1088/1742-6596/360/1/012040

Non-commutative holonomies in $2+1$ LQG and Kauffman's brackets

K Noui[1], **A Perez**[2] **and D Pranzetti**[3]

[1] Laboratoire de Mathématiques et Physique Théorique, 37200 Tours, France
[2] Centre de Physique Théorique, Campus de Luminy, 13288 Marseille, France
[3] Max-Planck-Institut für Gravitationsphysik AEI, Am Mühlenberg 1, 14476 Golm, Germany

E-mail: pranzetti@aei.mpg.de

Abstract. We investigate the canonical quantization of 2+1 gravity with $\Lambda > 0$ in the canonical framework of LQG. A natural regularization of the constraints of 2+1 gravity can be defined in terms of the holonomies of $A_{\pm} = A \pm \sqrt{\Lambda}e$, where the SU(2) connection A and the triad field e are the conjugated variables of the theory. As a first step towards the quantization of these constraints we study the canonical quantization of the holonomy of the connection $A_{\lambda} = A + \lambda e$ acting on spin network links of the kinematical Hilbert space of LQG. We provide an explicit construction of the quantum holonomy operator, exhibiting a close relationship between the action of the quantum holonomy at a crossing and Kauffman's q-deformed crossing identity. The crucial difference is that the result is completely described in terms of standard SU(2) spin network states.

1. Introduction

In his seminal work [1], Witten established the link between the Jones Polynomial, Chern-Simons theory and quantum gravity in 2+1 dimensions with $\Lambda \neq 0$. By defining a path integral quantization of the Chern-Simons theory with compact gauge Lie groups $SU(2)$, Witten provided a quantization of Euclidean gravity with a positive Λ and discovered a fascinating relation between knots invariants and the expectation values of Wilson loops observables in Chern-Simons theory, leading to a new covariant definition of the Jones polynomials. After this result, Turaev and Viro defined a state-sum model [2] showing the central role played by quantum groups in the construction of 3-manifolds invariants and knots polynomials. Turaev-Viro (TV) invariants can be viewed as a q-deformed version of Ponzano and Regge amplitudes. Moreover, the asymptotic of the vertex amplitudes (the quantum $6j$-symbol) has been shown to be related to the action of $2+1$ gravity with non vanishing cosmological constant in the WKB approximation [3]. All this, strongly motivates the idea that it should be possible to recover (in the context of LQG) the TV amplitudes as the physical transition amplitudes of 2+1 gravity with $\Lambda \neq 0$. This has been so far explicitly shown only in the simpler case for pure gravity with $\Lambda = 0$ [4]. Following this simpler case, in LQG it is natural to interpret the TV invariant as transition amplitudes between arbitrary pairs of spin-network states. For this to be the case, the TV amplitudes would have to be related to the classical $SU(2)$ spin networks states of the canonical theory. In contrast, the TV amplitudes are constructed from the combinatorics of q-deformed spin networks. This would imply that the understanding of the relationship between the TV invariants and quantum gravity requires the understanding of the dynamical interplay between classical spin-network states and q-deformed amplitudes. We shall find here some indications of how this relationship can arise.

2. Phase Space and Constraints

Let us first briefly recall the canonical structure of (Riemannian) gravity in 2+1 dimensions. The action of departure is $S(A,e) = \int_{\mathcal{M}} \mathrm{tr}\,[e \wedge F(A)] + \frac{\Lambda}{6}\mathrm{tr}\,[e \wedge e \wedge e]$, where $\Lambda \geq 0$, e is a cotriad field, and A is an $SU(2)$ connection. Assuming that the space time manifold has topology $\mathcal{M} = \Sigma \times \mathbb{R}$, and, upon the standard 2+1 decomposition, the phase space of the theory is parametrized by the pullback to Σ of A and e. In local coordinates we can express them in terms of the 2-dimensional connection A_a^i and the triad field e_a^i, where $a = 1, 2$ are space coordinate indices and $i, j = 1, 2, 3$ are $su(2)$ indices. The Poisson bracket among these is given by $\{A_a^i(x), e_b^j(y)\} = \epsilon_{ab}\delta^i_j\delta^{(2)}(x,y)$. The phase space variables are subjected to the first class local constraints

$$d_A e = 0 \quad \text{and} \quad F(A) + \Lambda\, e \wedge e = 0. \tag{1}$$

The basic kinematical observables are given by the holonomy of the connection and appropriately smeared functionals of the triad field e. Quantization of these (unconstrained) observables leads to an irreducible representation on a kinematical Hilbert space ($\mathcal{H}_{kin}$) spanned by spin-network states. The holonomy acts simply by multiplication while e acts as the derivative operator. Dynamics is defined by imposing the quantum constraints on the kinematical states. More precisely, the quantum constraint-equations of 2+1 gravity with $\Lambda \neq 0$ can be written as

$$\mathcal{G}\,[\alpha] \triangleright \Psi = \int_\Sigma \mathrm{Tr}[\alpha d_A e] \triangleright \Psi = 0 \quad \text{and} \quad C_\Lambda\,[N] \triangleright \Psi = \int_\Sigma \mathrm{Tr}\,[N\,(F(A) + \Lambda\, e \wedge e)] \triangleright \Psi = 0, \tag{2}$$

for all $\alpha, N \in \mathcal{C}^\infty(\Sigma, su(2))$. Quantization of the previous expression for the constraints turns out to be difficult due to the presence of non-linear functional terms of the basic fields which requires the introduction of a regularization. A possible way to construct the physical Hilbert space of the theory is to follow the example of [4] for the case $\Lambda = 0$. In fact, the main result of [4] is the definition of a path integral representation of the theory from the canonical picture reproducing the spin foam amplitudes of the Ponzano-Regge model, and this sets the bases for the extension of the analysis to the $\Lambda \neq 0$ case. Indeed, the key observation is that the curvature constraint in (2) can be quantized by first introducing a regulator consisting of a cellular decomposition Δ_Σ of Σ into plaquettes p and defining a new connection $A_\pm \equiv A \pm \sqrt{\Lambda} e$ so that we get

$$C_\Lambda\,[N] = \lim_{\epsilon \to 0} \sum_{p \in \Delta_\Sigma} \mathrm{Tr}\,[N_p\, W_p\,(A_\pm)] - \mathcal{G}\left[\pm\sqrt{\Lambda} N\right], \tag{3}$$

where $W_p(A_\pm) \in SU(2)$ is the Wilson loop computed in the fundamental representation. This provides a candidate background independent regularization of the curvature constraint $C_\Lambda[N]$. The quantization of the previous classical expression requires the quantization of the holonomy of $A_\pm$. More generally, as a first step towards the quantization of (3), we are now going to present the quantization of the holonomy h_λ of the general connection $A_\lambda \equiv A + \lambda e$ for $\lambda \in \mathbb{R}$, surveying the main steps of the more detailed analysis presented in [5].

3. Quantization of $h\,(A_\lambda)$

We are now going to explore the quantization of the kinematical observable $h_\eta\,[A_\lambda] = P\,\mathrm{e}^{-\int_\eta A + \lambda e}$ associated with a path $\eta \in \Sigma$, as operators on the $\mathcal{H}_{kin}$ of 2+1 LQG. In order to do this, let us first recall that in LQG there is a well-defined quantization of the e-field based on the smearing of e along one dimensional paths. More precisely, given a path $\eta^a(t) \in \Sigma$ one considers the (smeared) connection conjugate momentum $E(\eta) \equiv \int E^{ai}\tau_i n_a dt$, where $n_a \equiv \epsilon_{ab}\frac{d\eta^a}{dt}$ is the normal to the path. The quantum operator associated to $E(\eta)$ acts non trivially only on holonomies h_γ along a path $\gamma \in \Sigma$ that are transversal to η. It suffices to give its action on transversal holonomies that either end or start on η. The result is:

$$\hat{E}(\eta) \triangleright h_\gamma = \frac{1}{2}\hbar \begin{cases} o(p)\tau^i \otimes \tau_i h_\gamma & \text{if } \gamma \text{ ends at } \eta \\ o(p) h_\gamma \tau^i \otimes \tau_i & \text{if } \gamma \text{ starts at } \eta \end{cases}, \tag{4}$$

where $o(p)$ is the orientation of the intersection $p \in \Sigma$. Due to the tensorial form of the Poisson bracket between the connection A and the triad field e (inherited by the commutator in the quantum theory) the action of h_η on the vacuum simply creates a Wilson line excitation, i.e. it acts simply by multiplication by the holonomy of A along the path, namely $h_\eta[A_\lambda]|0\rangle = h_\eta[A]|0\rangle$.

Therefore, the simplest non-trivial example is the action of the holonomy of A_λ on a transversal Wilson loop in the fundamental representation. We define the quantization of h_η by quantizing each term in its series expansion in λ. Terms of order n have n powers of the e operators. The quantization of these products becomes potentially ill-defined due to factor ordering ambiguities. More precisely, let $\gamma, \eta : (0,1) \to \Sigma$ be two curves that cross each other *transversally*. Let us now study the action of η on γ. Developing in powers of λ the action $h_\eta(A_\lambda) \triangleright h_\gamma(A_\lambda)|0\rangle$, the generic coefficient of order p is:

$$\lambda^p \sum_{n\geq p}\sum_{m\geq p}(-1)^{m+n}\sum_{1\leq k_1<\cdots<k_p\leq n}\int_0^1 \mathrm{d}t_1\cdots\int_0^{t_{n-1}}\mathrm{d}t_n\int_0^1 \mathrm{d}s_1\cdots\int_0^{s_{m-1}}\mathrm{d}s_m$$
$$\left[A(\eta(t_1))\cdots E(\eta(t_{k_1}))\cdots E(\eta(t_{k_p}))\cdots A(\eta(t_n))\right] \triangleright A(\gamma(s_1))\cdots A(\gamma(s_m)) .$$

Ordering ambiguities now arise when computing the action of the derivation operators on the connection along γ. In fact, it is easy to see that only the terms containing p consecutive graspings Es acting on p consecutive As survive and this is where factor ordering ambiguities appear due to the non-commutativity of the grasping operators. We will see in a moment that relationship with Kauffman's bracket is found if one uses the so-called Duflo map. To lighten the presentation, let us now introduce a graphical notation for the quantum holonomy action

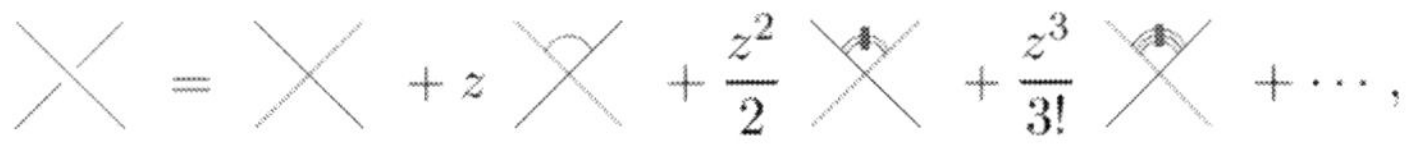
(5)

where $z = -io\hbar\lambda$ (here o is the crossing orientation coming from (4)), and the boxes, at this stage, denote a generic ordering prescription. At this point, inspired by [8], we introduce the Duflo isomorphism as a quantization map and explicitly work out the prescription leading, once the action (4) of the flux operator in LQG is also taken into account, to a crossing evaluation in accordance with the q-deformed binor identity. The Duflo map [9] (Q_D) is a generalization of the universal *quantization map* proposed by Harish-Chandra for semi-simple Lie algebras. The latter provides a prescription to quantize polynomials of commuting variables which after quantization acquire Lie algebra commutation relations. We write explicitly only the expression of the map Q_D for a pair of E fields since this is the only one we are going to use in the following, namely

$$\mathrm{Q}_D[E_j E_k] = \frac{1}{2}(\tau_j\tau_k + \tau_k\tau_j) + \frac{1}{6}\delta_{jk}. \tag{6}$$

In order to get the general form of the series (5) in the case where we use the LQG quantization of the flux operators, it suffices to write the first few terms. In the first order term, E acts as a left-invariant vector field on the portion of the holonomy which has the crossing as its source and as a right-invariant one on the other portion. The full result is just the same as in (5). In the second order diagram we have the action of two flux operators at the same point and therefore ordering ambiguities arise. In order to deal with them, we now use the prescription (6) induced by the Duflo map and, diagrammatically, for the second order term we have  $= \frac{1}{2} \;\; + \frac{1}{2} \;\; + \frac{1}{6} \;\; = \frac{1}{16}$. Therefore, the second order diagram is proportional to the order zero diagram. The third order term is consequently proportional to the first order one and so on. We get in this way the general expression for arbitrary order and perturbative expansion (5) can be exactly summed, yielding

$$h_\eta(A_\lambda) \triangleright h_\gamma(A_\lambda)\,|0\rangle = \quad = A \quad + A^{-1} \quad , \tag{7}$$

where $A = e^{\frac{i o \hbar\lambda}{4}}$. Therefore, the series expansion in powers of λ converges and leads to a simple expression for the crossing reproducing Kauffman's q-deformed binor identity for $q = \exp i\lambda/2$.

4. Discussion

We have shown that the ordering ambiguities arising in the quantization of the holonomy of A_λ in the fundamental representation can be sorted out by means of a simple quantization based on the Duflo map, leading to the Kauffman-like algebraic structure for the action of the quantum holonomy defining a crossing. The recovering of Kauffman's bracket related to the q-deformed crossing identity is a remarkable result since it was obtained starting from the standard $SU(2)$ $\mathcal{H}_{kin}$ of LQG. This result is promising in the road to finding a relationship between TV amplitudes and physical amplitudes in canonical LQG. In fact, recall that the value of d_q together with the deformed binor identity are the two ingredients for the combinatorial definition of the TV invariant. Despite the fact that, at the present kinematical level, loops still evaluate according to the classical $SU(2)$ recoupling theory, an intriguing indication that the implementation of dynamics could lead to the emergence of the quantum dimension for loops evaluation emerges from the study of the algebra of the operator (3), taken as a proposal for the regularized version of the curvature constraint in (2). More precisely, the classical constraint algebra dictates that this should be proportional to the Gauss constraint. If one performs this analysis, it is immediate to see that, besides the 'mildly' anomalous contributions found in [6] (see [7] for a possible way around this difficulty), there are stronger ones which do not annihilate gauge invariant states. Surprisingly, the latter anomalous terms happen to be proportional to $(A^2 + A^{-2} + \;\;)$, where represents the loop with no area in the fundamental representation $j = 1/2$. Thus the condition that an infinitesimal loop evaluates to the quantum dimension emerges from the constraint algebra. All this indicates that, even when we do not introduce a quantum group at any stage, and no dynamical constraint has been imposed yet, amplitudes such as the value of the quantum dimension and $q = A^2 = e^{\frac{i\hbar\lambda}{4}}$ naturally appear from our treatment. This indicates that perhaps a strict correspondence between LQG and the TV invariant can be established if one appropriately quantizes and imposes the curvature constraint (2). Quantum holonomies defined here might be the right tool along this direction.

References

[1] Witten E 1988 *Nucl. Phys. B* **311** 46; 1989 *Commun. Math. Phys.* **121** 351
[2] Turaev V G and Viro O Y 1992 *Topology* **31** 865
[3] Mizoguchi S and Tada T 1992 *Phys. Rev. Lett.* **68** 1795 (*Preprint hep-th/9110057*)
[4] Noui K and Perez A 2005 *Class. Quant. Grav.* **22** 1739 (*Preprint gr-qc/0402110*)
[5] Noui K, Perez A and Pranzetti D 2011 JHEP **10** 036 (*Preprint gr-qc/11050439*)
[6] Perez A and Pranzetti D 2010 *Class. Quant. Grav.* **27** 145009 (*Preprint gr-qc/10013292*)
[7] Pranzetti D 2011 *Class. Quant. Grav.* **28** 225025 (*Preprint gr-qc/11015585*)
[8] Sahlmann H and Thiemann T 2011 *J. Geom. Phys.* **61**, 1104-21 (*Preprint gr-qc/11011690*)
[9] Duflo M 1977 *Ann. scient. Ecole Norm. Sup.* **10**, 265288

Loops 11: Non-Perturbative / Background Independent Quantum Gravity IOP Publishing
Journal of Physics: Conference Series **360** (2012) 012041 doi:10.1088/1742-6596/360/1/012041

Regularized Hamiltonians and Spinfoams

Emanuele Alesci

Universität Erlangen, Institut für Theoretische Physik III, Lehrstuhl für Quantengravitation
Staudtstrasse 7, D-91058 Erlangen, EU

E-mail: alesci@theorie3.physik.uni-erlangen.de

Abstract. We review a recent proposal for the regularization of the scalar constraint of General Relativity in the context of LQG. The resulting constraint presents strengths and weaknesses compared to Thiemann's prescription. The main improvement is that it can generate the 1-4 Pachner moves and its matrix elements contain $15j$ Wigner symbols, it is therefore compatible with the spinfoam formalism: the drawback is that Thiemann anomaly free proof is spoiled because the nodes that the constraint creates have volume.

1. Introduction

The relation between Canonical [1] and Covariant [2] loop quantum gravity (LQG) [3] is still one of the main open issues of this approach to Quantum Gravity. The beauty of the first is that provides an anomaly free quantization of the Dirac algebra but it's difficult to extract the Physical Hilbert space from the dynamics; this problem is bypassed by the spinfoam formalism, where using the quantization of a constrained BF theory, one directly defines transition amplitudes [4, 3] using a vertex expansion of the spinfoam model. However these objects are defined for fixed complexes and is unclear how to recover the full diff invariant theory. Nevertheless these two approaches should describe the same theory but, in spite of the fact that the spinfoam formalism was originally conceived as an exponentiation of the canonical evolution [5], aimed to implement a Projector on the physical Hilbert Space, it is still not possible to formally relate the dynamics of the two, (see [6, 7]). Remarkably however in 3-d, the relation between the two formalisms has been clarified in [8] and in 4-d, the spinfoam *kinematics* matches the canonical one [9, 10]. One thus hopes to link the two approaches starting from the common kinematics as has been done in 3-d: Given a boundary state, the amplitude for a transition is given by a sum over histories of spinnetworks, based on 2-complexes, compatible with that state. This sequence is generated by elementary evolutions of spinnetworks taking places at the vertices of the 2-complex. To relate the dynamics one can then proceed looking at the Hamiltonian as the generator of these evolutions. Thiemann's Hamiltonian constraint [11, 12] is the natural candidate for this action, however acting on spinnetworks nodes it creates new "extraordinary links" joined to the original graph by "extraordinary nodes" without volume (Thiemann construction to this aim uses the AL version of the volume [13]): this feature, fundamental for the consistent quantization of the Dirac algebra, is nevertheless incompatible with the spinfoam description that in 4d, on a simplicial setting, requires a vertex amplitude based on a bulk 4-simplex: this implies that the Hamiltonian acting on a 4-valent node (dual to a tetrahedron) should produce four resulting 4-valent nodes: in other words it should be able to implement a 1-4 Pachner move. The natural question is then if it's possible to find an Hamiltonian that does; the answer is yes, and an example is a suitable modification of Thiemann's one appeared in [14].

2. Thiemann constraint and the new proposal

Here we consider only the "euclidean" Hamiltonian constraint $\mathcal{H} = -2\mathrm{Tr}[F \wedge e]$. Following Thiemann [11], and choosing units where $8\pi G c^{-3}\gamma = 1$, we can write $e^i_a(x) = \{A^i_a(x), V\}$, where V is the volume of an arbitrary region Σ containing the point x. Using this, and smearing the constraint with a lapse function $N(x)$, we have $\mathcal{H}[N] = \int_\Sigma d^3x\, N(x)\,\mathcal{H}(x) = -2\int_\Sigma N\, \mathrm{Tr}(F \wedge \{A, V\})$. In order to regularize this expression we review two options:

a) *Thiemann construction* [11, 12] based on a triangulation T of the manifold Σ into elementary tetrahedra with analytic edges. In this case we take a tetrahedron Δ of T, and a vertex v of this tetrahedron. Call the three edges that meet at v as s_i, $i = 1, 2, 3$ and denote a_{ij} the edge connecting the two end-points of s_i and s_j opposite to v; in this way s_i, s_j and a_{ij} form a triangle. Denote this triangle as $\alpha_{ij} := s_i \circ a_{ij} \circ s_j^{-1}$. Decompose the smeared Euclidean constraint into a sum of one term per each tetrahedron

$$\mathcal{H}[N] = \sum_{\Delta \in T} -2 \int_\Delta d^3x\, N\, \epsilon^{abc}\, \mathrm{Tr}(F_{ab}\{A_c, V\}) \,. \tag{1}$$

Define the classical *regularized* Hamiltonian constraint as $\mathcal{H}_T[N] := \sum_{\Delta\in T} \mathcal{H}^m_\Delta[N]$, with

$$\mathcal{H}^m_\Delta[N] := \frac{N(v)}{2N_m^2}\, \epsilon^{ijk}\, \mathrm{Tr}\left[h^{(m)}_{\alpha_{ij}} h^{(m)}_{s_k} \{h^{(m)-1}_{s_k}, V\}\right] , \tag{2}$$

where $h^{(m)}_{\alpha_{ij}}$ and $h^{(m)}_{s_k}$ are the holonomies, in representation m, around the triangle α_{ij} and along the segment s_k respectively, and $N_m^2 = \mathrm{Tr}[\tau^i\tau^i] = -(2m+1)m(m+1)$ with τ^i the (anti-hermitian) generators of $su(2)$. This expression converges to the Hamiltonian constraint (1) if the triangulation is sufficiently fine [15]. The expression (2) can finally be promoted to a quantum operator, since volume and holonomy have corresponding well-defined operators in LQG. The lattice spacing of the triangulation T then acts as a regularization parameter. Acting on a spin network state, the operator reduces to a sum over terms each acting on individual nodes. The continuum limit of the operator turns out to be trivial in the quantum theory. On diffeomorphism-invariant (bra) states the regulator dependence drops out trivially. The result [15, 14], in the action of the operator $\mathcal{H}^m_\Delta$ on trivalent nodes, $|v(j_i, j_j, j_k)\rangle \equiv |v_3\rangle$, whereas j_i, j_j, j_k are the spins of the adjacent edges e_i, e_j, e_k, with $l_0^2 = 16\pi\gamma l^2_{Planck}$, is given by:

$$\mathcal{H}^m_\Delta\, |v(j_i, j_j, j_k)\rangle = \frac{i l_0}{12 N_m^2}\, [\sum_{a,b} A^{(m)}(j_i, a|j_j, b|j_k) \quad\text{(diagram)}\quad + \mathit{Permutations}\,] \,. \tag{3}$$

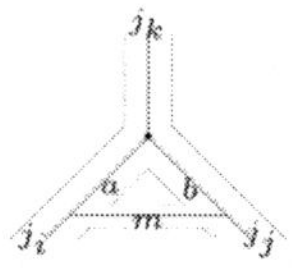

$\mathcal{H}^m_\Delta|v_3\rangle$ gives the original state with the new "extraordinary link" m between all the possible pairs of edges adjacent to the node, with amplitudes [14] given by cyclic permutations of arguments.

b) *A new operator* [14] is obtained if we consider a regularization based not only on triangulations T but also on the one-skeleton Γ, dual of T: Γ is a graph with nodes v in the center of the tetrahedra v of T, and straight links that cut the triangles of T. In this case we fix a tetrahedron v and one of its vertices, say s. Let s^a be the segment that joins the center of v to s, and u^a and v^a two of the sides of the triangle s opposite to the tetrahedron's vertex s. The volume of the tetrahedron can be written as $V = \sum_s \frac{1}{18}\epsilon_{abc} s^a u^b v^c$ where the sum is over the four vertices of the tetrahedron. Consider now the quantity

$$H_\Delta = \sum_s h^i_{\Delta_s}\, \mathrm{Tr}[\tau^i\, h_s^{-1}\{h_s, V\}] \tag{4}$$

Journal of Physics: Conference Series **360** (2012) 012041 doi:10.1088/1742-6596/360/1/012041

where $\triangle s$ is the triangle opposite to the vertex s. If A and e are constant on the tetrahedron, it is easy to see (for instance using coordinates in which the tetrahedron is regular) that this gives

$$H_\Delta = \sum_s F^i_{ab} u^a v^b s^c e^i_c = 18 \, \mathrm{Tr}(F_{ab} e_c) \epsilon^{abc} V = 18 \int_v \mathrm{Tr}(F \wedge e). \tag{5}$$

Therefore, using holomonies are in the representation m, we can replace (2) with

$$\mathcal{H}_\Delta[N] := \frac{N(v)}{36 N_m^2} \sum_s h^i_{\triangle_s} \, \mathrm{Tr}\Big[\tau^i h_s \{h_s^{-1}, V\}\Big] \tag{6}$$

Notice that the sum is over the four links emerging from v in T and $\triangle s$ is a triangle that joins the three points sitting on the three other links emerging from v. Notice also that this triangle and the center v define a tetrahedron, which we shall denote $\blacktriangle s$. It is then natural to replace the triangle regularization of the curvature with the tetrahedral one defined in [14]: the idea is to substitute $h^i_\triangle$ with a different object, denoted $h^i_\blacktriangle$, and which is defined by the *spin-network function* of the connection A, associated to the tetrahedron generated by *three* segments s_{01}, s_{02}, s_{03}, emerging from a point n, with one open link and explicitly defined as

$$h^i_\blacktriangle = c_m N_m \, i^{i\alpha\beta\gamma} \, i^{\delta\epsilon\zeta} \, i^{\theta\iota\kappa} \, i^{\lambda\mu\nu} \;\; h[s_{01}]_{\alpha\delta} \, h[s_{02}]_{\beta\kappa} \, h[s_{03}]_{\gamma\mu} \, h[s_{12}]_{\epsilon\theta} \, h[s_{23}]_{\iota\lambda} \, h[s_{31}]_{\nu\zeta} \, . \tag{7}$$

Here h are holomies in a representation of (integer) spin m. $i^{\alpha\beta\gamma}$ is the (unique) normalized 3-valent intertwiner between three representations m, and $i^{i\alpha\beta\gamma}$ is a normalized 4-valent intertwiner, with the first index in the adjoint representation and the other three in the representation m, satisfying $i^{i\alpha\beta\gamma} = i^{i\gamma\alpha\beta} = i^{i\beta\gamma\alpha}$ and $c_m^{-1} = i^{i\alpha\beta\gamma} \, i^{\alpha\epsilon\zeta} \, i^{\epsilon\beta\iota} \, i^{\lambda\gamma\zeta} \, i^{i\lambda\iota}$. The key property of $h^i_\blacktriangle$ is that in the limit in which the size of the tetrahedron is small, we have $h^i_\blacktriangle = h^i_\triangle$ where the triangle is the face of the tetrahedron opposite to the 4-valent node n.

With this observation we can replace $h^i_\triangle$ with $h^i_\blacktriangle$ in (6) and promote it to the quantum operator:

$$\hat{\mathcal{H}}.[N] := \frac{-i}{36 \cdot 8\pi\gamma l_p^2} \sum_v \frac{N(v)}{N_m^2} \sum_s \hat{h}^i_{\blacktriangle_s} \, \mathrm{Tr}\Big[\tau^i \hat{h}_s \hat{V} \hat{h}_s^{-1}\Big] \tag{8}$$

The action of the operator on a spin network state with support on a graph γ can then be defined, following [11], by choosing a regularizing triangulation T adapted to γ. Here one has to choose T such that γ is a subgraph of T^*. To analyze the new constraint we restrict our attention to 4-valent nodes. On a single 4-valent node, it is a sum over the four edges that emerge from the node, $H_\Delta \, |v_4\rangle = \sum_s \, H^s_\Delta \, |v_4\rangle$ with

$$\hat{H}^s_\Delta |v_4\rangle = N_m^2 c_m \sum_{c,d,e,f,g,h,k} d_c \, d_f \, d_g \, d_h \, d_k \, \frac{l_0^3}{4} \, V_{j_l, i_n}{}^{d,e}(j_i, j_j, j_k, m, c)$$

$$\times \begin{Bmatrix} j_l & d & 1 \\ m & m & c \end{Bmatrix} \; F_\times(m, d, e, f, g, h, k, j_i, j_j, j_k, j_l)$$

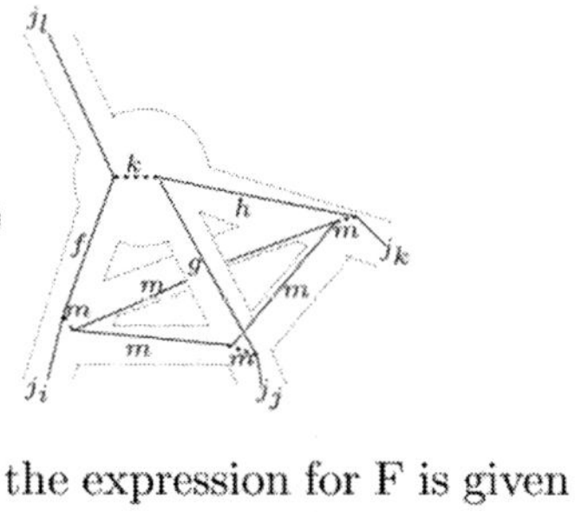

where $V_{j_l, i_n}{}^{d,e}$ is the volume of a non-gauge-invariant 5-valent node, the expression for F is given in [14] and remarkably contains 15j symbols.

Journal of Physics: Conference Series **360** (2012) 012041 doi:10.1088/1742-6596/360/1/012041

3. Conclusions

The main differences between the operator (8) and the quantum version of (6) are: 1) The curvature is computed on a surface that is properly *dual* to the direction indicated by e. Notice in fact that it is computed on a triangle that *surrounds* the direction of the link s. This is different from the old case; there, γ had to be a subgraph of the triangulation T itself, not its dual. 2) The new operator creates *three* new links instead than one. Therefore generically it transforms a 4-valent node into *four* nodes, thus implementing the 1-4 Pachner move [2]. This is precisely the action of the dynamics in the simplicial spinfoam models. 3) The nodes created are themselves 4-valent. Thus are "of the same kind" as the original node and therefore with volume. This is not the case as the old operator 4) Finally, when we compute matrix elements of this operator, we find $15j$ Wigner symbols, even if a way to recover from these one of the known spinfoam model is still under investigation.

Thiemann's proof that the Hamiltonian operator is anomaly free [11] does not go through with the new operator. The proof was indeed based on the fact that the nodes generated were rather "special" and had no volume, and therefore the Hamiltonian could not act again on them. On the other hand, this fact is sometime viewed as one of the unconvincing aspect [16] of the old construction, and this is partially corrected with the constraint considered here. (On different ways to address the issue, see in particular [17, 18].)

References

[1] A. Ashtekar and J. Lewandowski, 2004 *Class. Quant. Grav.* **21** R53
T. Thiemann, 2007 *Modern canonical quantum general relativity* (Cambridge: Cambridge University Press)
C. Rovelli, 2004 *Quantum Gravity* (Cambridge: Cambridge University Press).
[2] A. Perez, 2003 *Class. Quant. Grav.* **20** R43
[3] C. Rovelli, 2011 *Class. Quant. Grav.* **28** 114005
[4] L. Modesto and C. Rovelli, 2005 *Phys. Rev. Lett.* **95** 191301
E. Alesci and C. Rovelli, 2007 *Phys. Rev. D* **76** 104012
E. Alesci and C. Rovelli, 2008 *Phys. Rev. D* **77** 044024
E. Bianchi, E. Magliaro and C. Perini, 2009 *Nucl. Phys. B* **822** 245
[5] M. P. Reisenberger and C. Rovelli, 1997 *Phys. Rev. D* **56** 3490
[6] E. Alesci, K. Noui and F. Sardelli, 2008 *Phys. Rev. D* **78** 104009
[7] E. Alesci, T. Thiemann, A. Zipfel, 2001 Linking covariant and canonical LQG: New solutions to the Euclidean Scalar Constraint *Preprint* arXiv:1109.1290
[8] K. Noui and A. Perez, 2005 *Class. Quant. Grav.* **22** 1739
[9] J. Engle, E. Livine, R. Pereira and C. Rovelli, 2008 *Nucl. Phys. B* **799** 136
L. Freidel and K. Krasnov, 2008 *Class. Quant. Grav.* **25** 125018
[10] W. Kaminski, M. Kisielowski and J. Lewandowski, 2010 *Class. Quant. Grav.* **27** 095006
[11] T. Thiemann, 1998 *Class. Quant. Grav.* **15** 839
[12] T. Thiemann, 1998 *Class. Quant. Grav.* **15** 875
[13] A. Ashtekar and J. Lewandowski, 1998 *Adv. Theor. Math. Phys.* **1** 388
[14] E. Alesci and C. Rovelli, 2010 *Phys. Rev. D* **82** 044007
[15] M. Gaul and C. Rovelli, 2001 *Class. Quant. Grav.* **18** 1593
[16] R. Gambini, J. Lewandowski, D. Marolf and J. Pullin, 1998 *Int. J. Mod. Phys. D* **7** 97
J. Lewandowski and D. Marolf, 1998 *Int. J. Mod. Phys. D* **7** 299
L. Smolin, 96 The classical limit and the form of the Hamiltonian constraint in non-perturbative quantum general relativity *Preprint* arXiv:gr-qc/9609034.
[17] K. Giesel and T. Thiemann, 2007 *Class. Quant. Grav.* **24** 2465
[18] T. Thiemann, 2006 *Class. Quant. Grav.* **23**, 2211
B. Dittrich and T. Thiemann, 2006 *Class. Quant. Grav.* **23**, 1025
T. Thiemann, 2006 *Class. Quant. Grav.* **23** 2249

Journal of Physics: Conference Series **360** (2012) 012042 doi:10.1088/1742-6596/360/1/012042

Operator Spin Foams: holonomy formulation and coarse graining

Benjamin Bahr
Department of Applied Mathematics and Theoretical Physics
Wilberforce Road, Cambridge CB3 0WA, UK

Abstract. A dual holonomy version of operator spin foam models is presented, which is particularly adapted to the notion of coarse graining. We discuss how this leads to a natural way of comparing models on different discretization scales, and a notion of renormalization group flow on the partially ordered set of 2-complexes.

1. Introduction

In [1] a general class of models was presented, which are a generalization of current spin foam models (see [2] and references therein), called operator spin foams (OSF). Based on the work [3], these models are defined on a 2-complex κ, which can be seen as either embedded in a manifold (to depict a history of spin networks), or as abstract set of vertices, edges, faces, and boundary relations, like it would appear e.g. as term in a GFT expansion. They are quite general, containing the Ponzano-Regge- , Barrett-Crane- and EPRL-FK model (for $\gamma < 1$) as special cases.

In this article we will present a dual version of the OSF in terms of certain holonomies. Introducing a notion of coarse graining, we will also discuss how to compare models on different κ, leading to a notion of renormalization group flow.

2. Operator spin foams

Let κ be a finite, locally finite, oriented 2-complex, and G a compact Lie group. An operator spin foam consists of the following data: an assignment of irreducible, finite-dimensional, unitary representations ρ_f of G to faces f of κ. For each such assignment, denote the Hilbert space

$$\mathcal{H}_e \;=\; V_{\rho_{f_1}} \otimes \cdots V_{\rho_{fn}} \otimes V^*_{\rho_{f_{n+1}}} \otimes \cdots V^*_{\rho_{fm}} \tag{2.1}$$

for each edge e of κ, where $f_1, \ldots, f_n$ are the faces incident to e with agreeing orientation, and $f_{n+1}, \ldots, f_m$ the ones with opposite orientation (see figure 1). For each edge, assign an operator

$$P_e \;:\; \mathcal{H}_e \;\longrightarrow\; \mathcal{H}_e \tag{2.2}$$

Then, at each vertex v of κ, due to the index structure of the operators P_e, there is a canonical way of contracting the indices of all the P_e in order to obtain a number, which is denoted by tr_v. More generally, if κ has a non-empty boundary, which in [3] was defined as the graph γ consisting of all edges e in κ that are bounded by only one face, then the P_e are only assigned to

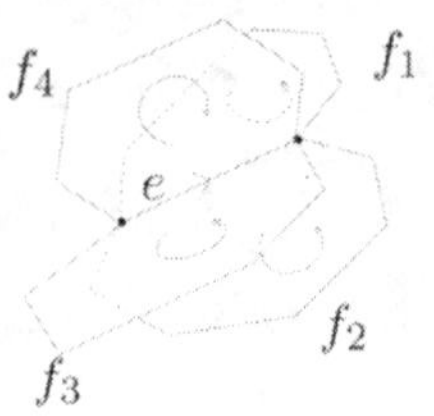

Figure 1. An edge e and four incident faces, with agreeing or opposite orientations.

edges *not* in γ. The index structure is such that contracting over all vertices *not* in γ leads to a linear form on $\mathcal{H}_\gamma$, the space of G-spin networks on γ (with representations assigned to each edge equal to the ρ_f of the unique face f touched by e).

$$Z(\kappa, \rho_f, P_e) = \prod_v \mathrm{tr}_v \left(\otimes_{e\supset v} P_e\right) \ : \ \mathcal{H}_\gamma \to \mathbb{C} \tag{2.3}$$

When the graph $\gamma = \gamma_i \sqcup \gamma_f^*$ is disconnected (where γ_f^* denotes the graph γ_f with all orientations reversed), then $Z(\kappa, \rho_f, P_e) : \mathcal{H}_i \to \mathcal{H}_f$ is interpreted as the transfer operator of the theory.

If P_e is chosen to commute with the group action on each $\mathcal{H}_e$, then the operator $Z(\kappa, \rho_f, P_e)$ becomes gauge-invariant, and $Z(\kappa, \rho_f, P_e)$ is invariant under trivially subdividing a face f by an additional edge. If the P_e are chosen to be self-adjoint, then $Z(\kappa, \rho_f, P_e)$ is invariant under change of an edge-orientation. If it furthermore commutes with the isomorphisms $V_\rho \to V_\rho^*$, it is also invariant under change of a face-orientation. If P_e is chosen to be an orthogonal projector, then $Z(\kappa, \rho_f, P_e)$ is invariant under changing κ by trivially subdividing an edge by an additional vertex.

3. Dual formulation

In the following we introduce a dual formulation of the OSF in terms of certain holonomies.[1] This also allows us to speak of the sum over all ρ_f. Denote the set of edges in κ by E and faces in κ by F. Then define the set

$$E \ltimes F \ := \ \left\{(e, f) \in E \times F \,\middle|\, e \subset \partial f\right\} \tag{3.1}$$

The dual formulation is given in terms of integral over $G^{E\ltimes F}$, i.e. using holonomies $h_{(e,f)}$ with $e \subset \partial f$. Define, for an edge, the function

$$\begin{aligned} C_e(h_{(e,f_1)}, \ldots, h_{(e,f_n)}) \ &:= \ \sum_{\{\rho_{f_k}\}} \left(\prod_{k=1}^m \dim \rho_f\right) \\ &\times \prod_{k=1}^n \rho_k(h_{e,f_k})^{m_k}{}_{n_k} \prod_{k=n+1}^m \rho_k^*(h_{e,f_k})_{m_k}{}^{n_k} \\ &\times (P_e)^{m_1\cdots m_n}{}_{m_{n+1}\cdots m_m, n_1\cdots n_n}{}^{n_{n+1}\cdots n_m} \end{aligned} \tag{3.2}$$

If the norm of the operators P_e does not grow too fast w. r. t. the ρ_f, then (3.2) exists as a distribution on $G^{E\ltimes F}$. For each face f consider a smooth class function $S_f(g) = S_f(hgh^{-1})$, and denote its Fourier coefficients by

$$\hat{S}_f(\rho) \ = \ \int_G dg\, S_f(g) \chi_\rho(g) \tag{3.3}$$

[1] This formulation is a generalization of the holonomy formulations given in [4] for the Barrett-Crane-, and in [5] for the EPRL-FK model.

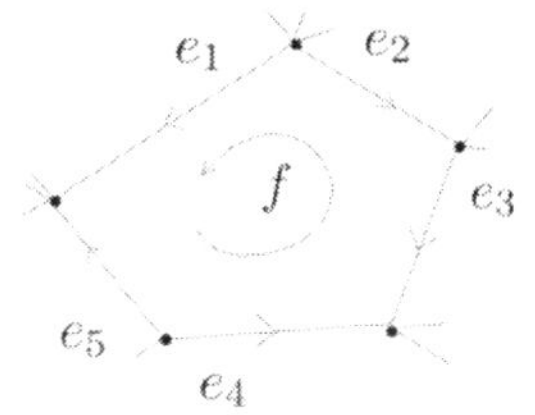

Figure 2. The holonomy g_f around a face f is given by an ordered product of $h_{(e,f)}$, $e \subset \partial f$. In this case $g_f = h_{(e_1,f)}h^{-1}_{(e_2,f)}h^{-1}_{(e_3,f)}h_{(e_4,f)}h^{-1}_{(e_5,f)}$

where χ_ρ denotes the character in the representation ρ. Finally, for a face f denote the holonomy around f by (see figure 2)

$$g_f \;:=\; \overrightarrow{\prod_{e\supset\partial f}} h^{\pm 1}_{(e,f)} \tag{3.4}$$

Then it can be shown that

$$\begin{aligned} Z[\kappa] \;&:=\; \int_{G^{E\ltimes F}} dh_{e,f} \prod_f S(g_f) \prod_e C_e(h_{(e,f_1),\ldots}) \\ &=\; \sum_{\{\rho_f\}} \left(\prod_f \hat{S}_f(\rho_f)\right) Z(\kappa, \rho_f, P_e) \end{aligned} \tag{3.5}$$

Note that formula (3.5), allows to write the sum over spin foam amplitudes as either a sum over representations or integral over $G^{E\ltimes F}$. The $\hat{S}_f$ here act as face amplitudes, and at the same time as regulators (if S_f are chosen to be smooth) to make the sum finite. Note that, in the case of BF theory, when all P_e are the projectors onto the gauge-invariant subspace of $\mathcal{H}_e$, one has (if all orientations of e and f_k agree) that

$$C_e^{(BF)}(h_{(e,f_1)},\ldots) \;=\; \prod_{k=1}^{m-1} \delta(h_{(e,f_k)}h^{-1}_{(e,f_{k+1})}) \tag{3.6}$$

so the integral over $G^{E\ltimes F}$ collapses to an integral over G^E. In this case, one formally also chooses $S_f(g) = \delta(g)$, so the edge amplitudes are given by $\hat{S}_f(\rho) = \dim\rho$, reproducing e.g. the known formulae for the Ponzano-Regge model.

The class of possible models which can arise by choosing different functions S_f and distributions C_e include not only BF theory, but also the BC- or the EPRL-FK-model (for $\gamma < 1$), or the finite group spin foam models introduced in [6].

Note that it is straightforward to include a boundary graph $\gamma \subset \kappa$, in which case the action of the operator $Z[\kappa]$ on a state ψ can be either written in terms of a sum over indices $\psi^{m\cdots}{}_{n\cdots}$, or as integration over a wave function $\psi(h_1,\ldots)$. The formalism can be easily extended to include boundary edges with more than one face touching it. Details will appear in a future publication.

4. Coarse graining

In quantum field theories and statistical systems, coarse graining is a method to construct a theory on a larger (macroscopic) scale from a theory on a finer (microscopic) scale. The result

Journal of Physics: Conference Series 360 (2012) 012042 doi:10.1088/1742-6596/360/1/012042

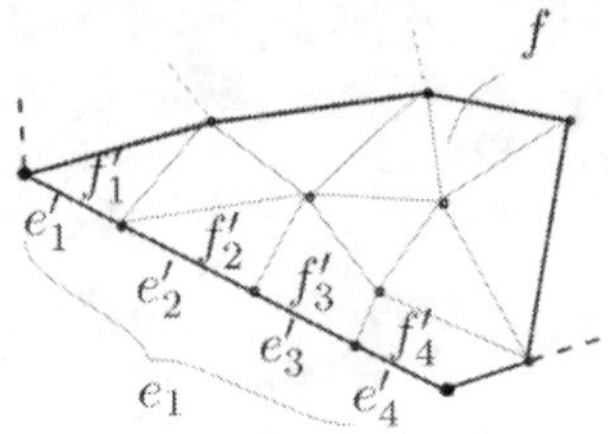

Figure 3. Microscopic holonomies $h'_{(e'_k,f'_k)}$ are composed to macroscopic holonomy $h_{(e,f)}$.

is an effective theory which describes only the behaviour of the long range degrees of freedom within the microscopic theory. By repeated coarse graining one can construct a series of theories, labeled by their characteristic (length or energy) scale. Within e.g. quantum field theory it is desirable to find so-called renormalizable theories. These have the same structure on all scales, differing only by a finite set of parameters, i.e. the coupling constants, which are allowed to depend on the scale.

We now introduce a notion of coarse graining for OSF. The motivation for this is two-fold: First of all, the OSF are candidates for physical theories which are truncated to finitely many degrees of freedom by the 2-complex κ. The setup is tailored to include background-independent theories, such as quantum gravity, so having a notion of coarse graining for these theories can address questions of renormalizability within this context. Secondly, it has been demonstrated that the discretization that lies at the foundation of Spin Foam models, breaks diffeomorphism-invariance [7]. It has also been observed in toy models that one can regain the correct diffeomorphism symmetry *on the discrete level* by following the renormalization trajectory to the fixed point [8]. It is therefore important to investigate renormalization in the Spin Foam context, in order to see whether diffeomorphism symmetry can be restored this way as well.

In the following we will think of the 2-complexes κ as being embedded in a manifold M, since this induces a natural partial ordering on the set of (semianalytic) 2-complexes. Consider two 2-complexes κ and κ' such that every edge/face in κ can be composed of edges/faces in κ'. We then write $\kappa \leq \kappa'$.

Let e be an edge in κ composed of edges e'_k in κ'. Define

$$H_{(e,f)} := h_{(e'_1,f'_1)} \dots h_{(e'_n,f'_n)} \tag{4.1}$$

where the faces f'_k are uniquely determined by the conditions of containing e'_k and being contained in f (see figure 3).

Two OSF models, one defined on κ, with functions S_f and C_e, and the other defined on κ', with functions $S'_{f'}$ and $C'_{e'}$ are said to be *cylindrically consistent* if

$$\mu(h_{(e,f)}) := \prod_f S_f(g_f) \prod_e C_e(h_{(e,f)}, \dots) \tag{4.2}$$

$$\stackrel{!}{=} \int dh'_{(e',f')} \prod_{f'} S'_{f'}(g_{f'}) \prod_{e'} C'_{e'}(h'_{(e',f')}, \dots)$$

$$\times \prod_{(e,f)} \delta\big(h^{-1}_{(e,f)} H_{(e,f)}\big)$$

The construction in (4.2) is such that any function ψ of the $h'_{(e',f')}$ which only depends on the combinations $H_{(e,f)}$ (4.1) have the same expectation values in the model defined on κ and the one on κ', i.e.

$$\int dh_{(e,f)}\mu\,\psi(h_{(e,h)}) \;=\; \int dh'_{(e',f')}\mu'\,\psi(H_{(e,f)}) \tag{4.3}$$

In this sense, the model on κ is a coarse graining of the model on κ', defining an effective theory for the macroscopic degrees of freedom. Some comments are in order:

- The setup can also be applied to abstract 2-complexes, in which case one has to decide what it means that $\kappa \leq \kappa'$. In [9] a notion of refinement has been considered where $\kappa \leq \kappa'$ if each edge/face in κ is already contained in κ'. Note that this is a slightly weaker condition than the one considered in this article.
- Eq. (4.2) can be viewed as renormalization group flow equations for OSF. If one parametrized the set of all OSF by $\{\lambda_n\}$, then (4.2) would be equations relating $\{\lambda_n\}$ and $\{\lambda'_n\}$ on different 2-complexes. Hence, unlike in background-dependent contexts, the $\{\lambda_n\}$ would not depend on just one length scale, but on all of κ. The RG flow would therefore not live on a single trajectory, but on the partially ordered set of all 2-complexes.[2] Note that for these the notion of limit still exists, so it is meaningful to ask where the $\{\lambda_n\}$ are flowing to some fixed point.
- If all the S_f and C_e are positive, then (4.3) is in fact the condition for cylindrical consistency for the measures $\mu\, dh_{(e,f)}$ and $\mu'\, dh'_{(e',f')}$. If these conditions are satisfied on all κ, then the measures can be combined to a measure on the projective limit of all the $G^{E\ltimes F}$, which can be seen as the phase space of the continuum theory. This is in complete analogy to the construction of the space of generalized connections in LQG. In this sense renormalizability is tied to the existence of a continuum limit.
- A priori it is not clear whether, given any OSG on κ', (4.2) is enough to uniquely determine the S_f and C_e on κ, or even that they exist at all. If not, then it might be necessary to go over to more general versions of the OSF, which also include non-localities [11].

5. Summary

We have discussed a holonomy formulation for the OSF models, utilizing holonomies $h_{(e,f)}$ labeled by pairs of edges e and faces f, with $e \subset \partial f$. These are related to the wedge holonomies. With this the OSF can be either formulated in terms of sum over representations or integration over group variables. In the latter form it gives rise to a natural coarse graining procedure, allowing it to compare models on different 2-complexes. We have also commented on how this implies that renormalization group flow for OSF is naturally defined along a partially ordered set, rather than a single scale, and discussed how cylindrical consistency of the models is closely connected to its continuum limit.

[1] B. Bahr, F. Hellmann, W. Kaminski, M. Kisielowski and J. Lewandowski, Class. Quant. Grav. **28** (2011) 105003 [arXiv:1010.4787 [gr-qc]].
[2] C. Rovelli, Class. Quant. Grav. **28** (2011) 114005 [arXiv:1004.1780 [gr-qc]].
[3] W. Kaminski, M. Kisielowski and J. Lewandowski, Class. Quant. Grav. **27** (2010) 095006 [arXiv:0909.0939 [gr-qc]].
[4] H. Pfeiffer, Class. Quant. Grav. **19** (2002) 1109 [gr-qc/0112002].
[5] E. Magliaro and C. Perini, arXiv:1010.5227 [gr-qc].
[6] B. Bahr, B. Dittrich and J. P. Ryan, arXiv:1103.6264 [gr-qc].
[7] B. Bahr and B. Dittrich, Class. Quant. Grav. **26** (2009) 225011 [arXiv:0905.1670 [gr-qc]].
[8] B. Bahr, B. Dittrich and S. Steinhaus, Phys. Rev. D **83** (2011) 105026 [arXiv:1101.4775 [gr-qc]].
[9] C. Rovelli and M. Smerlak, arXiv:1010.5437 [gr-qc].
[10] E. Manrique, R. Oeckl, A. Weber and J. A. Zapata, Class. Quant. Grav. **23** (2006) 3393 [hep-th/0511222].
[11] B. Dittrich, F. C. Eckert and M. Martin-Benito, arXiv:1109.4927 [gr-qc], B. Dittrich and F. C. Eckert, arXiv:1111.0967 [gr-qc].

[2] Ideas in this direction are in fact not new, see e.g. [10] for more on this issue.

Journal of Physics: Conference Series **360** (2012) 012043 doi:10.1088/1742-6596/360/1/012043

Scalar-Tensor theory as a singular subsector of $\Lambda(\phi)$ Plebanski gravity

David Beke

Faculty of Applied Sciences EA16, Gent University, Galglaan 2, 9000 Gent, Belgium

E-mail: david.beke@ugent.be

Abstract. It is shown that, in the absence of matter fields, the coupling of a scalar field to the non-chiral Plebanski action can be obtained by relaxing the trace component of the simplicity constraints. This is realized by considering a subclass of generalized theories, where a potential depending on invariants of the Lagrange multipliers is added to the Plebanski action. Generically such theories propagate eight degrees of freedom, but here the (singular) subclass with the potential only depending on the trace of the Lagrange multiplier is displayed as the Bergmann-Wagoner-Nordtvedt class of scalar-tensor theories.

1. Introduction

In the non-chiral Plebanski formalism, Euclidean general relativity is described by a set of $\mathfrak{so}(4)$ algebra valued bivectors $B^{IJ}_{\mu\nu} = -B^{JI}_{\mu\nu} = -B^{IJ}_{\nu\mu}$, where we use Greek letters $\mu, \nu, ... = 0, 1, 2, 3$ for space-time indices and capital Latin letters $I, J, ... = 0, 1, 2, 3$ for internal indices.[1] These bivectors are put into correspondence with a tetrad by imposing the constraints[2]

$$B^{IJ} \wedge B^{KL} - \frac{1}{24}\epsilon^{IJKL}\epsilon_{MNPQ}B^{MN} \wedge B^{PQ} = 0, \tag{1}$$

which are solved by

$$B^{IJ} = \pm(\star e \wedge e)^{IJ}, \tag{2}$$
$$B^{IJ} = \pm e^I \wedge e^J, \tag{3}$$

where $\star$ is the hodge dual in the Lie algebra. The correct dynamics is obtained by taking the kinetic part of the action to be BF:

$$S_{\text{Pl}}\left[B^{IJ}_{\mu\nu}, A^{IJ}_{\mu}, \phi_{IJKL}\right] = \int B^{IJ} \wedge F_{IJ}(A) - \frac{1}{2}\left(\phi_{IJKL} + \frac{\Lambda}{6}\epsilon_{IJKL}\right) B^{IJ} \wedge B^{KL}. \tag{4}$$

Here, the Lagrange multiplier ϕ^{IJKL} is subject to $\epsilon_{IJKL}\phi^{IJKL} = 0$, such that variation wrt it yields (1). Due to the cubic dependence on the fields, this action has proven useful in the study of quantum gravity (see e.g. [2] for a recent review).

[1] Taking $\mathfrak{so}(3,1)$-valued bivectors gives a description of the Lorentzian theory. We focus on the Euclidean case here, avoiding technicalities arising from reality conditions, which are needed to ensure one is working with a real Lorentzian metric. These will be discussed in [1].

[2] ϵ is the totally antisymmetric symbol, with $\epsilon^{0123} = 1$. We also use $\delta^{IJ}_{KL} = \delta^{[I}_{K}\delta^{J]}_{L}$.

A class of modifications, where the constraints (1) are relaxed by promoting Λ to a function of the scalar invariants of ϕ_{IJKL},

$$S_{\Lambda(\phi)} = \int B^{IJ} \wedge F_{IJ} - \frac{1}{2}\left(\phi_{IJKL} + \frac{\Lambda(\phi)}{6}\epsilon_{IJKL}\right) B^{IJ} \wedge B^{KL}, \tag{5}$$

has appeared in the literature as the gravitational part in the unification of gravity with Yang-Mills theory[3, 4]. A further motivation for studying these theories, is that it has been argued that this class should be closed under the renormalization group [5], suggesting they can be used as an effective action describing quantum gravity. However, such theories are generically plagued by a very basic problem, containing a massless graviton and a massive graviton, coupled to the Boulware-Deser ghost[6, 7, 8]. Here, a singular subclass avoiding these problems is presented. The dynamical degrees of freedom (DOFs) are a massless graviton and a scalar field, and the ghost mode can be avoided. A more detailed discussion will be given in [1].

2. Irreducible decomposition of the simplicity constraints

Using the isomorphism $\mathfrak{so}(4) \cong \mathfrak{so}(3) \oplus \mathfrak{so}(3)$, one can split B^{IJ} in its selfdual and anti-selfdual parts $B^{IJ} = P^{IJ}_{(+)i}B^i_{(+)} + P^{IJ}_{(-)i}B^i_{(-)}$, where the projectors are the 't Hooft symbols, explicitly given by $P^{IJ}_{(\epsilon)i} = \epsilon\delta_0^{[I}\delta_i^{J]} + \frac{1}{2}\epsilon_{0i}{}^{IJ}$. Using the parametrization of $\mathfrak{so}(3)$-valued bivectors [9, 10], B^{IJ} can be parametrized by two tetrads (e^I_μ and $\bar{e}^I_\mu$), two $SL(3,\mathbb{R})$ fields (b^i_j and $\bar{b}^i_j$) and $\eta = \pm 1$, $\bar{\eta} = \pm 1$, in the following way:

$$B^{IJ} = \eta P^{IJ}_{(+),i} b^i_j \Sigma^j_{(+)\mu\nu}[e^I_\mu] + \bar{\eta} P^{IJ}_{(-),i} \bar{b}^i_j \Sigma^i_{(-)\mu\nu}[\bar{e}^I_\mu]. \tag{6}$$

Here $\Sigma^i_{(\epsilon)} = P^i_{(\epsilon)IJ} e^I \wedge e^J$ is a basis for the space of two-forms (anti-)selfdual wrt $e^I_\mu e^J_\nu \eta_{IJ}$. [3]

The lefthandside of equation (1) is valued in $((\mathbf{1},\mathbf{0}) \oplus (\mathbf{0},\mathbf{1})) \otimes_{\text{symm}} ((\mathbf{1},\mathbf{0}) \oplus (\mathbf{0},\mathbf{1}))$ and vanishes upon contraction with ϵ_{IJKL}. Hence it is isomorphic to $(\mathbf{2},\mathbf{0}) \oplus (\mathbf{0},\mathbf{2}) \oplus (\mathbf{1},\mathbf{1}) \oplus (\mathbf{0},\mathbf{0})$. As such, (1) can be decomposed in its irreducible representations:[4]

$$\begin{aligned} 2P^{IJKL}_{(\mathbf{2},\mathbf{0})}{}_{ij} e\left(m^{ij} - \frac{1}{3}m\delta^{ij}\right) + 2P^{IJKL}_{(\mathbf{0},\mathbf{2})}{}_{ij}\bar{e}\left(\bar{m}^{ij} - \frac{1}{3}\bar{m}\delta^{ij}\right) \\ + P^{IJKL}_{(\mathbf{1},\mathbf{1})}{}_{ij} b^i_k \bar{b}^j_l \Sigma^k[e^I_\mu] \wedge \bar{\Sigma}[\bar{e}^I_\mu] + 2P^{IJKL}_{(\mathbf{0},\mathbf{0})}(em - \bar{e}\bar{m}) = 0, \end{aligned} \tag{7}$$

where $m^{ij} = b^i_k b^{jk}$, $e = \det(e^I_\mu)$ and $m = m^i{}_i$. The $(\mathbf{2},\mathbf{0})$, resp. $(\mathbf{0},\mathbf{2})$, components imply, since $\det(m) = \det(b)^2 = 1$, $m^{ij} = \delta^{ij}$, resp. $\bar{m}^{ij} = \delta^{ij}$. The solutions of the $(\mathbf{1},\mathbf{1})$ components are given by $\bar{e}^I_\mu \propto e^I_\mu$. From the scalar component it then follows that $\bar{e}^I_\mu = e^I_\mu$, such that

$$B^{IJ}_{\mu\nu} = \frac{\eta + \bar{\eta}}{2} e^I \wedge e^J + \frac{\eta - \bar{\eta}}{2}(\star e \wedge e)^{IJ}. \tag{8}$$

For $\eta = \bar{\eta}$, one obtains (3) which upon substitution in (4) results in $S = \pm \int e^I \wedge e^J \wedge F_{IJ}[A]$ which is a topological theory. For $\eta = -\bar{\eta}$, solution (2) and consequently the Einstein-Cartan (EC) action is recovered.

[3] One can check that both a rotation and a boost of the tetrad induce a rotation of Σ^i. The 36 components of $B^{IJ}_{\mu\nu}$ are thus decomposed in the 2×16 components of two tetrads and the 2×8 components of the fields with determinant 1, minus $2 \times (3+3)$ components because of rotational invariance in the $\mathfrak{so}(3)$ contractions.

[4] The isomorphisms between the tensors with $\mathfrak{so}(3)$ indices and those with $\mathfrak{so}(4)$ indices are given explicitly by $P^{IJKL}_{(\mathbf{2},\mathbf{0})}{}_{ij} = P^{IJ}_{(+)i}P^{KL}_{(+)j} - \frac{1}{3}\delta_{ij}P^{IJ}_{(+)k}P^{KL}_{(+)k}$, $P^{IJKL}_{(\mathbf{0},\mathbf{2})}{}_{ij} = P^{IJ}_{(-)i}P^{KL}_{(-)j} - \frac{1}{3}\delta_{ij}P^{IJ}_{(-)k}P^{KL}_{(-)k}$, $P^{IJKL}_{(\mathbf{1},\mathbf{1})}{}_{ij} = 2P^{(IJ|}_{(+)i}P^{|KL)}_{(-)j}$, $P^{IJKL}_{(\mathbf{0},\mathbf{0})} = \frac{1}{6}\delta^{IJKL}$.

3. Massive Scalar Tensor theory from modified constraints

A 'weak' relaxation of the constraints is to drop only one: the $(\mathbf{0},\mathbf{0})$ component of (7). In that case $\bar{e}^I_\mu$ and e^I_μ are not equal, but are conformally related to one another:

$$b^i_j = \bar{b}^i_j = \delta^i_j, \quad \bar{e}^I_\mu = \psi e^I_\mu, \tag{9}$$

such that

$$B^{IJ} = \frac{\eta + \bar{\eta}\psi^2}{2}(\star e \wedge e)^{IJ} + \frac{\eta - \bar{\eta}\psi^2}{2} e^I \wedge e^J. \tag{10}$$

Already at this level, one can expect to be dealing with a scalar-tensor theory, where the scalar degree of freedom results from the conformal factor relating the two metrics.

In the class of theories described by (5), this relaxation is achieved by choosing

$$\Lambda(\phi) = \Lambda_0(\mathrm{Tr}(\phi)). \tag{11}$$

As the Hessian $\frac{\partial^2 \Lambda}{\partial \phi_{IJKL} \partial \phi^{MNPQ}} = \Lambda_0''(\mathrm{Tr}(\phi))\delta^{IJKL}\delta_{MNPQ}$ is singular as an operator on $(\mathbf{2},\mathbf{0}) \oplus (\mathbf{0},\mathbf{2}) \oplus (\mathbf{1},\mathbf{1}) \oplus (\mathbf{0},\mathbf{0})$, this theory is not subject to the conclusions of [6] where it was shown that if this Hessian has maximal rank, (5) contains eight DOFs.

Taking Λ as in (11), varying the action (5) wrt ϕ_{IJKL} and taking the $(\mathbf{2},\mathbf{0})$, $(\mathbf{0},\mathbf{2})$ and $(\mathbf{1},\mathbf{1})$ components indeed yields (9). The $(\mathbf{0},\mathbf{0})$ component implies

$$\Lambda_0'(\mathrm{Tr}(\phi)) = \frac{-\delta_{IJKL} B^{IJ} \wedge B^{KL}}{\epsilon_{MNPQ} B^{MN} \wedge B^{PQ}} = -\frac{1-\psi^4}{2(1+\psi^4)} \tag{12}$$

which can be solved for $\mathrm{Tr}(\phi) = \phi(\psi)$ (there are multiple sectors if this solution is not unique). For a quadratic potential $\Lambda_0(\mathrm{Tr}(\phi)) = \frac{a}{16}\mathrm{Tr}^2(\phi)$, one finds a unique solution

$$\mathrm{Tr}(\phi) = -\frac{a}{4}\frac{1-\psi^4}{1+\psi^4}. \tag{13}$$

The equations resulting from varying (5) wrt the connection can be solved for the connection, and upon substitution in the action one obtains, up to a total derivative,

$$\int e \left[\frac{\eta - \bar{\eta}\psi^2}{2} R - 3\bar{\eta}\nabla_\kappa\psi\nabla^\kappa\psi - V_0(\psi)\left(1+\psi^4\right)\right], \tag{14}$$

$$V_0(\psi) = \Lambda_0(\phi(\psi)) + \frac{1}{2}\phi(\psi)\frac{1-\psi^4}{1+\psi^4}: \tag{15}$$

while the choice of Λ_0 does affect the potential terms, the kinetic terms are independent of it.

One can diagonalize the kinetic terms by making a conformal field redefinition of the metric $\hat{g}_{\mu\nu} = |\frac{\eta - \bar{\eta}\psi^2}{2}| g_{\mu\nu}$, such that the action is that of a scalar field minimally coupled to this metric. One obtains, up to total derivatives,

$$\int e \left[\pm R \mp \frac{6\eta\bar{\eta}}{(\eta - \bar{\eta}\psi^2)^2}\nabla_\mu\psi\nabla^\mu\psi - 4\frac{1+\psi^4}{(\eta - \bar{\eta}\psi^2)^2} V_0(\psi)\right], \tag{16}$$

where we have dropped the hats, and where the + sign corresponds to $\eta - \bar{\eta}\psi^2 > 0$ and the - sign to $\eta - \bar{\eta}\psi^2 < 0$.[5] Note that all sectors have three DOFs, both for $\eta = -\bar{\eta}$ and $\eta = \bar{\eta}$, which would correspond to the topological sector of (4). The sector with $\eta = -\bar{\eta}$, having the wrong

[5] Both signs correspond to different sectors of the theory. For brevity, we discuss the sector $\eta - \bar{\eta}\psi^2 > 0$.

sign of the kinetic term, is unstable. The signs of the kinetic terms do not pose problems when $\eta = \bar{\eta}$, and upon making a final field redefinition $\tilde{\psi} = \sqrt{6}\,\mathrm{arctanh}(\psi)$, the canonical form

$$S = \int eR - \nabla_\mu \tilde{\psi} \nabla^\mu \tilde{\psi} - V(\tilde{\psi}) \tag{17}$$

is obtained, with $V(\tilde{\psi}) = 4\cosh^4(\frac{\tilde{\psi}}{\sqrt{6}}) V_0(\tanh(\frac{\tilde{\psi}}{\sqrt{6}}))$. For the quadratic Λ_0 potential, this potential is bounded and given by

$$V(\tilde{\psi}) = -\frac{4}{a} \frac{\left(\sinh^4(\frac{\tilde{\psi}}{\sqrt{6}}) - \cosh^4(\frac{\tilde{\psi}}{\sqrt{6}})\right)^2}{\sinh^4(\frac{\tilde{\psi}}{\sqrt{6}}) - \cosh^4(\frac{\tilde{\psi}}{\sqrt{6}})} \tag{18}$$

In [1] it will be shown that the introduction of the cosmological constant and/or the Immirzi parameter affects this potential in such a way as to reproduce the scalar field dark matter (SFDM) phenomenology discussed in [11, 12].

4. Coupling to matter

To extract physical predictions, the coupling of matter to this scalar-tensor theory has to be specified. Tensor matter fields Q are coupled to (5) by adding a term $S[B, Q]$. If one is to recover minimally coupled matter in the Plebanski limit $\Lambda(\phi) \to \Lambda$, this term has to satisfy

$$S(\star e \wedge e, Q) = S_m(g_{\mu\nu}, Q), \tag{19}$$

with S_m the minimally coupled matter action. This condition guarantees that substituting (2) in the action yields matter minimally coupled to the EC action. We propose the following coupling term:[6]

$$S(B, Q) = S_m(\frac{\eta}{2} e^I_\mu e^J_\nu \delta_{IJ} - \frac{\bar{\eta}}{2} \bar{e}^I_\mu \bar{e}^J_\nu \delta_{IJ}, Q). \tag{20}$$

It is precisely this metric $\eta e^I_\mu e^J_\nu \delta_{IJ} - \bar{\eta} \bar{e}^I_\mu \bar{e}^J_\nu \delta_{IJ}$ which has been identified as the massless graviton DOF [7, 8]. In the scalar-tensor subclass this implies a coupling to (14) with a term

$$S_m(\frac{\eta - \bar{\eta}\psi^2}{2} g_{\mu\nu}, Q). \tag{21}$$

Making the conformal transformation as in (16), one finds

$$S = \int eR - \nabla_\mu \tilde{\psi} \nabla^\mu \tilde{\psi} - V(\tilde{\psi}) + S_m(g_{\mu\nu}, Q). \tag{22}$$

The scalar field is decoupled from matter fields, which is the coupling needed to obtain SFDM phenomenology.

5. Conclusion

It has been discussed that $\Lambda(\phi)$ Plebanski gravity, which generically has an unstable bi-metric nature, contains a subclass in which the two metrics are conformally related to one another. This class has been shown to be equivalent to a subclass of the Bergmann-Wagoner-Nordtvedt scalar-tensor theories. As a technical result, this tells that if one is studying gravity coupled to a scalar field, this coupling does not need to be done by hand: it suffices to relax the $(\mathbf{0}, \mathbf{0})$ components of the constraints. As a physical result, this shows that, if a more rigorous relation between $\Lambda(\phi)$ Plebanski gravity and effective quantum gravity actions can be established, the phenomenological consequences of scalar-tensor theories, such as SFDM and inflation[14], can turn out to have a quantum origin.

[6] This is not the only coupling term satisfying (19), e.g. the form proposed for Plebanski theory in [13] also achieves this. In [1] we will comment on other possible choices, which generically have stability issues not present in (22).

Acknowledgements

We are very happy to thank S. Speziale for cooperation in the work summarized here, and N. Van den Bergh for many stimulating discussions. The author is supported by the Research Foundation-Flanders (FWO).

References

[1] Beke D and Speziale S *in preparation*
[2] Rovelli C 2011 arXiv:1102.3660 [gr-qc]
[3] Smolin L 2009 *Phys. Rev. D* **80** 124017 arXiv:0712.0977 [hep-th]
[4] Torres-Gomez A and Krasnov K 2010 *Phys. Rev. D* **81** arXiv:0911.3793 [hep-th]
[5] Krasnov K 2006 arXiv:hep-th/0611182
[6] Alexandrov S and Krasnov K 2009 *Class. Quant. Grav.* **26** 055005 arXiv:0809.4763 [gr-qc]
[7] Speziale S 2010 *Phys. Rev. D* **82** 064003 arXiv:1003.4701 [hep-th]
[8] Beke D, Palmisano G and Speziale S arXiv:1112.4051 [hep-th]
[9] Krasnov K 2008 *Class. Quant. Grav.* **25** 025001 arXiv:gr-qc/0703002
[10] Freidel L 2008 arXiv:0812.3200 [gr-qc]
[11] Matos T and Ureña-López L 2000 *Class. Quant. Grav.* **17** L75 arXiv:astro-ph/0004332
[12] Sahni V and Wang L 2000 *Phys. Rev. D* **62** 103517 arXiv:astro-ph/9910097
[13] Tennie F and Wohlfarth M 2010 *Phys. Rev. D* **82** 104052 arXiv:1009.5595 [gr-qc]
[14] La D and Steinhardt P 1989 *Phys. Rev. Lett.* **62**(4) 376–378

Journal of Physics: Conference Series **360** (2012) 012044 doi:10.1088/1742-6596/360/1/012044

The time-oriented boundary states and the Lorentzian-spinfoam correlation functions

Eugenio Bianchi[1,2] **and You Ding**[1,3]

[1] Centre de Physique Théorique de Luminy, Case 907, F-13288 Marseille, France

E-mail: ebianchi@perimeterinstitute.ca, dingy@bjtu.edu.cn

Abstract. A time-oriented semiclassical boundary state is introduced to calculate the correlation function in the Lorentzian Engle-Pereira-Rovelli-Livine spinfoam model. The resulting semiclassical correlation function is shown to match with the one in Regge calculus in a proper limit.

To test the semiclassical behavior of the Engle-Pereira-Rovelli-Livine (EPRL) spinfoam amplitude [1], we compute the two-point correlation functions for the Penrose metric operator in the Lorentzian signature. The setting is the one introduced in [2] and developed in [3, 4]. In particular, the correlation functions are calculated on a *time-oriented* semiclassical state [4], which is peaked on the space-like boundary geometry of a Lorentzian 4-simplex.

1. The time-oriented boundary states

The boundary of a Lorentzian 4-simplex consists of five space-like tetrahedra a which meet at ten triangles (ab) $(a, b = 1...5$ and $a < b)$. Suppose all the time-like normals to the tetrahedra are outward-pointing, then the tetrahedra can be divided into two types: the time-like normals to the tetrahedra are either future-pointing or past-pointing. Each triangle, with the two adjoining tetrahedra, defines a *wedge*. These wedges are then classified into two types: it is called in [5, 6] *thick wedge* if the incident tetrahedra are of same pointing type, which means both future-pointing or both past-pointing, otherwise called *thin wedge*. Let us assign a quantity to the triangle (ab) to denote this classification:

$$\Pi_{ab} = \begin{cases} 0, & \text{thick wedge;} \\ \pi, & \text{thin wedge.} \end{cases} \tag{1}$$

Consider the complete graph Γ dual to the boundary of the 4-simplex, with five nodes a dual to tetrahedra and ten links (ab) dual to the corresponding meeting triangles. Consider the group $SU(2)$, the *spin network states* $|\Gamma, j_{ab}, i_a\rangle$ supported on this graph are given by coloring each link (ab) with an irreducible representation j_{ab} (a spin), and coloring each node a with an $SU(2)$ intertwiner i_a. These spin network states span the truncated $SU(2)$ Hilbert space $\mathcal{H}_\Gamma$ of loop quantum gravity (LQG). There is another (overcomplete) basis of $\mathcal{H}_\Gamma$, which

[2] Present address: Perimeter Institute, 31 Caroline St N, Waterloo ON, Canada N2L 2Y

[3] Present address: Department of Physics, Beijing Jiaotong University, 100044 Beijing, China

is *Lorentzian coherent spin network states* with nodes labeled by *Lorentzian Livine-Speziale coherent intertwiners*[4]:

$$|\Gamma, j_{ab}, \Upsilon_a(\vec{n})\rangle = \exp\,(-i \sum_{(a<b)} \Pi_{ab} j_{ab}) |\Gamma, j_{ab}, \Phi_a(\vec{n})\rangle. \tag{2}$$

Notation is as follows. On the right hand side, $\Phi_a(\vec{n})$ denotes the Euclidean coherent intertwiner between the representations $j_{ab}(b \neq a)$ [7],

$$\Phi_a(\vec{n}) = \int_{SU(2)} \mathrm{d}h \prod_{b\neq a} \langle j_{ab}, m_{ab}|h|j_{ab}, \vec{n}_{ab}\rangle \tag{3}$$

where the integral measure $\mathrm{d}h$ is the $SU(2)$ Haar measure and $|j_{ab}, \vec{n}_{ab}\rangle$ denotes the $SU(2)$ coherent state in the spin-j representation to the link (ab) of the boundary graph Γ; these coherent states $|j, \vec{n}\rangle$ are peaked on the geometry of a classical triangle: $\vec{n}$ are associated to unit-normals to triangles of a tetrahedron, and j areas of the triangles. We assume all the normals outward to the tetrahedron, which satisfy $j_1\vec{n}_1 + j_2\vec{n}_2 + j_3\vec{n}_3 + j_4\vec{n}_4 = 0$, thus we associate to each triangle (ab) normals $-\vec{n}_{ab}$ when a is target of the triangle and respectively $\vec{n}_{ba}$ when b is the source. Given a tetrahedron a, the *Lorentzian-geometry phase* $\exp\,(-i\sum_{b>a}\Pi_{ab}j_{ab})$ maps the Euclidean coherent intertwiners into the Lorentzian ones [4]:

$$\Upsilon_a(\vec{n}) = \exp\,(-i \sum_{b>a} \Pi_{ab} j_{ab}) \Phi_a(\vec{n}), \tag{4}$$

where the Lorentzian coherent intertwiner $\Upsilon(\vec{n})$ in the l.h.s. is defined by time reversing the past-pointing tetrahedra of the Euclidean coherent intertwiner $\Phi(\vec{n})$:

$$\Upsilon(\vec{n}) := \begin{cases} \Phi(\vec{n}), & \text{for future-pointing tetrahedra;} \\ T\Phi(\vec{n}), & \text{for past-pointing tetrahedra.} \end{cases} \tag{5}$$

Given a Euclidean coherent intertwiner in equation (3), the effect of time reversal T on the (past-pointing) coherent intertwiner is defined as

$$T\Phi_a(\vec{n}) := \int_{SU(2)} \mathrm{d}h \prod_{b\neq a} \langle j_{ab}, m_{ab}|h\widetilde{T}_{ab}|j_{ab}, \vec{n}_{ab}\rangle, \tag{6}$$

where "thickener" $\widetilde{T}_{ab}$ associated to triangle (ab) time-reverses thin wedges but unchanges thick wedges:

$$\widetilde{T}_{ab}|j_{ab}, \vec{n}_{ab}\rangle := \begin{cases} |j_{ab}, \vec{n}_{ab}\rangle, & \text{for thick wedge;} \\ (-1)^{j_{ab}}|j_{ab}, \vec{n}_{ab}\rangle, & \text{for thin wedge.} \end{cases} \tag{7}$$

The action (7) of the *thickener* $\widetilde{T}$ can be reexpressed as

$$\widetilde{T}_{ab}|j_{ab}, \vec{n}_{ab}\rangle = \exp(-i\Pi_{ab}j_{ab})|j_{ab}, \vec{n}_{ab}\rangle, \tag{8}$$

with Π_{ab} related to the Lorentzian geometry by equation (1), and thus equation (4) can be obtained by combining equations (5)-(8). The Lorentzian coherent spin network state $|\Gamma, j_{ab}, \Upsilon_a(\vec{n})\rangle$ with nodes labeled by Lorentzian coherent intertwiners $\Upsilon_a(\vec{n})$ is thus given by equation (2).

Before going to use the Lorentzian coherent spin network states to construct the boundary semiclassical states, we want to make clear two points: (i)The Lorentzian coherent intertwiner is obtained by time-reversing the past-pointing tetrahedra and unchanging the future-pointing ones; the effect of the time reversal on tetrahedra is given by thickener $\widetilde{T}_{ab}$ on each triangle (ab). The action (7) of thickener $\widetilde{T}_{ab}$ is defined by "time-reversing" thin wedges and unchanging thick wedges. The action of the time-reversal on the thin wedge is motivated from the time reversal $T|j,\vec{n}\rangle = (-1)^j|j,-\vec{n}\rangle$ in quantum mechanics. However, there is a slight difference between them of the minus sign for $\vec{n}$. This minus sign is supposed to keep normals outward-pointing after time reversal. (ii)The two phases from the intertwiner relation (4) and from the spin network relation (2) are different, in the sense that a is fixed in the former but free in the latter. In fact, the latter phase is product of the five phase in the former form. However, here, we call both of them *Lorentzian-geometry phase*, if no confusion arises.

Now let us come to the Lorentzian semiclassical state superposed by the Lorentzian coherent spin network states. Let ϕ_o label the simplicial extrinsic curvature, which is an angle associated to the triangle shared by the tetrahedra; a Lorentzian semiclassical state peaked both on intrinsic and extrinsic geometry can be given by a superposition of Lorentzian coherent spin network states:

$$|\Psi_o\rangle = \sum_{j_{ab}} \psi_{j_o,\phi_o}(j)|j,\Upsilon(\vec{n})\rangle\,, \tag{9}$$

with coefficients $\psi_{j_o,\phi_o}(j)$ given by a gaussian times a phase [2],

$$\psi_{j_o,\phi_o}(j) = \exp\Big(-i\sum_{ab}\gamma\phi_o^{ab}\,(j_{ab}-(j_o)_{ab})\Big)\times\exp\Big(-\sum_{ab,cd}\gamma\alpha^{(ab)(cd)}\,\frac{j_{ab}-(j_o)_{ab}}{\sqrt{(j_o)_{ab}}}\,\frac{j_{cd}-(j_o)_{cd}}{\sqrt{(j_o)_{cd}}}\Big)\,, \tag{10}$$

where the 10×10 matrix $\alpha^{(ab)(cd)}$ is assumed to be complex with positive definite real part. In the following we will use the Lorentzian semiclassical state (9) to calculate the Lorentzian two-point correlation function.

2. The EPRL correlation functions

The connected two-point correlation function G^{abcd}_{nm} on a semiclassical boundary state $|\Psi_o\rangle$ is defined as

$$G^{abcd}_{nm} = \langle E^a_n\!\cdot\! E^b_n\; E^c_m\!\cdot\! E^d_m\rangle - \langle E^a_n\!\cdot\! E^b_n\rangle\,\langle E^c_m\!\cdot\! E^d_m\rangle\,, \tag{11}$$

where $(E^a_n)_i$ is a flux operator through a surface f_{an} dual to the triangle between the tetrahedra a and n, parallel transported in the tetrahedron n. Here the dynamical expectation value of an operator $\mathcal{O}$ on the semiclassical state $|\Psi_o\rangle$ is defined via

$$\langle\mathcal{O}\rangle = \frac{\langle W|\mathcal{O}|\Psi_o\rangle}{\langle W|\Psi_o\rangle}\,, \tag{12}$$

where W is the EPRL spinfoam amplitude [1]. The correlation function (11) can be reexpressed in terms of group integral [4]:

$$G^{abcd}_{nm} = \frac{\sum_j \psi_j \int \mathrm{d}^4g\,\mathrm{d}^{10}z\; q^{ab}_n q^{cd}_m e^S}{\sum_j \psi_j \int \mathrm{d}^4g\,\mathrm{d}^{10}z\; e^S} - \frac{\sum_j \psi_j \int \mathrm{d}^4g\,\mathrm{d}^{10}z\; q^{ab}_n e^S}{\sum_j \psi_j \int \mathrm{d}^4g\,\mathrm{d}^{10}z\; e^S}\,\frac{\sum_j \psi_j \int \mathrm{d}^4g\,\mathrm{d}^{10}z\; q^{cd}_m e^S}{\sum_j \psi_j \int \mathrm{d}^4g\,\mathrm{d}^{10}z\; e^S}. \tag{13}$$

Notation is as follows. The coefficient ψ_j is short for $\psi_{j_o,\phi_o}(j)$ given in equation (10). The group integral d^4g is short for the Haar measure of $SL(2,\mathbb{C})^4$, one per each tetrahedron, and one redundant integral is removed to obtain the finite integral [8]. We consider an irreducible representation of the Lorentz group $SL(2,\mathbb{C})$ on the space $\mathcal{H}_{(k,p)}$ of homogeneous functions $f^j_m(z)^{(k,p)}$ on *complex projective line* $\mathcal{P}$, and the integral $d^{10}z$ is over one $\mathcal{P}$ per each triangle, up to a factor $-(\langle Z_{ab}, Z_{ab}\rangle\langle Z_{ba}, Z_{ba}\rangle)^{-1}$, with $Z_{ab} = g_a^\dagger z$ and $Z_{ba} = g_b^\dagger z$. The "action" S is given by

$$S(g,z) = \sum_{(a<b)} \left(j_{ab} \log \frac{\langle J\xi_{ab}, Z_{ab}\rangle^2 \langle Z_{ba}, \xi_{ba}\rangle^2}{\langle Z_{ab}, Z_{ab}\rangle\langle Z_{ba}, Z_{ba}\rangle} + i\gamma j_{ab} \log \frac{\langle Z_{ba}, Z_{ba}\rangle}{\langle Z_{ab}, Z_{ab}\rangle} - i\Pi_{ab} j_{ab} \right), \qquad (14)$$

with γ the Barbero-Immirzi parameter; the spinor $\xi \in \mathbb{C}^2$ is related to the coherent state $|j,\vec{n}\rangle$ in the sense that the corresponding $SU(2)$ group element $n(\xi)$ rotate the direction $(0,0,1)$ in sphere $\mathbb{S}^2$ into the $\vec{n}$ direction. The antipodal vector $-\vec{n}(\xi)$ can be associated to $J\xi$, i.e. $-\vec{n}(\xi) = \vec{n}(J\xi)$. The insertion q_n^{ab} is given by

$$q_n^{ab} = \gamma^2 j_{an} j_{bn} \frac{\langle \vec{\sigma} Z_{an}, \xi_{an}\rangle}{\langle Z_{an}, \xi_{an}\rangle} \cdot \frac{\langle \vec{\sigma} Z_{bn}, \xi_{bn}\rangle}{\langle Z_{bn}, \xi_{bn}\rangle}. \qquad (15)$$

The large-spin asymptotics of the correlation function (13) can be obtained via stationary phase approximation [3, 6], which is given by (for details, see [4]):

$$G^{abcd}_{nm}(\alpha) = (\gamma j_o)^3 (R^{abcd}_{nm}(\alpha) + O(\gamma)) \qquad (16)$$

with

$$R^{abcd}_{nm} = \frac{1}{\gamma^3 j_o^3} \sum_{p<q, r<s} Q^{-1}_{(pq)(rs)} \frac{\partial q_n^{ab}}{\partial j_{pq}} \frac{\partial q_m^{cd}}{\partial j_{rs}} \qquad (17)$$

and

$$Q_{(ab)(cd)} = -\frac{\gamma \alpha^{(ab)(cd)}}{\sqrt{(j_o)_{ab}}\sqrt{(j_o)_{cd}}} + S''_{\text{Regge}}. \qquad (18)$$

If we take the classical limit, introduced in [3], where the Barbero-Immirzi parameter is taken to zero $\gamma \to 0$, and the spin of the boundary state is taken to infinity $j \to \infty$, keeping the size of the quantum geometry $A \sim \gamma j$ finite and fixed, the two-point function (16) we obtain exactly matches the one obtained from Lorentzian Regge calculus [9].

Deriving the LQG correlation function at the level of a single spin foam vertex is certainly only a first step. Within the setting of a vertex expansion, an analysis of the LQG correlation function for an arbitrary number of spinfoam vertices is needed. It would be interesting to investigate the contribution of the γ-term to correlation functions when more than a single spin foam vertex is considered.

References

[1] Engle J, Livine E, Pereira R and Rovelli C 2008 *Nucl. Phys.* **B799** 136–149 (*Preprint* 0711.0146)
[2] Rovelli C 2006 *Phys. Rev. Lett.* **97** 151301 (*Preprint* gr-qc/0508124)
[3] Bianchi E, Magliaro E and Perini C 2009 *Nucl. Phys.* **B822** 245–269 (*Preprint* 0905.4082)
[4] Bianchi E and Ding Y 2011 (*Preprint* 1109.6538)
[5] Barrett J W and Foxon T J 1994 *Class. Quant. Grav.* **11** 543–556 (*Preprint* gr-qc/9310016)
[6] Barrett J W, Dowdall R J, Fairbairn W J, Hellmann F and Pereira R 2010 *Class. Quant. Grav.* **27** 165009 (*Preprint* 0907.2440)
[7] Livine E R and Speziale S 2007 *Phys. Rev.* **D76** 084028 (*Preprint* 0705.0674)
[8] Engle J and Pereira R 2009 *Phys. Rev.* **D79** 084034 (*Preprint* 0805.4696)
[9] Regge T 1961 *Nuovo Cim.* **19** 558–571

Singular topologies in the Boulatov model

Sylvain Carrozza
Max Planck Institute for Gravitational Physics, Albert Einstein Institute, Am Mühlenberg 1, 14476 Golm, Germany, EU
Laboratoire de Physique Théorique, CNRS UMR 8627, Université Paris XI, F-91405 Orsay Cedex, France, EU
E-mail: `sylvain.carrozza@aei.mpg.de`

Abstract. Through the question of singular topologies in the Boulatov model, we illustrate and summarize some of the recent advances in Group Field Theory.

1. Introduction

Group Field Theories[1] (GFTs) are genuine quantum field theories defined on group manifolds, especially relevant to quantum gravity in general, and in particular to the question of the continuum limit in Spin Foam models (SF). This is mainly because GFTs generate SF models as Feynman amplitudes in their perturbative expansions, thus bringing additional tools and concepts to define and understand the sum over foams. But interestingly, GFTs can also be seen as a generalization of matrix models [2] (and a particular class of tensor models), which were particularly successful as discrete models of 2d quantum gravity: a continuum regime could be found that matches exactly the continuum theory. This link with matrix models has recently triggered much progress in GFT and tensor models in general[2]. Focusing on the Boulatov model, dual to the Ponzano-Regge model for euclidean 3d quantum gravity, we illustrate some of these results and their consequences for our understanding of singular topologies in GFT.[3]

2. Boulatov model and singular topologies

2.1. Original model

The Boulatov model can be defined by the action:

$$S[\varphi] = \int [\mathrm{d}g]^3 \varphi(g_1, g_2, g_3)^2 + \lambda \int [\mathrm{d}g]^6 \varphi(g_1, g_2, g_3)\varphi(g_3, g_5, g_4)\varphi(g_5, g_2, g_6)\varphi(g_4, g_6, g_1)\,, \qquad (1)$$

where φ is a real field on $\mathrm{SU}(2)^3$, required to be invariant under the diagonal action of SU(2):

$$\forall h \in \mathrm{SU}(2), \qquad \varphi(hg_1, hg_2, hg_3) = \varphi(g_1, g_2, g_3). \qquad (2)$$

This field has the geometrical interpretation of a quantized triangle, the variable g_i being understood as the holonomy encoding parallel transport from the centre of the triangle to the

[1] See [1] for a recent review.
[2] See the review [6], and references therein.
[3] The present article is a summary of the motivations and results of the recent work [5], where more technical details can be found.

center of the edge i. In this interpretation, the interaction term in the action encodes the gluing of four quantum triangles along their boundary edges so as to form a tetrahedron, whereas the (trivial) kinetic term identifies two triangles. When formally expanding the (euclidean) path integral:

$$\mathcal{Z} = \int \mathrm{d}\mu(\varphi)\, \mathrm{e}^{-S[\varphi]} \tag{3}$$

in powers of λ, one gets Feynman amplitudes labelled by simplicial complexes. Their building blocks are tetrahedra, to which we associate the kernel of the interaction part of S, glued together through their boundary triangles. The propagator (kernel of the kinetic part of the action, supplemented by a projection on gauge invariant fields (2)) encodes these gluings. Alternatively, these simplicial complexes are represented by dual 3-stranded graphs, where a line of the graph corresponds to a triangle of the simplicial complex, and its three strands are dual to the edges of the same triangle. In a graph $\mathcal{G}$, one can identify a 2-complex structure (the faces being the closed chains of strands) and show that the GFT amplitude of a simplicial complex is equal to the Ponzano-Regge amplitude of its dual 2-complex:

$$\mathcal{A}^{\mathcal{G}} = \int [dh_t] \prod_e \delta\left(H_e\right) . \tag{4}$$

This formula is written in terms of simplicial data: to a triangle t is associated an holonomy h_t (representing parallel transport along the dual edge in the 2-complex)[4], e is an edge, and H_e is the holonomy around its dual face. It is defined as an ordered product of group elements h_t, associated to the triangles which contain the edge e: $H_e = \overrightarrow{\prod_{t \supset e}} h_t$.

As in ordinary quantum field theories, the amplitudes are divergent, so the path integral needs to be regulated. This is usually achieved via a regularization of the δ-function on SU(2), for instance by cutting-off high spins in its Peter-Weyl expansion:

$$\delta^{\Lambda}(g) \equiv \sum_{j \in \mathbb{N}/2, j \leq \Lambda} (2j+1)\chi^j(g)\,, \tag{5}$$

where χ^j are the characters of SU(2). Divergences can then be expressed in powers of the value of the regulated δ-function at the identity: $\delta^{\Lambda}(\mathbb{1}) \sim \Lambda^3$.

The simplicial complexes generated by the GFT represent (virtual) histories of discrete spaces (here triangulated surfaces), and are the structures from which we hope to recover known (and hopefully unknown) spacetime physics. On the way to the emergence of a continuum spacetime from GFT, with the full structure of differential manifold, it is reasonable to first focus on more primitive aspects, such as topology. Namely, we would like to show that in some regime of the GFT, the simplicial complexes that dominate the perturbative expansion are regular topological manifolds, i.e. topological spaces in which the neighborhood of any point has trivial topology (a point that fails to have this property is called a topological singularity). The problem is that GFTs generically generate very complicated topological spaces[5], with pointlike singularities as well as extended singularities (for example a one dimensional singular subspace) [7].

Two complementary strategies have been successfully used to control the singular topologies in the Boulatov model, which we describe in the following: the first consists in imposing suitable combinatorial conditions that restrict the class of simplicial complexes summed over in the Feynman expansion; and the second amounts to find scaling bounds showing that the remaining singular complexes are reasonably suppressed with respect to the leading order.

[4] h_t is also the auxiliary variable appearing in the gauge condition (2).
[5] The topological structure used is the obvious one, induced by the identifications of the boundaries of the tetrahedra, themselves having the topology of a ball in $\mathbb{R}^3$.

2.2. Colored model

Colored models were originally introduced [8] as a general prescription allowing to restrict, in a controlled way, the class of simplicial complexes generated by GFTs. In the Boulatov case, it amounts to replacing the real field φ by four complex fields φ_ℓ, indexed by a color label $\ell \in \{1, \cdots, 4\}$. The gauge condition (2) is unchanged, and the colored action is defined as:

$$S_{col.}[\varphi, \overline{\varphi}] = \int |\varphi|^2 + \lambda \int \varphi_1 \varphi_2 \varphi_3 \varphi_4 + \text{c.c.}, \tag{6}$$

where the convolutions of fields follow the same patterns as in the uncolored model (1). The only effect of the coloring is to reduce the combinatorial complexity of the Feynman graphs generated by the GFT. But, and this is essential, the amplitude of a simplicial complex generated by the colored model is also equal to the Ponzano-Regge partition function.

It can be shown that no simplicial complex with extended singularities is generated by the colored model [8]. There are still singular simplicial complexes appearing in the expansion, but they have only pointlike singularities (located at the vertices): this type of topological spaces, which are regular except on a discrete set, are called pseudomanifolds. Their analysis will be reviewed in the next section.

But before that, we need to mention two additional and spectacular properties of the colored model, that are not available in the uncolored version. The first is that the cut-off Λ can be used to define a new perturbative expansion [9], exact analog of the $1/N$ expansion in matrix models, that was the essential ingredient missing so far to reproduce their success in higher dimensions. It allowed to prove that, upon a suitable rescaling of the coupling constant $\lambda \to \lambda/\sqrt{\delta^\Lambda(\mathbb{1})}$, the partition function can be expanded as [10]:

$$\mathcal{Z} = [\delta^\Lambda(\mathbb{1})]^2 \mathcal{Z}_0(\lambda\overline{\lambda}) + \mathcal{O}([\delta^\Lambda(\mathbb{1})]), \tag{7}$$

where $\mathcal{Z}_0$ contains the contributions of a simple subclass of triangulations of the 3-sphere. Therefore, leading order contributions are not only manifolds, but also have trivial topology.

The second breakthrough [4] (that was again only made possible by the introduction of colors) is the realization of simplicial diffeomorphism symmetry, defined at the level of the amplitudes, as a symmetry of the GFT action itself. Interestingly for us, this symmetry acts on vertices of the triangulations, that is exactly where potential singularities are located.

3. Vertex representation

3.1. Vertex variables

As far as the symmetries of the Boulatov model are concerned, the natural variables in a quantum triangle $\varphi(g_1, g_2, g_3)$ are the $G_{ij} \equiv g_i^{-1} g_j$ [3, 4]. It is therefore natural to look for a formulation of the model in which the fields are defined in such variables. Thanks to the gauge invariance (2), this transformation can actually be performed, and was extensively studied in [5]. One obtains a theory with four complex fields $\psi_\ell(G_u, G_v, G_w)$, and the same structure of colored action (6), but with two notable differences: a) the gauge condition (2) is traded for a closure condition, encoded by a distributional factor $\delta(G_u G_v G_w)$ in the propagator; b) the interaction term encodes the tetrahedral geometry through identifications of vertices common to different triangles (as opposed to edges in the usual formulation), which is reflected in a stranded structure with 3-valent interactions.

The main advantage of this formulation is that, for any color ℓ, it allows to rewrite the amplitudes as integrals of the form:

$$\mathcal{A}^\mathcal{G} \propto \int [\mathrm{d}G] \left(\prod_{b \in \mathcal{B}_\ell} \prod_{v \in V_b} \delta\left(H_{v,b}\right) \right) \left(\prod_{t_\ell} \delta\left(G_u^{t_\ell} G_v^{t_\ell} G_w^{t_\ell}\right) \right), \tag{8}$$

where: $b \in \mathcal{B}_\ell$ are closed triangulated surfaces around vertices of color ℓ, called bubbles; $v \in V_b$ is a vertex of such a surface; $H_{v,b}$ is the holonomy around the vertex v in b; and t_ℓ is summed over all the triangles of color ℓ in the simplicial complex.

This expression as a very nice geometrical interpretation, describing the effective physics of the field of color ℓ: the bubble terms represent triangulated boundaries of effective flat cells (first factor), consistently glued through propagators of color ℓ (second factor).

3.2. Optimal scaling bounds

In the cut-off theory (and with the scaling of λ ensuring the existence of a large Λ expansion), the explicit factorization of bubble contributions (8) allows to prove the following scaling bounds:

$$\mathcal{A}^{\mathcal{G}} = \mathcal{O}\left([\delta^\Lambda(\mathbb{1})]^{2-2\sum_{b\in\mathcal{B}_\ell} g_b} \right) , \tag{9}$$

where g_b is the genus of the (closed and orientable) triangulated surface b. Remarkably $g_b \in \mathbb{N}$, and $g_b = 0$ if and only if b is a triangulation of the sphere. But precisely, a given vertex in the simplicial complex associated to $\mathcal{G}$ is topologically regular whenever the dual bubble b is spherical. So (9) is a bound on singular topologies, that was moreover proven to be optimal [5]: for any integers $(g_1, \cdots, g_n)$, there exists a simplicial complex which has exactly n bubbles of color ℓ and genera $(g_1, \cdots, g_n)$, and whose amplitude diverges like $[\delta^\Lambda(\mathbb{1})]^{2-2\sum_{i=1}^n g_i}$. This bound is indeed sharper than what was previously known, since any singular topology will have an amplitude in at most $\mathcal{O}(1)$, and not only $\mathcal{O}([\delta^\Lambda(\mathbb{1})])$ as can be deduced from (7).

4. Conclusion and outlook

In this short article, we briefly reviewed the question of singular topologies in the Boulatov model, which is in tension with the discrete spacetime interpretation. This gives an interesting point of view on recent developments in GFT. For instance, a colored model is required both as a way to get rid of very singular simplicial complexes, but also to realize simplicial symmetries as field symmetries, which in turn are essential to understand the left-over singular topologies. The optimal bounds arising from the vertex reformulation also justify by themselves the rescaling of λ used in the Λ expansion [9]: this is the unique scaling for which they can be interpreted as a hierarchical set of bounds that tame the contribution of singularities. Finally, they allow to identify the precise order in Λ at which the interpretation of the Boulatov model as a sum over discrete spacetimes ceases to make sense. The good news is that for the leading order (in $[\delta^\Lambda(\mathbb{1})]^2$), as well as all corrections in $[\delta^\Lambda(\mathbb{1})]^\gamma$ with $\gamma > 0$, this interpretation is available.

This analysis can be generalized to 4d topological models [11]. Whether this applies also to 4d gravity models needs to be investigated.

References

[1] D. Oriti, in *Foundations of space and time*, G. Ellis, J. Murugan (eds.), Cambridge University Press, Cambridge (2011). [arXiv:1110.5606 [hep-th]].
[2] F. David, Nucl. Phys. B257, **45** (1985); P. Ginsparg, 'Matrix models of 2-d gravity', [arXiv: hep-th/9112013]; P. Di Francesco, P. H. Ginsparg and J. Zinn-Justin, Phys. Rept. **254**, 1 (1995), [arXiv: hep-th/9306153].
[3] F. Girelli and E. R. Livine, Class. Quant. Grav. **27**, 245018 (2010) [arXiv:1001.2919 [gr-qc]].
[4] A. Baratin, F. Girelli and D. Oriti, Phys. Rev. D **83**, 104051 (2011) [arXiv:1101.0590 [hep-th]].
[5] S. Carrozza and D. Oriti, arXiv:1104.5158 [hep-th].
[6] R. Gurau and J. P. Ryan, arXiv:1109.4812 [hep-th].
[7] R. Gurau, Class. Quant. Grav. **27**, 235023 (2010) [arXiv:1006.0714 [hep-th]].
[8] R. Gurau, Commun. Math. Phys. **304**, 69 (2011) [arXiv:0907.2582 [hep-th]].
[9] R. Gurau, Annales Henri Poincare **12**, 829 (2011) [arXiv:1011.2726 [gr-qc]].
[10] V. Bonzom, R. Gurau, A. Riello and V. Rivasseau, Nucl. Phys. B **853**, 174 (2011) [arXiv:1105.3122 [hep-th]].
[11] S. Carrozza, D. Oriti, to appear.

Journal of Physics: Conference Series **360** (2012) 012046 doi:10.1088/1742-6596/360/1/012046

Holomorphic Simplicity Constraints for 4d Riemannian Spinfoam Models

Maïté Dupuis[1], **Etera R. Livine**[2]

[1] Institute for Theoretical Physics III, University of Erlangen-Nürnberg, Erlangen, Germany.
[2] Laboratoire de Physique, ENS Lyon, CNRS-UMR 5672, 46 Allée d'Italie, Lyon 69007, France.

E-mail: dupuis@theorie3.physik.uni-erlangen.de, etera.livine@ens-lyon.fr

Abstract.
Starting from the reformulation of the classical phase space of Loop Quantum Gravity in terms of spinor variables and spinor networks, we build coherent spin network states and show how to use them to write the spinfoam path integral for topological BF theory in terms of Gaussian integrals in the spinors. Finally, we use this framework to revisit the simplicity constraints reducing topological BF theory to 4d Riemannian gravity. These holomorphic simplicity constraints lead us to a new spinfoam model for quantum gravity whose amplitudes are defined as the evaluation of the coherent spin networks.

It was recently shown that the classical phase space for Loop Quantum Gravity on a fixed graph Γ can be elegantly parameterized in terms of spinor networks [1, 2, 3, 4, 5, 6, 7]. These spinor networks get quantized and lead to the spin network states of Loop Quantum Gravity. This spinor formalism provides a clear interpretation of these states as discrete geometries. Here we use it to define coherent spin networks and build the spinfoam amplitudes for topological BF theory as Gaussian integrals over the spinor variables [9]. This allows to revisit the simplicity constraints reducing topological BF theory to 4d Riemannian gravity and to construct a spinfoam model for 4d Riemannian gravity based on new holomorphic simplicity constraints [8, 9].

1. Classical spinor networks and semi-classical spinnetworks

Spin network states for Loop Quantum Gravity can be seen as the quantization of classical spinor networks. These spinor networks are defined on a given graph Γ and consist in spinor variables $z_e^v \in \mathbb{C}^2$ attached to the edges and vertices of Γ and satisfying certain closure and matching constraints.

Let us first focus on a given vertex v of Γ. For each edge e attached to v, we define a spinor $|z_e\rangle$ and its dual spinor, which are two-dimensional complex vectors living in the fundamental representation of SU(2):

$$|z_e\rangle = \begin{pmatrix} z_e^0 \\ z_e^1 \end{pmatrix} \in \mathbb{C}^2, \quad \langle z_e| = \begin{pmatrix} \bar{z}_e^0 & \bar{z}_e^1 \end{pmatrix}, \quad \text{and} \quad |z_e] = \varsigma|z_e\rangle = \begin{pmatrix} -\bar{z}_e^1 \\ \bar{z}_e^0 \end{pmatrix}. \tag{1}$$

The spinor space is naturally endowed with the canonical Poisson bracket $\{z_e^a, \bar{z}_e^b\} = -i\delta_{ab}$. The spinors determine 3-vectors $\vec{V}(z_e) \in \mathbb{R}^3$, or $\vec{V}_e$ for short, through their projection on Pauli matrices, $\vec{V}(z_e) = \langle z_e|\vec{\sigma}|z_e\rangle$. This vector $\vec{V}_e$ entirely determines the original spinor z_e up to a global phase. Its components form a $\mathfrak{su}(2)$ algebra, $\{V_e^i, V_e^j\} = 2\epsilon^{ijk}V_e^k$, and actually generates SU(2) transformation on the spinor, $|z_e\rangle \rightarrow g\,|z_e\rangle$ with $g \in \mathrm{SU}(2)$. We now impose the closure constraints $\sum_e \vec{V}_e = 0$ at the vertex v, which generate a global SU(2) invariance on the spinors z_e. It reads in terms of the spinors as:

$$\sum_e |z_e\rangle\langle z_e| = \frac{1}{2}\sum_e \langle z_e|z_e\rangle \mathbb{I}. \tag{2}$$

SU(2)-invariant observables are easily identified as scalar products between spinors as $E_{ef} = \langle z_e|z_f\rangle$ and $F_{ef} = [z_e|z_f\rangle$ [3, 4, 6]. The modulus of these quadratic functionals give back the usual scalar products $\vec{V}_e \cdot \vec{V}_f$, or more precisely $|E_{ef}|^2 = |\vec{V}_e||\vec{V}_f| + \vec{V}_e\cdot\vec{V}_f$ and $|F_{ef}|^2 = |\vec{V}_e||\vec{V}_f| - \vec{V}_e\cdot\vec{V}_f$.

To form a spinor network on the graph Γ, we glue together the structures defined at each vertex. We now have spinors z_e^v around each vertex v, or equivalently two spinors $z_e^{s,t}$ for each edge attached e to its source and target vertices, $s(e)$ and $t(e)$. These spinors satisfy the closure constraints at each vertex v and new matching constraints on each edge [2, 6]:

$$\langle z_e^s|z_e^s\rangle = \langle z_e^t|z_e^t\rangle \quad \text{or equivalently} \quad |\vec{V}_e^s| = |\vec{V}_e^t|\,. \tag{3}$$

These generate $U(1)$-transformations on the spinors, $z_e^{s,t} \rightarrow e^{\pm i\theta} z_e^{s,t}$. The reduced phase space defined as the symplectic quotient $(\mathbb{C}^2)^{2E}//\mathrm{SU}(2)^V//\mathrm{U}(1)^E$ is exactly the phase space of Loop Quantum Gravity on the graph Γ. This is achieved explicitly by the reconstruction of the holonomy along the edges from the spinors [2, 6, 7]. The geometrical interpretation of these spinor networks is as a set of polyhedra (dual to each vertex) glued together by area-matching of their faces (but not shape-matching).

2. Quantization and Coherent Spin Networks

The quantization is straightforward and the spinor components are raised to annihilation and creation operators of harmonic oscillators:

$$|z_e\rangle = \begin{pmatrix} z_e^0 \\ z_e^1 \end{pmatrix} \rightarrow \begin{pmatrix} a_e \\ b_e \end{pmatrix}, \qquad \langle z_e| = \begin{pmatrix} \bar{z}_e^0 & \bar{z}_e^1 \end{pmatrix} \rightarrow \begin{pmatrix} a_e^\dagger & b_e^\dagger \end{pmatrix}, \tag{4}$$

with $[a_e, a_e^\dagger] = [b_e, b_e^\dagger] = 1$ and $[a_e, b_e] = 0$ for all e. This leads to Schwinger representation of SU(2) in terms of a couple of harmonic oscillators. The components of the vector $\vec{V}_e$ become the $\mathfrak{su}(2)$ generators and the spin j_e of the corresponding irreducible SU(2)-representation is given by fixing the total energy $E_e = \widehat{|\vec{V}(z_e)|} = (a_e^\dagger a_e + b_e^\dagger b_e) = 2j_e$. At the quantum level, the closure constraints impose SU(2)-invariance at each vertex v, while the matching constraints impose that the spins at the source and target vertices of each edge are equal. At the end, we recover spin network states, with SU(2)-representation j_e on each edge e and intertwiners at each vertex v, that is an SU(2)-invariant state in the tensor product of the SU(2)-representations j_e living on the edges e of v.

Next we introduce coherent spin networks peaked on each point of the spinor phase space. Starting with intertwiners, we define coherent intertwiners[1] as the group averaging of the standard coherent states for the harmonic oscillators labeled by the spinors $\{z_e\}$ [8, 9]:

$$||\{z_e\}\rangle \equiv \int dg\, g \triangleright e^{\sum_e z_e^0 a_e^\dagger + z_e^1 b_e^\dagger}|0\rangle, \tag{5}$$

[1] There is a clear relation between these coherent states and the LS coherent intertwiners used in the definition of the EPRL-FK spinfoam models [10, 11]. The interested reader can find details in [4, 9].

where the vacuum state $|0\rangle$ is the vacuum of the harmonic oscillators. These coherent states transform nicely under $\mathrm{U}(N)$ (see [4, 9] for details) and provide a decomposition of the identity on the Hilbert space $\mathcal{H}_N$ of N-valent intertwiners:

$$\mathbb{I}_{\mathcal{H}_N} = \frac{1}{\pi^{2N}} \int [d^4 z_e]^N e^{-\langle z_e|z_e\rangle} ||\{z_e\}\rangle\langle\{z_e\}||. \tag{6}$$

Then we glue these coherent intertwiners to form a full coherent spin network on the graph Γ. We have intertwiners $||\{z_e^v\}_{e\ni v}\rangle$ for each vertex $v \in \Gamma$ where the set of spinors $\{z_e^v\}$ satisfy the closure and matching constraints. The coherent quantum state on Γ is defined as the tensor product of these intertwiner states, $\psi_{\{z_e^v\}} = \bigotimes_v ||\{z_e^v\}_{e\ni v}\rangle$. $\psi_{\{z_e^v\}}$ is truly labeled by points in the reduced phase space and its evaluation on the group elements $\{g_e\} \in \mathrm{SU}(2)^E$ reads:

$$\psi_{\{z_e^v\}}(g_e) = \int [dh_v]\, e^{\sum_e [z_e^{s(e)}|h_{s(e)}^{-1}\, g_e\, h_{t(e)}|z_e^{t(e)}\rangle}. \tag{7}$$

This expression is explicitly SU(2)-invariant at every vertex v and holomorphic in the spinors z_e^v's. These coherent spin networks are semi-classical states peaked on the discrete geometry defined by their classical spinor labels [8, 9].

3. Spinfoam amplitudes for BF theory in terms of spinors

At this stage we have all the tools necessary to re-write the spinfoam path integral for 4d BF theory with SU(2) as gauge group in terms of spinors and holomorphic coherent intertwiners. Generic spinfoams are defined on arbitrary 2-complexes, describing the time evolution of graphs. Representations (resp. intertwiners) are put on faces (resp. edges) on the 2-complexes and the spinfoam amplitude is defined as the integral over all possible data of the product of local amplitudes associated to each vertex of the 2-complex. The neighborhood of a vertex defined by the incoming edges and faces dressed with intertwiners and representations is described by a boundary graph and spin network states and the local vertex amplitude is defined as the evaluation of this boundary spin network (at the identity). Here, for simplicity's sake, we will only consider 4d triangulations made of 4-simplices glued together along tetrahedra, with the 2-complex defined as the dual 2-skeleton and the spinfoam vertices as dual to the 4-simplices. Considering on a single 4-simplex σ and its dual boundary graph, we attach coherent intertwiners to each dual vertex (i.e tetrahedron) and we define the 4-simplex amplitude as the evaluation of the corresponding coherent spin network i.e $\psi_{\{z_\Delta^\tau\}}(g_\Delta) \equiv \psi_{\{z_e^v\}}(g_e)$ defined by (7) evaluated at $g_e = \mathbb{I}$. τ label the five tetrahedra of the 4-simplex and are dual to the vertices v of the boundary graph, while Δ denote the ten triangles dual to the edges e of the boundary graph. Then the 4-simplices are glued together using the resolution of the identity on the space of intertwiners. There is nevertheless an ambiguity in the integration measure for the spinors and we fix it by requiring that the spinfoam amplitude for the SU(2) BF theory be topological, which gives a Gaussian integral over the spinor variables [9]:

$$Z[\mathcal{M}] = \int \prod_{\tau,\Delta\in\tau} \frac{e^{-\langle z_\Delta^\tau|z_\Delta^\tau\rangle} d^4 z_\Delta^\tau}{\pi^2} \prod_\Delta \mu(z_\Delta^{\tau(\Delta)}) \prod_\sigma \int \prod_{\tau\in\sigma} dh_\tau^\sigma\, e^{\sum_{\Delta\in\sigma} [\varsigma^{\epsilon_{s(\Delta)}^\sigma} z_\Delta^{s(\Delta)}|(h_{s(\Delta)}^\sigma)^{-1} h_{t(\Delta)}^\sigma|\varsigma^{\epsilon_{t(\Delta)}^\sigma} z_\Delta^{t(\Delta)}\rangle}$$

where the measure factor $\mu(z) = (\langle z|z\rangle - 1)$ is inserted for a single tetrahedron $\tau(\Delta)$ for each triangle Δ (the origin of the dual plaquette) and the signs ϵ_τ^σ register the orientations of the tetrahedra with respect to each 4-simplex. Using the decomposition of the δ-distribution of SU(2) as an integral over spinors,

$$\delta(g_1..g_n) = \int \frac{e^{-\langle z|z\rangle} d^4 z}{\pi^2} \mu(z)\, e^{\langle z|g_1..g_n|z\rangle} = \int \frac{\prod_i^n e^{-\langle z_i|z_i\rangle} d^4 z_i}{\pi^{2n}} \mu(z_1)\, e^{\langle z_1|g_1|z_2\rangle}..e^{\langle z_n|g_n|z_1\rangle}, \tag{8}$$

we show that the definition above reproduces exactly the discretized path integral for BF theory:

$$Z[\mathcal{M}] = \int \prod_{\sigma,\tau\in\sigma} dh^{\sigma}_{\tau} \prod_{\Delta} \delta(\overrightarrow{\prod}_{\sigma\ni\Delta}(h^{\sigma}_{s(\Delta)})^{-1}h^{\sigma}_{t(\Delta)})\,. \tag{9}$$

4. Holomorphic simplicity constraints for the Euclidean case

We now go from the topological BF theory for Spin(4) $\sim \mathrm{SU}_L(2)\times\mathrm{SU}_R(2)$ to 4d Riemannian gravity within the spinorial framework. The simplicity constraints are imposed at level of each intertwiner (or tetrahedron or edges of the dual 2-complex). Working at a vertex v with now two sets of spinors $\{z^{v,L}_e\}$ and $\{z^{v,R}_e\}$ independently satisfying the closure constraint, we introduce new simplicity constraints (where we have dropped the index v):

$$\forall e,f \quad F^L_{ef} = \rho^2 F^R_{ef} \qquad \text{i.e.} \quad \forall e,f \quad [z^L_e|z^L_f\rangle = \rho^2[z^R_e|z^R_f\rangle \tag{10}$$

where ρ is a fixed parameter related to the Immirzi parameter γ. At the classical level, these holomorphic simplicity constraints with the closure constraints imply the usual quadratic simplicity constraints (by taking their modulus square). They are also equivalent to the linear simplicity constraints proposed in [12], i.e the existence of a 4-vector normal to all the bivectors $B_e + \gamma \star B_e$ where B_e is defined by its self-dual and anti-self dual parts $(\vec{V}(z^L_e), \vec{V}(z^R_e))$ [9]. At the quantum level, the coherent intertwiner $||\{z_e\}\rangle$ diagonalizes the $\hat{F}$ annihilation operators, $\hat{F}_{ij}\,||\{z_e\}\rangle = [z_i|z_j\rangle\,||\{z_e\}\rangle$. Thus we solve exactly the holomorphic simplicity constraints by the coherent intertwiners $|\{z_e\}\rangle_\rho \equiv ||\{\rho z_e\}\rangle_L \otimes ||\{z_e\}\rangle_R$ with $z^L_e = \rho z_e = \rho z^R_e$ [8, 9]. Then we define a new spinfoam model solving exactly the holomorphic simplicity constraints by its vertex amplitude given by the evaluation of the coherent spin network on the boundary 4-simplex graph obtained by gluing these coherent simple intertwiners $|\{z_e\}\rangle_\rho$:

$${}_\rho\mathcal{A}_\sigma(z^\tau_\Delta) = \psi_{\{\rho z^\tau_\Delta\}}(\mathbb{I})\psi_{\{z^\tau_\Delta\}}(\mathbb{I}) = \int [dh_\tau]^5\, e^{\sum_{\Delta\in\sigma}\rho^2[z^{s(\Delta)}_\Delta|h^L_{s(\Delta)}{}^{-1}h^L_{t(\Delta)}|z^{t(\Delta)}_\Delta\rangle[z^{s(\Delta)}_\Delta|h^R_{s(\Delta)}{}^{-1}h^R_{t(\Delta)}|z^{t(\Delta)}_\Delta\rangle}\,. \tag{11}$$

The full spinfoam amplitude is obtained by gluing these vertex amplitudes and integrating over the spinors with a Gaussian measure. The main difference with the EPRL-FK spinfoam model is that we do have quantum states solving the holomorphic simplicity constraint operators. The "drawback" is that the diagonal simplicity constraints are not strongly imposed and we do not work with simple representations satisfying $j^L_e = j^R_e$, but with Gaussian wave-packets peaked on this relation.

Finally, the reformulation of spinfoam amplitudes in terms of spinors presents the advantage of a clear link with coherent spin network states and an expression directly as the path integral of a discrete action in terms of spinors and holonomies. We hope it could be an interesting starting point for the study of the renormalization and coarse-graining of the spinfoam amplitudes.

References

[1] Freidel L and Speziale S 2010 *Phys. Rev.* **D82** 084040 (*Preprint* 1001.2748)
[2] Freidel L and Speziale S 2010 *Phys. Rev.* **D82** 084041 (*Preprint* 1006.0199)
[3] Freidel L and Livine E R 2010 *J. Math. Phys.* **51** 082502 (*Preprint* 0911.3553)
[4] Freidel L and Livine E R 2011 *J.Math.Phys.* **52** 052502 (*Preprint* 1005.2090)
[5] Borja E F, Diaz-Polo J, Garay I and Livine E R 2010 *Class. Quant. Grav.* **27** 235010 (*Preprint* 1006.2451)
[6] Borja E F, Freidel L, Garay I and Livine E R 2011 *Class.Quant.Grav.* **28** 055005 (*Preprint* 1010.5451)
[7] Livine E R and Tambornino J 2011 (*Preprint* 1105.3385)
[8] Dupuis M and Livine E R 2011 *Class.Quant.Grav.* **28** 085001 (*Preprint* 1006.5666)
[9] Dupuis M and Livine E R 2011 (*Preprint* 1104.3683)
[10] L. Freidel and K. Krasnov 2008 *Class.Quant.Grav.* **25** 125018 (*Preprint* 1006.0199)
[11] E.R. Livine and S. Speziale 2008 *Europhys.Lett.* **81** 50004 (*Preprint* 1006.0199)
[12] J. Engle, R. Pereira and C. Rovelli 2007 *Phys.Rev.Lett.* **99** 161301 (*Preprint* 0705.2388)
[13] J. Engle, E.R. Livine, R. Pereira and C. Rovelli 2008 *Nucl.Phys.B* **799** 136-149 (*Preprint* 1006.0199)

Journal of Physics: Conference Series **360** (2012) 012047 doi:10.1088/1742-6596/360/1/012047

Canonical formalism for simplicial gravity[1]

Philipp A Höhn

Institute for Theoretical Physics, Universiteit Utrecht, Leuvenlaan 4, NL-3584 CE Utrecht, The Netherlands

E-mail: p.a.hohn@uu.nl

Abstract. We summarise a recently introduced general canonical formulation of discrete systems which is fully equivalent to the covariant formalism. This framework can handle varying phase space dimensions and is applied to simplicial gravity in particular.

1. Motivation and Goal

Research in simplicial gravity[2] has mainly been carried out in the covariant setting which directly involves an action principle, while efforts to construct a suitable canonical formalism have been few and far between. For Regge Calculus [2], a specific discretization of General Relativity, first attempts to construct a canonical formulation [3] relied on discretized space, but a continuous (infinitesimal) time evolution generated by a set of constraints. A consistent canonical formulation, however, should reproduce exactly the covariant solutions following from the action, which in simplicial gravity implies a discrete time evolution that can thus no longer be generated by a set of constraints employing a bracket structure. Rather, a well–defined set of evolution moves is required to generate such a discrete time evolution. The role of the constraints (if existent) in such a framework, in turn, is reduced to generating (infinitesimal) transformations of the phase space variables which under certain conditions correspond to gauge symmetries of the (discrete) action. Based on ideas put forward in [4, 5, 6], the first canonical formulation (for Regge triangulations) which reproduces exactly the covariant theory and encompasses a discrete version of a hypersurface deformation algebra appeared in [7]. The drawback of this first consistent canonical formulation is, however, that the evolution moves employed in the discrete time evolution belong to a special class [5] that preserve both the topology and connectivity of the 'spatial'[3] triangulated hypersurfaces which carry the phase space variables. This formalism is therefore only applicable to a very special class of triangulations with preserved phase space dimension. Instead, in order to deal with arbitrary triangulations (of fixed 'spatial' topolgy), a general class of evolution moves as well as a new canonical framework is necessary.

Such a general canonical formalism for discrete theories which can handle varying phase space dimensions and constraints and, furthermore, is fully equivalent to the covariant formulation has been introduced recently [8]. In particular, it has been applied to (Euclidean) Regge Calculus with arbitrary 'spatial' hypersurfaces.

[1] Based on a talk at the conference *Loops '11* in Madrid on May 24th 2011 [1].

[2] We refer to discrete gravity theories based on simplices and triangulations.

[3] For simplicity, we work in Euclidean signature. Hence, we write 'spatial' in quotation marks.

Journal of Physics: Conference Series **360** (2012) 012047 doi:10.1088/1742-6596/360/1/012047

Here we want to summarise how one may treat the general situation where the 'spatial' lattice evolves/changes (akin to the situation with graphs in Loop Quantum Gravity) [8]. Specifically, this implies that the numbers of degrees of freedom may vary from step to step, a feature which we must be able to incorporate into our formalism. For several physical situations (e.g. expanding or contracting universes) and their numerical implementation it may be advantageous to be able to adapt the discretization density in time. But also in the light of several approaches to quantum gravity—especially Loop Quantum Gravity which entails changing spatial graphs and the simplicial Spin Foam models—it seems fruitful to be able to handle evolving lattices.

2. Evolution in discrete 'multi-fingered' or 'bubble' time

The central idea behind the general canonical evolution scheme for triangulations is to glue (or remove) a single D–simplex, to (or from) a $(D-1)$–dimensional triangulated hypersurface Σ_k at each elementary step counted by $k \in \mathbb{Z}$ (see figure 1); gluing moves correspond to forward evolution, while removal moves correspond to backward evolution. Accordingly, the triangulated hypersurface evolves in a discrete 'multi–fingered' or 'bubble' time through the full D–dimensional solution—in complete analogy to canonical General Relativity. The nice

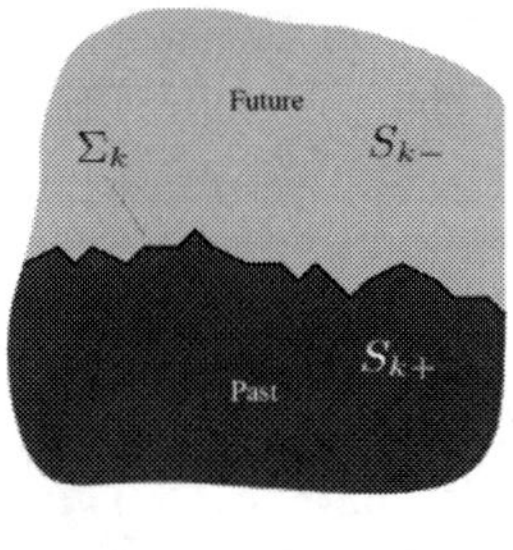

Figure 1. Hypersurface Σ_k separating 'past' and 'future' region at step k.

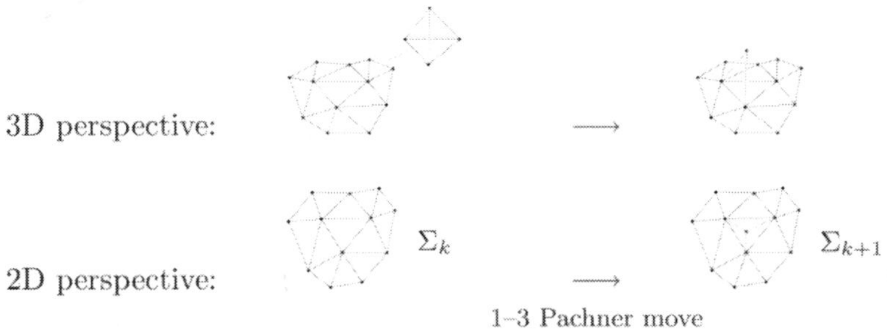

Figure 2. Gluing a single tetrahedron onto a single triangle in the 2D boundary hypersurface of some 3D bulk triangulation.

feature of these D–dimensional gluing or removal moves is that they have the interpretation of $(D-1)$–Pachner moves [9] within the $(D-1)$–hypersurface. For instance, figure 2 shows the 3D example of gluing a single tetrahedron onto a single triangle in a 2D hypersurface. From the 2D perspective, this move amounts to a subdivision of the triangle into three new ones: this is a so–called 1–3 Pachner move in 2D. It turns out that all other allowed gluing and removal moves in 3D and 4D also amount to Pachner moves in the hypersurfaces [8]. Pachner moves are an elementary class of moves (they only involve a fixed number of simplices) which are ergodic piecewise linear homeomorphisms (one can map between any finite triangulations of the same topology by finite sequences of these moves). These Pachner moves are therefore the general evolution moves which we will henceforth use in our construction of canonical simplicial gravity.

The implementation of these gluing and removal moves requires the action to be additive. This is, in fact, the case in Regge Calculus on which we therefore put our focus hereafter.[4] Recall that the lengths of the edges are the configuration variables of Regge Calculus. Thus, when attempting to implement the Pachner moves in a canonical formalism, one, in general, encounters the following two 'problems' (which can already be diagnosed for the 1–3 move in figure 2):

(a) subsets of variables coincide at different steps, i.e. $\Sigma_{k+1} \cap \Sigma_k \neq \emptyset$, and

(b) numbers of variables differ (phase space dimension varies) from step to step.

The canonical formalism clearly ought to appropriately address and 'solve' these 'problems'.

[4] But note that the present formalism is applicable to any discrete theory with additive action [8].

3. Discrete Legendre transformation and implementation of the Pachner moves

In order to define discrete Legendre transformations, we must group up the D-simplices counted by $k \in \mathbb{Z}$ into a choice of fat slices which we label by $n \in \mathbb{Z}$ (see figure 3). Lengths of edges in Σ_n are denoted by l^e_n, while all lengths which are internal between Σ_n and Σ_{n-1} are assigned to step n and denoted by l^i_n.[5] The key ingredient is to use the action contribution of fat slice n, $S_n(l^e_n, l^i_n, l^{e'}_{n-1})$, as a 'generating function' of first type, defining the conjugate momenta via a pair of discrete Legendre transformations $\mathbb{F}^{\pm}$ from the direct product of configuration manifolds $\mathcal{Q}_{n-1} \times \mathcal{Q}_n$ to the two phase spaces $\mathcal{P}_{n-1}, \mathcal{P}_n$ at steps $n-1, n$, respectively, [10, 8]

$$
\begin{aligned}
{}^{+}p^n_e &:= \frac{\partial S_n}{\partial l^e_n} & \qquad {}^{-}p^{n-1}_e &:= -\frac{\partial S_n}{\partial l^e_{n-1}} \\
{}^{+}p^n_i &:= \frac{\partial S_n}{\partial l^i_n} & \qquad {}^{-}p^{n-1}_i &:= -\frac{\partial S_n}{\partial l^i_{n-1}} = 0.
\end{aligned}
\tag{1}
$$

The last equation yields a set of constraints which follows from the fact that S_n does not depend

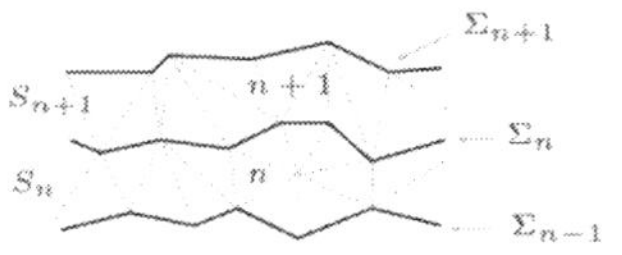

Figure 3. Group up simplices into a choice of fat slices.

Figure 4. The 1–4 and 4–1 Pachner moves.

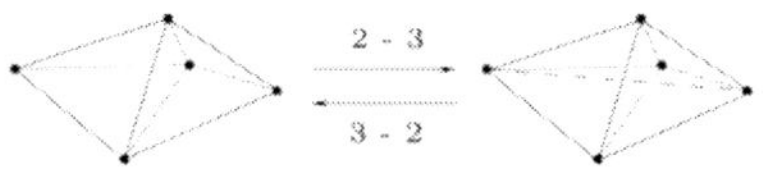

Figure 5. The 2–3 and 3–2 Pachner moves.

on l^i_{n-1}. In the general case (e.g. due to varying numbers of edges) the Legendre transformations will fail to be isomorphisms and only map onto subspaces of $\mathcal{P}_{n-1}$ and $\mathcal{P}_n$ [8]. We call the image of $\mathbb{F}^-$ the *pre-constraint surface* and the image of $\mathbb{F}^+$ the *post-constraint surface*.

From a similar transformation, using S_{n+1} as the generating function, one finds ${}^{-}p^n_e = -\frac{\partial S_{n+1}}{\partial l^e_n}$. Hence, the equations of motion for the l^e_n, i.e. $\frac{\partial S_n}{\partial l^e_n} + \frac{\partial S_{n+1}}{\partial l^e_n} = 0$, imply a *momentum matching* $p^n_e := {}^{+}p^n_e = {}^{-}p^n_e$. We will therefore subsequently drop the $+,-$ indices. The same applies to internal variables l^i, however, now the equations of motion, $\frac{\partial S}{\partial l^i} = 0$, imply $p_i = 0$, i.e. equations of motion for internal lengths actually appear as canonical constraints.

Returning to the gluing moves counted by k (henceforth, we restrict ourselves to gluing moves) and assuming $\Sigma_k = \Sigma_n$, it turns out [8] that at each k the previous momenta translate into $p^k_e = \frac{\partial S_{k+}}{\partial l^e_k}$, where $S_{k+} = \sum_{k'=1}^{k} S_{\sigma_{k'}}$ ($S_{\sigma_{k'}}$ action of k'-th simplex) is the action up to step k (see figure 1). Since the action is additive, Pachner moves require a *momentum updating* [8],

$$
p^{k+1}_e = p^k_e + \frac{\partial S_{\sigma_{k+1}}}{\partial l^e_k}. \tag{2}
$$

This construction defines our discrete time evolution.

We are now ready to solve our 'problems'. We can solve 'problem' (a) by mapping the lengths of all edges occurring in both Σ_k and Σ_{k+1} to themselves, $l^e_{k+1} = l^e_k$, and by updating the momenta according to (2). 'Problem' (b) can be solved by a natural phase space extension: we can formally 'add' new or old variables l^n_k, l^o_k of edges occurring only 'to the future' or only 'to the past' of hypersurface Σ_k to $\mathcal{P}_k$, noting that these variables come with constraints $p^k_n = 0 = p^k_o$ because of the equations of motion. For the example of the 1–3 Pachner move (figure 2), we can use the action of the new tetrahedron, $S_\tau(l^n_{k+1}, \ldots)$, as a type one generating function with trivial dependence on l^n_k because the new edges $n = 1, 2, 3$ do not occur at k, in order to consistently

[5] For simplicity, we neglect any boundary variables.

transform their momenta, $\{p_n^k = \frac{\partial S_\tau}{\partial l_k^n} = 0\} \mapsto \{p_n^{k+1} = \frac{\partial S_\tau}{\partial l_{k+1}^n}\}$. The first equations are three examples of *pre–constraints*, while the last three equations constitute *post–constraints* (S_τ only depends on quantities of Σ_{k+1}). If this recipe is applied correctly, all Pachner moves in 3D and 4D can be implemented as canonical transformations on (a suitably extended) phase space [8].

For 4D simplicial gravity there are four types of Pachner moves [8]. The 1–4 move (figure 4) introduces a new vertex and four new edges which, as in the case of the 1–3 move, come with four *post–constraints* $p_n^{k+1} - \frac{\partial S_\sigma}{\partial l_{k+1}^n} = 0$ that are automatically satisfied after the move. The 2–3 move (figure 5) introduces one edge equipped with a *post–constraint*, but renders one triangle internal. In 4D these two moves are the only moves which introduce new edges into the triangulation. However, since these moves do not involve any equations of motion (no edges become internal) all new edge lengths can *a priori* be freely chosen. This implies that, with the 2–3 move, one can generate an *a priori* free curvature because in 4D Regge Calculus internal triangles carry the deficit angles and the new deficit angle depends on the new length l_{k+1}^n. The inverse moves, i.e. the 4–1 and 3–2 moves (figures 4 and 5), on the other hand, annihilate edges o from Σ_k and come with non–trivial *pre–constraints* $p_o^k + \frac{\partial S_\sigma}{\partial l_k^o} = 0$ which need be satisfied *prior* to the move.

The *post–constraints* form an abelian Poisson algebra, but *a priori* do not generate gauge transformations [8], which in Regge Calculus correspond to vertex displacements in the bulk that leave the action invariant [6, 7]. Rather, they reflect the lack of information in a given hypersurface about the full 4D Regge solution; given initial data, non–uniqueness of solutions arises. *A posteriori* the first or second class nature of the *post–constraints* of Σ_k depends on the *pre–constraints*: if no complete *constraint matching* at k takes place, some constraints turn into second class and the 'unmatched' *pre–constraints* may fix some (or all) *a priori* free lengths of the 1–4 and 2–3 moves [8, 11]. Those lengths which eventually remain free constitute gauge parameters. However, gauge symmetry is generically broken in the presence of curvature [6, 7].

4. Conclusions and Outlook

We devised a general canonical framework for simplicial gravity by employing gluings/removals of single simplices as evolution moves that can be interpreted as Pachner moves within the hypersurfaces [8]. The Pachner moves can be implemented as canonical transformations via a natural phase space extension that is controlled by constraints which are equations of motion. By using the action as a generating function, the canonical framework generates general solutions of the covariant formalism and is thus equivalent. Likewise, a quantization should presumably yield a direct connection between the canonical framework and the path integral. The application of the present scheme to linearized 4D Regge gravity, as well as a more detailed discussion and classification of the constraints will appear in a separate paper [11].

Acknowledgments The author thanks Bianca Dittrich for collaboration on this topic.

References

[1] slides at `http://loops11.iem.csic.es/loops11/index.php?option=com_content&view=article&id=117`
[2] T. Regge, Nuovo Cim. **19** (1961) 558
[3] T. Piran and R. M. Williams, Phys. Rev. D **33** (1986) 1622; J. L. Friedman and I. Jack, J. Math. Phys. **27** (1986) 2973
[4] R. Gambini and J. Pullin, Phys. Rev. Lett. **90** (2003) 021301; C. Di Bartolo, R. Gambini, R. Porto and J. Pullin, J. Math. Phys. **46** (2005) 012901
[5] J. Barrett, M. Galassi, W. Miller, R. Sorkin, P. Tuckey and R. Williams, Int. J. Theor. Phys. **36** (1997) 815
[6] B. Bahr and B. Dittrich, Class. Quant. Grav. **26** (2009) 225011
[7] B. Dittrich, P. A. Höhn, Class. Quant. Grav. **27** (2010) 155001
[8] B. Dittrich, P. A. Höhn, (*Preprint* arXiv:1108.1974)
[9] U. Pachner, Abh. Math. Sem. Univ. Hamburg **57** (1986) 69; Europ. J. Combinatorics **12** (1991), 129-145
[10] J. E. Marsden, M. West, *Acta Numerica* **10**, pp. 357-514 (CUP 2001)
[11] B. Dittrich, P. A. Höhn, *to appear*

Dynamics of Loop Quantum Gravity encoded in graphs

Marcin Kisielowski[1,2]
Based on work done in collaboration with prof. Jerzy Lewandowski and Jacek Puchta.

[1] Instytut Fizyki Teoretycznej, Uniwersytet Warszawski, ul. Hoża 69, 00-681 Warszawa (Warsaw), Polska (Poland)
[2] St. Petersburg Department of Steklov Mathematical Institute RAS, 27, Fontanka, 191023 St. Petersburg, Russia

Abstract. We present an approach to spin foam models, where the 2-complex is constructed from more basic building blocks. Those building blocks are graphs which we equip with certain additional structure. We show how the spin foam amplitude can be read directly from this diagrammatic notation.

1. Introduction

One could interpret spin foam models as procedures for constructing discretized path integrals for quantum gravity, and for many other theories. In the case of models of quantum gravity the foams are 2-complexes having one vertex in the center of each 4-simplex, one edge intersecting each tetrahedron and one face intersecting each triangle of simplicial decomposition of a manifold (such foams we call simplicial foams) [1]. A spin foam is a labelled foam obtained by assigning to each face an irreducible representation of a given group (e.g. SU(2) group) and to each edge a pair of intertwiners (or equivalently an operator [2]). In this interpretation spin foams contain all data needed to calculate transition amplitudes between states of quantized discrete gravity.

However, spin foam models of quantum gravity may be also interpreted as histories of states of Loop Quantum Gravity – the states of *full* quantum gravity. In this case, the class of possible foams is broader. An example of such a spin foam model is the Engle-Pereira-Rovelli-Livine model [3] extended in [4] to oriented linear 2-cell complexes. The extended model allows more varied structures of a vertex neighbourhood (see [4]) and full variety of boundary graphs (a boundary of a foam in this class can be an arbitrary graph). We looked for an optimal class of foams compatible with an extension of this type – a class which is just broad enough to have the desired properties discussed above (but not too broad) and does not use auxiliary linear structures which are not compatible with diffeomorphism invariance. The solution we found is naturally expressed in terms of certain diagrams, which we call graph diagrams [5, 6]. We found a coloring of a graph diagram which is equivalent to a coloring of the corresponding foam. A graph diagram together with a coloring is called an operator spin network diagram (OSN diagram). All information needed to calculate the transition amplitudes between Loop Quantum Gravity states (in general – the boundary transition amplitude [7]) is encoded in the OSN diagrams. Therefore, one does not need to reconstruct foams and may use only OSN diagrams. Since it

Journal of Physics: Conference Series **360** (2012) 012048 doi:10.1088/1742-6596/360/1/012048

is easier to imagine the diagrams than the foams, we expect OSN diagrams to become a useful technical tool.

This article is structured as follows: In section 2 we introduce graph diagrams and their coloring, and we comment on the correspondence between OSN diagrams and spin foams. In section 3 we present a formula for the corresponding spin foam operator [2] which involves only the data from the OSN diagram.

2. Graph diagrams and operator spin network diagrams

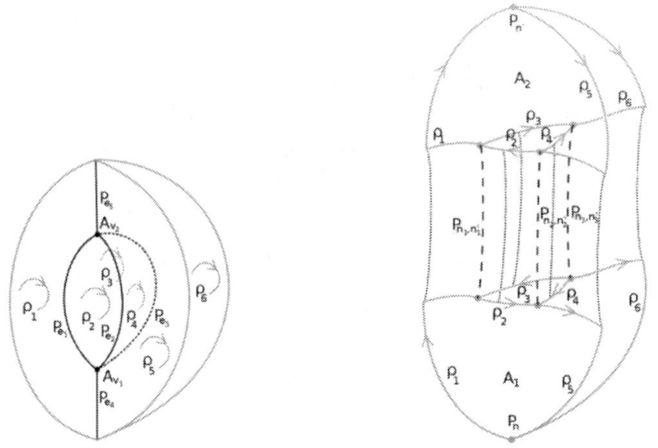

(a) Spin foam

(b) Operator spin network diagram

Figure 1: A spin foam and an equivalent operator spin network diagram

A **graph diagram** $(\mathcal{G}, \mathcal{R})$ consists of a set $\mathcal{G}$ of oriented graphs $\{\Gamma_1, ..., \Gamma_N\}$, and a family $\mathcal{R}$ of relations. One of the relations is called a *node relation.* It is a symmetric relation in the set of nodes of the graphs. Each node n is either in relation with precisely one $n' \neq n$ or it is unrelated (such node is called a boundary node). In order to be related, two nodes need to have equal number of incoming / outgoing links (with possible closed links counted twice). One defines a bijective map between all the incoming / outgoing links of Γ_I at n, with the outgoing / incoming links of $\Gamma_{I'}$ at n'. Two links identified with each other by the bijection are called to be in relation $\mathcal{R}^{(n,n')}_{\text{link}}$ at the pair of nodes n, n' (this relation is symmetric). A link of Γ_I / $\Gamma_{I'}$ which intersects n / n' twice, emerges in the relation twice: once as incoming and once as outgoing. A family of relations $\mathcal{R}^{(n,n')}_{\text{link}}$ is called a *link relation.* An example of a graph diagram is depicted on figure 1b (omit a coloring A, ρ and P at this point), node relations are depicted by dashed lines and link relations are depicted by dotted lines.

To each graph in a graph diagram corresponds a foam which is an image of a homotopy of this graph to a point. It has one internal vertex and its boundary is the input graph. This foam may be considered to be an elementary building block of the 2-complex corresponding to the graph diagram. Node and link relations code the way in which we glue together these building blocks. Detailed construction of a foam from a graph diagram is given in [6]. Each graph in the graph diagram is a boundary of a suitable neighbourhood of a vertex in the corresponding 2-complex – the so called vertex graph [4]. In this sense graph diagrams define a natural class of spin foams compatible with the extension of the EPRL model presented in [4].

Graph diagrams can be colored in a way which corresponds to a coloring of the corresponding foam. A colored diagram is called an **operator spin network diagram**

$(\mathcal{G} = \{\Gamma_1, ..., \Gamma_n\}, R, \rho, P, A)$. The coloring is the following.

- Links are colored with representations ρ acting in a Hilbert space $\mathcal{H}_\rho$. If two links are in link relation the corresponding representations are equivalent.
- For each node define a Hilbert space $\mathcal{H}_n = \mathrm{Inv}\left(\bigotimes_i \mathcal{H}^*_{\rho_i} \otimes \bigotimes_j \mathcal{H}_{\rho_j}\right) \subset \left(\bigotimes_i \mathcal{H}^*_{\rho_i} \otimes \bigotimes_j \mathcal{H}_{\rho_j}\right)$, where i / j labels the links incoming / outgoing at n.
  Pairs of nodes $\{n, n'\}$ in the node relation are colored with an operator
  $$P_{n,n'} \in \mathcal{H}_n \otimes \mathcal{H}_{n'}.$$
  Boundary nodes (those not related with any other node) are colored with an operator
  $$P_n \in \mathcal{H}_n \otimes \mathcal{H}^*_n.$$
- Each graph Γ_I is colored with a tensor
  $$A_{\Gamma_I} \in \left(\bigotimes_n \mathcal{H}_n\right)^* \tag{1}$$
  where n runs through the nodes of Γ_I. We call it a contractor.

Figure 1 may clarify the correspondence between colorings of foams and diagrams. For every link, there is a face colored with the representation equivalent to the one assigned to a link. For every pair of nodes in relation there is an edge with both endpoints being internal vertexes of the corresponding foam, and for every boundary node there is an edge which has one endpoint being internal vertex and one being boundary vertex. The coloring with contractors is a new idea we introduced to include the SU(2) spin foam model constructed from EPRL vertex [8, 9] (and possible other models). The contractor is a generalization of a standard vertex contraction [4, 2], i.e. it generalizes the contractor A^{Tr} implicitly used in operator spin foam formalism [2]:

$$A^{\mathrm{Tr}}_\Gamma = \bigotimes_l \mathrm{Tr}_l \in \bigotimes_l \mathcal{H}^*_{\rho_l} \otimes \mathcal{H}_{\rho_l} \equiv \left(\bigotimes_n \mathcal{H}_n\right)^*,$$

where l runs through links of the graph Γ.

3. The spin foam operator

There is a natural contraction

$$\left(\bigotimes_{\Gamma_I} A_{\Gamma_I}\right) \lrcorner \left(\bigotimes_n P_n\right), \tag{2}$$

where Γ_I ranges the set of graphs and n ranges the set of boundary nodes and pairs of nodes related by the node relation (for more details on the contraction see [6]).

Operator (2) is a spin foam operator of the spin foam corresponding to the graph diagram. In (2) every face amplitude and boundary edge amplitude is equal 1. We now generalize this formula to include general boundary edge and face amplitudes.

We redefine the operators P_n corresponding to boundary nodes. Since boundary nodes correspond to boundary vertexes and links intersecting these nodes correspond to boundary edges (the second correspondence is 2-1), the redefinition of P_n allows us to include general boundary edge amplitudes. We define

$$A_n = \sqrt{\prod_l A_l}, \tag{3}$$

where l ranges the set of links intersecting the node n. The terms A_l become boundary edge amplitudes when the spin foam is reconstructed. The square root in formula (3) comes from the fact that each link is counted twice. We redefine operators P_n by setting:

- $\tilde{P}_n = A_n P_n$ for boundary nodes,
- $\tilde{P}_n = P_n$ when $n = \{n_1, n_2\}$.

The node relation and the link relation lead to an equivalence relation in the set of links of the graphs. The resulting equivalence relation, which we call face relation, carries information about faces of the corresponding 2-complex, and allows to introduce face amplitude without direct reference to that 2-complex. The face relation $\sim$ is described as follows:

- Each link is in relation with itself.
- Two different links l and l' are in relation when they are in the link relation.
- It may happen that a link is in the link relation with two other (different) links (because every link has two ends). In general there may be a sequence of links $(l_1, l_2, \ldots, l_m)$ such that l_i is in the link relation with l_{i+1}, $i \in \{1, \ldots, m-1\}$ and it is natural to assign a single amplitude to a set $\{l_1, \ldots, l_m\}$. As a result we define l and l' to be in a face relation when there exists l'' such that l is in the link relation with l'', and l'' is in the link relation with l.

The face relation induces division of the set of links into equivalence classes. We assign an amplitude (a complex number) to f and denote it by A_f. It becomes a face amplitude in the reconstructed spin foam. Denote by $\mathfrak{L}/\sim$ a set of those equivalence classes. The spin foam operator is:

$$P = \left(\prod_{f\in\mathfrak{L}/\sim} A_f\right)\left(\bigotimes_I A_I\right) \lrcorner \left(\bigotimes_n \tilde{P}_n\right). \tag{4}$$

It is equal to an operator assigned to the corresponding spin foam [6]. As anticipated in the introduction, this formula involves only objects defined by an operator spin network diagram and no reference to the resulting spin foam is needed.

Summary. Graph diagrams describe a class of foams which we suggest to use in spin foam models of quantum gravity [6]. The data from the operator spin network diagram is sufficient to calculate the spin foam operator – no reference to the corresponding spin foam is needed. Therefore this formalism is independent from the 2-complex approach. We expect it to be a useful technical tool, as the diagrams are easier to imagine than the 2-complexes.

Acknowledgements. Marcin Kisielowski acknowledges financial support from the project International PhD Studies in Fundamental Problems of Quantum Gravity and Quantum Field Theory of Foundation for Polish Science, cofinanced from the programme IE OP 2007-2013 within European Regional Development Fund. The work was also partially supported by the grants N N202 104838, and 182/N-QGG/2008/0 (PMN) of Polish Ministerstwo Nauki i Szkolnictwa Wyższego.

References

[1] Baez J (2000) Lect.Notes Phys.**543**,25-94, (*Preprint* arXiv:gr-qc/9905087v1)
[2] Bahr B, Hellmann F, Kamiński W, Kisielowski M, Lewandowski J (2010) Class.Quant.Grav.**28**,105003,2011 (*Preprint* arXiv:1010.4787v1)
[3] Engle J, Livine E, Pereira R, Rovelli C (2008) Nucl.Phys.**B799**,136-149 (*Preprint* arXiv:0711.0146v2)
[4] Kamiński W, Kisielowski M, Lewandowski J (2010) Class.Quantum Grav.**27**,095006, arXiv:0909.0939v2
[5] Hellmann F (2011) PhD Thesis, University of Nottingham (*Preprint* arXiv:1102.1688v1)
[6] Kisielowski M, Lewandowski J, Puchta J (2011) *Preprint* arXiv:1107.5185 [gr-qc]
[7] Rovelli C (2004) *Quantum Gravity*, (Cambridge: Cambridge University Press)
[8] Rovelli C (2010) Class.Quant.Grav.**28**,114005,2011 (*Preprint* arXiv:1004.1780v4)
[9] Bianchi E, Regoli D, Rovelli C (2010) Class.Quant.Grav.**27**,185009,2010 (*Preprint* arXiv:1005.0764v1)
You Ding, Muxin Han, Rovelli C (2011) Phys.Rev.**D83**,124020,2011 (*Preprint* arXiv:1011.2149v2)
Rovelli C (2011) (*Preprint* arXiv:1102.3660v3)

Journal of Physics: Conference Series **360** (2012) 012049 doi:10.1088/1742-6596/360/1/012049

Effective action for EPRL/FK spin foam models

Aleksandar Miković[1,2] **and Marko Vojinović**[2]

[1]Departamento de Matemática, Universidade Lusófona de Humanidades e Tecnologias
Av. do Campo Grande, 376, 1749-024 Lisboa, Portugal

[2]Grupo de Fisica Matemática da Universidade de Lisboa
Av. Prof. Gama Pinto, 2, 1649-003 Lisboa, Portugal

E-mail: amikovic@ulusofona.pt, vmarko@cii.fc.ul.pt

Abstract. We show that a natural modification of the EPRL/FK vertex amplitude gives a finite spin foam model whose effective action gives the Einstein-Hilbert action in the limit of large spins and arbitrarily fine spacetime triangulations. The first-order quantum corrections can be easily computed and we show how to calculate the higher-order corrections.

1. Introduction

The EPRL/FK spin foam models [1, 2] represent a class of spin foam models whose boundary states are the Loop Quantum Gravity spin networks, while the vertex amplitudes are constructed such that all the discretized constraints of the Plebanski action are imposed. The partition function for an EPRL/FK spin foam model can be written as

$$Z = \sum_j \int d\vec{n} \prod_f W_f(j_f) \prod_v W_v(j_f, \vec{n}_{lf}) \,, \tag{1}$$

where j_f are $SU(2)$ spins assigned to the faces f of the dual 2-complex K for a spacetime triangulation, $\vec{n}_{lf}$ are the Livine-Speziale normal vectors, one for each edge l and the adjacent face f of K, see [3]. The face amplitude $W_f(j)$ can be chosen to be $2j_f + 1$, see [4], while W_v is the vertex amplitude, which can be represented as a certain linear combination of the Lorentz group $15j$-symbols, see [1, 2] for details.

The classical limit of the quantum gravity theory defined by the path integral (1) can be determined by computing the corresponding effective action. The effective action can be expanded as the classical action plus the quantum corrections, which are proportional to the powers of $\hbar$, so that in the limit $\hbar \to 0$ one will obtain the classical action. One way to do this is to calculate the graviton correlation functions in the large-spin limit, see [5, 6]. However, this approach has a drawback, since it requires the knowledge of infinitely many correlation functions in order to construct the corresponding effective action. On the other hand, the effective action can be computed directly [7], which is much easier to do. We will sketch this computation here.

In order to calculate the effective action, the partition function has to be finite. The convergence of (1) can be studied by introducing a real deformation parameter p such that $Z \to Z_p$, where

$$W_v \to \frac{W_v}{\prod_{f\in v}(2j_f+1)^p} \,,$$

Journal of Physics: Conference Series **360** (2012) 012049 doi:10.1088/1742-6596/360/1/012049

and W_f stays the same, see [8]. One also needs to know the large-spin asymptotics of W_v, which is given by

$$W_v(j,\vec{n}) \approx \frac{1}{V(j)}\left[N_+(j)e^{i\alpha S_{vR}(j)} + N_-(j)e^{-i\alpha S_{vR}(j)}\right], \tag{2}$$

where $V(j)$ is a homogeneous function of order 12, $N_\pm(j)$ are homogeneous functions of order zero, $\alpha = \gamma$ or $\alpha = 1$, depending on whether the vertex boundary constitutes a Lorentzian or Euclidean Regge geometry, respectively. γ is the Barbero-Immirzi parameter and S_{vR} is the Regge action for the vertex v. If $(j,\vec{n})$ at a vertex do not form a Regge geometry, then $W_v(\lambda j,\vec{n})$ falls off faster that any power of $1/\lambda$, see [9, 10].

One also needs the asymptotics for the case when some of the j are big and other are small, but it is not difficult to see that in any case $|W_v(j,\vec{n})|$ is a bounded function. Then it can be shown that Z_p is absolutely convergent for $p > 1$, see [8]. Another complication which has to be resolved before calculating the effective action is that the asymptotics (2) is of the $\cos(S_R)$ type, instead of the e^{iS_R} type. This can be resolved by redefining the vertex in the following way

$$W_v \to A_v = \frac{1}{\prod_{f\in v}(2j_f+1)^p}\,\frac{W_v + \sqrt{W_v^2 - 4N_+N_-/V^2}}{2N_+}. \tag{3}$$

Then

$$A_v(j,\vec{n}) \approx \frac{1}{\prod_{f\in v}(2j_f+1)^p}\,\frac{e^{i\alpha S_{vR}(j)}}{V(j)}, \tag{4}$$

when all j are large.

If $N_+ \neq N_-$ one can obtain the exponential asymptotics of the vertex amplitude by making a linear combination of W and its complex conjugate [11]. Another way is to use the fact that W is a linear combination of $15j$-symbols, and to change the coefficients such that the exponential asymptotics is achieved, see [12].

2. Effective action

An effective action Γ can be defined for a spin foam model by adapting the the background field method from quantum field theory (QFT), see [7]. In the case of the EPRL/FK models one obtains

$$e^{i\Gamma(j,\vec{n})} = \sum_{j'}\int d\vec{n}'\prod_f W_f(j+j')\prod_v A_v(j+j',\vec{n}+\vec{n}'), \tag{5}$$

where $(j,\vec{n})$ are the background variables and $j' \in \mathbb{N}/2$. This formula is valid in the semiclassical limit, i.e. when $j \to \infty$ and the Barbero-Immirzi parameter γ is fixed. One can then employ the asymptotic formula (4) for the vertex amplitude in the definition of the effective action (5), since the background spins j are all large. It is known that the background configurations $(j,\vec{n})$ which form a Regge geometry have a dominant contribution to the path integral (1) for large spins, see [13, 14]. In that case $(j,\vec{n}) = (j(L),\vec{n}(L))$, where L is a set of lengths for the edges of the spacetime triangulation, and one can show by using (4) and the standard Gaussian integral formulas, see [7], that

$$\Gamma(j,\vec{n}) = S_R(L) + \sum_f c_f(p)\log j_f + \frac{1}{2}\operatorname{Tr}\log\tilde{S}_R'' + O(1). \tag{6}$$

Journal of Physics: Conference Series **360** (2012) 012049 doi:10.1088/1742-6596/360/1/012049

$S_R(L)$ denotes the Regge action for the spacetime triangulation and $c_f(p) = 1-(p+12)n_f$, where n_f is the number of vertices of a face f. The matrix $\tilde{S}''_R$ is given by

$$\tilde{S}''_R = \begin{pmatrix} \tilde{S}''_{R\,jj} & S''_{R\,jn} \\ S''_{R\,jn} & S''_{R\,nn} \end{pmatrix},$$

where S''_{jj}, S''_{jn} and S''_{nn} denote the matrices formed from the corresponding second-order partial derivatives of $S_R = \gamma \sum_f j_f \Theta_f(j, \vec{n})$ evaluated at $j = j(L)$ and $\vec{n} = \vec{n}(L)$. The elements of the matrix $\tilde{S}''_{jj}$ are given by

$$\tilde{S}''_{R\,ff'} = S''_{R\,ff'} - i\frac{c_f}{j_f^2}\delta_{ff'}.$$

The notation $O(1)$ in (6) means that

$$\Gamma(\lambda j, \vec{n}) = \lambda\tilde{\Gamma}_0(j,\vec{n}) + (\log\lambda)\tilde{\Gamma}_1(j,\vec{n}) + \tilde{\Gamma}_2(j,\vec{n}) + \lambda^{-1}\tilde{\Gamma}_3(j,\vec{n}) + \cdots, \tag{7}$$

so that the first term in (6) is of $O(j)$, the second and the third term in (6) are of $O(\log j)$, the next term is of $O(1)$ and so on. Hence the leading term for $j \to \infty$ is the Regge action, which means that it is the classical limit of the theory. If the triangulation of the spacetime is fine enough, the Regge action can be approximated by the Einstein-Hilbert action as $S_R \approx S_{EH}/2$, and one recovers general relativity in the limit of smooth geometry [7].

The second term in (6) comes from the face amplitude and the regularization factor in the vertex amplitude, while the third term is a discretization of the trace-log term from QFT.

3. Higher-order corrections

The method used in [7] to calculate the effective action was based on the "on-shell" background, i.e. the background $(j, \vec{n})$ was chosen such that $(j, \vec{n})$ formed an "on-shell" Regge geometry. In this case $\vec{n} = \vec{n}(L_0)$ and $j = j(L_0)$ where L_0 is a set of edge lengths which satisfy the Regge action equations of motion. After the calculation was done, the L_0 background was replaced with a background L, where the lengths in the set L do not have to satisfy the Regge equations.

In QFT it is known that such an approach is accurate only up to the first-order in $\hbar$, since the exact formula for the effective action is given by

$$e^{\frac{i}{\hbar}\Gamma(\phi)} = \int \mathcal{D}h \exp\left(\frac{i}{\hbar}S(\phi+h) - \frac{i}{\hbar}\int dx\, \frac{\delta\Gamma}{\delta\phi(x)}h(x)\right), \tag{8}$$

see [15]. The formula (8) is an integro-differential equation, which can be solved perturbatively as

$$\Gamma = S + \hbar\Gamma_1 + \hbar^2\Gamma_2 + \cdots, \tag{9}$$

and one does not need to use an on-shell background ϕ in the calculation.

The exact QFT formula (8) can be applied to an EPRL/FK spin foam model as

$$e^{i\Gamma(j,\vec{n})} = \sum_{j'}\int d\vec{n}' \prod_f W_f(j+j') \prod_v W_v(j+j', \vec{n}+\vec{n}')\, e^{-i\left\langle \frac{\partial\Gamma}{\partial j}j' + \frac{\partial\Gamma}{\partial\vec{n}}\vec{n}'\right\rangle}, \tag{10}$$

where

$$\left\langle \frac{\partial\Gamma}{\partial j}j' + \frac{\partial\Gamma}{\partial\vec{n}}\vec{n}'\right\rangle = \sum_f \frac{\partial\Gamma}{\partial j_f}j'_f + \sum_{fl}\frac{\partial\Gamma}{\partial\vec{n}_{fl}}\vec{n}'_{fl},$$

is the spin-foam analog of the higher-loop correction term from (8).

The background $(j,\vec{n})$ will be chosen to be a Regge background $(j(L),\vec{n}(L))$, where the lengths L are off-shell, i.e. do not have to satisfy the Regge equations of motion. If the background is chosen such that it does not form a Regge geometry, then we expect that $e^{i\Gamma(j,\vec{n})}$ will be exponentially suppressed, i.e.

$$\left| e^{i\Gamma(j,\vec{n})} \right|_{j\neq j(L),\vec{n}\neq\vec{n}(L)} \ll \left| e^{i\Gamma(j,\vec{n})} \right|_{j=j(L),\vec{n}=\vec{n}(L)} .$$

The analog of the $\hbar$-expansion (9) is the formula (7), which we will write as

$$\Gamma = S_R + \Gamma_1 + \Gamma_2 + \cdots + \Gamma_n + \cdots ,$$

where $S_R = O(j)$, $\Gamma_1 = O(\log j)$ and $\Gamma_n = O(j^{-n+2})$ for $n \geq 2$. By using the techniques from [7] it is not difficult to show that Γ_1 is the same as the $O(\log j)$ term from (6), while the differences between the formula (5) and the formula (10) will appear for $n \geq 2$ terms.

4. Conclusions

The effective action approach from QFT is quite useful for exploring the semiclassical limit of spin foam models of quantum gravity since the calculations in this approach are much easier than the calculations in the other approaches. The effective action approach can also tell us what kind of vertex amplitude we should choose in order to obtain a spin foam model whose classical limit is GR.

The first-order quantum corrections have two terms, see (6). The third term in (6) is a discrete analog of the usual trace-log term from QFT, while the second term is specific for spin foam models. It will be important to find out what is the form of these quantum corrections in the smooth geometry limit, since these corrections will determine the first-order back-reaction effects for black holes and the cosmological evolution. The higher-order corrections can be calculated from the equation (10).

It would be interesting to apply the effective action approach to the Regge model, and compare the results to the results obtained in the spin foam case.

Acknowledgments

This work has been partially supported by FCT project PTDC/MAT/099880/2008. MV was also supported by the FCT grant SFRH/BPD/46376/2008.

References

[1] Engle J, Livine E, Pereira R and Rovelli C 2008 *Nucl. Phys.* B **799** 136
[2] Freidel L and Krasnov K 2008 *Class. Quant. Grav.* **25** 125018
[3] Livine E R and Speziale S 2007 *Phys. Rev.* D **76** 084028
[4] Bianchi E, Regoli D and Rovelli C 2010 *Preprint* arXiv:1005.0764
[5] Mamone D and Rovelli C 2009 *Class. Quant. Grav.* **26** 245013
[6] Rovelli C and Zhang M 2011 *Preprint* arXiv:1105.0566
[7] Miković A and Vojinović M 2011 *Class. Quant. Grav.* **28** 225004
[8] Miković A and Vojinović M 2011 *Preprint* arXiv:1101.3294
[9] Barrett J W, Dowdall R J, Fairbairn W J, Hellmann F and Pereira R 2009 *Preprint* arXiv:0907.2440
[10] Barrett J W, Dowdall R J, Fairbairn W J, Gomes H, Hellmann F and Pereira R 2010 *Preprint* arXiv:1003.1886
[11] Miković A and Vojinović M 2011 *Preprint* arXiv:1103.1428
[12] Engle J 2012 *Preprint* arXiv:1201.2187
[13] Conrady F and Freidel L 2008 *Phys. Rev.* D **78** 104023
[14] Magliaro E and Perini C 2011 *Europhys. Lett.* **95** 30007
[15] Nair V P 2005 *Quantum Field Theory: A Modern Perspective* (New York: Springer Science + Business Media, Inc.)

Journal of Physics: Conference Series **360** (2012) 012050 doi:10.1088/1742-6596/360/1/012050

Einstein-Regge equations in spinfoams[1]

Claudio Perini
Institute for Gravitation and the Cosmos, Physics Department, Penn State, University Park, PA 16802-6300, USA

E-mail: claude.perin@gmail.com

Abstract. We consider spinfoam quantum gravity on a spacetime decomposition with many 4-simplices, in the double scaling limit in which the Immirzi parameter γ is sent to zero (flipped limit) and the physical area in Planck units (γ times the spin quantum number j) is kept constant. We show that the quantum amplitude takes the form of a Regge-like path integral and enforces Einstein equations in the semiclassical regime. In addition to quantum corrections which vanish when the Planck constant goes to zero, we find new corrections due to the discreteness of geometric spectra which is controlled by the Immirzi parameter.

1. Introduction

Spinfoams [2, 3] are a tentative covariant quantization of general relativity. They provide transition amplitudes between quantum states of 3-geometry, in the form of a Misner-Hawking sum over virtual geometries. In the so-called 'new models' [4, 5] considered here, these virtual states are the ones of canonical loop quantum gravity, the $SU(2)$ spin-network states, a remarkable feature that promotes the spinfoam framework to a tentative path integral representation of Hamiltonian loop quantum gravity.

A major open problem is to understand how the semiclassical expansion of the spinfoam Feynman path integral reproduces classical Einstein equations. We expect something similar to the formal semiclassical approximation of the gravitational path integral

$$\int Dg_{\mu\nu} e^{\frac{i}{\hbar}S_{EH}(g_{\mu\nu})} \sim e^{\frac{i}{\hbar}S_{EH}(g^0_{\mu\nu})} \tag{1}$$

in the classical limit $\hbar \to 0$, where S_{EH} is the Einstein-Hilbert action for general relativity, evaluated on the r.h.s. on the classical solution $g^0_{\mu\nu}$ determined by the boundary conditions on the metric field $g_{\mu\nu}$. In [6, 7] the author (in collaboration with E. Magliaro) have proposed a tentative solution to this problem that has a compelling interpretation in terms of the interplay between *two* physical scales in loop quantum gravity: the Planck length and the minimal physical length of loop gravity. To simplify the presentation, we work here in Euclidean signature.

2. Spinfoam amplitudes: statement of the semiclassical regime

A dual simplicial 2-complex σ is a collection of vertices, faces, and edges which are dual to 4-simplices, triangles and tetrahedra respectively. The 2-complex is decorated in the following way. For each face f there is an associated discrete area $a_f = \gamma\hbar j_f(j_f+1)$, with j_f an integer

[1] Based on a talk by the author at the conference *Loops '11* in Madrid on May 24th 2011 [1].

Journal of Physics: Conference Series **360** (2012) 012050 doi:10.1088/1742-6596/360/1/012050

spin. Faces are oriented and bounded by a cycle of edges e. The positive numer γ is the Barbero-Immirzi parameter. Here we consider $0 < \gamma < 1$. Each edge bounding a face has a source vertex $s(e)$ and a target vertex $t(e)$, where source/target is relative to the orientation of the face. To each edge let us associate $SU(2)$ elements n_{ef} (f runs over the faces meeting at the edge e) and two source/target $Spin(4) \simeq SU(2) \times SU(2) \sim SO(4)$ gauge group variables $g_{e,s(e)}, g_{e,t(e)}$. The variables n_{ef} can also be interpreted as unit vectors $\vec{n}_{ef}$ in $\mathbb{R}^3$, up to a phase ambiguity, by saying that n is a rotation that brings a reference direction to the direction of $\vec{n}$.

The spinfoam amplitude [4, 5] for a 2-complex σ without matter is given by

$$Z(\sigma) = \sum_{\{a_f\}} \int dg_{ve} \int d\vec{n}_{ef} \, \exp\left(\frac{S}{\hbar} + \frac{S'}{\gamma\hbar}\right), \tag{2}$$

where S and S' are gauge-invariant actions. We restrict our analysis to the nondegenerate part of the integral (the 4-volume of any 4-simplex cannot be arbitrarily small, that is smaller than a fixed $\delta > 0$) as done in [8].

Now we state the semiclassical regime we are interested in as follows. We introduce two physical scales: the first one is the length scale λ_{QG} of *quantum gravity*, that can be identified with the Planck length, the second one is the scale λ_{LQG} of *loop quantum geometry*, that is the scale where we can 'see' the discreteness of spacetime, according to the predictions of loop quantum gravity. We have

$$l_{QG} \equiv l_P, \quad l_{LQG} \equiv \sqrt{\gamma}\, l_P. \tag{3}$$

The semiclassical regime of the spinfoam path-integral we are looking for is a two-scale problem expressed by the following relation:

$$l \gg l_{QG} \gg l_{LQG}, \tag{4}$$

where l is the typical linear scale of each 4-simplex in the spacetime triangulation. Formally, we study this regime by letting $\hbar$ go to zero and at the same time letting γ be small. This is done in a two-step stationary phase approximation as explained in the following.

3. The flipped $\gamma \to 0$ limit

The first step is the flipped limit expansion $\gamma \to 0$, small Immirzi parameter, with physical area a kept constant ('flipped' is named after the 'flipped' spinfoam model [9] following a suggestion by C. Rovelli). In other words, restoring the spin quantum numbers, we consider large spins in such a way that

$$j \to \infty, \ \gamma \to 0, \ j\gamma = const. \tag{5}$$

and γj is the macroscopic physical area in Planck area units. Notice that the Immirzi parameter controls the spacetime discreteness. In particular it controls the area gap and the spacing between area eigenvalues, thus the limit (5) is a continuum limit for the area operator.

Because the coefficient of S' in (2) is large, we use the extended stationary phase method to study the 'flipped' regime and compute the main contribution to the path integral. The action S' is complex with nonpositive real part $\operatorname{Re} S' \leq 0$, so we expect that the main contribution can come only from the configurations such that $\operatorname{Re} S' = 0$. Splitting the group $SO(4)$ in selfdual and antiselfdual parts ($+$ and $-$), this condition is equivalent to the glueing equations

$$g^{+}_{ev}\vec{n}_{ef} = -g^{+}_{e'v}\vec{n}_{e'f} \tag{6}$$

$$g^{-}_{ev}\vec{n}_{ef} = -g^{-}_{e'v}\vec{n}_{e'f} \tag{7}$$

Journal of Physics: Conference Series **360** (2012) 012050 doi:10.1088/1742-6596/360/1/012050

where e, e' are adjacent edges in the face f, sharing the vertex v. Furthermore, we require that the critical points are stationary points, because otherwise the rapidly oscillating phase would suppress the amplitude. Varying S' with respect to $SO(4)$ group variables, and evaluating at a critical point we get the closure condition

$$\delta_{g_{ev}} S'\big|_{crit.} = 0 \longrightarrow \sum_{f\in e} a_f \vec{n}_{ef} = 0, \tag{8}$$

namely the closure relation for the tetrahedron dual to the edge e. The variations with respect to the unit vectors $\vec{n}_{ef}$ and with respect to the areas a_f do not impose further restrictions, once the previous conditions are imposed.

It can be shown (see [7] and also the more recent [10]) that the glueing and closure equations imply that a nondegenerate critical configuration $(j_f, \vec{n}_{ef}, g_{ev})$ describe a 4-dimensional spacetime Regge triangulation. More precisely, the equivalence classes of parity-related nondegenerate critical points are in one-to-one correspondence with Regge triangulations. The areas of the triangulation are given by a_f, the 3-dimensional normals to the faces of the tetrahedra are specified by $\vec{n}_{ef}$, and the Levi-Civita connection is coded in the holonomies $g_{e,v}$. The simplicial metric tensor can be built out of these on-shell variables.

Curvature is distributional and concentrated on triangles. Moreover, the evaluation of the full action at a nondegenerate critical point with globally consistent parity gives the Regge action

$$S + \frac{1}{\gamma} S'|_{crit.} = iS_{\text{Regge}} = i\sum_f a_f \Theta_f \tag{9}$$

of general relativity, where the deficit angle Θ_f codes the curvature concentrated on the triangle dual to the face f. The other parity-related, possibly non globally consistent critical points yield a generalization of the Regge action (see also [11]). Each of them contibutes to the small γ approximation of the path integral. In general this is very difficult to compute explicitly and depends on the simplicial 2-complex chosen, but we expect it to take the form of a sum over the set $\mathcal{C}$ of critical points, with some induced measure μ. The result is

$$Z(\sigma) = \int_{\mathcal{C}} d\mu(a_f, n_{ef}, g_{ev}) e^{iS_{\text{Regge}}} + \text{p.r.} + \mathcal{O}(\gamma) \tag{10}$$

where p.r. are the parity-related terms, and $\mathcal{O}(\gamma)$ denotes the γ-corrections to the partition function that come from the next orders of the 'flipped' asymptotic approximation.

Thus the spinfoam amplitude reduces to a Regge-like quantum gravity path integral [12] given by formula (10), up to γ-corrections. The latter take into account the effect of quantum discreteness of volumes and areas of loop quantum gravity. This is a remarkable feature of the 'flipped' limit $\gamma \to 0$. The result resonates with the recent computations of the spinfoam graviton propagator [13]. In particular, the leading order graviton propagator $G(x, y)$ in the $\hbar$ expansion shows the same kind of γ-corrections,

$$G(x, y) = \frac{R + \gamma X + \gamma^2 Y}{|x - y|^2} + \hbar\text{-corr.} \tag{11}$$

and only in the limit $\gamma \to 0$ with physical area kept constant the 2-point function matches with the one computed in Regge gravity (the term proportional to the matrix R in (11)). It also matches with a suitable projection of the standard propagator of perturbative gravity on flat background. The γ-corrections in (11) were also shown to break the parity symmetry of the propagator. Interestingly, similar corrections were found by Bojowald in the cosmological context [14].

4. The semiclassical $\hbar \to 0$ limit

Starting from (10), we can parametrize the integral using length variables. Indeed given that the nondegenerate critical points correspond to Regge triangulations, there must exist an assignment of lengths l_s to the sides s of the triangles such that the areas a_f coincide with the areas computed out of the lengths: $a_f = a_f(l_s)$. Let us parametrize the set of critical points $\mathcal{C}$ with the side lengths l_s of the triangulation, and rewrite (10) as

$$Z(\sigma) = \int d\tilde{\mu}(l_s) e^{\frac{i}{\hbar} S_R(l_s)} + \text{p.r.} + \mathcal{O}(\gamma), \tag{12}$$

where the Regge action is now viewed as a function of the lengths.

Now we perform the second step of our semiclassical expansion which is the usual classical limit obtained formally in the asymptotic regime $\hbar \to 0$, as in a WKB expansion. Let us concentrate on the first term in (12). In the classical limit the stationary lengths of the Regge action dominate the integral. Using also the Schlafli identity, which tells us that the variation of the deficit angles do not contribute to the total variation of the action, these are the solutions of the Regge equations [15]

$$\sum_f \frac{\partial a_f}{\partial l_s} \Theta_f = 0, \tag{13}$$

a simplicial version of the Einstein equations in vacuum, $R_{\mu\nu} = 0$, namely of the vanishing of the Ricci tensor. The simplicial Einstein equations impose a relation between the deficit angles of different faces. Notice that there may be semiclassical contributions from the parity-related terms, for which a geometric interpretation is less transparent to us. Suppose the physical problem we are studying is such that there is only one solution of the Regge equations compatible with the boundary data; then by the stationary phase method the path integral (12) goes like

$$Z(\sigma) \sim e^{iS_{\text{Regge}}} + \text{p.r.} + \mathcal{O}(\gamma), \qquad \hbar \to 0 \tag{14}$$

where S_{Regge} is now evaluated on the solution of (13), and thus can be interpreted as the Hamilton-Jacobi function. The last equation is a realization of the semiclassical approximation (1) in spinfoams, and shows that in the two-scale limit considered here the Einstein-Regge equations correctly emerge.

We leave the study of the degenerate configurations as an open problem.

[1] Slides at `http://loops11.iem.csic.es/loops11/index.php?option=com_content&view=article&id=108`
[2] Baez J C 1998 *Class. Quant. Grav.* **15** 1827–1858 (*Preprint* gr-qc/9709052)
[3] Reisenberger M and Rovelli C 2000 (*Preprint* gr-qc/0002083)
[4] Engle J, Livine E, Pereira R and Rovelli C 2008 *Nucl. Phys.* **B799** 136–149 (*Preprint* 0711.0146)
[5] Freidel L and Krasnov K 2008 *Class. Quant. Grav.* **25** 125018 (*Preprint* 0708.1595)
[6] Magliaro E and Perini C 2011 (*Preprint* 1105.0216)
[7] Magliaro E and Perini C 2011 *Europhys.Lett.* **95** 30007 (*Preprint* 1108.2258)
[8] Conrady F and Freidel L 2008 *Phys. Rev.* **D78** 104023 (*Preprint* 0809.2280)
[9] Engle J, Pereira R and Rovelli C 2008 *Nucl.Phys.* **B798** 251–290 (*Preprint* 0708.1236)
[10] Han M and Zhang M 2011 (*Preprint* 1109.0500)
[11] Barrett J W and Foxon T 1994 *Class.Quant.Grav.* **11** 543–556 (*Preprint* gr-qc/9310016)
[12] Rocek M and Williams R M 1981 *Phys.Lett.* **B104** 31
[13] Bianchi E, Magliaro E and Perini C 2009 *Nucl. Phys.* **B822** 245–269 (*Preprint* 0905.4082)
[14] Bojowald M 2001 *Class.Quant.Grav.* **18** L109–L116 (*Preprint* gr-qc/0105113)
[15] Regge T 1961 *Nuovo Cim.* **19** 558–571

Journal of Physics: Conference Series **360** (2012) 012051 doi:10.1088/1742-6596/360/1/012051

Shape of spinfoams encoded in graphs

J Puchta

Instytut Fizyki Teoretycznej, Uniwersytet Warszawski, ul. Hoża 69, 00-681 Warszawa (Warsaw), Polska (Poland)

Centre de Physique Theorique de Luminy, Case 907, Luminy, F-13288 Marseille, France

E-mail: jpa@fuw.edu.pl

Abstract. As the EPRL model was evolving into its present form, a precise identification of the family of complexes (foams) used in the definition of the spinfoams has been abandoned. Most often mentioned possibilities were: simplicial complexes, linear complexes or locally linear complexes. We have solved that problem and we present a new compact characterization of a general foam. Our characterization leads also to a new type of diagrammatic notation directly from which a spinfoam amplitude can be written. This allows one to consider the diagrammatic notation as an alternative to standard 2-complex approach.

1. Motivation

The spinfoam models provide a covariant description of dynamics of Loop Quantum Gravity. The framework defines a way to calculate transition amplitudes between boundary spinnetwork states. Since the states are labeled by graphs, the key object in spinfoam theories are 2-complexes considered as time-histories of spinnetworks.

First of all, the only foams considered were 2-complexes dual to triangulations of a 4-manifold (see, e.g., [2]). Then became clear that the spinfoam amplitude can be calculated for arbitrary vertex [3]. It turned out that to calculate a vertex amplitude one needs data that can be expressed as a spinfoam, i.e. the intersection of a colored 2-complex with a little 3-sphere centered at the considered vertex.

To calculate the spinfoam amplitude for a given boundary spinnetwork state one has to sum the amplitudes given by all 2-complexes of a given boundary (or calculate the amplitude of a maximally refined 2-complex, which is equivalent [4]). However the precise class of 2-complexes that one should consider in this calculation has not been defined yet.

We propose such a class. Our idea is based on an observation: given any spinfoam one can decompose it into a set of spinnetworks (one per each vertex - see [3, 5]) with some relations between nodes and links of these spinnetworks - representing the way in which vertices are connected by edges and faces. Thus let us reverse this procedure and say that each relevant spinfoam is given by some set of graphs (spinnetworks) with their nodes and links connected by some relations. Our class will consist of all 2-complexes that can be obtained by such construction.

The key issue is now to prove that for every such initial data one can construct a 2-complex. The catch is in the relations between nodes and links: from spinfoams one can obtain some of them, but can one get all the 2-complexes? We will prove it by giving a detailed construction of a 2-complex from the set of (arbitrarily related) graphs, which is the core of the talk.

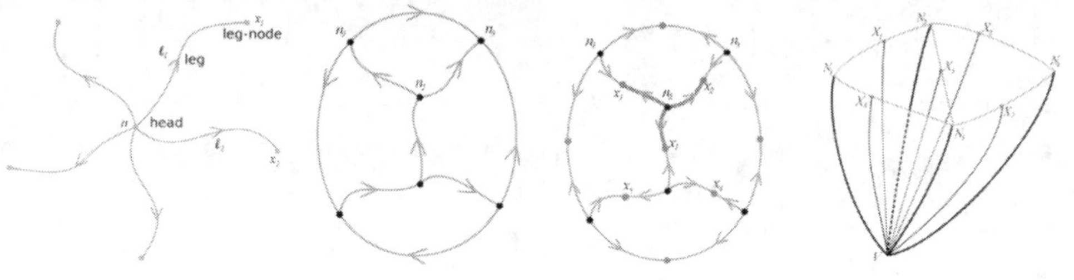

(a) A squid (b) A graph (c) A squid-graph (d)

Figure 1. The idea of squid-graphs. The squid λ_{n_2} is highlighted in blue at (c). Some edges must be reoriented when passing from (b) to (c), but since one considers spinnetworks, the information is not lost - it passes to the coloring. (d) A one-vertex foam obtained by shrinking the boundary squid-graph (green).

The problem is treated in detail (and extended) in the paper of Marcin Kisielowski, Jerzy Lewandowski and JP [1], especially in section 3. Moreover, as one can find in section 2 of [1], our construction allows to rewrite the spinfoam amplitude in a transparent form of a contraction of some projection operators with some tensors (see also [3]), what is the subject of the complementary talk of Marcin Kisielowski.

2. Construction

Let us make a remark: the paper [5] defines some operations on spinfoams, i.e. edge- and face-splitting and edge/face reorientation. In what follow we consider equivalent any two 2-complexes that differ only by these operations.

2.1. *Ingredients - graph diagrams*

The construction is started from the following ingredients (called a *graph diagram*):

- A set of graphs $A = \{\gamma_1, \ldots, \gamma_N\}$.
- A family of homeomorphic maps $\phi_{n \to n'}$: $\mathcal{D}_n \to \mathcal{D}_{n'}$, where $\mathcal{D}_n$ is a small neighborhood of a node n of one of graphs. Each node appears *at most* once, either as n or as n' in the family.

The maps $\phi_{n \to n'}$ indicates the gluing of pieces of 2-complex. It is important which link ℓ incident to n is glued to which link ℓ' incident to n'. That is why the maps are required to be homeomorphic.

2.2. *Squid-graphs*

The following technical definition of *squid-graph* makes further steps (gluings) more transparent. Let us define a *squid* as a graph consisting of a selected node, called *head*, and a number of nodes, called *leg-nodes*, where every one is connected only with the *head* by a single link, called *leg* - see Fig. 1(a).

A sufficiently small neighborhood of each node of a graph is a squid. Thus given a graph γ we define its decomposition into set of squids $\mathcal{S}_\gamma$ as follows:

(i) split each link ℓ of γ into two links by putting a node x_ℓ on it.

(ii) Each node n_i of the old graph γ becomes a head of one squid λ_{n_i}.

(iii) Each link $\tilde{\ell}$ of the new graph $\tilde{\gamma}$ has one end x_ℓ and one end n_i. Each such link becomes a leg of the squid λ_{n_i} - see Fig. 1(c).

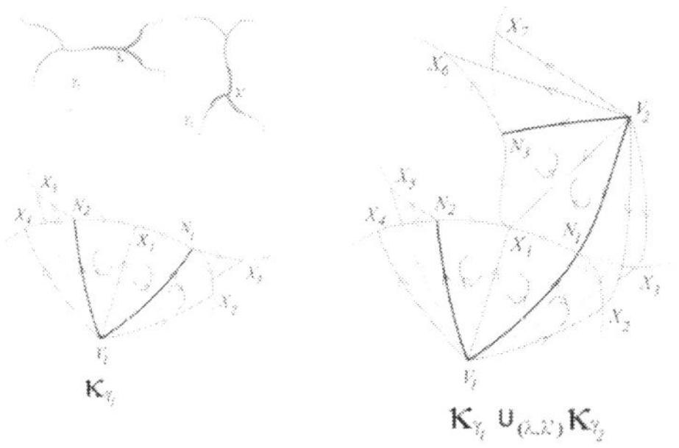

Figure 2. Gluing of two 2-complexes along a squid. First figure presents two fragments of graphs, that will be glued. Below there is a fragment of 1-vertex foam before gluing (the second one looks analogous). The figure at the right hand side presents the 2-complex after gluing. All possible types of edges are visible here.

The pair $(\tilde{\gamma}, \mathcal{S}_\gamma)$ will be called a *squid-graph*. As the first step of our construction let us construct a squid-graph $(\tilde{\gamma}, \mathcal{S}_\gamma)$ out of each graph $\gamma \in A$. It is straightforward to see that each map $\phi_{n\to n'}$ defines a morphism of squids $\Phi_{\lambda_n,\lambda_{n'}}$, i.e. a map saying which leg of λ_n is identified with which leg of $\lambda_{n'}$.

2.3. *One-vertex foams*
Given any graph γ one can construct a 2-complex κ_γ being a neighborhood of a vertex with structure given by the graph. The prescription is precisely the one used to obtain the spinnetwork used in evaluation of the vertex amplitude [3] but reversed (see Fig. 1(d)):

(i) Draw the graph γ on a 3-sphere of radius 1.

(ii) Shrink the sphere to a point. The tracks of links of γ give faces of κ_γ, the tracks of nodes of γ are internal edges of κ_γ, the middle point of the ball is the only internal vertex of κ_γ.

We call κ_γ a *one-vertex foam*. The boundary 1-complex of κ_γ is the graph γ itself. Thus if γ has a squid-graph structure, the boundary of κ_γ inherit the squid-decomposition.

Let us construct a one-vertex foam κ_γ out for each $\gamma \in A$ and take their disjoint union $\kappa_0 := \bigsqcup_{\gamma\in A} \kappa_{\tilde{\gamma}}$.

2.4. *Gluing along pair of squids*
Given any 2-complex κ with boundary decomposed into squids and any map $\Phi_{\lambda,\lambda'}$ being morphism between two boundary squids let us define gluing κ along λ and λ' as follows:

- All the components of κ (i.e. faces, edges, vertices, relations between them) remain unchanged, but,
- the edges and vertices that belong to λ or λ' are identified according to $\Phi_{\lambda,\lambda'}$ - see Fig. 2,
- the set of boundary squids is reduced by its subset $\{\lambda, \lambda'\}$.

The resulting 2-complex is called $\kappa/\Phi_{\lambda,\lambda'}$.

2.5. *Induction*
Gluing along any two pairs of squids commutes (see App. A of [1]). Thus having a set of maps $\{\Phi_{\lambda_1,\lambda_1'}, \ldots, \Phi_{\lambda_E,\lambda_E'}\}$ one may glue κ_0 along them in arbitrary order obtaining the same unique 2-complex: $\kappa_0 \to \kappa_0/\Phi_{\lambda_1,\lambda_1'} \to \ldots \to (\ldots(\kappa_0/\Phi_{\lambda_1,\lambda_1'})/\ldots)/\Phi_{\lambda_E,\lambda_E'} =: \kappa_{final}$ which is the one we are looking for.

3. Properties of obtained 2-complexes
Below we present a brief characterization of the 2-complexes that might be obtained using our prescription.

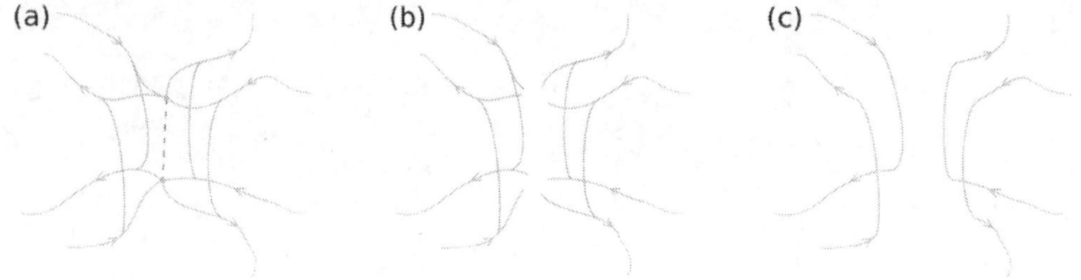

Figure 3. A procedure of building the boundary graph of a graph diagram. First remove a squid containing both glued nodes - (a)→(b), then connect the remaining links according to the map Φ - (b)→(c).

3.1. Edges
Edges of three types appear in our 2-complexes (see Fig. 2):

(i) VN edges being traces of heads of squids. They are always internal edges, each of them is at least 2-valent (in general they are 3-valent or more).
(ii) VX edges are traces of leg-nodes of squids. They are always internal and 2-valent edges, thus they are called *removable* (see face-splitting in [5]).
(iii) NX edges are legs of squids. They are either boundary edges (if the squid to which they belong was not glued) or internal 2-valent (thus *removable*) edges (if the squid was glued to another squid).

3.2. Boundary
The gluing procedure induces the following simple algorithm to find the boundary graph of the 2-complex κ_{final}:

- Given a map $\Phi_{\lambda,\lambda'}$ remove the squids λ and λ' from the graphs they belong.
- Glue the remaining links of the graphs according to the map $\Phi_{\lambda,\lambda'}$ - see Fig. 3.
- Repeat it for each of maps $\Phi_{\lambda_i,\lambda'_i}$.

3.3. Faces
Each face that appears in the complex κ_{final} is a triangle with one vertex being the middle of one of the balls (V), the second one being a head of one of squids (N) and the third one being a leg-node of the squid (X). However, some edges are trivial and thus removable. Removing of this edges results in obtaining the so-called generalized faces being several triangles glued. Each of such generalized faces is topologically a disc glued onto a skeleton. However, in some cases gluing onto the skeleton may cause the face to have topology more complicated than a disc. An example is shown at Fig. 4.

4. Summary and outlook
We have proven that given any graph diagram one can reconstruct a 2-complex out of it. We claim that all such 2-complexes constitute the class that should be used when considering spinfoams. The class is wide enough to contain 2-complexes with faces of nontrivial topology. However, detailed study and characterization of the class is still needed. Moreover our construction is compatible with the decomposition of the spinfoam amplitude into projection

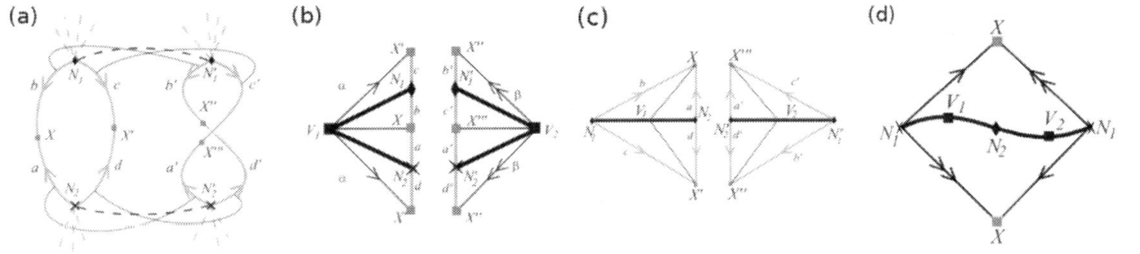

Figure 4. An example of obtaining a face with topology of a projective plane. (a) A fragment of a graph diagram. (b) Fragments of one vertex foams, cut along edges for transparency. (c) The same fragments of one vertex foams, now without the cuts. (d) The face with all removable edges erased. The bold line $N_1V_1N_2V_2N_1$ is built of the edges (the 1-skeleton of 2-complex) onto which the face is glued.

operators and tensors called contractors, what may simplify some calculations based on complicated (e.g. knoted) spinfoams.

Acknowledgments

We thank the organizers of Loops 11 Conference for the possibility to give this talk. There is not enough room here to thank all the other people properly, the paper [1] list them all.

The work was realized in the project "International PhD Studies in Fundamental Problems of Quantum Gravity and Quantum Field Theory" of Foundation for Polish Science, cofinanced from the program IE OP 2007-2013 within European Regional Development Fund.

References

[1] Kisielowski M, Lewandowski J and Puchta J (2012) Class. Quant. Grav. **29**, 015009
[2] Engle J, Livine E, Pereira R and Rovelli C (2008) Nucl. Phys. B **799**, 136-149
[3] Kamiński W, Kisielowski M and Lewandowski J (2010) Class. Quant. Grav. **27**, 095006
[4] Rovelli C and Smerlac M (2010) *In quantum gravity, summing is refining* (Preprint 10105437v3 [gr-qc])
[5] Bahr B, Hellmann F, Kamiński W, Kisielowski M and Lewandowski J (2011) Class. Quant. Grav. **28**, 105003

Journal of Physics: Conference Series **360** (2012) 012052 doi:10.1088/1742-6596/360/1/012052

Non-commutative dual representation for quantum systems on Lie groups

Matti Raasakka

Max Planck Institute for Gravitational Physics (Albert Einstein Institute), Am Mühlenberg 1, 14476 Potsdam, Germany, EU

E-mail: `matti.raasakka@aei.mpg.de`

Abstract. We provide a short overview of some recent results on the application of group Fourier transform to quantum mechanics and quantum gravity. We close by pointing out some future research directions.

1. Introduction

The group Fourier transform is an integral transform from functions on a Lie group to functions on a non-commutative dual space. It was first formulated for functions on SO(3) [1, 2], later generalized to SU(2) [2–4] and other Lie groups [5, 6], and is based on the quantum group structure of Drinfel'd double of the group [7, 8]. The transform provides a unitarily equivalent representation of quantum systems with a Lie group configuration space in terms of non-commutative algebra of functions on the classical dual space, the dual of the Lie algebra. As such, it has proven useful in several different ways. Most importantly, it provides a clear connection between the quantum system and the corresponding classical one, allowing for a better physical insight into the system. In the case of quantum gravity models this has turned out to be particularly helpful in unraveling the geometrical content of the models, because the dual variables have an intuitive interpretation as classical geometrical quantities.

We will first review the general formulation of the group Fourier transform, and then mention its recent applications to quantum mechanics on $SO(3)$, and to Loop Quantum Gravity and spin foam models. We will close by pointing out some future reseach directions.

2. Group Fourier transform

Let $\mathcal{G}$ be a (finite dimensional) Lie group, and $\mathfrak{g}^*$the dual of the Lie algebra $\mathfrak{g}$ of $\mathcal{G}$. The group Fourier transform is an isometry between $L^2(\mathcal{G}, \mathrm{d}g)$ and a non-commutative function space $L^2_\star(\mathfrak{g}^*, \mathrm{d}X)$ defined in the following way [6]: We choose a function $E : \mathcal{T}^*\mathcal{G} = \mathcal{G} \times \mathfrak{g}^* \to U(1)$, $(g, X) \mapsto E_g(X)$, the non-commutative plane wave, such that we may decompose the delta distribution on $\mathcal{G}$ in terms of the Fourier modes in $\mathfrak{g}^*$ as $\frac{1}{\kappa^d}\delta_e(g) = \int_{\mathfrak{g}^*} \frac{\mathrm{d}X}{(2\pi\hbar)^d} E_g(X)$. Here $\mathrm{d}X$ the Lebesgue measure on $\mathfrak{g}^*$, $d = \dim \mathcal{G}$, $\kappa \in \mathbb{R}_+$ is a deformation parameter, which determines the physical dimensions on $\mathcal{G}$, and $\hbar$ is the Planck constant. We then define a non-commutative $\star$-product via the relation $E_g(X) \star E_h(X) \equiv E_{gh}(X)$. The $\star$-product is extended to functions on $\mathfrak{g}^*$ through linearity by defining the group Fourier transform of $\phi \in L^2(\mathcal{G}, \mathrm{d}g)$ as $\tilde{\phi}(X) = \int_{\mathcal{G}} \kappa^d \mathrm{d}g\ E_g(X)\phi(g)$, where $\mathrm{d}g$ is the (left invariant) Haar measure on $\mathcal{G}$. Due to above properties, the inverse transform is obtained as $\phi(g) = \int_{\mathfrak{g}^*} \frac{\mathrm{d}X}{(2\pi\hbar)^d} E_g(X) \star \tilde{\phi}(X)$. We denote the image of $L^2(\mathcal{G}, \mathrm{d}g)$ under the transform as $L^2_\star(\mathfrak{g}^*, \mathrm{d}X)$, since by requiring $E_{g^{-1}}(X) = \overline{E_g(X)}$ the transform is an isometry from $L^2(\mathcal{G}, \mathrm{d}g)$ to $L^2_\star(\mathfrak{g}^*, \mathrm{d}X)$, i.e.,

$$\int_{\mathcal{G}} \kappa^d \mathrm{d}g\, \overline{\phi(g)}\phi'(g) = \int_{\mathfrak{g}^*} \frac{\mathrm{d}X}{(2\pi\hbar)^d} \overline{\tilde{\phi}(X)} \star \tilde{\phi}'(X) \ . \quad (1)$$

Accordingly, we obtain a unitarily equivalent non-commutative dual representation of fields living on $\mathcal{G}$ via the group Fourier transform.

The choice for the explicit form of the non-commutative plane waves $E_g(X)$ is not obvious. Typically the plane waves are of the form $E_g(X) = \exp(i \sum_i Z_i(g) \cdot X^i)$, where $Z_i(g)$ are some coordinates on the group manifold. In particular, [1, 2, 9, 10] use the coordinates $Z^i(g) = -\frac{i}{2} \operatorname{tr}(g\sigma^i)$ for $SO(3)$, where the trace is taken in the fundamental spin-$\frac{1}{2}$-representation and σ^i are the Pauli matrices. For $SU(2)$ the construction must be modified, as for example in [2–4], since the above coordinates are two-to-one in this case, $Z^i(g^{-1}) = Z^i(-g)$. In [2,3] an extra parameter was introduced to make the transform one-to-one for $SU(2)$. For an exponential Lie group the canonical choice for coordinate functions Z^i on $\mathcal{G}$ can be obtained from the inverse of the exponential map $\exp : \mathfrak{g} \simeq \mathbb{R}^d \to \mathcal{G}$, $Z \mapsto \exp Z$, for a portion of $\mathfrak{g}$ containing the origin, where the exponential map is one-to-one. With this choice we have the natural inner product $\cdot : \mathfrak{g} \times \mathfrak{g}^* \to \mathbb{R}$ to use for the plane waves $E_g(X) = \exp(iZ(g) \cdot X)$.[1]

The non-commutative variables $X \in \mathfrak{g}^*$ can be seen to correspond to the classical dual cotangent space variables in the way of deformation quantization: The canonical Lie derivative operators $\hat{X}_i = -i\frac{\hbar}{\kappa}\mathcal{L}_i$, where $\mathcal{L}_i$ are Lie derivatives with respect to a basis in the Lie algebra, satisfy the commutation relations $[\hat{X}_i, \hat{X}_j] = -i\frac{\hbar}{\kappa}c_{ij}{}^k \hat{X}_k$, where $c_{ij}{}^k$ are the structure constants of $\mathfrak{g}$. These commutators correctly reflect the classical Poisson algebra of the phase space $\mathcal{T}^*\mathcal{G}$ arising from the symplectic structure of the cotangent bundle. If as usual we have $\mathcal{L}_i E_e(X) = i\frac{\kappa}{\hbar} X_i$, then the non-commutative plane wave is the generating function for $\star$-polynomials of coordinates of the non-commutative dual space, $\hat{X}_{i_1} \cdots \hat{X}_{i_n} E_g(X) = E_g(X) \star X_{i_1} \star \cdots \star X_{i_n}$, and we recover the Lie algebraic commutation relations for the non-commutative coordinates $[X_i, X_j]_\star = X_i \star X_j - X_j \star X_i = -i\frac{\hbar}{\kappa} c_{ij}{}^k X_k$. Here we see that the quantity $(\hbar/\kappa)$ controls the non-commutativity, but it is κ^{-1}, which is the deformation parameter associated to the $\star$-product in this case, since $\hbar$ plays a distinct role [10]. In particular, κ determines the physical dimensions of the group, present even in the classical theory, whereas $\hbar$ is the parameter controlling the quantum fluctuations. Therefore there are two distinct limits of a quantum theory with the phase space $\mathcal{T}^*\mathcal{G}$ to be considered, the classical limit $\hbar \to 0$ and the commutative limit $\kappa^{-1} \to 0$. In the commutative limit the Lie algebra commutators vanish, and the group coincides essentially with the tangent space $\mathfrak{g} \simeq \mathbb{R}^d$ at the unit element.

3. Phase space path integral for quantum mechanics on $SO(3)$

Taking advantage of the group Fourier transform, we obtain a dual non-commutative representation of quantum mechanics on $\mathcal{G}$ in terms of a set of states $\{|X\rangle \mid X \in \mathfrak{g}^*\}$ by setting $\langle g|X\rangle := E_g(X)$, where Dirac notation is used. These states constitute a basis with respect to the non-commutative $\star$-product [6, 10]. We may then write down the first order phase space path integral. As usual, we have a Hamiltonian operator $\hat{H}$ generating the time-evolution of the system. Then, the propagator for the quantum system in $\mathcal{G}$-basis reads $\langle g', t'|g, t\rangle := \langle g'|e^{-i(t'-t)\hat{H}/\hbar}|g\rangle$. We then introduce time-splitting $(t'-t) \equiv \epsilon N$ and insert

[1] In fact, since the $\star$-product may be defined only under integration, we only need $\mathcal{G}$ to be weakly exponential, i.e., the image of the exponential map to be dense in $\mathcal{G}$. (See [11] and references therein for details on exponential Lie groups.) If this is not the case, it may be possible to generalize to several coordinate patches as in [3].

resolutions of identity $\hat{\mathbb{1}} \equiv \int_{\mathcal{G}} \kappa^d \mathrm{d}g \; |g\rangle\langle g|$ and $\hat{\mathbb{1}} \equiv \int_{\mathfrak{g}^*} \frac{\mathrm{d}X}{(2\pi\hbar)^d} \; |X\rangle \star \langle X|$ for each timestep. To obtain the path integral, we take the limits $\epsilon \to 0$, $N \to \infty$, while $\epsilon N = (t' - t)$. For the path integral to satisfy the Schrödinger equation, only the linear order in ϵ must be taken into account [12], and we may approximate $\langle X_k|e^{-i\epsilon\hat{H}/\hbar}|g_k\rangle \approx e^{-i\epsilon H_\star(g_k, X_k)/\hbar} \star \langle X_k|g_k\rangle$, where $\langle g|\hat{H}|X\rangle =: \langle g|X\rangle \star H_\star(g, X)$. Further, using the properties of the plane waves, we arrive at the discrete timestep expression for the path integral

$$\langle g', t'|g, t\rangle = \lim_{\substack{\epsilon \to 0 \\ N \to \infty}} \left[\prod_{k=1}^{N-1} \int_{\mathcal{G}} \kappa^d \mathrm{d}g_k\right] \left[\prod_{k=0}^{N-1} \int_{\mathfrak{g}^*} \frac{\mathrm{d}X_k}{(2\pi\hbar)^d}\right] \exp\left\{\frac{i}{\hbar} \sum_{k=0}^{N-1} \epsilon \left(\frac{1}{\epsilon} V_k \cdot X_k - H_{q,k}\right)\right\}, \quad (2)$$

where $V_k := Z(g_k^{-1} g_{k+1})$, $H_{q,k} := H_q(g_k, X_k)$, $Z^i(g)$ are the coordinates on $\mathcal{G}$ used in the plane waves, and $H_q(g, X) := [\omega(-i\partial_X^i)]^{-1} H_\star(g, X)$ is *the quantum corrected Hamiltonian* for the system [6]. Here $\omega(Z^i(g))\mathrm{d}^d Z \equiv \mathrm{d}g$, where $\mathrm{d}^d Z$ is the Lebesgue measure on the coordinates Z^i. These quantum corrections to the classical Hamiltonian, which are due to the non-commutative structure of the phase space, are crucial for the propagator to satisfy the Schrödinger equation [10].

For the special case of $\mathcal{G} = \mathbb{R}^d$ with the coordinates $Z^i(g)$ obtained via the inverse exponential map (2) agrees with the usual expressions for the first order path integral. In [10] the stationary phase approximation was studied for $SO(3)$ with coordinates $Z^i(g) = -(i/2)\,\mathrm{tr}(g\sigma^i)$ and found to yield the correct classical equations of motion in the limit $\hbar \to 0$. This further confirms the identification of the non-commutative dual variables as corresponding to the classical canonically conjugate variables. Also, it was shown that for the case of a free particle on $SO(3)$ the quantum corrections agree with the well-established results [12–14]. In the limit $\kappa \to \infty$ the path integral coincides with the one for $\mathcal{G} = \mathbb{R}^d$ as the diameter of the group becomes infinite. Thus the first order path integral facilitates access to semi-classical as well as 'semi-commutative' approximations for quantum systems on Lie groups.

4. Non-commutative metric variables for quantum gravity

The group Fourier transform has recently provided a better insight into the geometrical content of quantum gravity models. The configuration space of Loop Quantum Gravity (LQG), for example, which corresponds to a graph Γ representing the quantum geometry of a spatial slice of spacetime, is $SO(3)^E/SO(3)^V$, where E and V are the numbers of edges and vertices in Γ [15]. Here the $SO(3)^E$ part corresponds to parallel transports along the edges of the embedded graph, and $SO(3)^V$ corresponds to local gauge transformations at the vertices[2]. From the canonical analysis of LQG it can be seen that the dual variables to the parallel transports along edges of Γ taking values in $SO(3)$ are area bivectors valued in $\mathfrak{so}(3)$ associated to faces dual to the edges of Γ. In [16] the group Fourier transform for $SO(3)$ was used to formulate the metric representation for the LQG state space $L^2(SO(3)^E/SO(3)^V)$. In this case the non-commutative variables of the dual space $L^2_\star((\mathfrak{so}(3)^*)^E)$ to $L^2(SO(3)^E)$ correspond to these area bivectors. Integrating out the gauge degrees of freedom imposes geometrical closure constraints on the bivectors, so that the faces bound polyhedra dual to the vertices of Γ. Therefore the non-commutative metric representation obtained via the group Fourier transform provides a precise connection between LQG and simplicial geometry.

Group field theories are quantum field theoretical models with field arguments taking values in Lie groups, which give a covariant path integral formulation of quantum gravity as a sum over

[2] Usually the group $SU(2)$, the double cover of $SO(3)$, is used for the group variables associated to the edges of the graph. However, we will restrict to $SO(3)$, since the construction of the non-commutative metric representation for LQG has been carried out only in this case. The generalization to $SU(2)$ should be straightforward by using one of the already existing transforms for $SU(2)$.

spin foam geometries. (See e.g. [17,18] for further details.) Using the group Fourier transform, a metric representation for 3d group field theory, the Boulatov model, which corresponds to a path integral quantization of discrete 3d BF theory, was formulated in [5,9,19]. (For spin foam models the corresponding non-commutative methods were considered in [20], and the relation between group field theory models and deformed κ-Poincaré symmetry via group Fourier transform was also considered in [21].) Importantly, the non-commutative metric representation led to the realization of diffeomorphism symmetry in the colored 3d model [22], which however due to the deformed coproduct of the symmetry algebra still needs to be properly implemented on the quantum level in the framework of braided quantum field theory [23] as pointed out in [9,19]. Another important feature of the metric representation is that the simplicity constraints on the bivector variables, taking the 4d BF theory to general relativity, can be imposed directly and in a geometrically clear way on the classical phase space variables [24,25].

5. Future directions

The group Fourier transform was recently generalized [6] to other Lie groups besides copies of $SO(3)$ and $SU(2)$, which opens up the possibility for formulating the non-commutative metric representation of 4d Lorenzian spin foam models. This should allow for a host of developments in the 4d case, as it has done in the 3d case. In particular, the metric representation should allow for a convenient imposition of the simplicity constraints at the level of the classical phase space variables, and a better understanding of the geometrical symmetries of the models. Also, the study of the semi-classical limit of the models is expected to be greatly facilitated by the non-commutative methods, since no excursion to the elaborate representation theory and special functions is necessary. In addition to these interesting possibilities, the commutative limit can also be used to develop convenient approximations to the full model. A pursuit in this direction of research is currently under way.

References

[1] Freidel L and Livine E 2006 *Class. Quant. Grav.* **23** 2021 (*Preprint* arXiv:hep-th/0502106)
Freidel L and Livine E 2006 *Phys. Rev. Lett.* **96** 221301 (*Preprint* arXiv:hep-th/0512113)
[2] Freidel L and Majid S 2008 *Class. Quant. Grav.* **25** 045006 (*Preprint* arXiv:hep-th/0601004)
[3] Joung E, Mourad J and Noui K 2009 *J. Math. Phys.* **50** 052503 (*Preprint* arXiv:0806.4121)
[4] Dupuis M, Girelli F and Livine E 2011 *Preprint* arXiv:1107.5693
[5] Livine E 2009 *Class. Quant. Grav.* **26** 195014 (*Preprint* arXiv:0811.1462)
[6] Raasakka M 2011 *Preprint* arXiv:1111.6481
[7] Majid S 2000 *Foundations of Quantum Group Theory* (Cambridge: Cambridge University Press)
[8] Batista E and Majid S 2003 *J. Math. Phys.* **44** 107 (*Preprint* arXiv:hep-th/0205128)
[9] Baratin A and Oriti D 2010 *Phys. Rev. Lett.* **105** 221302 (*Preprint* arXiv:1002.4723)
[10] Oriti D and Raasakka M 2011 *Phys. Rev.* D **84** 025003 (*Preprint* arXiv:1103.2098)
[11] Moskowitz M and Sacksteder R 2008 *Math. Res. Lett.* **15** 1197
[12] DeWitt B 1957 *Rev. Mod. Phys.* **29** 377
[13] Chaichian M and Demichev A 2001 *Path Integrals in Physics, Volume I* (London: IOP Publishing)
[14] Marinov M S and Terentyev M V 1979 *Fortsch. Phys.* **27** 511
[15] Thiemann T 2007 *Modern Canonical Quantum General Relativity* (Cambridge: Cambridge University Press)
[16] Baratin A, Dittrich B, Oriti D and Tambornino J 2011 *Class. Quant. Grav.* **28** 175011 (*Preprint* arXiv:1004.3450)
[17] Freidel L 2005 *Int. J. Theor. Phys.* **44** 1769 (*Preprint* arXiv:hep-th/0505016)
[18] Oriti D 2011 *Foundations of Space and Time*, ed J Murugan *et al* (Cambridge: Cambridge University Press)
[19] Girelli F and Livine E 2010 *Class. Quant. Grav.* **27** 245018 (*Preprint* arXiv:1001.2919)
[20] Bonzom V and Livine E 2009 *Phys. Rev.* D **79** 064034 (*Preprint* arXiv:0812.3456)
[21] Girelli F, Livine E and Oriti D 2010 *Phys. Rev.* D **81** 024015 (*Preprint* arXiv:0903.3475)
[22] Baratin A, Girelli F and Oriti D 2011 *Phys. Rev.* D **83** 104051 (*Preprint* arXiv:1101.0590)
[23] Oeckl R 2001 *Comm. Math. Phys.* **217** 451 (*Preprint* arXiv:hep-th/9906225)
[24] Baratin A and Oriti D 2011 *Preprint* arXiv:1108.1178
[25] Baratin A and Oriti D 2011 *Preprint* arXiv:1111.5842

Journal of Physics: Conference Series **360** (2012) 012053 doi:10.1088/1742-6596/360/1/012053

Spinfoams: summing or refining?

Matteo Smerlak (in collaboration with Carlo Rovelli)
Centre de Physique Théorique, Campus de Luminy Case 907, 13288 Marseille cedex 9, France
E-mail: smerlak@cpt.univ-mrs.fr

Abstract. We consider two alternative prescriptions for the continuum limit in spinfoam gravity – *summing* over all foams vs. *refining* a single foam –, and argue that, under certain conditions, they are the same.

The spinfoam program is based on truncations of the quantum gravitational field on finite 2-complexes, aka foams. How does one take the *continuum limit* in this context? Should one *sum* over all foams, as in perturbative quantum field theory? Or should one *refine* a single foam, as in lattice gauge theory? Both alternatives have been considered in the spinfoam literature [1, 2, 3, 4], and – to this day – neither has won the consensus. Here, we argue that they may turn out to be equivalent: summing and refining could be the two faces of the same limit [5].

At the heuristic level, there are good reasons to believe that such a coincidence could arise. Indeed, recall that spinfoam quantum gravity operates without a background metric, for it is meant to be the quantum theory *of the metric itself.* This means that each foam is *both* a lattice, viz. a discretization of spacetime, and a Feynman diagram, viz. a virtual history of field quanta.[1]

To make this intuition precise, the key mathematical concept is that of *net.* Let X be a (Hausdorff) topological space. A *net* N is an X-valued function on a *directed set* S, i.e. a partially ordered set in which any two elements have an upper bound. A net is *convergent to* $N_\infty \in X$, if for any neighborhood U of N_∞, there is $s_U \in S$ such that $N(s) \in U$ whenever $s \geq s_U$. Thanks to the directedness of S, the *limit* $\lim_S N = N_\infty$, when it exists, is unique.

Define a *foam* as a triple $\Gamma = (V, E, F)$ with V a finite set of *vertices*, E a set of ordered pairs of vertices $e = \big(s(e), t(e)\big)$, called *edges*, and F a finite set of *faces*. Here, we call *face* a finite sequence of edges $f = (e_1^{\epsilon_1}, ..., e_n^{\epsilon_n}, ..., e_{n_f}^{\epsilon_{n_f}})$, where $t(e_n^{\epsilon_n}) = s(e_{n+1}^{\epsilon_{n+1}})$, $\epsilon_n = \pm 1$ and $n_f + 1 := 1$. Say that two foams are *isomorphic* if there exists a bijective map between the vertex sets V_1 and V_2 respecting all adjacency relations, and call *u-foam* an isomorphism class $[\Gamma]$ of foams (u stands for "unlabeled").

It is not hard to see that the (discrete) set $\mathcal{F}$ of all u-foams with a given boundary forms a directed set with respect to the relation defined by $[\Gamma_1] \leq [\Gamma_2]$ if Γ_1 can be embedded into Γ_2. In a nutshell, this is because, given any two foams, it is always possible to attach to them as many vertices, edges and faces as necessary to embed them both into a third, finer, foam.

A *spinfoam model* $\mathcal{Z}$ can then be thought of as a net over $\mathcal{F}$, with values in a Hilbert space $\mathcal{H}$ of "quantum states". This observation suggests an obvious definition for the *refinement limit* of $\mathcal{Z}$, namely the limit $\mathcal{Z}_\infty = \lim_{\mathcal{F}} \mathcal{Z}$. What is the relation between this limit and the sum $\sum_{[\Gamma]\in\mathcal{F}} \mathcal{Z}([\Gamma])$ (assuming both exist)?

[1] See e.g. [6] for more details on the physical interpretation of spinfoam gravity.

Published under licence by IOP Publishing Ltd

Journal of Physics: Conference Series **360** (2012) 012053 doi:10.1088/1742-6596/360/1/012053

As a rule, spinfoam amplitudes $\mathcal{Z}([\Gamma])$ are defined as sums over colorings σ of the faces of a foam $\Gamma \in [\Gamma]$ by *spins*, i.e. unitary irreducible representations of some compact Lie group. To each such coloring is associated an amplitude $z(\Gamma, \sigma)$, and

$$\mathcal{Z}([\Gamma]) = \sum_{\sigma \in \mathrm{Col}(\Gamma)} z(\Gamma, \sigma), \tag{1}$$

where $\mathrm{Col}(\Gamma)$ is the set of face-colorings of Γ. We can also consider the alternative definition

$$\mathcal{Z}^*([\Gamma]) = \sum_{\sigma \in \mathrm{Col}^*(\Gamma)} z(\Gamma, \sigma), \tag{2}$$

where now $\mathrm{Col}(\Gamma)$ denotes the set of face-colorings of Γ with trivial representations excluded (non-zero spins).

The relationship between the continuum limit $\mathcal{Z}_\infty$ and the sum of $\mathcal{Z}([\Gamma])$ over all u-foams with a fixed boundary is then the following: under a general condition on the amplitude $\mathcal{Z}_\Gamma(\sigma)$, which we detail in the next paragraph, we have

$$\mathcal{Z}([\Gamma]) = \sum_{[\Gamma'] \leq [\Gamma]} \mathcal{Z}^*([\Gamma']), \tag{3}$$

and hence, passing to the limit (assuming it exists),

$$\mathcal{Z}_\infty = \sum_{[\Gamma] \in \mathcal{F}} \mathcal{Z}^*([\Gamma]). \tag{4}$$

This is our central result. It states that refining foams in the model $\mathcal{Z}$ is the same as summing over foams in the model $\mathcal{Z}^*$. (Note that, due to the peculiar net structure, all foams appear in (4) only in the following sense: every u-foam $[\Gamma]$ appears in one finite sum (3) whose value can be chosen arbitrarily close to $\mathcal{Z}_\infty$.

The condition on $z(\Gamma, \sigma)$ for the relation (4) to hold is one of *cylindrical consistency*. First, observe that each coloring σ of a foam Γ comes with a multiplicity, related to the symmetries of Γ. The *multiplicity* $|\sigma|_\Gamma$ of a coloring $\sigma \in \mathrm{Col}(\Gamma)$ is the number of colorings σ' such that $\sigma = \sigma' \circ \phi$, with ϕ an automorphism of Γ. Then we say that the amplitude

$$a(\Gamma, \sigma) := |\sigma|_\Gamma \, z(\Gamma, \sigma), \tag{5}$$

is *cylindrically consistent* if $a(\Gamma, \sigma) = a(\Gamma', \sigma')$ when (Γ', σ') is a *trivial extension* of (Γ, σ), that is when Γ is a subfoam of Γ', σ and σ' coincide on the faces of Γ and σ' is trivial on the other faces of Γ'.

Under this condition, the proof of (3) is straightforward. First, observe that the subfoams of Γ index a partition of $\mathrm{Col}(\Gamma)$, in which each each class is made of the trivial extensions of a given subfoam $\Gamma' \subset \Gamma$:

$$\mathrm{Col}(\Gamma) = \bigsqcup_{\Gamma' \subset \Gamma} \mathrm{Col}^*(\Gamma'). \tag{6}$$

This implies that

$$\mathcal{Z}([\Gamma]) = \sum_{\Gamma' \subset \Gamma} \left(\sum_{\sigma' \in \mathrm{Col}^*(\Gamma')} |\sigma'|_\Gamma^{-1} a(\Gamma, \sigma') \right). \tag{7}$$

Second, check that we have

$$|\sigma'|_\Gamma = |\sigma'|_{\Gamma'} N_{\Gamma', \Gamma}, \tag{8}$$

with $N_{\Gamma',\Gamma}$ the number of subfoams of Γ isomorphic to Γ'. Third, use cylindrical consistency to get

$$\mathcal{Z}([\Gamma]) = \sum_{\Gamma' \subset \Gamma} N^{-1}_{\Gamma',\Gamma} \mathcal{Z}^*([\Gamma']) \tag{9}$$

and conclude to (3).

This result in principle applies equally well to the Euclidean and Lorentzian spinfoam models of quantum gravity [6] – assuming convergence issues can be suitably addressed, of course.

Acknowledgements

This work has benefited from critical remarks by Jerzy Lewandowski and Aristide Baratin. I also thank Jose Antonio Zapata for drawing my attention to his work on the continuum limit of spinfoam models.

References

[1] Zapata J A 2002 *Journ. Math. Phys.* **43** 12
[2] Reisenberger M P and Rovelli C 2000 *Class. Quant. Grav.* **18** 32
[3] Oriti D 2009 *Approaches to Quantum Gravity - toward a new understanding of space, time, and matter* (*Preprint* 0607032)
[4] Rivasseau V 2011 (*Preprint* 1103.1900v1)
[5] Rovelli C and Smerlak M 2010 (*Preprint* 1010.5437)
[6] Rovelli C 2004 *Quantum Gravity* (Cambridge University Press)

Journal of Physics: Conference Series **360** (2012) 012054 doi:10.1088/1742-6596/360/1/012054

Local gauge theory and coarse graining

José A. Zapata
Centro de Ciencias Matemáticas, UNAM, 58190 Morelia, Mexico
E-mail: zapata@matmor.unam.mx

Abstract. Within the discrete gauge theory which is the basis of spin foam models, I study the problem of macroscopically faithful coarse graining. Macroscopic data is identified; it contains the holonomy evaluation along a discrete set of loops and the homotopy classes of certain maps. When two configurations share this data they are related by a local deformation. The interpretation is that such configurations differ by "microscopic details". In many cases the homotopy type of the relevant maps is trivial for every connection; two important cases in which the homotopy data is composed by a set of integer numbers are: (i) a two dimensional base manifold and structure group $U(1)$, (ii) a four dimensional base manifold and structure group $SU(2)$. These cases are relevant for spin foam models of two dimensional gravity and four dimensional gravity respectively. This result suggests that if spin foam models for two-dimensional and four-dimensional gravity are modified to include all the relevant macroscopic degrees of freedom –the complete collection of macroscopic variables necessary to ensure faithful coarse graining–, then they could provide appropriate effective theories at a given scale.

1. Connections (modulo gauge) in terms of local holonomy variables

Let (E, π, M) be a principal G-bundle over a d-dimensional base space M and $\mathcal{A}_\pi$ be the space of connections on it.

In order to introduce a concept of measuring scale and locality we use a smooth triangulation Δ of M; alternatively, we can use a cellular decomposition that refines a smooth triangulation. This decomposition is inherited by the bundle.

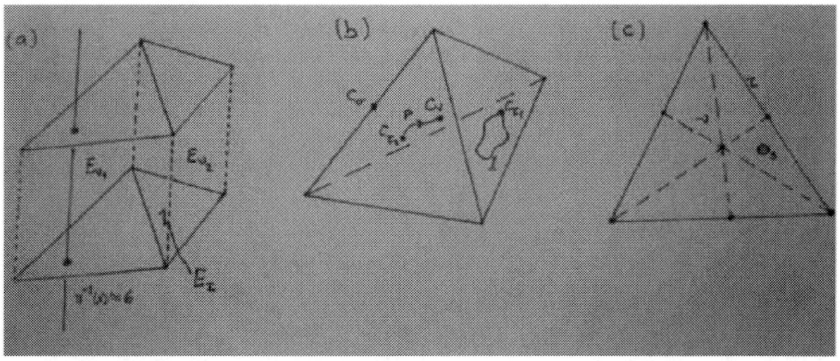

Figure 1. (a) The smooth G-bundles π_{ν_1} and π_{ν_2} intersect at the smooth G-bundle π_τ.
(b) Paths and loops in $\mathcal{P}(\nu)$.
(c) $\mathrm{Sd}(\nu)$, the baricentric subdivision of ν. We also show s, the support of B in the example.

Journal of Physics: Conference Series **360** (2012) 012054 doi:10.1088/1742-6596/360/1/012054

Definition 1 *Decomposition of the bundle (see figure 1.a):*

- *The principal G-bundle over M, $(E_\Delta, \pi_\Delta, M_\Delta)$, is called Δ-smooth if its restriction to π_ν^{-1} for each simplex ν of the triangulation is the restriction of a smooth bundle to an open neighborhood of π_ν^{-1}.*
- *$\mathcal{A}^\infty_{\pi_\tau}$ is the space of smooth connections in the bundle $\pi_\tau = \pi|_{\tau\in(|\Delta|,\phi)}$ (restrictions to π_τ of connections which are smooth in an open neighborhood of the subbundle π_τ of π, where τ is a simplex in the triangulation).*
- *$\bar{\times}_{\tau\in(|\Delta|,\phi)}\mathcal{A}^\infty_{\pi_\tau}$ is the subset of the cartesian product defined by the compatibility condition $\sigma \subset \tau \Rightarrow A_\sigma = A_\tau|_\sigma$ for every two simplices of the triangulation. We will write $\mathcal{A}^{\Delta\text{-}\infty}_\pi = \bar{\times}_{\tau\in(|\Delta|,\phi)}\mathcal{A}^\infty_{\pi_\tau}$.*
- *A Δ-smooth bundle map is a homeomorphism of the total space $\tilde{f} : E \to E$ which preserves each subbundle π_τ and is the restriction to π_τ of a smooth bundle map on an open neighborhood of it. Clearly a Δ-smooth bundle map acts on connections by pull back and sends $\mathcal{A}^{\Delta\text{-}\infty}_\pi$ to itself.*
- *$\mathcal{G}^{\Delta\text{-}\infty}_{\star,\pi}$ is the group of Δ-smooth bundle equivalence maps which consists of Δ-smooth bundle maps which induce the identity map on the base space M. We will write $\mathcal{A}/\mathcal{G}^{\Delta\text{-}\infty}_{\star,\pi} = \mathcal{A}^{\Delta\text{-}\infty}_\pi/\mathcal{G}^{\Delta\text{-}\infty}_{\star,\pi}$.*

Now I present local holonomy data for the bundle and the connection. It is the application of a construction of Barrett and Kobayashi to the subbundles over the simplices.

In simplex ν of the triangulation choose its baricenter $C\nu$ as base point, and consider the space $\mathcal{L}(C\nu, \nu)$ of piecewise smooth oriented $C\nu$-based loops modulo reparametrization and modulo retracing. It turns out that $\mathcal{L}(C\nu, \nu)$ is a group. Once we have identified G with $\pi^{-1}(C\nu)$ the holonomy of a given connection $A \in \mathcal{A}^\infty_{\pi_\nu}$ A_π around a loop $l \in \mathcal{L}(C\nu, \nu)$ is a group element $\mathrm{Hol}(l, A) \in G$. Moreover, if we fix the connection the holonomy map gives us a group homomorphism

$$\mathrm{Hol}_A \equiv \mathrm{Hol}(\cdot, A) : \mathcal{L}(C\nu, \nu) \to G.$$

Gauge transformations act on holonomies only at the base point and the action is by conjugation.

Holonomy evaluations are specified by points of the space $\mathcal{A}/\mathcal{G}_{\star,\pi_\nu}$, the space of connections modulo the group of gauge transformations whose restriction to the fiber $\pi^{-1}(C\nu)$ is the identity. Holonomy evaluations characterize the bundle over ν (modulo bundle equivalence) and the connection (modulo gauge); this is an application of a theorem of Barrett [1].

Theorem 1 (Barrett, Kobayashi) *Let $\mathrm{H}^\nu : \mathcal{L}(C\nu, \nu) \to G$ be a group homomorphism which is smooth*[1]*. Then, there is a differentiable principal G-bundle (E_ν, π_ν, ν), a point $b \in \pi^{-1}(C\nu)$, and a connection $A \in \mathcal{A}_{\pi_\nu}$ such that $\mathrm{H}^\nu = \mathrm{Hol}_A$. The bundle and the connection are unique up to a bundle equivalence transformation.*

The theorem does not only state the existence of the bundle and the connection; its proof is constructive. Here is a sketch of the main idea. The total space is given by

$$E_\nu = G \times \mathcal{P}_{C_\nu} / \sim_{\mathrm{Hol}^\nu}$$

where $\mathcal{P}_{C_\nu}$ is the space of paths in ν (modulo reparametrization and retracing) whose source is C_ν. The idea behind the equivalence relation is that a pair $(g, p) \in G \times \mathcal{P}_{C_\nu}$ gives a point on the fiber over $t(p)$, the target of p; g specifies a point on the fiber at the beginning of the path and it is parallel transported to the end of the path. One only has to acknowledge that

[1] The definition of smoothness is axiom H3 in Barrett's article, stating that any smooth finite dimensional family of loops leads to a smooth submanifold in G after composition with H^ν. See [1] for a detailed explanation.

Journal of Physics: Conference Series **360** (2012) 012054 doi:10.1088/1742-6596/360/1/012054

(g_1, p_1) and (g_2, p_2) determine the same point if $t(p_1) = t(p_2)$ and the parallel transports agree, $\mathrm{PT}_{p_1} g_1 = \mathrm{PT}_{p_2} g_2$. This last condition can also be written as $g_1 \mathrm{H}^\nu(p_2^{-1} \circ p_1) = g_2$.

If this theorem is applied to each simplex of the smooth triangulation, the connection and the bundle will be described by means of holonomies. However, a collection of such holonomy evaluations for each simplex will be related to a bundle and a connection on it only if certain compatibility conditions hold. The pasting of descriptions over the different simplices needs extra data. The extra data glues the bundle over a simplex to the bundle over any simplex that intersect it.

Consider two simplices of the smooth triangulation such that $\tau \subset \nu$. Thus, for each point $x \in \tau$ the above construction gives two constructions of π_x: $E_\tau|_x$ and $E_\nu|_x$. The glue for these bundles is again parallel transport. There are two groups $\mathcal{L}(C\tau, \tau)$ and $\mathcal{L}(C\nu, \nu)$; and there are also paths in ν between $C\tau$ and $C\nu$, $\mathcal{P}^{\tau}_{C_\nu}$. The three spaces can be considered together forming a semigroup; two of its elements can be composed only if the target of the first is the same as the source of the second. Consider all the subsimplices of ν, $\tau_i \subset \nu$. The semigroup of paths in ν which have sources and targets at C_ν or any of the C_{τ_i} will be called $\mathcal{P}(\nu)$ (see figure 1.b). Notice that this semigroup contains $\mathcal{L}(C\nu, \nu)$ and $\mathcal{L}(C\tau_i, \tau_i)$ for each $\tau_i \subset \nu$. Once the fibers over C_ν and all the C_{τ_i}s have been identified with G, parallel transport is described by assignment of group elements, and parallel transport is a semigroup homomorphism

$$\mathrm{PT}^\nu : \mathcal{P}(\nu) \to G.$$

PT^ν contains all the information in H^ν and each of the H^{τ_i}; in addition, it induces a further equivalence relation that glues the bundle over ν with the bundles over each τ_i. For example, $(g_1, p_1) \in E_\nu|_x$ is equivalent to $(g_2, p_2) \in E_\tau|_x$ if $g_1 \mathrm{PT}^\nu(p_2^{-1} \circ p_1) = g_2$.

One can extend this construction to glue the bundles over all the simplices of the smooth triangulation to obtain a local version of theorem of Barrett and Kobayashi. The ingredients are $\mathcal{P}(\Delta)$ -the semigroup associated to the smooth triangulation of the manifold- and a parallel transport map which will be called simply PT. $\mathcal{P}(\Delta)$ is generated by composing paths (modulo reparamerization and retracing) contained in some simplex of the triangulation and with source and target being baricenters of simplices of the triangulation. PT is required to be a Δ-smooth semigroup homomorphism. The smoothness condition means that its restriction to paths contained in any simplex is the restriction of a map on paths contained in an open neighborhood of the simplex which is smooth in the sense of Barrett [1]. A detailed study of the abelian case was presented at [2].

Theorem 2 *Let M be a manifold, Δ a smooth triangulation of it, and let* $\mathrm{PT} : \mathcal{P}(\Delta) \to G$ *be a Δ-smooth semigroup homomorphism. Then, there is a Δ-smooth G-bundle (E, π_Δ, M), a collection of points $\{b_\nu \in \pi^{-1}(C\nu)\}_{\nu \in \Delta}$, and a connection $A \in \mathcal{A}_{\pi_\Delta}$ such that* $\mathrm{PT} = \mathrm{PT}_A$. *The bundle and the connection are unique up to a bundle equivalence transformation.*

2. Macroscopic variables and faithful coarse graining

The last theorem describes the bundle and the connection from a parallel transport map. Since this theorem also applies to manifolds with boundary, it extends Theorem 1 gluing the bundles over each simplex using the parallel transport data of paths that link the centers of different simplices.

Notice that since bundles over discs are trivial and simplices have the topology of the disc, Theorem 1 applied to each simplex constructs a trivial bundle (and a connection on it); that is, the bundle over a given simplex is essentially independent of the parallel transport map used to build it. Different parallel transport maps may yield inequivalent bundles only because they glue these *standard building blocks* in inequivalent ways chosen by the parallel transport data.

Journal of Physics: Conference Series **360** (2012) 012054 doi:10.1088/1742-6596/360/1/012054

Each $\mathcal{P}(\nu)$ is decomposed into subsemigroups that intersect each other as dictated by the simplicial structure of triangulation

$$\nu_1 \cap \nu_2 = \tau \Rightarrow \mathcal{P}(\nu_1) \cap \mathcal{P}(\nu_2) = \mathcal{P}(\tau).$$

This property lets us treat one simplex at a time, and see the non trivial gluing just be seeing how the bundle over a simples is glued to the bundles over its boundary faces (its lower dimensional subsimplices).

If $\tau \subset \partial\nu$ then Theorem 2 provides π_ν, π_τ and a gluing map $I_{\tau\nu} : \pi_\tau \to \pi_\nu$ determined by PT^ν. We will show that there is macroscopic data that can be extracted from PT^ν assuring that if $\mathrm{Data}_\Delta(\mathrm{PT}^\nu_1) = \mathrm{Data}_\Delta(\mathrm{PT}^\nu_2)$ then $I_{\tau\nu} : \pi_\tau \to \pi_\nu(\mathrm{PT}^\nu_1)$ is equivalent to $I_{\tau\nu} : \pi_\tau \to \pi_\nu(\mathrm{PT}^\nu_2)$. A realization of this goal in the abelian case was presented at [2].

Below is an example of inequivalent gluings. It shows that $\mathrm{Data}_\Delta(\mathrm{PT}^\nu)$ must include some data a part from a set of parallel transport evaluations along a discrete collection of paths and loops.

Example:
Let $G = U(1)$ and $M \sim S^2$ with Δ a smooth triangulation of it; also let $\nu \in \Delta$ be a triangle and $\tau \in \Delta$ be one of its one dimensional faces. The baricentric subdivision of ν, $\mathrm{Sd}(\nu)$, is shown in figure 1.c.

In S^2 we have a parallel transport map PT_1 such that PT^ν_1 is a flat parallel transport (all holonomies are the identity). In an appropriate local trivialization over ν, PT^ν_1 would be described by a connection one form $A_1 = 0$. We have a second parallel transport map PT_2 such that for every $\delta \in \Delta$ such that $\delta \neq \nu$ we have $\mathrm{PT}^\delta_2 = \mathrm{PT}^\delta_1$.

Let us use the trivialization over ν according to which $A_1 = 0$. PT^ν_2 described in the same trivialization is characterized by a connection one form A_2 determined by a curvature two form B_2 which is zero everywhere except for the compact support $s \subset \nu$ shown in figure 1.c. The size of $B_2|_s$ is adjusted in such a way that $\int_D B_2 = 2\pi$, where D is the triangle in $\mathrm{Sd}(\nu)$ containing s.

All the gluings between bundles over simplices determined by PT_2 agree with those determined by PT_1 except for the gluing of the bundles over ν and τ and ν and ν' (the triangle which shares τ with ν). However, the induced total bundles $\pi_{\Delta,1}$ and $\pi_{\Delta,2}$ are inequivalent since their Euler numbers differ by 1.

Notice that the parallel transport maps PT_1 and PT_2 agree when evaluated on any path contained in the 1-skeleton of $\mathrm{Sd}(\Delta)$. Thus, this discrete set of parallel transport evaluations is not capable of detecting the difference between PT_1 and PT_2. It is clear that given any other choice of discrete set of paths, with an appropriate choice of s we can fabricate a parallel transport map PT_2 which agrees with PT_1 when restricted to the chosen paths while inducing an inequivalent bundle.

The gluing between π_ν and π_τ is a change of trivialization of π_τ, from the one constructed using PT^τ to the one constructed using PT^ν. As any other change of trivialization it is described by an assignment $g : \tau \to U(1)$. In the case of the PT_1 gluing the assignment is $g(x) = \mathrm{id}$ for all $x \in \tau$, while the PT_2 gluing (which needs to agree at $\partial\tau$) winds around $U(1)$ once. For a study of the abelian case see [2].

Now we describe (up to scale Δ) $[\{\pi_\nu \to_A \pi_\tau\}_{\tau\subset\partial\nu}(\text{mod. equivalence}), A \in \mathcal{A}/\mathcal{G}_{\pi_\nu}]$ in terms of parallel transport maps. The macroscopic data needed for such description is $\mathrm{Data}_\Delta(\mathrm{PT}^\nu) = (\mathrm{PT}^\nu_\Delta, W^\nu)$ where

$$\mathrm{PT}^\nu_\Delta : \mathcal{P}_\Delta(\nu) \to G$$

is the restriction of PT^ν to the discrete semigroup of paths $\mathcal{P}_\Delta(\nu) \subset \mathcal{P}(\nu)$ consisting of paths that fit in the 1-skeleton of $\mathrm{Sd}(\nu)$. And,

$$W^\nu \in \{\text{Homotopy class of map } T_{\partial\nu,\tau} : \tau \to G, \text{ with } T|_{\partial\tau} \text{ fixed}\}_{\tau\subset\nu},$$

where the gluing map $T_{\partial\nu,\tau}$ is constructed form PT^ν following a variation of Barrett's construction [1]. The result is the following:

Theorem 3 (Faithful coarse graining) *The data $\mathrm{Data}_\Delta(\mathrm{PT}^\nu)$ characterizes*

- *the gluing maps $\{\pi_\nu \to_{\mathrm{PT}^\nu} \pi_\tau\}_{\tau\subset\partial\nu}$ up to equivalence and*

- *the connection modulo gauge* $A \in \mathcal{A}/\mathcal{G}_{\pi_\nu})$ *up to a microscopical deformation;*

where $A, A' \in \mathcal{A}/\mathcal{G}_{\pi_\nu})$ *are said to agree up to a microscopical deformation if they can be deformed to the same (singular) semigroup homomorphism by a homotopy of semigroup homomorphisms which preserves* $Data_\Delta(\mathrm{PT}^\nu)$.

In the cases relevant for two-dimensioanal and four-dimensional euclidean gravity ([dim$M$ = $2, G = U(1)$] and [dim$M = 4, G = SU(2)$] respectively) W^ν is characterized by a set of integers, one per codimension one face of ν. For many other cases, like the one relevant for three-dimensional euclidean gravity ([dim$M = 3, G = SU(2)$]), W^ν is trivial.

This result suggests that if the current spin foam models for two-dimensional and four-dimensional gravity are modified to include all the relevant macroscopic degrees of freedom –the complete collection of macroscopic variables necessary to ensure faithful coarse graining–, then they could provide appropriate effective theories at a given scale.

A spin foam model for two-dimensional gravity which incorporates extra data regarding the bundle structure was given in [3]. In the same reference a naive spin foam model which ignored the bundle data was shown to yield unphysical results. An explicit relation between the work presented here and [3] will be presented elsewhere.

Acknowledgements
This work was partially supported by CONACyT grant 80118.

3. References

[1] J. W. Barrett, "Holonomy and path structures in general relativity and Yang-Mills theory," Int. J. Theor. Phys. **30**, 1171 (1991).
S. Kobayashi and K. Nomizu, "Foundations of differential geometry," John Wiley & Sons Inc., New York (reprint 1996)
[2] H. G. Diaz-Marin and J. A. Zapata, "Curvature function and coarse graining," J. Math. Phys. **51**, 122307 (2010) [arXiv:1101.3818 [hep-th]].
[3] D. Oriti, C. Rovelli and S. Speziale, "Spinfoam 2-d quantum gravity and discrete bundles," Class. Quant. Grav. **22**, 85 (2005) [gr-qc/0406063].

Journal of Physics: Conference Series **360** (2012) 012055 doi:10.1088/1742-6596/360/1/012055

Loop Quantum Brans-Dicke Theory

Xiangdong Zhang and Yongge Ma

Department of Physics, Beijing Normal University, Beijing 100875, China

E-mail: zhangxiangdong@mail.bnu.edu.cn; mayg@bnu.edu.cn

Abstract. The loop quantization of Brans-Dicke theory (with coupling parameter $\omega \neq -\frac{3}{2}$) is studied. In the geometry-dynamical formalism, the canonical structure and constraint algebra of this theory are similar to those of general relativity coupled with a scalar field. The connection dynamical formalism of the Brans-Dicke theory with real $su(2)$-connections as configuration variables is obtained by canonical transformations. The quantum kinematical Hilbert space of Brans-Dicke theory is constituted as that of loop quantum gravity coupled with a polymer-like scalar field. The Hamiltonian constraint is promoted as a well defined operator to represent quantum dynamics. This formalism enable us to extend the scheme of non-perturbative loop quantum gravity to the Brans-Dicke theory.

1. Introduction

In the past 25 years, loop quantum gravity (LQG), a background independent approach to quantize general relativity (GR), has been widely investigated [1, 2, 3, 4]. Recently, this non-perturbatively loop quantization procedure has been generalized to the metric $f(\mathcal{R})$ theories [5, 6]. In fact, modified gravity theories have recently received increased attention in issues related to "dark Universe" and non-trivial tests on gravity beyond GR. Besides $f(\mathcal{R})$ theories, a well-known competing relativistic theory of gravity was proposed by Brans and Dicke in 1961 [7], which is apparently compatible with Mach's principle. To represent a varying "gravitational constant", a scalar field is non-minimally coupled to the metric in Brans-Dicke theory (BDT). On the other hand, since 1998, a series of independent observations implied that our universe is currently undergoing a period of accelerated expansion [8]. These results have caused the "dark energy" problem in the framework of GR. It is reasonable to consider the possibility that GR is not a valid theory of gravity on a galactic or cosmological scale. The scalar field in BDT of gravity is then expected to account for "dark energy". Furthermore, a large part of the non-trivial tests on gravity theory is related to Einstein's equivalence principle (EEP) [9]. There exist many local experiments in solar-system supporting EEP, which implies the metric theories of gravity. Actually, BDT is a representative metric theory, which has been received most attention. Thus it is interesting to see whether this class of metric theories of gravity could be quantized nonperturbatively. Note that the metric $f(\mathcal{R})$ theories are equivalent to the special kind of BDT with the coupling parameter $\omega = 0$ and some non-vanishing potential of the scalar field [10]. In this work, for simplicity consideration, we only consider BDT with coupling parameter $\omega \neq -\frac{3}{2}$. The connection formalism of BDT is derived from its geometrical dynamics. Based on the resulted connection dynamical formalism, we then quantize the BDT by extending the nonperturbative quantization procedure of LQG in the way similar to loop quantum $f(\mathcal{R})$ gravity. Throughout the paper, we use Greek alphabet for spacetime indices, Latin alphabet a,b,c,..., for spatial indices, and i,j,k,..., for internal indices.

2. Classical and Quantum Aspects of Brans-Dicke Theory

The original action of Brans-Dicke theory reads

$$S(g) = \frac{1}{2}\int_\Sigma d^4x\sqrt{-g}[\phi\mathcal{R} - \frac{\omega}{\phi}(\partial_\mu\phi)\partial^\mu\phi], \quad (1)$$

where we set $8\pi G = 1$, and $\mathcal{R}$ denotes the scalar curvature of spacetime metric $g_{\mu\nu}$. The Hamiltonian analysis of BDT can be found in Refs. [11, 12]. By doing 3+1 decomposition of the spacetime, the four-dimensional scalar curvature can be expressed as

$$\mathcal{R} = K_{ab}K^{ab} - K^2 + R + \frac{2}{\sqrt{-g}}\partial_\mu(\sqrt{-g}n^\mu K) - \frac{2}{N\sqrt{h}}\partial_a(\sqrt{h}h^{ab}\partial_b N), \quad (2)$$

where K_{ab} is the extrinsic curvature of a spatial hypersurface Σ, $K \equiv K_{ab}h^{ab}$, R denotes the scalar curvature of the 3-metric h_{ab} induced on Σ, n^μ is the unit normal of Σ and N is the lapse function. By Legendre transformation, the momenta conjugate to the dynamical variables h_{ab} and ϕ are defined respectively as

$$p^{ab} = \frac{\partial\mathcal{L}}{\partial\dot{h}_{ab}} = \frac{\sqrt{h}}{2}[\phi(K^{ab} - Kh^{ab}) - \frac{h^{ab}}{N}(\dot{\phi} - N^c\partial_c\phi)], \quad (3)$$

$$\pi = \frac{\partial\mathcal{L}}{\partial\dot{\phi}} = -\sqrt{h}(K - \frac{\omega}{N\phi}(\dot{\phi} - N^c\partial_c\phi)), \quad (4)$$

where N^c is the shift vector. The resulted Hamiltonian of BDT can be derived as a liner combination of constraints as $H_{total} = \int_\Sigma d^3x(N^aV_a + NH)$, where the smeared diffeomorphism and Hamiltonian constraints read respectively

$$V(\overrightarrow{N}) = \int_\Sigma d^3xN^aV_a = \int_\Sigma d^3xN^a\left(-2D^b(p_{ab}) + \pi\partial_a\phi\right), \quad (5)$$

$$H(N) = \int_\Sigma d^3xN\Big[\frac{2}{\sqrt{h}}\left(\frac{p_{ab}p^{ab} - \frac{1}{2}p^2}{\phi} + \frac{(p-\phi\pi)^2}{2\phi(3+2\omega)}\right) + \frac{1}{2}\sqrt{h}(-\phi R + \frac{\omega}{\phi}(D_a\phi)D^a\phi + 2D_aD^a\phi)\Big]. \quad (6)$$

Here the condition $\omega \neq -\frac{3}{2}$ was assumed. Lengthy but straightforward calculations show that the constraints comprise a first-class system similar to GR. Since the geometric canonical variables (h_{ab}, p^{ab}) of BDT are the same as those of $f(\mathcal{R})$ theories [6], we can use the same canonical transformations of $f(\mathcal{R})$ theories to obtain the connection dynamical formalism of BDT. To this aim, we let

$$\tilde{K}^{ab} = \phi K^{ab} + \frac{h^{ab}}{2N}(\dot{\phi} - N^c\partial_c\phi). \quad (7)$$

The new geometric variables are $E^a_i = \sqrt{h}e^a_i$ and $A^i_a = \Gamma^i_a + \gamma\tilde{K}^i_a$, where e^a_i is the triad such that $h_{ab}e^a_ie^b_j = \delta_{ij}$, $\tilde{K}^a_i \equiv \tilde{K}^{ab}e^i_b$, Γ^i_a is the spin connection determined by E^a_i, and γ is a nonzero real number. It is clear that our new variable A^i_a coincides with the Ashtekar-Barbero connection [13, 14] when $\phi = 1$. The only non-zero Poisson bracket among the new variables reads $\{A_a^j(x), E_k^b(y)\} = \gamma\delta_a^b\delta_k^j\delta(x,y)$. Now, the phase space of BDT consists of conjugate pairs (A_a^i, E_j^b) and (ϕ, π), with the additional Gaussian constraint $\mathcal{G}_i = \mathscr{D}_a E_i^a \equiv \partial_a E_i^a + \epsilon_{ijk} A_a^j E^{ak}$, which justifies A_a^i as an $su(2)$-connection. The original vector and Hamiltonian constraints can be respectively written up to Gaussian constraint as

$$V_a = \frac{1}{\gamma} F_{ab}^i E_i^b + \pi\partial_a\phi, \tag{8}$$

$$\begin{aligned} H &= \frac{\phi}{2}\left[F_{ab}^j - (\gamma^2 + \frac{1}{\phi^2})\varepsilon_{jmn}\tilde{K}_a^m\tilde{K}_b^n\right]\frac{\varepsilon_{jkl}E_k^aE_l^b}{\sqrt{h}} \\ &+ \frac{1}{3+2\omega}\left(\frac{(\tilde{K}_a^iE_i^a)^2}{\phi\sqrt{h}} + 2\frac{(\tilde{K}_a^iE_i^a)\pi}{\sqrt{h}} + \frac{\pi^2\phi}{\sqrt{h}}\right) + \frac{\omega}{2\phi}\sqrt{h}(D_a\phi)D^a\phi + \sqrt{h}D_aD^a\phi, \end{aligned} \tag{9}$$

where $F_{ab}^i \equiv 2\partial_{[a}A_{b]}^i + \epsilon_{kl}^i A_a^k A_b^l$ is the curvature of A_a^i. All the constraints are of first class. The total Hamiltonian can be expressed as a linear combination $H_{total} = \int_\Sigma \Lambda^i\mathcal{G}_i + N^aV_a + NH$.

Based on the connection dynamical formalism, the nonperturbative loop quantization procedure can be straightforwardly extended to the BDT. The kinematical structure of BDT is the same as that of $f(\mathcal{R})$ theories [5, 6]. The kinematical Hilbert space of the system is a direct product of the Hilbert space of geometry and that of scalar field, $\mathcal{H}_{\rm kin} := \mathcal{H}_{\rm kin}^{\rm gr} \otimes \mathcal{H}_{\rm kin}^{\rm sc}$, with the orthonormal spin-scalar-network basis $T_{\alpha,X}(A,\phi) \equiv T_\alpha(A) \otimes T_X(\phi)$ over some graph $\alpha \cup X \subset \Sigma$. Here α and X consist of finite number of curves and points respectively in Σ. The basic operators are the quantum analogue of holonomies $h_e(A) = \mathcal{P}\exp\int_e A_a$ of connections, densitized triads smeared over 2-surfaces $E(S,f) := \int_S \epsilon_{abc}E_i^a f^i$, point holonomies $U_\lambda = \exp(i\lambda\phi(x))$ [15], and scalar momenta smeared on 3-dimensional regions $\pi(R) := \int_R d^3x\pi(x)$. Note that the whole construction is background independent, and the spatial geometric operators of LQG, such as the area [16], the volume [17] and the length operators [18, 19] are still valid here. As in LQG, it is straightforward to promote the Gaussian constraint $\mathcal{G}(\Lambda)$ to a well-defined operator [2, 4]. Its kernel is the internal gauge invariant Hilbert space $\mathcal{H}_G$ with gauge invariant spin-scalar-network basis. Since the diffeomorphisms of Σ act covariantly on the cylindrical functions in $\mathcal{H}_G$, the so-called group averaging technique can be employed to solve the diffeomorphism constraint [3, 4]. Thus we can also obtain the desired diffeomorphism and gauge invariant Hilbert space $\mathcal{H}_{Diff}$ for the BDT.

Now, we come to implement the Hamiltonian constraint (9) at quantum level. In order to compare the Hamiltonian constraint of BDT with that of $f(\mathcal{R})$ theories in connection formalism, we write Eq. (9) as $H(N) = \sum_{i=1}^7 H_i$. It is easy to see that the terms H_1, H_2, H_7 just keep the same form as those in $f(\mathcal{R})$ theories, and the H_3, H_4, H_5 terms are also similar to the corresponding terms in $f(\mathcal{R})$ theories. Here differences are only reflected by the coefficients. Now we come to the completely new term, $H_6 = \int_\Sigma d^3xN\frac{\omega}{2\phi}\sqrt{h}(D_a\phi)D^a\phi$. We can introduce well-defined operators ϕ, ϕ^{-1} as in Ref. [6]. By the same regularization techniques as in Refs. [6, 20], we triangulate Σ in adaptation to some graph α underling a cylindrical function in $\mathcal{H}_{\rm kin}$ and reexpress connections by holonomies. The corresponding regulated operator $\hat{H}_6^\varepsilon$ can act on a basis vector $T_{\alpha,X}$ over some graph $\alpha \cup X$. It is easy to see that the action of $\hat{H}_6^\varepsilon$ on $T_{\alpha,X}$ is graph changing. It adds a finite number of vertices at $t(s_I(v)) = \varepsilon$ for edges $e_I(t)$ starting from each high-valent vertex of α. As a result, the family of operators $\hat{H}_6^\varepsilon(N)$ fails to be weakly convergent when $\varepsilon \to 0$. However, due to the diffeomorphism covariant properties of the triangulation, the limit operator can be well defined via the so-called uniform Rovelli-Smolin topology induced by diffeomorphism-invariant states Φ_{Diff}. It is obviously that the limit is independent of ε. Hence the regulators can be removed. We then have

$$\begin{aligned}\hat{H}_6 \cdot T_{\alpha,X} &= \sum_{v\in V(\alpha)} \frac{2^{17}N(v)\omega}{3^6\gamma^4(i\lambda_0)^2(i\hbar)^4E^2(v)}\hat{\phi}^{-1}(v)\\ &\times \sum_{v(\Delta)=v(\Delta')=v} \epsilon(s_Ls_Ms_N)\epsilon^{LMN}\hat{U}^{-1}_{\lambda_0}(\phi(s_{s_L(\Delta)}))[\hat{U}_{\lambda_0}(\phi(t_{s_L(\Delta)}))-\hat{U}_{\lambda_0}(\phi(s_{s_L(\Delta)}))]\\ &\times \mathrm{Tr}(\tau_i\hat{h}_{s_M(\Delta)}[\hat{h}^{-1}_{s_M(\Delta)},(\hat{V}_v)^{3/4}]\hat{h}_{s_N(\Delta)}[\hat{h}^{-1}_{s_N(\Delta)},(\hat{V}_v)^{3/4}])\\ &\times \epsilon(s_Is_Js_K)\epsilon^{IJK}\hat{U}^{-1}_{\lambda_0}(\phi(s_{s_I(\Delta')}))[\hat{U}_{\lambda_0}(\phi(t_{s_I(\Delta')}))-\hat{U}_{\lambda_0}(\phi(s_{s_I(\Delta')}))]\\ &\times \mathrm{Tr}(\tau_i\hat{h}_{s_J(\Delta')})[\hat{h}^{-1}_{s_J(\Delta')},(\hat{V}_v)^{3/4}]\hat{h}_{s_K(\Delta')}[\hat{h}^{-1}_{s_K(\Delta')},(\hat{V}_v)^{3/4}])\cdot T_{\alpha,X}. \end{aligned} \tag{10}$$

Thus the total Hamiltonian constraint operator $\hat{H}(N) = \sum_{i=1}^{7}\hat{H}_i$ is well defined in $\mathcal{H}_G$. Furthermore, master constraint programme can be introduced for BDT to avoid possible quantum anomaly and find the physical Hilbert space [11].

3. Conclusions

With the key observation that LQG is based on its $su(2)$-connection dynamical formalism which can be derived via canonical transformations from the geometric dynamics, the $su(2)$-connection dynamics of BDT is obtained. Thus LQG has been successfully extended to the BDT by coupling to a polymer-like scalar field. The quantum kinematical structure of BDT is the same as that of loop quantum $f(\mathcal{R})$ theories. Hence the important physical result that both the area and the volume are discrete remains valid for quantum BDT. While the dynamics of BDT is more general than that of $f(\mathcal{R})$ theories, the Hamiltonian constraint can still be promoted to a well-defined operator in $\mathcal{H}_G$. Hence the classical BDT can be non-perturbatively quantized. Therefore, besides GR and $f(\mathcal{R})$ theories, LQG method is also valid for the BDT of gravity.

Acknowledgments

We thank the organizers of Loops 11 conference for the financial support to our attendance. This work is supported by NSFC (Grant No.10975017) and the Fundamental Research Funds for the Central Universities.

References

[1] C. Rovelli, Quantum Gravity, (Cambridge University Press, 2004).
[2] T. Thiemann, Modern Canonical Quantum General Relativity, (Cambridge University Press, 2007).
[3] A. Ashtekar and J. Lewandowski 2004 Class. Quant. Grav. **21** R53
[4] M. Han, Y. Ma and W. Huang 2007 Int. J. Mod. Phys. D **16** 1397
[5] X. Zhang and Y. Ma 2011 Phys. Rev. Lett. **106** 171301
[6] X. Zhang and Y. Ma 2011 Phys. Rev. **D 84** 064040
[7] C. Brans and R. H. Dicke 1961 Phys. Rev. **124** 925
[8] J. Friemann, M. Turner, D. Huterer 2008 Ann. Rev. Astron. Astrophys. **46** 385
[9] C. M. Will 2006 Living Rev. Rel. **9** 3
[10] T. P. Sotiriou and V. Faraoni 2010 Rev. Mod. Phys. **82** 451
[11] X. Zhang and Y. Ma 2011 Phys. Rev. **D 84** 104045
[12] G. J. Olmo and H. Sanchis-Alepuz 2011 Phys. Rev. **D 83** 104036
[13] A. Ashtekar 1986 Phys. Rev. Lett. **57** 2244
[14] J. Barbero 1995 Phys. Rev. **D 51** 5507
[15] A. Ashtekar, J. Lewandowski, H. Sahlmann 2003 Class. Quant. Grav. **20** L11
[16] C. Rovelli and L. Smolin 1995 Nucl. Phys. B **442** 593
[17] A. Ashtekar and J. Lewandowski 1998 Adv. Theor. Math. Phys. **1** 388
[18] T. Thiemann 1998 J. Math. Phys. **39** 3372
[19] Y. Ma, C. Soo, J. Yang 2010 Phys. Rev. **D 81** 124026
[20] M. Han and Y. Ma 2006 Class. Quant. Grav. **23** 2741

Journal of Physics: Conference Series **360** (2012) 012056 doi:10.1088/1742-6596/360/1/012056

Braided Topology and the Emergence of Matter

Sundance Bilson-Thompson

School of Chemistry and Physics, University of Adelaide, SA 5005, Australia

E-mail: `sundance.bilson-thompson@adelaide.edu.au`

Abstract. In recent years, the possibility that particles may be modelled by topological structures occurring in the discrete spacetime of certain quantum gravity theories has been investigated. We present an overview of this research program, based on a hypothetical substructure of quarks and leptons. Under appropriate dynamical rules, such topological structures are capable of interacting with each other, and we show how the ability to interact may be combined with a mapping to fermion and boson states of the Standard Model.

1. Introduction

During the 1970s and 1980s several models were proposed that treated quarks and leptons as composite fermions, formed from a common set of subcomponents or "preons" (after the nomenclature used in [1]). One particularly interesting example was proposed independently and simultaneously by Harari and Shupe [2, 3]. This model postulates two kinds of preons (called "rishons" in Harari's terminology), one carrying an electric charge of $+e/3$ (where e is the charge on the proton), the other neutral, that combine into triplets. Permutations of the triplets with a mix of neutral and charged rishons are interpreted as colour charge. The resulting list of possible triplets consists of a single neutral triplet (interpreted as the ν_e), three states with charge $+e/3$ (the $\overline{d}$), three states with charge $+2e/3$ (the u), and one state with charge $+e$ (the e^+). The model also incorporates equivalent combinations formed from charged and neutral anti-rishons to create the e^-, $\overline{u}$, d, and $\overline{\nu_e}$. This accounted for the precise ratios of lepton and quark electric charges, and the correspondence between colour charge and fractional electric charge. Gauge bosons were postulated to be formed form a mix of rishons and anti-rishons.

This report discusses a modification of the Harari/Shupe model, dubbed the Helon Model [4], originally described in 2005, along with subsequent work [5] which has developed the idea that topological structures in a discrete spacetime may play the role of preons, combining to form emergent fermions and bosons consistent with those of the Standard Model.

In the notation of the Helon Model, the positively-charged preons are denoted by H_+, the negatively-charged anti-preons by H_-, and the neutral preons by H_0. There are no distinct neutral anti-preons. The construction of the fermions of the first generation is then as shown in the table of Figure 1 (quarks at top, leptons in the bottom row). Notice immediately that without a neutral anti-preon there is no explicit construction of the anti-neutrino.

In Loop Quantum Gravity and related approaches to canonical quantum gravity, discrete spacetime structure may be represented by a network or graph, with nodes dual to volumes and links dual to areas. The simplest graphs to consider are trivalent. In the case of a non-zero cosmological constant, the links become framed (i.e. acquire a width) which allows them to carry twists. When the graph is embedded within a manifold, the links may braid around each

Journal of Physics: Conference Series **360** (2012) 012056 doi:10.1088/1742-6596/360/1/012056

$H_+H_+H_0\,(u_B)$	$H_+H_0H_+\,(u_G)$	$H_0H_+H_+\,(u_R)$
$H_0H_0H_+\,(\overline{d_B})$	$H_0H_+H_0\,(\overline{d_G})$	$H_+H_0H_0\,(\overline{d_R})$
$H_-H_-H_0\,(\overline{u_B})$	$H_-H_0H_-\,(\overline{u_G})$	$H_0H_-H_-\,(\overline{u_R})$
$H_0H_0H_-\,(d_B)$	$H_0H_-H_0\,(d_G)$	$H_-H_0H_0\,(d_R)$
$H_+H_+H_+(e^+)$	$H_0H_0H_0\,(\nu_e)$	$H_-H_-H_-(e^-)$

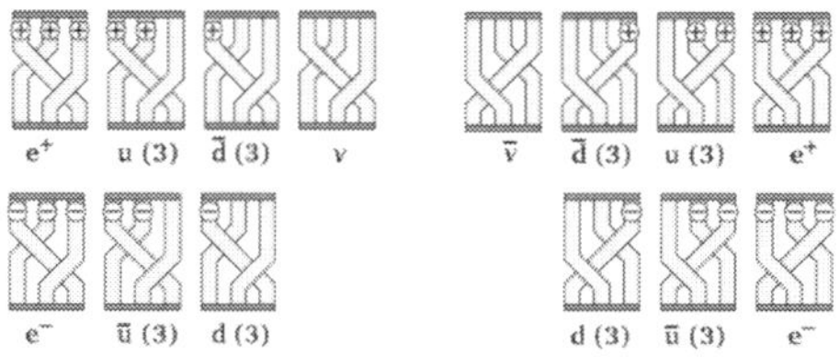

Figure 1. The fermions formed by adding zero, one, two or three charges (i.e. twists) to a neutral braid, where (3) denotes that there are three possible permutations of charge position. We are using a basic braid for illustrative purposes, however arbitrary braiding is possible, in principle. For a given charge magnitude there are twice as many charged states as neutral states.

other, however many different topological structures (braids) may occur for a graph with a given number of links and nodes. The braids therefore appear to have no effect on the volume and area of the corresponding spacetime. By treating the links in a graph as preons, it is possible to use the Helon Model to map the braids onto the first generation of fermions of the Standard Model. In this context untwisted links are equated to H_0s, with left-handed and right-handed twisted links equated to H_-s and H_+s respectively. A braid on three strands then corresponds to the triplets of preons occurring in the Harari/Shupe model. In this construction all fermions may be viewed as a basic neutrino "framework" to which electric charges are added.

Given a neutral braid we may construct its mirror image. We will interpret the distinction between a braid and its mirror image as a difference in handedness. If we add positive or negative twists to braids of either handedness, as in Figure 1, we construct topological states that map to the left-handed and right-handed charged fermions exactly once. However, as noted above there is no neutral anti-preon. Hence we obtain only a single left-handed neutral braid, and a single right-handed neutral braid, which we identify as the neutrino and anti-neutrino respectively.

If a braid is connected to a larger network ("the Universe") at only one end, which we may think of as the bottom, and the other end is attached only to the braid itself, then the top of the braid is simply a trivalent node. In this case it is possible to deform a braid on three strands to obtain a structure which carries only twisting. Such a braid is simply a series of crossings of the first (leftmost) leg over/under the second, and crossings of the second leg over/under the third (rightmost) leg. When a trivalent node is flipped over along the axis connecting the centre of the node with the crossing of two links, as in Fig. 2, the crossing is undone and a series of twists (through $\pm\pi$) are induced on the three links leading out of the node. In [6] it was shown that by undoing each crossing in a braid in sequence, one obtains a series of three uncrossed strands with twists a, b, and c on the first, second, and third strands respectively (a, b, c are multiples of π). The triple $[a, b, c]$ defines an equivalence class of framed braids. Any two braids which may be reduced to the same unbraided state fall within the same topological equivalence class. Therefore for braids within networks it is not different braids, but rather different equivalence classes of braids that correspond to the various fermions. This process of undoing crossings will only work in the trivalent case (flipping a node with four or more legs may undo the crossing of two strands but will induce a crossing on the others).

We can model the electroweak interactions between quarks and leptons in terms of braid products. The product of braids in the Artin braid [7] group is taken by joining the bottom of the first strand in the first braid to the top of the first strand in the second braid, and likewise with each of the subsequent strands, to form a single larger braid. If the first braid consists of n crossings and the second braid consists of m crossings, their product consists of at most $n + m$ crossings (some crossings may cancel with each other, resulting in fewer crossings in the product). When we take the product of two framed braids, the twists on the joined

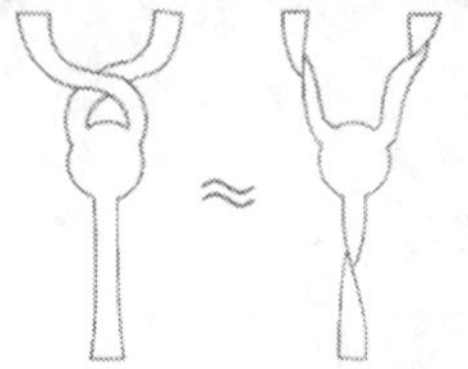

Figure 2. Flipping a trivalent node over uncrosses the links emerging from it, but induces twists on the links.

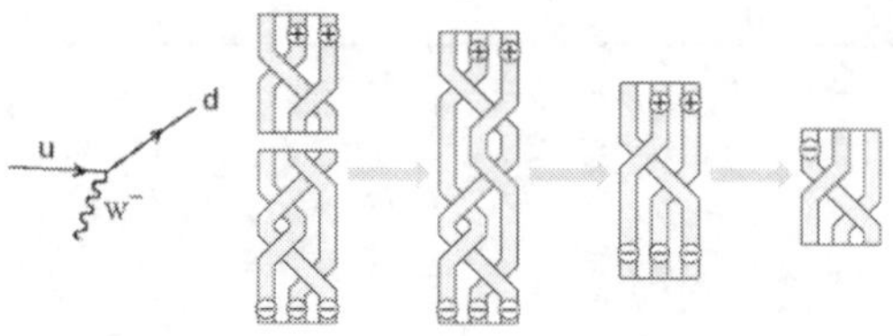

Figure 3. Interactions between fermions and bosons (left) can be represented by taking the product of framed braids (right).

strands may combine together or cancel. To illustrate this, consider a braid composed of two H_+s (strands with right-handed twists), and one H_0, and another braid composed of three H_-s. The first braid corresponds in charge to an up quark, the second corresponds to a W^- boson. When the product of these braids is taken, the resultant braid is composed of a single H_- and two H_0s. This corresponds to a down quark. This interaction is illustrated in Fig. 3. Weak interactions can not change the location of the "odd-man-out" helon within a braid (this would change quark colour) hence the substructure of the electroweak bosons must be braids with a trivial permutation of the strand ordering between the top and bottom. In general several braids including the identity braid (three strands with no crossings) may fulfill this requirement, with different braid structures corresponding to bosons with different helicities.

2. Interactions in networks

If the fermions and bosons of the SM truly emerge from an appropriate theory of quantum gravity, we must move from abstract braids to considering the interactions of braids embedded within networks. In [5] it was shown that under the standard evolution moves of LQG, braids in a network will not decay, but neither will multiple braids combine or a single braid split into several. It seems that a new evolution move is necessary to allow interactions to occur.

Reassuringly, an approach to interactions which does not require a new evolution move was described by Wan and Smolin in [8]. Here it was shown that *tetravalent* networks evolving under the dual Pachner moves [9] allow certain braided structures to interact in a manner reminiscent of the interactions of bosons and fermions. The Pachner moves allow one tetravalent node to become four nodes, and vice-versa, and allow a pair of tetravalent nodes connected by a single link to become three nodes arranged on the vertices of a triangular ring. These networks are realised as spherical nodes, connected by tubular links. This format can be made to correspond with the framed networks discussed throughout the rest of this paper by observing that if one slices such tubular networks in half, down opposite sides of each link, the resulting surfaces are tetravalent framed networks, and tetravalent nodes may be viewed as pairs of trivalent nodes.

It was shown in [8] that there exists a certain class of braids, which can be merged with other braids, by a process illustrated schematically in Fig. 4. The two braids are moved through the network until they are connected by a common link. The two nodes in between the pair of braids are then replaced by a ring, via a Pachner move. This ring is manipulated through the crossings of one braid, to form a tetrahedral structure of four nodes at one end of a new braid, composed of the crossings of the two original braids. The four nodes are then reduced to a single node via a Pachner move. Such interactions are, of course, reversible, and have a natural interpretation in terms of a boson being absorbed or emitted by a fermion.

While these results suggested that interactions may more naturally be achieved in the tetravalent case than the trivalent case, it was not clear whether there was a direct mapping of

Journal of Physics: Conference Series **360** (2012) 012056 doi:10.1088/1742-6596/360/1/012056

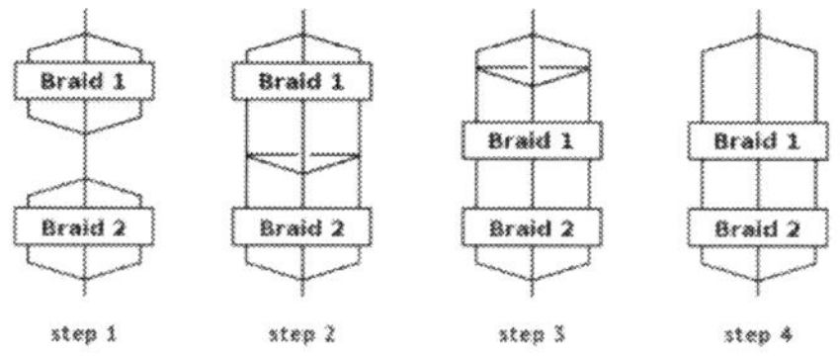

Figure 4. Schematic representation of two braids in a tetravalent network merging.

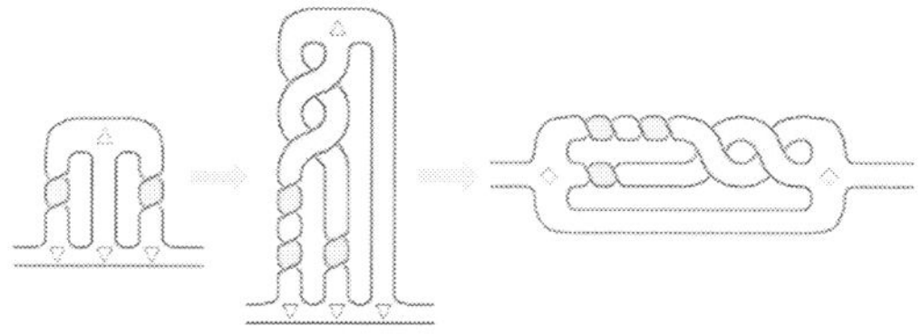

Figure 5. Identifying braids in the trivalent ($\triangle$=node) and tetravalent ($\diamond$=node) format.

tetravalent braids to fermion and boson states, as provided in the trivalent case by the Helon Model. Also, as noted above, the number of distinct braids is not reduced in the tetravalent case, as it was in the trivalent case. If we are to avoid having an unrealistically large list of braids (and hence fermion species), we must limit the type of allowed braids in the tetravalent case, or find a way to combine tetravalent braids into equivalence classes, as was done in the trivalent case. We now describe how the latter possibility can be achieved.

We start with a trivalent braid, and reduce it to its pure twist form. It is then possible to use node-flipping to eliminate all the twists on one strand (the rightmost, say). This introduces new twists and crossings on the other two strands, but that is not problematic. Once there is no twisting on one strand, one of the nodes at the "bottom" of the braid may be moved along this strand, and merged with the node at the "top" of the braid, to form a tetravalent node. The remaining two nodes at the bottom of the braid are likewise merged to form a tetravalent node, as in Fig 5. In this way we can associate a tetravalent braid to each trivalent braid. This allows us (as in the trivalent case) to use the Helon Model to map equivalence classes of braids to particle states, in the tetravalent setting where interactions have been shown to occur.

3. Discussion and Unresolved Issues

A more complete discussion of the Helon Model can be found in [4]. While there remain several issues the Helon Model has yet to address, including the asymmetry of the weak interaction, the origin of mass, and the nature of Cabbibo-mixing, the work presented here goes some way towards resolving similarly significant problems. It now appears that interactions between braids in networks can be achieved, without sacrificing the mapping between braid states and particle states, and without the need to arbitrarily invent new evolution moves, by the approach we have described, which combines the best features of the tetravalent and trivalent schemes. This brings the prospect of emergent, interacting matter from discrete spacetime a step closer.

Acknowledgments

The author is supported by a Ramsay Postdoctoral Fellowship at the University of Adelaide, and wishes to thank Derek Leinweber and Max Lohe for their encouragement.

References

[1] J. C. Pati, A. Salam, *Phys. Rev.* **D10**, 275 (1974).
[2] H. Harari, *Phys. Lett.* **B86**, 83 (1979).
[3] M. Shupe, *Phys. Lett.* **B86**, 87 (1979).
[4] S. O. Bilson-Thompson, [hep-ph/0503213] (2005).
[5] S. O. Bilson-Thompson, F. Markopoulou, L. Smolin, [hep-th/0603022] *Class. Quant. Grav.* **24**, 3975 (2007).
[6] S. Bilson-Thompson, J. Hackett, L. Kauffman, [arXiv:0903.1376] *J. Math. Phys.* **50**, 113505 (2009).
[7] See e.g. Kunio Murasugi, *Knot Theory and its Applications*, Birkhauser (1996), or any equivalent text.
[8] L. Smolin and Y. Wan, [arXiv:0710.1548] *Nucl. Phys.* **B796**, 331 (2008).
[9] U. Pachner, *Abh. Math. Sem. Univ. Hamburg* **57**, 69 (1986).

Journal of Physics: Conference Series **360** (2012) 012057 doi:10.1088/1742-6596/360/1/012057

Observable consequences of quantum gravity: Can light fermions exist?

A. Eichhorn [1]

Theoretisch-Physikalisches Institut, Friedrich-Schiller-Universität Jena, Max-Wien-Platz 1, D-07743 Jena, Germany

Abstract. Any theory of quantum gravity must ultimately be connected to observations. This demand is difficult to be met due to the high energies at which we expect the quantum nature of gravity to become manifest. Here we study, how viable quantum gravity proposals can be restricted by investigating the interplay of gravitational and matter degrees of freedom. Specifically we demand that a valid quantum theory of gravity must allow for the existence of light (compared to the Planck scale) fermions, since we observe these in our universe. Within the effective theory framework, we can thus show that UV completions for gravity are restricted, regardless of the details of the microscopic theory. Specialising to asymptotically safe quantum gravity, we find indications that universes with light fermions are favoured within this UV completion for gravity.

1. How to connect quantum gravity to observations

In quantum gravity, the goal is the construction of an internally consistent theory, which can then be tested against experiment. Since the typical scale of quantum gravity effects is generally expected to be $\mathcal{O}(M_{\rm Planck})$, devising doable experiments that test the quantum nature of gravity is highly challenging. Hence it is considerably simpler to *indirectly* test quantum gravity: Clearly the observed low-energy properties of matter must be compatible with a quantum theory of gravity, if it is to describe our universe. A particular quantity that can be sensitive to quantum gravity fluctuations is constituted by fermion masses. Within the Standard Model, these arise from chiral symmetry breaking (χSB), either induced by the strong interactions, or by the Higgs sector. In a similar fashion, quantum gravity fluctuations might induce strong correlations in the fermion sector and lead to χSB. In this case, the induced fermion masses would naturally be $\mathcal{O}(M_{\rm Planck})$, analogously to QCD, where the fermion masses induced by χSB are comparable to $\Lambda_{\rm QCD}$. Since our universe contains fermions which are considerably lighter than the Planck scale, we conclude that any viable theory of quantum gravity must evade such a mechanism.

Here, we will show that this requirement allows to put restrictions on quantum gravity theories. We first investigate generic UV completions for gravity by making use of the framework of effective field theories: Irrespective of the UV completion for gravity, an effective parameterisation of quantum gravity fluctuations in terms of metric fluctuations should hold on scales up to a UV scale $\Lambda \lesssim M_{\rm Planck}$. (This allows, e.g. to compute quantum corrections to the Newtonian potential, see, e.g. [1].) Then the UV completion for gravity determines the values of the couplings in the effective theory on the UV cutoff scale Λ. A Renormalisation Group (RG)

[1] astrid.eichhorn@uni-jena.de

study allows to investigate, whether the predicted values for the couplings are compatible with the existence of light fermions on lower scales.

A parameterisation of quantum gravity fluctuations in terms of metric fluctuations is also central to a specific UV completion for gravity, namely the asymptotic-safety scenario [2], where the gravitational degrees of freedom are carried by the metric up to arbitrarily high momentum scales. The far UV is then dominated by an interacting fixed point in the running couplings. Evidence for the existence of this fixed point has been collected, e.g. in [3]; for reviews see [4].

The main difference between these two settings is, that in the latter case the requirement of an interacting fixed point determines the UV behaviour at arbitrarily high momenta. In the former setting, our analysis extends over a finite range of scales, only. The values of the couplings at the finite UV-scale Λ are then not restricted by a fixed-point requirement, and are determined by the microscopic theory.

2. How to investigate the existence of light fermions

Within the framework of the functional RG, χSB is signalled by divergent four-fermion couplings. This connection arises as follows: Introducing composite bosonic degrees of freedom, schematically $\phi \sim \bar{\psi}\psi$, in the path integral, allows to rewrite a four-fermion coupling $\lambda\left(\bar{\psi}\psi\right)^2 \sim \phi\bar{\psi}\psi + \frac{1}{\lambda}\phi^2$. Within the bosonic picture, the onset of spontaneous symmetry breaking is signalled by the bosonic mass going to zero. Thus $\lambda \to \infty$ signals the onset of spontaneous χSB in the purely fermionic language, which we will use here[2].

To investigate, whether the four-fermion couplings diverge, we study their β functions. We use a functional RG equation, the Wetterich-equation [5] for the scale-dependent effective action Γ_k, which is the generating functional of one-particle irreducible correlators that include all fluctuations from the UV down to the infrared (IR) scale k. At $k=0$, Γ_k coincides with the standard effective action $\Gamma = \Gamma_{k=0}$. The scale derivative of Γ_k is given by

$$\partial_t \Gamma_k = \frac{1}{2}\mathrm{STr}\{[\Gamma_k^{(2)} + R_k]^{-1}(\partial_t R_k)\}. \tag{1}$$

Here, $\partial_t = k\,\partial_k$; $\Gamma_k^{(2)}$ is the second functional derivative of Γ_k with respect to the fields, and R_k is an IR regulator function. The supertrace STr contains a trace over the spectrum of the full propagator (for reviews, see [6]). This equation is exact, and is applicable in perturbative as well as non-perturbative settings, since its derivation does *not* rely on the existence of a small parameter. Practical calculations do however usually require truncations of all possible operators in the effective action to a, typically finite, subset.

In choosing an appropriate truncation it is crucial that we include all four-fermion couplings compatible with the symmetries, since any of these might diverge and induce χSB. We thus study a Fierz-complete basis of $\mathrm{SU(N}_f)_L \times \mathrm{SU(N}_f)_R$ symmetric four-fermion interactions, where N_f is the number of fermions, and $i,j = 1, ..., \mathrm{N}_f$.

$$\begin{aligned} \Gamma_{k\,\mathrm{F}} &= \frac{1}{2}\int d^4x\,\sqrt{g}\,\big[\bar{\lambda}_-(k)(V-A) + \bar{\lambda}_+(k)(V+A)\big], \qquad (2)\\ \text{where } V &= \left(\bar{\psi}^i\gamma_\mu\psi^i\right)\left(\bar{\psi}^j\gamma^\mu\psi^j\right), \qquad A = -\left(\bar{\psi}^i\gamma_\mu\gamma^5\psi^i\right)\left(\bar{\psi}^j\gamma^\mu\gamma^5\psi^j\right). \end{aligned}$$

Generically, the β functions of the dimensionless renormalised couplings $\lambda_\pm = \frac{k^2\bar{\lambda}_\pm}{Z_\psi}$ read [7]

$$\beta_{\lambda_\pm} = (2+\eta_\psi)\lambda_\pm + a\,\lambda_\pm^2 + b\,\lambda_\pm\lambda_\mp + c\,\lambda_\mp^2 + d\lambda_\pm + e, \tag{3}$$

[2] The introduction of composite bosonic degrees of freedom becomes necessary only for an investigation of the symmetry-broken regime. A purely fermionic formulation suffices to detect the onset of symmetry breaking, which is all we will be interested in here.

where $\eta_\psi = -\partial_t \ln Z_\psi$ is the fermionic anomalous dimension related to the wave-function renormalisation, see below. Herein the first term arises from dimensional (and anomalous) scaling and reflects the perturbative non-renormalisability of four-fermion couplings in four dimensions. The quadratic contributions follow from a purely fermionic two-vertex diagram. A tadpole contribution $\sim d\lambda_\pm$ may also exist, as well as a $\lambda_\pm$-independent part $\sim e$ resulting from the coupling to other fields, e.g. to the metric, thus being decisive for the question of χSB.

The crucial observation is that whenever fixed points exist, initial conditions for the RG flow can be found such that chiral symmetry breaking is avoided, see fig. 1.

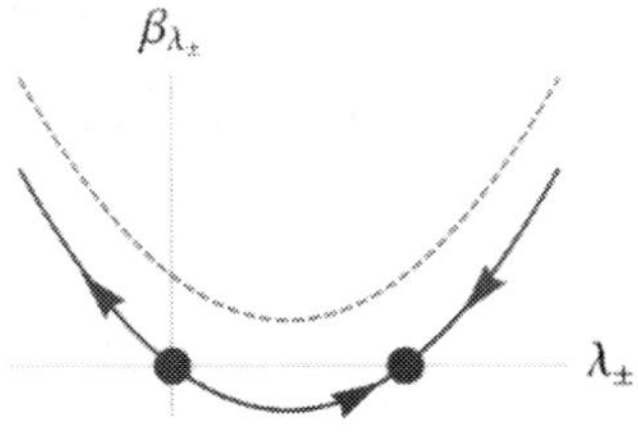

Figure 1. In the absence of further interactions, the parabola-type β function (solid blue line) exhibits two fixed points, the Gaußian one at $\lambda_\pm^* = 0$ and a non-Gaußian one at $\lambda_\pm^* = \lambda_{\pm,\mathrm{cr}} > 0$. Arrows indicate the RG flow towards the IR, and initial conditions to the left of $\lambda_\pm = 0$ lead to χSB. Further interactions can shift the parabola (red dashed line) such that no fixed points exist. Such a scenario typically occurs in QCD-like theories [8, 9, 10].

To evaluate the gravitational contribution to the β functions, we include terms carrying dynamics for gravitational as well as fermionic degrees of freedom in the following truncation

$$\Gamma_k = \frac{Z_N(k)}{16\pi G_N}\int d^4x\sqrt{g}(-R + 2\bar{\lambda}(k)) + \Gamma_{k\,\mathrm{gf}} + \int d^4x\sqrt{g}\, iZ_\psi \bar{\psi}^i\gamma^\mu\nabla_\mu\psi^i + \Gamma_{k\,\mathrm{F}}, \tag{4}$$

Herein the bare Newton constant G_N is related to the dimensionless renormalised Newton coupling $G(k) = k^2 Z_N(k) G_N$ and we have included a cosmological constant $\bar{\lambda}(k) = k^2\lambda(k)$. $\Gamma_{k\,\mathrm{gf}}$ denotes a standard gauge-fixing and ghost term in Landau deWitt background field gauge, see, e.g. [11]. Our truncation is motivated by similar truncations in QCD, which allow for a determination of the critical temperature for χSB [12]. The β functions for the Newton coupling and the cosmological term have been analysed in this truncation (with $Z_\psi = 1$) in [13].

3. Restricting UV completions for gravity

The main question is whether quantum gravity fluctuations lead to fixed-point annihilation in the β functions. As our main result we observe that this is not the case. We find that for any $\{G > 0, \lambda, N_f\}$ the system of β functions $\beta_{\lambda_\pm}$ shows four fixed points (for details of the calculation, see [7]). Thus metric fluctuations do not contribute strongly to $\lambda_\pm$-independent terms in Eq. (3), which could lead to fixed-point annihilation. Instead, we find that gravity fluctuations mainly enhance anomalous scaling terms $\sim \lambda_\pm$. Thus our first conclusion is that, although gravity is an attractive force, metric fluctuations do not automatically induce strong fermionic correlations and lead to χSB and bound state formation.

Thus χSB is a question of *initial conditions* for the RG flow, cf. fig. 1. Here lies the main difference between the asymptotic-safety scenario and other UV completions for gravity: In the former, the system has to sit on a fixed point in the far UV. If there were no fixed points in $\beta_{\lambda_\pm}$, asymptotic safety could not constitute a UV completion for gravity with dynamical fermions. Here, we have four fixed points with different universal properties at our disposal, which all define a valid UV completion. The number of relevant couplings (connected to the free parameters of the theory) is either zero, one or two, depending on the fixed point. Thus these fixed points constitute UV completions with varying predictive power. Their existence entails that metric fluctuations do not directly induce χSB and the scenario is compatible with observations.

In the case of other UV completions for gravity, there is no fixed-point requirement, and in fact the initial conditions for the RG flow can lie anywhere in the space of all couplings. Here, the crucial observation is that unbroken chiral symmetry restricts the allowed initial conditions, see fig. 2. We thus conclude that the existence of light fermions puts restrictions on any UV completion for gravity.

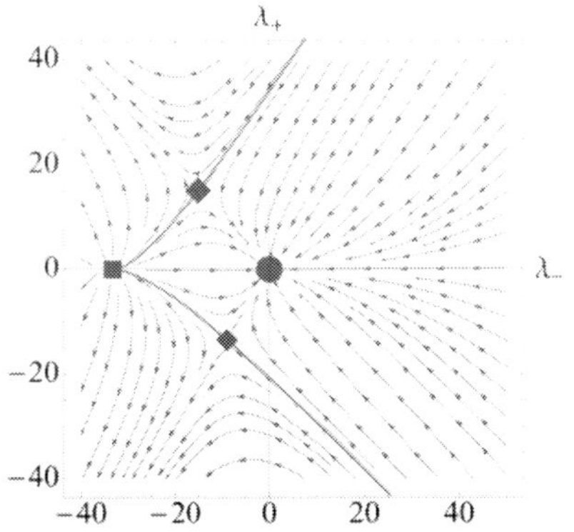

Figure 2. Flow towards the IR in the λ_+, λ_--plane for $\eta_N = 0, \eta_\psi = 0, G = 0.1, \lambda = 0.1$ and $N_f = 6$. For initial values to the right of the red lines the chiral system is in the universality class of the (shifted) Gaußian fixed point. Any microscopic theory that would put the effective quantum field theory to the left of the red lines would generically not support light fermions; thus the initial conditions are restricted to lie in the basin of attraction of the Gaußian fixed point.

In conclusion, we have found - within a truncation of the full RG equations - that, although gravity is an attractive force, even strong metric fluctuations do not directly lead to χSB and bound state formation for fermions. This result applies to asymptotically safe quantum gravity, as well as other UV completions, where we show that the requirement of unbroken χSB can be used to restrict the space of couplings.

Acknowledgments
I thank Holger Gies for helpful comments on this manuscript. This work was supported by the DFG-Research Training Group "Quantum- and Gravitational Fields" (GRK 1523/1).

[1] J. F. Donoghue, Phys. Rev. Lett. **72**, 2996 (1994) [arXiv:gr-qc/9310024].
[2] S. Weinberg, *In *Hawking, S.W., Israel, W.: General Relativity*, 790-831.*
[3] M. Reuter, Phys. Rev. D **57**, 971 (1998) [arXiv:hep-th/9605030]; O. Lauscher and M. Reuter, Phys. Rev. D **65**, 025013 (2002) [arXiv:hep-th/0108040]; M. Reuter and F. Saueressig, Phys. Rev. D **65**, 065016 (2002) [arXiv:hep-th/0110054]; P. Fischer and D. F. Litim, Phys. Lett. B **638**, 497 (2006) [arXiv:hep-th/0602203]; A. Codello, R. Percacci and C. Rahmede, Annals Phys. **324**, 414 (2009) [arXiv:0805.2909 [hep-th]]; D. Benedetti, P. F. Machado and F. Saueressig, Mod. Phys. Lett. A **24**, 2233 (2009) [arXiv:0901.2984 [hep-th]]; A. Eichhorn, H. Gies and M. M. Scherer, Phys. Rev. D **80**, 104003 (2009) [arXiv:0907.1828 [hep-th]]; E. Manrique, M. Reuter and F. Saueressig, arXiv:1006.0099 [hep-th]; M. Niedermaier, Nucl. Phys. B **833**, 226 (2010); E. Manrique, S. Rechenberger, F. Saueressig, Phys. Rev. Lett. **106**, 251302 (2011). [arXiv:1102.5012 [hep-th]].
[4] M. Niedermaier and M. Reuter, Living Rev. Rel. **9**, 5 (2006); M. Reuter and F. Saueressig, arXiv:0708.1317 [hep-th]; R. Percacci, In *Oriti, D. (ed.): Approaches to quantum gravity* 111-128. arXiv:0709.3851 [hep-th]; D. F. Litim, arXiv:0810.3675 [hep-th]; arXiv:1102.4624 [hep-th].
[5] C. Wetterich, Phys. Lett. B **301**, 90 (1993).
[6] J. Berges, N. Tetradis and C. Wetterich, Phys. Rept. **363** (2002) 223 [hep-ph/0005122]; J. Polonyi, Central Eur. J. Phys. **1** (2003) 1. [hep-th/0110026]; J. M. Pawlowski, Annals Phys. **322** (2007) 2831 [arXiv:hep-th/0512261]; H. Gies, arXiv:hep-ph/0611146; B. Delamotte, arXiv:cond-mat/0702365; O. J. Rosten, arXiv:1003.1366 [hep-th].
[7] A. Eichhorn and H. Gies, New J. Phys. **13**, 125012 (2011) [arXiv:1104.5366 [hep-th]].
[8] K. i. Kondo, S. Shuto and K. Yamawaki, Mod. Phys. Lett. A **6**, 3385 (1991).
[9] H. Gies, J. Jaeckel and C. Wetterich, Phys. Rev. D **69**, 105008 (2004) [arXiv:hep-ph/0312034].
[10] H. Gies and J. Jaeckel, Eur. Phys. J. C **46**, 433 (2006) [arXiv:hep-ph/0507171].
[11] A. Eichhorn and H. Gies, Phys. Rev. D **81**, 104010 (2010) [arXiv:1001.5033 [hep-th]].
[12] J. Braun and H. Gies, Phys. Lett. B **645**, 53 (2007) [arXiv:hep-ph/0512085]; JHEP **0606**, 024 (2006) [arXiv:hep-ph/0602226].
[13] R. Percacci and D. Perini, Phys. Rev. D **67**, 081503 (2003) [arXiv:hep-th/0207033]; Phys. Rev. D **68**, 044018 (2003) [arXiv:hep-th/0304222].

Journal of Physics: Conference Series **360** (2012) 012058 doi:10.1088/1742-6596/360/1/012058

The Coupling of Shape Dynamics to Matter

H Gomes

Department of Physics, University of California, Davis, CA, 95616

E-mail: gomes.ha@gmail.com

Abstract. Shape Dynamics (SD) is a theory dynamically equivalent to vacuum General Relativity (GR), which has a different set of symmetries. It trades refoliation invariance, present in GR, for local 3-dimensional conformal invariance. This contribution to the Loops 11 conference addresses one of the more urgent questions regarding the equivalence: is it possible to incorporate normal matter in the new framework? The answer is yes, in certain regimes. We present general criteria for coupling and apply it to a few examples.The outcome presents bounds and conditions on scalar densities (such as the Higgs potential and the cosmological constant) not present in GR.

1. Introduction

Shape Dynamics (SD) [1, 2] is a classical theory equivalent to pure gravity in which the spacetime picture that underlies GR is replaced by an intrinsically spatial picture. In particular, invariance under spacetime refoliations is traded for invariance of the theory under local spatial conformal transformations that preserve the total spatial volume. The equivalence is obtained by manipulating the constraint structure of General relativity, but can only be achieved if one first uses the Stuckelberg mechanism to extend the original phase space. Much like other known uses of extended phase space[1] the use of extended phase space for obtaining SD seems unavoidable. The basic conditions required for the symmetry trading is the existence of two (the original and the "new") symmetries which are maximally symplectic (i.e. gauge fix one another). Motivations for SD are many, and discussed in [1, 2], but a "uniqueness" type argument comes about only when considering field quantization. In fact, even before field quantization, in the absence of a true Hamiltonian (as in generally covariant systems), the fact that we have two symmetries gauge-fixing one another automatically implies that both symmetries are kept in the BRST formalism. Work on the construction of SD through "symmetry doubling" will be reported on shortly.

One of the upsides of the symmetry trading is that the constraint structure of SD becomes much simpler than that of GR, and in fact qualifies as a (infinite-dimensional) Lie algebra, as opposed to the Lie algebroid structure of GR. Shape dynamics is furthermore left with a unique (and single) Hamiltonian constraint to generate global time translation, yielding a tool to attack those problems of time related to refoliation invariance. However the trade-off implies that this global Hamiltonian is non-local. This holistic aspect of the theory makes it more

[1] We just mention two examples: the original Stueckelberg trick [3], which introduces a scalar field into an Abelian gauge theory making it massive but preserving gauge invariance, and the BRST construction, which adds ghosts and ghost momenta to the original phase space in order to apply cohomology tools to understand the space of gauge invariant functionals.

Journal of Physics: Conference Series **360** (2012) 012058 doi:10.1088/1742-6596/360/1/012058

difficult to incorporate matter as an "add on" to the theory. Instead, one must input the matter Hamiltonian already at the base level of the construction of SD, in the linking theory.

The problem of incorporating different interactions to the theory are of utmost importance, as it is through matter that one may obtain most phenomenological predictions from the theory, particularly those that might differ from its general relativistic counterparts. It turns out that the construction of a SD extension with standard matter content can be obtained from the appropriate GR theory by straightforward application of the construction principle given in [1]. This paper is based on [4].

2. Preliminaries

To make this paper sufficiently self-contained, we briefly review the construction principle for equivalent gauge theories, in particular the construction of Shape Dynamics in vacuum. The following procedure is a summary of that contained in [1]. Basically, the procedure in the concrete case of General Relativity consists in first extending phase space to accommodate extra redundant symmetries obtained through "conformalization". This does *not* mean that the extended theory is in any way conformally invariant. Then we apply two different particular gauge fixings and follow the Dirac analysis to obtain two dual theories with distinct symmetry properties: GR and Shape Dynamics.

2.1. Pure Shape Dynamics

Let us now briefly review the construction of Shape Dynamics as a theory equivalent to ADM gravity on a compact Cauchy surface Σ without boundary. For a more thorough review see [1]. We start with the standard ADM phase space $\Gamma_{ADM} = \{(g,\pi) : g \in \text{Riem}, \pi \in T^*_g(\text{Riem})\}$, where Riem denotes the set of Riemannian metrics on the above defined 3-manifold Σ, and the usual first class ADM constraints, i.e. the scalar constraints $S(x) = \frac{\pi^{ab}\pi_{ab} - \frac{1}{2}\pi^2}{\sqrt{g}} - \sqrt{g}R$ and momentum constraints $H^a(x) = \pi^{ab}_{;b}(x)$ thereon. We extend the ADM phase space with the phase space of a scalar field $\phi(x)$ and its canonically conjugate momentum density $\pi_\phi(x)$, which we introduce as additional first class constraints $Q(x) = \pi_\phi(x) \approx 0$. The system is thus merely a trivial embedding of the original ADM onto the extended phase space. Now use the canonical transformation T_ϕ generated by the generating functional $F = \int d^3x \left(g_{ab}e^{4\hat{\phi}}\Pi^{ab} + \phi\Pi_\phi\right)$, where $\hat{\phi}(x) := \phi(x) - \frac{1}{6}\ln\langle e^{6\phi}\rangle_g$ using the mean $\langle f\rangle_g := \frac{1}{V}\int d^3x\sqrt{|g|}f(x)$ and 3-volume $V_g := \int d^3x\sqrt{|g|}$. These "means" and volumes technically appear as a manner to yield dynamically equivalent theories. They force the appearance of a global Hamiltonian in SD.

At this point we have the first class set of constraints $T_\phi S(x)$, $T_\phi H(x)$ and $T_\phi Q(x) = \pi_\phi(x) - 4(\pi - \langle\pi\rangle\sqrt{g})(x)$. We then perform the phase space reduction for the gauge fixing condition $\pi_\phi(x) = 0$. The crucial technical steps come once one has imposed such a gauge-fixing, also sometimes called "best-matching constraint". For it is necessary for the symmetry trading to work that this gauge-fixing be completely second class wrt the set of original constraints one wants to trade for, and first class wrt all the other constraints. Furthermore one must show that the variable conjugate to π_ϕ, i.e. ϕ can be set such that that same original set of constraints can be strongly solved for (solved for in the entire phase space).

After some algebra, we find that all but one of the $T_\phi S(x)$ constraints can be solved for $\phi = \phi_o$, and this one leftover constraint yields the global Hamiltonian of SD. The end result is the following set of first class constraints, which define Shape Dynamics

$$
\begin{aligned}
H_{SD} &= T_{\phi_o}S(N_o)\\
H(\xi) &= \int d^3x\pi^{ab}\mathcal{L}_\xi g_{ab}\\
Q(\rho) &= \int d^3x\rho\left(\pi - \langle\pi\rangle\sqrt{|g|}\right),
\end{aligned}
\tag{1}
$$

Journal of Physics: Conference Series **360** (2012) 012058 doi:10.1088/1742-6596/360/1/012058

where $\phi_o(g,\pi)$ is such that $H_{SD}=0$ combined with $\phi=\phi_o(g,\pi)$ is equivalent to $T_\phi S(x)=0$ (at the surface $\pi_\phi=0$). Moreover, N_o denotes the CMC lapse function, $\pi=\pi^{ab}g_{ab}$ and $\langle\pi\rangle=\frac{1}{V}\int d^3x\pi$. The constraints $H(\xi)$ generate spatial diffeomorphisms and the constraints $Q(\rho)$ generate conformal transformations that leave the total spatial volume invariant.

The main issue that forbids or allows the trading is that the gauge-fixed constraints arising from the Stueckelberg mechanism (best-matching constraints) are first-class amongst themselves (and possibly also with respect to other constraints that one does not want to disturb, e.g. 3-diffeomorphisms) and second class solely with respect to the constraint that one wants to trade for. Thus the construction of a linking theory is non-generic. The condition that (a subset of) the best matched original constraints can be uniquely solved for the auxiliary variables ϕ is indispensable for the existence of a linking theory.

3. General Procedure

3.1. What constrains the coupling to the conformal factor?

When we try to couple different fields to gravity, we will have to face basically one question: how does one scale fields? What exponent is viable in the conformal coupling of a scalar field, $\psi\to e^{\alpha\hat\phi}\psi$? This is an important issue because if the scaling is not correct we could encounter two difficult obstructions.

The first obstruction is that if we are dealing with a field that possesses some kind of gauge symmetry it might not be possible to find a constraint $\mathcal{Q}$ that is first class with respect to the gauge constraint. The second is that the conserved charge (which we call D) implicit inside $\mathcal{Q}$ that defines the foliation might depend on the field. As a subset case, there might be an even worse consequence if the field in question possesses some gauge symmetry (like electromagnetism). For then D (which gives the foliation) could turn out to depend on the gauge potential. This is possible in spite of the existence of a field-dependent D that is first class with respect to the gauge generator of the field. Indeed with any other choice of scaling than the one chosen in the text, this possibility is realized with any Yang-Mills type interaction for non "neutral coupling". Neutral coupling is the particular choice of *empty* scaling $\alpha=0$, i.e. non-gravitational fields have conformal weight zero. This means that fields are only scaled in the sense that they are "carried along" by the scaling of the spatial metric.

3.2. Explicit coupling

We follow the construction principle for pure Shape Dynamics and suppose we have a gravity-matter system on a compact manifold without boundary, with constraints

$$\begin{array}{rcl} H(\xi) &=& \int d^3x\,(\pi^{ab}(\mathcal{L}_\xi g)_{ab}+\pi^A(L_\xi\phi)_A)\\ S(N) &=& \int d^3x\left(\frac{1}{\sqrt{|g|}}\pi^{ab}G_{abcd}\pi^{cd}-(R-2\Lambda)\sqrt{|g|}+H_{\text{matter}}(g_{ab},\phi_A,\pi^A)\right)N(x)\\ G^\alpha(\lambda_\alpha) &=& \int d^3x G^\alpha(g_{ab},\phi_A)\lambda_\alpha\sqrt{g}\end{array} \tag{2}$$

where we denote the non-gravitational configuration degrees of freedom collectively by ϕ_A and their canonically conjugate momenta π^A. We assume that the matter Hamiltonian H_{matter} does neither contain π^{ab} nor any spatial derivatives of g_{ab}, which seems to be an assumption that is realized in nature. We furthermore assume the constraint associated to internal gauge symmetries not to be a functional of the metric momenta. This is a condition realized in all matter fields studied, and which greatly simplifies treatment. As an example, consider Yang-Mills: the constraint is given by $\nabla_a\bar{E}^a_i=\frac{1}{\sqrt{g}}\partial_a\sqrt{g}\bar{E}^a_i=\frac{1}{\sqrt{g}}\partial_aE^a_i$ where $\bar{E}^a_i\sqrt{g}=E^a_i$. Thus $G_i=\nabla_a\bar{E}^a_i\to e^{-6\hat\phi}G_i$. In the cases of interest that we have studied, this is always the case, so that $T_\phi G^\alpha_i\propto G^\alpha_i\approx0$.

Journal of Physics: Conference Series **360** (2012) 012058 doi:10.1088/1742-6596/360/1/012058

To construct a linking theory we extend phase space by a scalar $\phi(x)$ with canonically conjugate momentum density $\pi_\phi(x)$ and impose the additional first class constraint $Q(x) = \pi_\phi(x)$, which eliminates the phase space extension on shell. To perform canonical best matching for volume preserving conformal transformations, we specify the conformal weight of matter fields by neutral coupling. We thus use the generator

$$F = \int d^3x \left(e^{4\phi} g_{ab}\Pi^{ab} + \phi\Pi_\phi + \phi_A \Pi^A \right) \tag{3}$$

to implement the conformal Kretschmannization. This acts non-trivially only on the gravitational variables. The resulting canonical transformations have $Q(\rho)$ and (respectively) $H(\xi)$ transform weakly into

$$\begin{array}{rcl} Q(\rho) & = & \int d^3x \rho \left(\pi_\phi - 4\left(\pi - \langle\pi\rangle \sqrt{|g|} \right) \right) \\ H(\xi) & = & \int d^3x \left(\pi^{ab}(L_\xi g)_{ab} + \pi_\phi \mathcal{L}_\xi \phi + \pi^A (\mathcal{L}_\xi \phi)_A \right). \end{array} \tag{4}$$

After imposing the best-matching condition $\pi_\phi(x) = 0$ we always get an equation of the form:

$$\{T_\phi S(N), \pi_\phi(x)\} = T_\phi \left[-\frac{3}{2} S(x) + 2(\Delta_{\text{matter}} N(x) - \langle \Delta_{\text{matter}} N \rangle) \right] \tag{5}$$

In the general case one has to compute explicitly the entire bracket $\{T_\phi H_{\text{matter}}, \pi_\phi\}$ and work out the appropriate form of the Δ_{matter} to check its invertibility. Let us here restrict to the simpler case occurring when the matter Hamiltonian does not contain spatial derivatives of the metric tensor nor metric momenta. In this case one gets:

$$\Delta_{\text{matter}} := \nabla^2 - \frac{\pi\langle\pi\rangle}{4\sqrt{g}} - R + \frac{1}{2\sqrt{|g|}} \left(\frac{\delta H_{\text{matter}}}{\delta g_{ab}} g_{ab} + \frac{3}{2} H_{\text{matter}} \right) \tag{6}$$

On the constraint surface $T_\phi S = 0$ and $\mathcal{Q} = 0$, the end result is equivalent to taking

$$\Delta_{\text{matter}} \approx (\nabla^2 - \frac{1}{12}\langle\pi\rangle^2) - \frac{\sigma^{ab}\sigma_{ab}}{g} + \frac{1}{2\sqrt{g}} \left(\frac{\delta H_{\text{matter}}}{\delta g_{ab}} g_{ab} - \frac{1}{2} H_{\text{matter}} \right) \tag{7}$$

Thus in this case the criterium for invertibility of the operator rests on:

$$\frac{1}{2\sqrt{g}} \left(\frac{\delta H_{\text{matter}}}{\delta g_{ab}} g_{ab} - \frac{1}{2} H_{\text{matter}} \right) \le \sqrt{g}\frac{1}{12}\langle\pi\rangle^2 + \frac{\sigma^{ab}\sigma_{ab}}{g} \tag{8}$$

in particular if

$$\frac{\delta H_{\text{matter}}}{\delta g_{ab}} g_{ab} - \frac{1}{2} H_{\text{matter}} \le 0 \tag{9}$$

under the previously assumed conditions the field can always be included in our model. A useful form of (9) for the specific case of homogeneous scaling is given by the following: if $H_{\text{matter}} \to e^{n\phi} H_{\text{matter}}$ scales with power n, we have the equivalent condition (assuming H_{matter} is positive):

$$n - 2 \le 0 \tag{10}$$

In some cases the inequality (8) cannot be attained, in others it implies a bound on the density of the fields, and yet in others it is always obeyed, since (9) is valid. Examples of the first kind are obtained from fields with four-dimensional conformal coupling, examples of the second kind are obtained by adding a mass potential term to the Hamiltonian, of the form $\psi^2\sqrt{g}$, and of the third are Yang-Mills and the massless scalar. However, as soon as this restriction is imposed, one can use the implicit function theorem to complete the construction of SD for these matter models.

Journal of Physics: Conference Series **360** (2012) 012058 doi:10.1088/1742-6596/360/1/012058

4. Conclusions

As a theory possessing 3-dimensional conformal invariance and unique surfaces of simultaneity, Shape Dynamics seems very counter-intuitive in the light of the teachings of Relativity. Furthermore, the existence of fields with mass would seem to undermine conformal invariance. In this paper we have shown that these prejudices are not justified. Applying "neutral conformal coupling" to the fields - so that they only feel space stretching but have otherwise no independent scaling - our method preserves the fields' canonical constraints for all cases of interest. Furthermore, for all such cases the matter Hamiltonian contains neither metric momenta nor metric spatial derivatives, enabling us to apply our bound (8). In such cases, the bound gives a straightforward criterium telling us whether a given field can or cannot be encompassed by Shape Dynamics. All the constituent fields automatically obey the bound except for the Higgs potential and the cosmological constant, whose value must be bounded. We point out the important fact that the deSitter solution does not obey the bound (8), hence it is not a solution for SD [2], and we thus have a first phenomenologically verifiable deviation from standard GR. This work makes the construction of a phenomenologically viable cosmological model imperative for the development of Shape Dynamics.

Acknowledgments

This work was supported in part by the U.S. Department of Energy under grant DE-FG02-91ER40674.

[1] H. Gomes and T. Koslowski, "The Link between General Relativity and Shape Dynamics," arXiv:1101.5974 [gr-qc].
[2] H. Gomes, S. Gryb and T. Koslowski, "Einstein gravity as a 3D conformally invariant theory," Class. Quant. Grav. **28** (2011) 045005 [arXiv:1010.2481 [gr-qc]].
[3] E. C. G. Stueckelberg, Helv. Phys. Acta **11**, 225 (1938).
[4] H. Gomes and T. Koslowski, "Coupling shape dynamics to matter gives spacetime," arXiv:1110.3837 [gr-qc].

[2] We thank Sean Gryb for pointing this out to us. The statement that the model does not obey the bound implies that the dual theory does not gauge fix the refoliations, and we are left with more than one "global Hamiltonians" in the dual theory. So one could wonder if this ambiguity is in fact not due to the high symmetry content of deSitter space. To this effect we note that we could always make the $\sigma^{ab}\sigma_{ab}$ term in (8) small enough so that no symmetry is present and yet it still breaks the bound.

Journal of Physics: Conference Series **360** (2012) 012059 doi:10.1088/1742-6596/360/1/012059

Shape Dynamics[1]

Tim Koslowski

E-mail: tkoslowski@perimeterinstitute.ca
Perimeter Institute for Theoretical Physics
31 Caroline Street, Waterloo, Ontario N2L 2Y5, Canada

Abstract. General Relativity can be reformulated as a geometrodynamical theory, called Shape Dynamics, that is not based on spacetime (in particular refoliation) symmetry but on spatial diffeomorphism and local spatial conformal symmetry. This leads to a constraint algebra that is (unlike General Relativity) a Lie algebra, where all local constraints are linear in momenta and may thus be quantized as vector fields on the geometrodynamic configuration space. The Hamiltonian of Shape Dynamics is complicated but admits simple expressions whenever spatial derivatives are negligible.

Shape Dynamics [1] (for motivations see [2]) is a canonical geometrodynamic theory that is equivalent General Relativity, but with spacetime refoliation invariance traded for local spatial conformal invariance, implementing Mach's idea of relative size. In other words: there are two different theory spaces that contain the dynamics of gravity, both defined on the same metric field content but with two distinct symmetry principles. Knowing the theory space is important because detail of a theory space can alter properties of effective field theories, in particular nonperturbative renormizability.

Spacetime Versus Spatial Symmetry Abandoning spacetime refoliation invariance for a spatial symmetry is not just a change of symmetry, but also abandons the beautiful Hamiltonian construction principle for General Relativity of Hojman, Kuchař and Teitelboim [3]. The starting point of their work is spacetime diffeomorphism symmetry and in particular its generators, the Lie-algebra of spacetime vector fields. These yield the Dirac-algebra of hypersurface deformations when pulled back to a Riemannian hypersurface:

$$[H(u), H(v)] = H([u,v]]), \ \ [H(u), S(N)] = S(\mathcal{L}_u N), \ \ [S(N), S(M)] = H(N\nabla M - M\nabla N), \quad (1)$$

where spacetime vectors have been decomposed into spatial parts u and v tangent to the hypersurface and refoliation parts N and M. Then [3] proves a representation theorem of the Dirac algebra on the phase space of geometrodynamics, where the fundamental canonical pair is the spatial metric $g_{ab}(x)$ and its canonically conjugate momentum density $\pi^{ab}(x)$. Assuming time reversibility and a strong notion of locality, one finds that the generators $H(u)$ and $S(N)$ have to take the form of the ADM constraints

$$H(u) = \int d^3x \pi^{ab} \mathcal{L}_u g_{ab}, \ \ S(N) = \int d^3x N \left(\frac{1}{\sqrt{|g|}} \pi^{ab} G_{abcd} \pi^{cd} - (R - 2\Lambda)\sqrt{|g|} \right), \quad (2)$$

where G_{abcd} denotes the DeWitt supermetric. This is precisely the form of constraints that one obtains from the Legendre transform of the Einstein–Hilbert action. In other words one

[1] Based on a talk given at Loops'11 (Madrid)

finds the canonical form of the Einstein–Hilbert action directly from spatial diffeomorphism and refoliation symmetry. The price of spacetime symmetry however is the complicated bracket $[S(N), S(M)] = H(N\nabla M - M\nabla N)$ which involves structure functions and a representation of $S(N)$ that is non-linear in the momenta. Both complications are associated with refoliation invariance and are thus eliminated if refoliation invariance is traded for spatial conformal invariance, whose generator is linear in the momenta and whose algebra is simple.

Canonical Best Matching Canonical best matching, which is a Hamiltonian realization of Barbour's Lagrangian best matching, is a construction that implements a configuration space symmetry in a dynamical system. Consider a dynamical system $(\Gamma, \{.,.\}, \{\chi^\mu\}_{\mu\in\mathcal{M}})$ on phase space Γ with Poisson bracket $\{.,.\}$ and first class constraints[2] $\{\chi_\mu\}$ and denote the configuration variables by $\{q_i\}_{i\in\mathcal{I}}$. We want to implement the action of a symmetry group $\mathbb{G}$ acting as $q_i \mapsto Q_i(q,\phi)$, where we choose the group parameters $\{\phi_\alpha\}_{\alpha\in\mathcal{A}}$ such that $\phi_\alpha = 0$ denotes the unit element. The canonical best matching procedure consists of three steps:
1. Extend the original phase space Γ by $T^*(\mathbb{G})$ equipped with the canonical symplectic structure[3], denote the momenta canonically conjugate to ϕ_α by π^α. We retain the original system by adjoining to the set of first class constraints the additional constraints $\{\pi^\alpha\}_{\alpha\in\mathcal{A}}$.
2. Implement $\mathbb{G}$-symmetry through a canonical transformation T generated by $F = Q_i(q,\phi)P^i + \phi_\alpha\Pi^\alpha$, where capitals denote the transformed variables and p^i is canonically conjugate to q_i. This turns the first class constraints π^α into a conserved charge that commutes with all $T\chi_\mu$

$$T : \pi^\alpha \mapsto \pi^\alpha - \pi_i^\alpha(q)p^i. \tag{3}$$

3. Impose the "best matching" (or "pure gauge") condition $\pi^\alpha = 0$. This may lead to three different types of dynamical systems:
(a) π^α is first class. This means that the system was already $\mathbb{G}$-symmetric.
(b) π^α gauge-fixes (some of) the χ_μ. We will see in a moment that this allows us to trade symmetries.
(c) π^α generates secondary constraints; in this case one has to follow the Dirac procedure to analyze the system. No general statements about the system are possible in this case.
The generic case is of course a mixture of the three prototype cases (a-c).

Linking Gauge Theories Let us now suppose canonical best matching yields case (b). Then we define a "Linking Gauge Theory" on $\Gamma \times T^*(\mathbb{G})$ with first class constraints $T\chi_\mu, T\pi^\alpha$. The constraints can be split into three disjoint sets:
(1) The constraints $T\chi_\mu$ that are gauge fixed by π^α. These are locally solvable for ϕ_α by the implicit function theorem, so they are of the form $\chi^1_\alpha = \phi_\alpha - \phi^o_\alpha(q,p)$.
(2) The constraints $T\pi^\alpha$, which are of the form $\chi^\alpha_2 = \pi^\alpha - \pi^\alpha_i(q)p^i$.
(3) The rest of the $T\chi_\mu$, which can be shown to not be invertible for ϕ_α, i.e. these constraints χ^3_ν weakly commute with the π^α because we assumed best matching yields a theory of type (b).
We can now implement two partial gauge-fixing conditions for this linking theory:

The first is $\phi_\alpha = 0$, which gauge fixes only the χ^α_2 constraints, whose Lagrange multipliers are set to vanish. We can thus perform phase space reduction by setting $(\phi_\alpha, \pi^\alpha) = (0, \pi^\alpha_i(q)p^i)$ and removing the constraints χ^α_2. Using $q_i = Q_i(q,0)$ and the fact that the π^α do not occur in the other constraints, we find that the constraints reduce to the set χ_μ after phase space reduction, the reduced phase space coincides with Γ and the Dirac bracket coincides with the Poisson bracket on reduced phase space, so we have recovered the gauge theory before best matching.

The second partial gauge fixing condition is $\pi^\alpha = 0$. In this case the constraints $T\chi^1_\alpha$ drop out and the phase space reduction $(\phi_\alpha, \pi^\alpha) = (\phi^o_\alpha, 0)$ again yields Γ as the reduced phase space and the Dirac bracket again coincides with the Poisson bracket. The remaining constraints

[2] We include the Hamiltonian in the set of first class constraints as an energy conservation constraint.
[3] For simplicity we consider only commutative groups at this point, but a similar treatment holds more generally.

Journal of Physics: Conference Series **360** (2012) 012059 doi:10.1088/1742-6596/360/1/012059

are not necessarily independent of ϕ_α, but because the $T\chi^3_\mu$ can not be solved for ϕ_α they are weakly equivalent to χ^3_μ and $\pi^\alpha_i(q)p^i$.

We can directly check that the two partially gauge-fixed theories are equivalent by imposing the χ^1_α and χ^α_2 respectively as additional partial gauge fixing conditions, so the two theories have the same initial value problem and the same equations of motion. We have thus traded the constraints χ^1_α of the first theory for the constraints $\pi^\alpha_i(q)p^i$ of the second theory without changing the phase space or Poisson bracket. If the two sets of constraints generate two different sets of gauge symmetries, then we have traded one set of gauge symmetries for another.

Pure Shape Dynamics Brabour's interpretation of Mach's principle suggests that gravity should exhibit spatial conformal invariance. This can be introduced by canonical best matching. However, to retain an evolution generator, we have to restrict the conformal transformations, which we choose to be volume preserving. We thus introduce a scalar ϕ with conjugate momentum density π_ϕ and apply the canonical transform generated by

$$F = \int d^3x \left(g_{ab} e^{4\hat{\phi}} \Pi^{ab} + \phi \Pi_\phi \right) \tag{4}$$

to ADM gravity (equation 2). Here g_{ab} denotes the spatial metric, π^{ab} its canonically conjugate momentum density, capitals denote transformed quantities, and we have used the volume preserving conformal factor $\hat{\phi}(x) = \phi(x) - \frac{1}{6}\ln\langle e^{6\phi}\rangle$ and the mean $\langle f \rangle = \frac{\int d^3x \sqrt{|g|} f}{\int d^3y \sqrt{|g|}}$. This yields the conserved charge $Q(x) = \pi_\phi(x) - 4(\pi(x) - \langle\pi\rangle\sqrt{|g|(x)})$ in terms of $\pi = g_{ab}\pi^{ab}$ and $\langle\pi\rangle = \frac{\int d^3x \pi}{\int d^3y \sqrt{|g|}}$. It turns out that imposing $\pi_\phi(x) = 0$ yields a theory of type (b), so we have a linking theory between General Relativity and a theory which we call Shape Dynamics. The conserved charge $Q(x)$ becomes a spatial conformal constraint $C(x) = \pi(x) - \langle\pi\rangle\sqrt{|g|(x)}$ that preserves the total spatial volume, while the ADM-diffeomorphisms $H(u)$ remain unchanged. All but one of th ADM scalar constraints are gauge fixed; the remaining one becomes the Shape Dynamics Hamiltonian H_{SD}. H_{SD} is defined by the normalization $\langle e^{6\hat{\phi}}\rangle = 1$ and

$$TS(x) = H_{SD}\sqrt{|g|(x)} e^{6\hat{\phi}(x)}, \tag{5}$$

which simplifies on the constraint surface $C(x) = 0$ to a modified Lichnerowicz–York equation, which can be proven to have unique solutions $\hat{\phi}(x)$. The total Shape Dynamics Hamiltonian is

$$\mathbf{H} = \mathcal{N} H_{SD} + H(u) + C(\rho), \tag{6}$$

where $\mathcal{N}$ is a single global Lagrange multiplier and $\rho(x)$ a local Lagrange multiplier. We see that all local constraints are linear in momenta at the expense of a nonlocal Hamiltonian H_{SD}.

The Lichnerowicz–York equation (5) obstructs a straightforward derivation of H_{SD}, although one can prove the existence of H_{SD}. However, this is **not** more complicated than General Relativity, where the analogous equation has to be solved in the initial value problem. Despite this, it is clear that the nonlocal Hamiltonian is a challenge for Shape Dynamics, but this seems to be an acceptable price to pay for its advantages compared with General Relativity:

1. All (local) constraints are linear in momenta and generate vector fields on the geometrodynamic configuration space.
2. The constraint algebra is a Lie algebra. The following brackets can be derived from the linking theory through phase space reduction:

$$\begin{array}{rclrclrcl} \{C(\rho_1), C(\rho_2)\} &=& 0 & \{C(\rho), H_{SD}\} &=& 0 & \{H(u), H_{SD}\} &=& 0 \\ \{H(u), C(\rho)\} &=& C(\mathcal{L}_u\rho) & \{H(u), H(v)\} &=& H([u,v]). & & & \end{array} \tag{7}$$

3. Shape Dynamics is contained in a different theory space than General Relativity, which may change nonperturbative renormalizability.

The usual construction of effective field theories uses a theory space (i.e. field content, symmetries and locality) as well as dimensional analysis. We adopt V and $\langle\pi\rangle$ in addition to local variables as SD-local. We can then mimic the usual construction of effective Hamiltonians: Write down all low-dimensional SD-local field monomials in the linking theory that are invariant under diffeomorphisms and banal transformations (i.e. invariant under $g_{ab} \to T g_{ab}|_{\phi=\sigma}, \pi^{ab} \to T\pi^{ab}|_{\phi=\sigma}, \phi \to \phi - \sigma$) that preserve the total spatial volume. Then determine the conformal field ϕ as a function of g_{ab}, π^{ab} by requiring local spatial conformal invariance.

Explicit Shape Dynamics There are at least three instances where the Shape Dynamics Hamiltonian can be explicitly constructed:
1. Using the fact that the solution to (5) is homogeneous if all coefficient functions are homogeneous and that the normalization condition implies that $\hat{\phi}$ vanishes if ϕ is homogeneous, one finds the Shape Dynamics Hamiltonian to coincide with the temporal gauge Hamiltonian $H(N \equiv 1)$ on the restricted phase space

$$\Gamma_r = \left\{ (g_{ab}, \pi^{ab}) \in \Gamma_{ADM} : \frac{\sigma^{ab}\sigma_{ab}}{|g|} = \langle \frac{\sigma^{ab}\sigma_{ab}}{|g|} \rangle, R = \langle R \rangle \right\}, \tag{8}$$

where $\sigma^{ab} = \pi^{ab} - \frac{1}{3}\pi g^{ab}$. Notice that we can use $R = \langle R\rangle$ as a gauge fixing for local conformal symmetry, so Γ_r contains *three out of four* local degrees of freedom of Shape Dynamics.
2. In the strong gravity (BKL) limit, one can neglect the spatial derivatives, so (5) becomes an algebraic equation which can be solved giving the nonlocal Hamiltonian

$$H_{SD}^{BKL} = V \left(\langle \sqrt{\sigma^{ab}\sigma_{ab}} \rangle^2 - \frac{1}{6}\langle \pi \rangle^2 + 2\Lambda \right). \tag{9}$$

This may be a good starting point for quantization, because in light of the BKL-conjecture [5] one expects the effects of quantum gravity to be most important when this Hamiltonian becomes exact.
3. In 2+1 dimensions on the sphere and torus one can solve (5) explicitly, because Γ_r contains all physical degrees of freedom in this case. The Shape Dynamics Hamiltonian turns out to coincide with the reduced phase space Hamiltonian for these topologies and one can thus explicitly Dirac quantize Shape Dynamics in this case.

Summary 1. Shape Dynamics is a geometrodynamic theory that is equivalent to General Relativity, but refoliation invariance is traded for local spatial conformal invariance.
2. Linking Theories (and canonical best matching) provide a general mechanism for trading gauge symmetries, leaving the possibility that further equivalent formulations of General Relativity may exist.
3. The Shape Dynamics Hamiltonian exhibits nonlocality for 1/4 of the physical degrees of freedom.
4. There exists a construction principle for effective Hamiltonians based on the symmetries of Shape Dynamics.
5. There are instances where Shape Dynamics can be explicitly constructed (in particular: strong gravity, 2+1 torus and sphere).
6. Since all local constraints are linear in momenta, one can perform Dirac quantization on the 2+1 torus and investigate the large-CMC-volume/CFT-correspondence.

Acknowledgments
I thank J. Barbour for helpful comments on the draft. Research at the Perimeter Institute is supported in part by the Government of Canada through NSERC and by the Province of Ontario through MEDT.

[1] J. Barbour: arXiv:1105.0183, H. Gomes, S. Gryb and T. Koslowski: Class. Quant. Grav. **28** (2011) 045005 arXiv:1010.2481, H. Gomes and T. Koslowski: arXiv:1101.5974, T. Budd, T. Koslowski: [arXiv:1107.1287].

[2] E. Anderson, J. Barbour, B. Z. Foster, B. Kelleher and N. O. Murchadha: Class. Quant. Grav. 22, 1795 (2005) [arXiv:gr-qc/0407104], J. Barbour and N. O. Murchadha: arXiv:1009.3559.

[3] S. Hojman, K. Kuchař, C. Teitelboim: Ann. Phys. **96** 88-135 (1976)

[4] J. York: Phys. Rev. Lett. **26**, 1656 (1971), Phys. Rev. Lett. **28**, 1028 (1972), J. Math. Phys. **14**, 456 (1973), N. Ó Murchadha, J. York: Phys. Rev. D **10**, 428 (1974), Phys. Rev. D **10**, 437 (1974), J. Math. Phys. **14**, 1551 (1973)

[5] I. M. Khalatnikov and E. M. Lifshitz: Phys. Rev. Lett., 24:7679, (1970), V. A. Belinsky, I. M. Khalatnikov, and E. M. Lifshitz: Adv. Phys., 19:525573, (1970), Adv. Phys., 31:639667, (1982)

Journal of Physics: Conference Series **360** (2012) 012060 doi:10.1088/1742-6596/360/1/012060

Longitudinal and transverse relativity of spacetime locality in Planck-scale-deformed phase spaces

Niccoló LORET †*

Leonardo BARCAROLI †

Giacomo ROSATI †*

† Dipartimento di Fisica, Università di Roma "La Sapienza", P.le A. Moro 2, 00185 Roma, EU
* INFN, Sez. Roma1, P.le A. Moro 2, 00185 Roma, EU

E-mail: niccolo.loret@roma1.infn.it , giacomo.rosati@roma1.infn.it

Abstract. We summarize some of the results we obtained in arXiv:1006.2126 (Physical Review Letters 106, 071301), arXiv:1102.4637 (Physics Letters B 700, 150-156) and in arXiv:1107.3334, giving complementary characterizations of the relativity of spacetime locality that affects certain Planck-scale-deformed phase-space constructions.

1. Preliminaries

These notes are based on our works [1, 2, 3]. Our aim is to put together the suggestions we derived about relative locality manifestations from deformed simmetries scenarios in two complementary regimes: longitudinal and transverse one.

It is valuable for our purposes to observe that "relative locality" is a natural next step in a logical line on relativistic theories that starts with the relativity of rest (Galilean Relativity) and then, accommodates "relative simultaneity" (Special Relativity). The relativity of rest and the absoluteness of time hosted by Galilean relativity famously leads to a structureless law of composition of velocities: $\vec{v} \oplus \vec{\delta v} = \vec{v} + \vec{\delta v}$. The observer-independence of the speed of light, and the relativity of simultaneity, lead to a more complex (nonlinear, noncommutative, nonassociative) law of composition of velocities, the special-relativistic law

$$\vec{v}_a \oplus \vec{v}_b = \frac{\vec{v}_a + \vec{v}_b}{1 + \frac{\vec{v}_a \vec{v}_b}{c^2}} + \frac{1}{c^2} \frac{\gamma_a}{1+\gamma_a} \frac{\vec{v}_a \wedge (\vec{v}_a \wedge \vec{v}_b)}{1 + \frac{\vec{v}_a \vec{v}_b}{c^2}} ,$$

where $\gamma_a = 1/\sqrt{1 - |\vec{v}_a|^2/c^2}$ is the Lorentz factor. In quantum-gravity research one finds motivation for endowing a specific momentum scale, essentially the Planck scale, with a special role, and it is conceivable [4, 5] that this momentum scale, here denoted with $|\ell|^{-1}$, be an absolute relativistic scale, to which the laws of transformation among observers could adapt in the context of a novel relativistic theory of "Doubly Special Relativity (DSR) type" [4, 5].

It is emerging that these sorts of constructions may require [1, 2] a relativity of spacetime locality in the same sense that the observer-independence of the speed-of-light scale requires the relativity of simultaneity. And, just like special relativity required a nonlinear law of composition of velocities, these DSR constructions require a non-linear law of composition of momenta $(p \oplus q)_a \neq (p+q)_a$, of the type

$$p_a \oplus dq_a = p_a + U(p)^b_a dq_b = p_a + dq_a + \Gamma^{bc}_a p_b dq_c + ... \qquad (1)$$

Published under licence by IOP Publishing Ltd

Journal of Physics: Conference Series **360** (2012) 012060 doi:10.1088/1742-6596/360/1/012060

where we denoted with Γ_a^{bc} the feature $\Gamma_a^{bc} = -\frac{\partial}{\partial p_a}\frac{\partial}{\partial q_b}(p \oplus q)_c|_{q,p=0}$, implicitly referring to the fact that this may be viewed [6] as an affine connection on momentum space.

The liaison between our approach (which we are here going to summarise) and the one based on the study of geometry of momentum space has been investigated in [7], where a first insight of interaction vertexes in a deformed simmetries scenario is given.

We shall here not rely explicitly on the geometry of momentum space and on the form of the composition law for momenta, since we do not need the full characterization of the shape of interactions to present longitudinal and transverse effect of relative locality. Instead we summarize some of the results obtained in Refs. [1] and [3], of the non-interactive case, giving usefully complementary characterizations of the relativity of spacetime locality that affects certain descriptions of free particles in Planck-scale-deformed phase-space constructions.

In this paper from now on we will formalize the features of obstruction of measurability and modified momenta composition rule with κ-Poincaré Hopf algebra. The algebric sector of κ-Poincaré can be interpreted as a deformation of Poincaré Lie algebra with $\ell \sim 1/M_P$ (where M_P is the Planck mass) the parameter of deformation. It is, therefore, always possible to rely on the existence of a "classical" limit by turning off this deformation ($\ell \to 0$). In this context moreover it is easier to give a simple interpretation of relative locality phenomena on physical observables and thus implement a phenomenology.

We adopt units such that the speed-of-light scale is 1 ($c = 1$).

2. Longitudinal effects

Let us start, in this section, by summarizing the main results of Ref. [1, 2], where a relativistic description of momentum dependence of the speed of massless particles was given within a classical-phase-space construction inspired by properties of the κ-Poincaré Hopf algebra [8, 9, 10].

At the quantum level the κ-Poincaré algebra is intimately related to the κ-Minkowski noncommutative spacetime [9, 10]. For classical-phase-space constructions indeed one can consider standard Minkowski spacetime coordinates in combination with a description of relativistic transformations inspired by properties of the κ-Poincaré Hopf algebra. So we assume trivial Poisson brackets for the spacetime coordinates, $\{x, t\} = 0$. One then codifies [1, 11] boost transformations in terms of Poisson brackets with the translation generators, with deformation parameters α, β, γ, as follows [1]:

$$\{\mathcal{N}_i, \Omega\} = P_i - \alpha\ell\Omega P_i \quad , \qquad \{P_i, \Omega\} = 0 \tag{2}$$

$$\{\mathcal{N}_i, P_j\} = \Omega\delta_{ij} + \ell\left((1+\gamma-\alpha)\Omega^2 + \beta\vec{P}^2\right)\delta_{ij} - \ell\left(\gamma+\beta-\tfrac{1}{2}\right)P_iP_j \ , \tag{3}$$

where Ω, P_i and $\mathcal{N}_i$ are respectively time traslation, space traslation and boost geneartors.

It is easy to check that (2),(3) satisfy all Jacobi identities with $\{x, t\} = 0$ and standard symplectic structure. And from (2), (3) one finds the following deformed on-shell relation:

$$\mathcal{C}_\ell = \Omega^2 - \vec{P}^2 + \ell(2\gamma\Omega^3 + (1-2\gamma)\Omega\vec{P}^2) \ . \tag{4}$$

By a standard Hamiltonian analysis one finds [1, 2] from (4) that, in particular, the worldlines of massless particles ($\mathcal{C}_\ell = 0$) are governed by

$$(x - x_0)_i = (1 - \ell|\vec{P}|)\frac{P_i}{|\vec{P}|}(t - t_0) \ . \tag{5}$$

The fact that the speed of massless particles here depends on momentum[1] is the main intriguing feature of this relativistic framework, and was the subject of several investigations (see, *e.g.*, Refs. [11, 12, 13, 14]), including some which established the presence of longitudinal relative locality [1, 2, 15, 16].

[1] The coefficients of the terms $\ell\Omega^3$ and $\ell\Omega\vec{P}^2$ in $\mathcal{C}_\ell$ were arranged in Ref. [1] just so that this speed law for massless particles, $1 - \ell|p|$, would be produced. This is how from the more general two-parameter case $\mathcal{C}_\ell = \Omega^2 - \vec{P}^2 + \ell(\gamma'\Omega^3 + \gamma''\Omega\vec{P}^2)$ one arrives at the one-parameter case considered here and in Ref. [1]: $\mathcal{C}_\ell = \Omega^2 - \vec{P}^2 + \ell(2\gamma\Omega^3 + (1-2\gamma)\Omega\vec{P}^2)$.

Journal of Physics: Conference Series **360** (2012) 012060 doi:10.1088/1742-6596/360/1/012060

The relativity of locality is immediately seen upon studying the covariance under α, β, γ-deformed boosts of the worldlines (5). One finds [1] that an infinitesimal deformed boost with rapidity vector ξ_i acts as follows:

$$P'_i = P_i - \xi_i\Omega - \ell\xi_i(\beta P^2 + (1+\gamma-\alpha)\Omega^2) - \ell\xi_k\left(\frac{1}{2}-\beta-\gamma\right)P^kP_i \quad (6)$$

$$t' = t - \xi_i x^i - \ell\xi_i(\alpha t P^i + 2(1+\gamma-\alpha)x^i\Omega) \quad (7)$$

$$x'_i = x_i - \xi_i t + \ell(\alpha\xi_i t\Omega + 2\beta\xi_k x^k P_i) - \ell\left(\gamma+\beta-\frac{1}{2}\right)(\xi_k P^k x_i + \xi_i x_k P^k) \quad (8)$$

We can now specialize these (6), (7) and (8) to the case of boosts along the direction x, which is the direction of motion of the particle according to (5), and it then follows that when a given observer Alice has the particle on the worldline (5) an observer boosted with respect to Alice along the x direction sees the particle on the worldline

$$(x' - x'_0)_i = (1 - \ell|\vec{P}'|)\frac{P'_i}{|\vec{P}'|}(t' - t'_0) \; , \quad (9)$$

indeed consistently with the relativistic nature of the framework.

And these results on the action of boosts also allow one to expose the relativity of spacetime locality. We first recall [1] that the absoluteness of spacetime locality amounts to the property that when one observer establishes that two events coincide then all other observers agree that those two events coincide.
To see that this is not the case in the framework under consideration, it suffices to consider two worldlines of massless particles with different energy, a "soft" one with momentum $p^{(s)}$ and a "hard" one with momentum $p^{(h)}$, emitted simultaneously from the origin of an observer Alice, so that we have a coincidence of (emission) events. A distant observer Bob at rest with respect to Alice would still describe the two emission events as coincident, but, using our prescription for boosting, a third observer Camilla, purely boosted with respect to Bob, would describe the two emission events as non-coincident. Details on the derivation of this relative-locality effect can be found in Ref. [1], which focused on the case of a Bob-Camilla boost parallel to the Alice-Bob translation direction. And for that case Ref. [1] exposed a "longitudinal relative locality" in the sense that the nonlocality attributed to the emission events by distantly boosted observer Camilla was found to be along the direction connecting Camilla and the relevant pair of events.

3. Transverse effects

Our next task is to summarize the results on "transverse relative locality" of Ref. [3]. Again we want to first map the worldline (5), attributed to observer Alice, to the description of an observer Bob, purely translated with respect to Alice along the x direction, but then we are interested in an observer Camilla purely boosted with respect to Bob along a direction y, orthogonal to x. Specializing our boost actions to the case of a boost purely in the y direction[2], we arrive at this Camilla worldline, which we characterize in the x, y plane [3]:

$$y^C(x^C) = -\xi_y\left(1 - \left(\alpha-\beta-\gamma-\frac{1}{2}\right)\ell p\right)x^C - \ell\xi_y a p \quad (10)$$

where ξ_y is the boost parameter. Here we notice two main features that characterize this result with respect to the corresponding result that applies in the special-relativity limit ($\ell \to 0$):

(I) when $\ell\xi_y a p$ is within the reach of available experimental sensitivities it will be appreciated that the worldline does not cross Camilla's spatial origin, a feature we shall find convenient to label as "shift";

(II) when $\ell\xi_y p$ is within the reach of available angular resolutions (and $\alpha - \beta - \gamma - \frac{1}{2} \neq 0$) the angle

[2] Notice that it may be useful to also note that the coordinate velocity for massless particles transforms under our boosts according to

$$\{\mathcal{N}_j, (1-\ell|\vec{P}|)\frac{P_i}{|\vec{P}|}\} = \left(1 - \ell\left(\alpha-\beta-\gamma-\frac{1}{2}\right)|\vec{P}|\right)\left(\delta_{ij} - \frac{P_iP_j}{|\vec{P}|^2}\right) - \ell\delta_{ij}|\vec{P}|.$$

in the x, y plane by which Camilla sees the arrival of the particle is momentum dependent, which is a feature that may be labeled "dual-gravity lensing", following the terminology introduced in Ref. [17], where an analogous feature was encountered in working with the relative-locality curved-momentum-space framework of Ref. [6].
Both of these features in general contribute to the transverse relative locality: if one considers a simultaneous emission at Alice of two massless particles of different momentum our distantly transversely boosted observer Camilla ends up describing the two emission events as non-coincident and the lack of coincidence occurs along the y direction (a direction orthogonal to the one connecting the observers Alice and Camilla).

Particularly striking are the effects of dual-gravity lensing: Alice sends out simultaneously two particles toward Camilla along the same direction, but Camilla ends up detecting them along non-parallel directions forming an angle [3]

$$\theta \simeq \xi_y(\alpha - \beta - \gamma - \frac{1}{2})\,\ell(p^{(h)} - p^{(s)}),$$

where notably the angle depends linearly on the difference of the momenta $(p^{(h)} - p^{(s)})$.

Incidentally it is noteworthy that the feature of dual-gravity lensing exposed in Ref. [17] was proportional to the sum of the energies(/momenta) of the two particles whose wordlines were experiencing lensing. It was already clear from Ref. [17] that this result of dependence on the sum of energies had only been checked within a very specific setup for the derivation, including definite choices among the many possible chains of interactions that could be considered in the interacting-particle framework there being studied. The fact that Ref. [3], although within the limitations of a theory of free particles, found dual-gravity-lensing effects proportional to the difference of the energies(/momenta) of the two particles whose wordlines experience lensing can provide encouragement for the search of other chains of interactions, in which the difference of energies governs the dual-gravity lensing.

References

[1] G. Amelino-Camelia, M. Matassa, F. Mercati and G. Rosati (2011), *Phys. Rev. Lett.* **106**, 071301.
[2] G. Amelino-Camelia, N. Loret and G. Rosati (2011), *Phys. Lett. B* **700**, 150 (*Preprint* arXiv:1102.4637).
[3] G. Amelino-Camelia, L. Barcaroli, N. Loret, *Preprint* arXiv:1107.3334v1 [hep-th].
[4] G. Amelino-Camelia (2002), Int. J. Mod. Phys. **D11**, 35.
[5] G. Amelino-Camelia (2001), AIP Conf. Proc. **589**, 137.
[6] G. Amelino-Camelia, L. Freidel, J. Kowalski-Glikman L. Smolin (2011), *Phys. Rev. D* **84**, 084010 (*Preprint* arXiv:1101.0931[hep-th]).
[7] G. Amelino-Camelia, M. Arzano, J. Kowalski-Glikman, G. Rosati, G. Trevisan (2011), *Preprint* arXiv:1107.1724v1 [hep-th].
[8] J. Lukierski, A. Nowicki and H. Ruegg, (1992), *Phys. Lett. B* **293**, 344.
[9] S. Majid and H. Ruegg (1994), *Phys. Lett. B* **334**, 348.
[10] J. Lukierski, H. Ruegg, W.J. Zakrzewski, (1995), *Ann.Phys.* **243**, 90 (*Preprint* arXiv:hep-th/9312153v1).
[11] M. Daszkiewicz, K. Imilkowska and J. Kowalski-Glikman (2004), *Phys. Lett. A* **323**, 345 (*Preprint* hep-th/0304027,).
[12] G. Amelino-Camelia, F. D'Andrea and G. Mandanici (2003), *JCAP* **0309**, 006.
[13] S. Mignemi (2003), *Phys. Lett. A* **316**, 173.
[14] S. Ghosh and P. Pal (2007), *Phys. Rev. D* **75**, 105021.
[15] L. Smolin (2010), *Preprint* arXiv:1007.0718.
[16] M. Arzano, J. Kowalski-Glikman (2011), *Class.Quant.Grav.***28**, 105009 (*Preprint* arXiv:1008.2962 [hep-th]).
[17] L. Freidel, L. Smolin, (2011), *Preprint* arXiv:1103.5626.

Journal of Physics: Conference Series **360** (2012) 012061 doi:10.1088/1742-6596/360/1/012061

Quantum Geometry Phenomenology: Angle and Semiclassical States

Seth A. Major

Department of Physics, Hamilton College, Clinton NY 13323 USA

E-mail: smajor@hamilton.edu

Abstract. The phenomenology for the deep spatial geometry of loop quantum gravity is discussed. In the context of a simple model of an atom of space, it is shown how purely combinatorial structures can affect observations. The angle operator is used to develop a model of angular corrections to local, continuum flat-space 3-geometries. The physical effects involve neither breaking of local Lorentz invariance nor Planck scale suppression, but rather only involve the combinatorics of $SU(2)$ recoupling. Bhabha scattering is discussed as an example of how the effects might be observationally accessible.

1. Introduction

Due to the extreme physics defined by the Planck scale, quantum gravity phenomenology is only possible with "lever arms". For instance when local Lorentz symmetry is broken the lever arm bringing the Planck scale within reach of observation is the magnificent sensitivity of particle physics in the effective field theory framework to the breaking space-time symmetries [1, 2, 3, 4]. This paper reviews the combinatoric lever arm suggested in [5] and contrasts this model with semi-coherent states peaked on discrete classical geometries in flat space. The model's lever arm is intrinsic to the discrete geometry of an atom of spatial geometry. As such it is an example of a phenomenology arising not from breaking local Lorentz invariance but rather from the fundamentally combinatoric description of spatial geometry in LQG. The combinatorics of an atom of geometry, the spin network node, suggests a dimensionless "shape" parameter that serves as an expansion parameter for the corrections to classical, flat-space continuum angular geometry. An operational view is taken in which (angle) measurements determine the local deep geometry and which contrasts with results from semi-coherent states that are peaked on classical polyhedra. The extent to which local geometry can be re-constructed from observational data is discussed.

2. Angle Operator

The angle operator, originally defined in [6], may be conveniently cast into the combinatorial framework of loop quantum gravity. The angle operator is defined at a node; incident links are partitioned into three sets. One may visualize the partitioning as arising from three regions in the surface dual to the node. Three left-invariant gravitational field operators L_1^i, L_2^i, and L_3^i are associated to these partitions. In terms of these field operators the quantum angle operator between dual surfaces 1 and 2 is

$$\hat{\theta}_{(12)} := \arccos \frac{L_1^i L_2^i}{|L_1|\,|L_2|}, \tag{1}$$

in which $|L| = \sqrt{L^2}$. The sum $\sum_{k=1}^{3} L_k^i$ vanishes due to gauge invariance and exhaustive partitions. This partitioning of links incident to n gives a preferred intertwiner basis. These are given by trivalent tree graphs that connect in an "intertwiner core". Deriving the spectrum of the angle operator of equation (1) is a simple exercise in angular momentum algebra [6]. There are two aspects of the continuum angular distribution that are hard to model, small angles are sparse and the distribution of values is weighted toward large angles. The asymmetry persists even when the spins are large [7, 8].

It is convenient to visualize the action of the angle operator on classical polyhedra, with faces dual to the incident links. The closed dual surface of the node, topologically S^2, is partitioned into three regions, S_1, S_2, and S_3, such that all the regions S_k are simply connected. The intertwiner core then represents a decomposition of the polyhedron with "internal areas" determined by $\vec{n}$. Although this picture is convenient, states may be deeply quantum and so our intuitive classical notions of face adjacency and dihedral angles may not apply. In the Discussion I ask how much of this classical information semiclassical states can determine.

The notation is as follows. Twice the sum of the representations on the links incident to the node in partition C_k is denoted by the "flux" s_k also denoted $\vec{s}$. In the dual surface picture this is the flux of spin through the respective surfaces, roughly the face areas. The quantities $n_k = 2j_k$, the internal areas, uniquely specify the intertwiner core states $|\vec{n}\rangle$.

3. Combinatorial phenomenology

It was apparent in the numerical studies of [7] that the asymmetry in the spectrum shifted the angular distribution for $\vec{n}$ at fixed $\vec{s}$, $p_{\vec{s}}(\vec{n})$, away from the usual $\sin\theta$ distribution of angles in three dimensional flat space. To recover this it was necessary in these studies to take large fluxes, and, in particular $1 \ll s_j \ll s_3$, $j = 1, 2$. Fluxes $\vec{s}$ that satisfy these relations are called "semi-classical fluxes".

There is another reason why the we might wish to consider nodes with large spin. Of course most physical processes we currently consider, such as scattering events, are "local" on the scale of the theory being tested. But in terms of the quantum geometry the scales are very large, typically many orders of magnitude above the Planck scale. In the volume operator likely to be relevant for the combinatorial framework, the volume scales as the (total flux)$^{3/2}$. The scaling with volume can be used to define an effective length ℓ_s and an effective energy $M_s = M_{Pl}/\sqrt{s}$. A surprising result in [7, 8] suggests – but does not predict – that to model the correct distribution of angles in 3-space, the total fluxes are 10^{32} giving an effective length scale of about 10^{-19} m, a perhaps not altogether hopeless scale.

To model an atom of 3-geometry suppose that the probability measure on the space of intertwiners is uniform. Further assume that all incident links to the node are spin-1/2, monochromatic and "simple". The combinatorics of the model can be solved analytically for semi-classical fluxes. For large flux $\vec{s}$ the normalized probability distribution is given by $p_{\vec{s}}(\vec{n}) \simeq \prod_{i=1}^{3} \frac{n_i}{s_i} \exp\left(-\frac{n_i^2}{2s_i}\right)$. Accessible measurements of the atom include 3-volume, (roughly) determined by the total flux, and angle, determined by the states $|\vec{n}\rangle$ of the intertwiner core. The fluxes $\vec{s}$ determine a mixed state, $\rho_{\vec{s}} = \sum_{\vec{n}} p_{\vec{s}}(\vec{n}) P_{|\vec{n}\rangle}$, where $P_{|\vec{n}\rangle}$ is the projector on the orthonormal basis of the intertwiner core. The sum is over the admissible 3-tuple of integers $\vec{n}$ such that $n_i \leq s_i$. In the discrete case the projector is $P_{|\vec{n}\rangle} = |\theta_I\rangle\langle\theta_I|$ where $|\theta_I\rangle = \sum_{\vec{n}} c_{\theta_I}(\vec{n})\,|\vec{n}\rangle$.

Journal of Physics: Conference Series **360** (2012) 012061 doi:10.1088/1742-6596/360/1/012061

The probability of finding the angle eigenvalue θ_I in the mixed state $\rho_{\vec{s}}$ is

$$\text{Prob}(\theta = \theta_I; \rho_{\vec{s}}) = \text{tr}\,(\rho_{\vec{s}} P_{\theta_I}) = \sum_{\vec{n}} p_{\vec{s}}(\vec{n}) |\langle n \mid \theta_I \rangle|^2 \equiv p_{\vec{s}}(\theta). \tag{2}$$

This procedure can be used to calculate $p_{\vec{s}}(\theta)$ in the continuum approximation. For semi-classical fluxes and a value of the measured angle θ, now taking continuous values, within an interval $\Delta\theta = (\theta - \delta\theta, \theta + \delta\theta)$ the geometric probability distribution is

$$\text{Prob}(\theta \in \Delta\theta; \hat{\rho}_s) = \text{tr}\left(\hat{\rho}_s \hat{E}_{\Delta\theta}\right) = \int d^3n P_{\vec{s}}(\vec{n}) \langle \vec{n} \mid \hat{E}_{\Delta\theta} \mid \vec{n} \rangle \tag{3}$$

where $\hat{E}_{\Delta\theta}$ is the projector onto the interval $\Delta\theta$. Geometrically it projects the state onto a (thickened) surface in $\vec{n}$-space given by $\theta(n) \in \Delta\theta$. Taking the limit $\delta\theta \to 0$ gives continuum approximation to equation (2)

$$P_{\vec{s}}(\theta) := \int d^3n\, p_{\vec{s}}(\vec{n}) |c_\theta(\vec{n})|^2 \delta\left(\theta - \theta(n)\right). \tag{4}$$

The integration of equation (4) is straightforward [5]. The key step in the calculation is the identification of the "shape parameter" $\epsilon := \sqrt{s_1 s_2}/s_3$ that measures the asymmetry in the distribution of angles.

The resulting measure, when expressed in terms of Legendre polynomials and to $O(\epsilon^3)$, is [5]

$$\rho_\epsilon(\theta) \simeq \sin\theta \left(1 - \frac{8}{\pi} P_1(\cos\theta)\epsilon + \frac{3}{2} P_2(\cos\theta)\epsilon^2\right). \tag{5}$$

The affect of the modified distribution of polar angles is that the 'shape' of space is altered by the combinatorics of the vertex; the local angular geometry differs from flat 3-space. While these effects would be in principle observable at any flux, the results here are valid for semi-classical flux. In this model the total flux $s = \sum_i s_i$ determines the 3-volume of the spatial atom and thus an effective mesoscopic length scale, $\ell_s = \sqrt{s}\ell_P$, greater than the fundamental discreteness scale of ℓ_P. So while the the shape parameter ϵ is free of the Planck scale, the effective length scale, determined by total flux s, is tied to the discreteness scale of the theory.

4. Example: Scattering

If the scale ℓ_s of the spatial atom is large enough then the underlying geometry would be accessible to observations of particle scattering. To see how the combinatorial effects might be manifest I'll briefly discuss combinatoric corrections to Bhabha scattering. This process is convenient because the e^+e^- scattering process involves "point-like" fundamental particles and for the practical reason that data is readily available. We are used to using scattering data to learn about the effective field theory at a certain energy scale. But the angular data is also a measurement of (much more) local geometry on some effective scale. The modification of the local angular geometry of space - the shape - leads to a modified solid angle. This is the dominant effect.

Short distance modifications to QED may be expressed in the Drell parameterization [11] corresponding to a short range potential added to the Coulomb potential. The local, discrete geometry manifests itself through the combinatorial corrections to the local geometry. For instance the Bhabha scattering cross section is

$$\left(\frac{d\sigma}{d\Omega}\right) / \left(\frac{d\sigma}{d\Omega}\right)_{QED} = 1 \mp \left(\frac{3s}{\Lambda_\pm^2}\right)\left(\frac{\sin^2\theta}{3 + \cos^2\theta}\right)\left(1 + \frac{8}{\pi}\cos(\theta)\epsilon + \dots\right). \tag{6}$$

It remains to be seen whether a derivation along the lines of [11] that incorporates shape corrections yields this correction. A comparison between the model and the data (see [5]) shows that the shape correction reduces power at small angles and increases power at large angles.

5. Discussion

This paper reviews a new quantum gravity phenomenology, one that explores effects arising from combinatorial structures in the deep spatial quantum geometry of LQG [5]. The example model used here relies on the combinatorics of a specific discrete model of spatial geometry, that of a single atom of spatial geometry, the spin network node. This model shows that potentially observable effects of quantum geometry need not be tied to (obvious) violations of local Lorentz symmetry and that a scale above the fundamental scale of the theory can arise out of combinatorial effects.

We can compare this model to the semi-coherent states of Levine and Speciale [12], peaked on geometry and spread on curvature. These group averaged states $|j\underline{\hat{n}}\rangle$ are built from $SU(2)$ coherent states and are peaked on states satisfying closure at large spin. They are thus peaked on classical (oriented) geometries at large spin. Relatively moderate "large" spins are required, e.g. $j \sim 100$. The basic coherent states are constructed from N spins (or face areas) and N directions. These directions are preserved in the averaging $\frac{\langle j\underline{\hat{n}}|\cos\theta_{ij}|j\underline{\hat{n}}\rangle}{\langle j\underline{\hat{n}}|j\underline{\hat{n}}\rangle} \simeq \hat{n}_i \cdot \hat{n}_j$ at (moderately) large spin. Thus the LS states, constructed from $3N-3$ data, reproduce classical angles in 3-space at moderate flux, as long as N is large and the classical directions $\underline{\hat{n}}$ are chosen uniformly. In this case the LS states exhibit no lever arm.

Taking an operational point of view the accessible measurements are an effective volume and a set of angles. From the spin geometry theorem [9, 10] we know that classical angles in Euclidean 3-space exist as long as the relative uncertainties in angle are small. Suppose for the moment that the mesoscopic quantum geometry is homogeneous. In the course of a scattering experiment angles and uncertainties are determined. Two questions arise: Can the resulting angular distribution data be described as a set of angles of a classical N-sided polyhedron? What sort of corrections might we expect for an angular semi-classical state? We know that the answer to the first question is negative. Lacking the face areas, angular data are too few to uniquely specify a classical polyhedron geometry. To answer the second question, we need to specify a (class of) closed coherent states peaked on angles, not on classical polyhedra with a fixed number of faces. From the spin geometry theorem we know that this is possible with small relative uncertainties, for large spins. The scattering data then can be used to determine whether, on the effective scale, classical 3 dimensional angular geometry exists. Specifying the only angular data for such states reduces the requirements and the combinatorics may provide a lever arm as above.

References

[1] T. Jacobson, S. Liberati , D. Mattingly arXiv:hep-ph/0110094; *Phys. Rev.* **D 66** (2002) 081302 arXiv:hep-ph/0112207.
[2] T. Konopka and S. Major *New J. Phys.* (2002) **4** 57 arXiv: hep-ph/0201184.
[3] T. Jacobson, S. Liberati, D. Mattingly *Annals Phys.* **321** (2006) 150-196 arXiv:astro-ph/0505267v2.
[4] S. Liberati and L. Maccione *Ann. Rev. Nucl. Part. Sci.* **59** (2009) 245-267 arXiv:0906.0681.
[5] S. Major. "Shape in an Atom of Space: Exploring quantum geometry phenomenology" *Class. Quantum Grav.* **27** (2010) 225012 arXiv:1005.5460.
[6] S. Major, *Class. Quant. Grav.* **16** (1999) 3859 arXiv:gr-qc/9905019.
[7] M. Seifert, "Angle and Volume Studies in Quantized Space" B.A. Thesis, Swarthmore College, Swarthmore, 2001 arXiv:gr-qc/0108047.
[8] S. Major and M. Seifert, *Class. Quant. Grav.* **19** (2002) 2211 arXiv:gr-qc/0109056.
[9] Roger Penrose, "Angular momentum: An approach to combinatorial space-time" in *Quantum Theory and Beyond* edited by T. Bastin (Cambridge University Press, Cambridge, 1971).
[10] John P. Moussouris, "Quantum Models as Space-time Based on Recoupling Theory" D.Phil. Thesis, Oxford University, Oxford, 1983.
[11] S. Drell *Ann. Phys. (NY)* **4** (1958) 75.
[12] E. Livine and S. Speziale,*Phys. Rev.* **D 76** (2007) 084028 arXiv:0705.0674.

Journal of Physics: Conference Series **360** (2012) 012062 doi:10.1088/1742-6596/360/1/012062

Shape Dynamics and AdS/CFT

Flavio Mercati

Departamento de Física Teórica, Universidad de Zaragoza, Zaragoza 50009, Spain and School of Mathematical Sciences, University of Nottingham, Nottingham, UK

E-mail: `flavio.mercati@gmail.com`

Abstract. I present a semi-classical correspondence between three-dimensional euclidean conformal field theory and the large-volume dynamics of gravity with positive cosmological constant in 3+1 dimensions. The correspondence is based on a theory of gravity called Shape Dynamics, whose solutions are a subset of those of General Relativity, namely those that can be foliated with constant mean extrinsic curvature. Shape Dynamics makes it possible to solve exactly for all the local degrees of freedom, while in GR this is possible only approximately, within a derivative expansion, due to the non-linearity of the scalar constraint.

1. Introduction

Shape Dynamics (SD)[1] is a new theory of gravity, equivalent to General Relativity (GR), but which admits a much smaller set of solutions, *i.e.* those that can be foliated with constant mean extrinsic curvature leaves. Moreover, its symmetries are different from those of GR: invariance under 4-dimensional refoliation is traded for local spatial conformal transformations that preserve the 3-volume of every leaf. The new symmetry is of gauge type, in the sense that, in the Hamiltonian formulation of the theory, it generates a constraint which is linear in the momenta. This resolves the issues associated with the quadratic constraint that appears in the Arnowitt–Deser–Misner (ADM) formulation of GR. The price for this is having a nonlocal Hamiltonian constraint, which however constrains only a single global degree of freedom. Also, the volume is a physical (non-gauge) degree of freedom, and its evolution is responsible for the expansion of the universe.

In this contribution to the Loops 11 conference I consider a perturbative construction of SD near the infinite-volume fixed point of its dynamics. This approximation scheme is found to be particularly suitable for the study of the AdS/CFT correspondence, and brings out an important advantage of SD over GR that makes it possible to prove the correspondence at the semiclassical level. This could prove crucial for further developments of the AdS/CFT idea. This advantage is the possibility to solve *all* local constraints of SD, while this was not possible in GR without a derivative expansion, and prevented the correspondence from being rigorously proven. This note is based on the results reported in [6].

2. AdS/CFT

The AdS/CFT correspondence was originally conjectured by Maldacena [7], and then made precise by Witten [8]. Consider the partition function of quantum gravity on a background

[1] Shape Dynamics was created by J. Barbour, Ó Murchadha and collaborators during the last decade [1, 2, 3] as a way to make explicit the hidden Machian structure of GR. It was put into complete form by H. Gomes, S. Gryb and T. Koslowski [4, 5]

manifold Σ with a boundary $\partial\Sigma$; it depends on the boundary values $\tilde{\phi}_i$ of every field ϕ_i that propagates in the bulk

$$\Psi(\tilde{\phi}_i) = \int_{\phi_i|_{\partial\Sigma}=\tilde{\phi}_i\ldots} \Pi_j \mathcal{D}\phi_j \ldots e^{iS(\phi_i)} . \tag{1}$$

The conjecture states that the dependence of Ψ on $\tilde{\phi}_i$ equals the generating functional of correlation functions of a CFT living on $\partial\Sigma$, in which the boundary values of the fields $\tilde{\phi}_i$ act as sources for the primary operators $\mathcal{O}_j$ of the CFT

$$Z_{CFT}(\tilde{\phi}_i \ldots) = \left\langle \exp \int_{\partial\sigma} \mathcal{O}_j \, \phi_j \right\rangle . \tag{2}$$

In a semi-classical, WKB-like approximation, the wavefunctional of quantum gravity is given by the on-shell value of the gravity action: $\Psi \simeq \alpha \, e^{iS_{\text{on-shell}}}$, where $S_{\text{on-shell}}$ can be calculated as the solution to the Hamilton-Jacobi equation for GR. This gives, on the CFT side, the generating functional of connected graphs of composite operators.

The correspondence was initially proposed for string theory in 5-dimensional AdS space and $\mathcal{N} = 4$ Super-Yang-Mills (SYM) theory in 4 dimensions at the conformal point [7]. Later a number of similar correspondences in different dimensions and for different field theories have been proposed. String theory in AdS space is still incompletely understood, but at low energy the theory reduces to supergravity with an AdS ground state coupled to Kaluza–Klein modes. On the CFT side this corresponds to the large N and strong 't Hooft coupling regime of SYM, which is relevant for certain non-perturbative aspects of QCD. So in the AdS/CFT context the question is how can one reconstruct the bulk spacetime out of CFT data and, conversely, given a bulk spacetime, what properties of the dual CFT can one read off? Both sides of the correspondence suffer from infinities: infrared divergences on the gravity side and ultraviolet divergences on the CFT side. Thus, one needs to renormalize the theory to obtain a correspondence between finite quantities. Skenderis [9] has shown how to systematically perform this renormalization, reconstructing perturbatively the bulk metric and a solution of the other fields out of CFT data.

In [6] we have investigated the potential of SD for going beyond the approximations made in these calculations, and we have shown that with SD one avoid a derivative expansion, and can calculate the counterterms needed to regularize the effective action, solving all of the constraints simultaneously. It follows that the wavefunctional of SD is invariant under diffeomorphisms and volume preserving conformal transformations. The correspondence thus implies that the CFT partition function is also invariant under diffeomorphisms and volume-preserving conformal transformations also away from the fixed point. These results have been obtained in 3+1 dimensions, and with positive cosmological constant (dS). The extension to arbitrary dimensions, signatures and values of the cosmological constant is under way.[2]

3. Shape Dynamics

Shape Dynamics is an Hamiltonian theory whose canonical variables are the three-metric g_{ab}, a scalar field ϕ, and their conjugate momentum densities, π_{ab} and π_ϕ. The Hamiltonian is [4, 5]

$$H_{SD} = \int d^3x \left(\pi^{ab} \mathcal{L}_\xi g_{ab} + \rho(x) \left(\pi - \langle \pi \rangle \sqrt{g} \right) \right) + N \, \mathcal{H}_{\text{gl}}, \tag{3}$$

where $\pi = \pi^{ab} g_{ab}$, $\mathcal{L}_\xi$ is the Lie derivative with respect to the vector field $\xi_a(x)$, which is a Lagrange multiplier, together with $\rho(x)$ and N. R denotes the 3-dimensional Ricci scalar, and

[2] The 2+1 dimensional case with the topology of the 2-torus has already been considered by T. Koslowski and T. Budd, [10] and the full quantum wavefunctional of SD has been calculated exactly, as there are no local degrees of freedom in this case.

Λ the cosmological constant. $\mathcal{H}_{\rm gl}$ is a global constraint which is defined through the solution of the following equations for $\hat{\phi}$:

$$\mathcal{H}_{\rm gl} = \frac{e^{-12\hat{\phi}}}{g}\left(\pi^{ab}\pi_{ab} + \frac{\langle\pi\rangle^2}{3}g - \frac{2}{3}\langle\pi\rangle\,\pi\,\sqrt{g}\right) - \frac{\langle\pi\rangle^2}{6} + 2\Lambda - e^{-4\hat{\phi}}\left(R(g) + 8\,e^{-\hat{\phi}}\nabla^2 e^{\hat{\phi}}\right) \quad (4)$$

$$\left\langle e^{6\hat{\phi}}\right\rangle = 1 \quad (5)$$

where $\langle f\rangle = \int d^3x\sqrt{g}\,f(x)/V$, and $V = \int d^3x\sqrt{g}$. $\mathcal{H}_{\rm gl}$ is a spatial constant, and to determine it one should solve Eq. (4) for $\hat{\phi}$ simultaneously with Eq. (5), and then plug the solution into Eq. (4). As (4) is an elliptic equation, there are conditions to be satisfied in order for its solution to be unique [11, 12]. In the case we are considering, with no matter and a cosmological constant, these conditions amount to $\Lambda \geq 0$. The constraint $\pi^{ab}_{;b}(x) = 0$, imposed by the variation of the Lagrange multiplier $\xi_a(x)$, is the diffeomorphism constraint of the ADM formulation of GR [13]. The constraint imposed by the variation of $\rho(x)$ enforces invariance under volume-preserving conformal transformations (VPCT), which is the symmetry that this theory has in exchange for refoliation invariance. This symmetry is much more tractable than refoliation invariance, as it is generated by a linear constraint, and for the present proposal will prove an edge that this theory has over GR. It can be shown [4, 5] that the following (partial) gauge-fixing: $\rho(x) = 0$, reduces the theory to GR in ADM variables, but, due to the conformal constraint $\pi = \langle\pi\rangle\sqrt{g}$, only with a particular choice of the lapse function. Recall that π is the trace of the extrinsic curvature [13], therefore the lapse is such that we have a foliation with constant mean extrinsic curvature (CMC). The solutions of SD are equivalent only to the solutions of GR which are CMC foliable.

One can isolate the 3-volume V and its conjugate momentum P from the other degrees of freedom with a canonical transformation $(g_{ab};\pi^{ab}) \to (V, \bar{g}_{ab}; P, \bar{\pi}^{ab})$ and fixing a reference volume $V_0 = \int d^3x\sqrt{\bar{g}}$. This allows, for large volumes $V \gg V_0$, to solve Eq. (4) and (5) perturbatively, through an expansion in powers of V_0/V. It is convenient to express the solution in the "Yamabe" gauge, defined as that gauge in which the scalar curvature is spatially constant $R(x) = \tilde{R} = const$ [6]. The solution is:

$$\mathcal{H}_{\rm gl} = \left(2\Lambda - \frac{3}{8}P^2\right) - \left(\frac{V_0}{V}\right)^{2/3}\tilde{R} + \left(\frac{V_0}{V}\right)^2\left\langle\frac{\bar{\pi}^{ab}\bar{\pi}_{ab}}{\bar{g}}\right\rangle + \mathcal{O}\left[\left(\frac{V}{V_0}\right)^{-8/3}\right]. \quad (6)$$

We see that the infinite-volume limit is an attractor of the dynamics, and when approached all the degrees of freedom "freeze out" apart from V, which keeps growing at constant speed [14].

The expression (6) combined with the conformal constraints allows us to make the connection with CFT. Using $P = \frac{2}{3}\langle\pi\rangle$, solving $\mathcal{H}_{\rm gl} = 0$ for $\langle\pi\rangle$, and adding the result to $D \approx 0$, we obtain $\pi/\sqrt{g} = \pm c$, where c is a spatial constant and commutes with all conformal constraints. Thus the VPCT constraint reduces to a conformal constraint. At infinite volume the theory has full conformal symmetry.

4. Hamilton–Jacobi equation and effective action

The Hamilton–Jacobi (HJ) method allows one to calculate, from the Hamiltonian, the value of the action evaluated on classical solutions of the theory. The HJ equation can be obtained from (6) by making the substitutions $P \to \frac{\delta S}{\delta V}$, $\pi^{ab} \to \frac{\delta S}{\delta g_{ab}}$, where $S = S(g_{ab}, \alpha^{ab})$ is the HJ functional; it depends on the metric g_{ab} and parametrically on the separation constants α^{ab}. These integration constants are symmetric tensor densities of weight 1. To obtain a complete integral of the HJ equation, $S_{(0)}$ can be taken in the form $S_{(0)} = \int d^3x\,\alpha^{ab}g_{ab}$. The linear constraints require α^{ab} to be transverse and with covariantly constant trace. The leading

Journal of Physics: Conference Series **360** (2012) 012062 doi:10.1088/1742-6596/360/1/012062

order HJ equation determines the value of the trace of α^{ab}. This restricts the freely specifiable components of α^{ab} to be the freely specifiable momentum data in York's conformal approach [15] to the initial value problem of GR.

For this paper, I will restrict to separation constants with vanishing transverse traceless part. These conditions are compatible with asymptotic (in time) dS space, which has maximally symmetric CMC slices.[3] There are two solutions with opposite sign. The first three terms are

$$S_{\pm} = \pm\sqrt{\frac{16}{3}\Lambda}\, V \mp \sqrt{\frac{3}{\Lambda}} \int d^3x\sqrt{\bar{g}}\tilde{R}\sqrt[3]{\frac{V}{V_0}} \pm \left(\frac{3}{\Lambda}\right)^{3/2} \int d^3x\sqrt{\bar{g}}\left(\frac{3}{8}\tilde{R}^2 - \tilde{R}^{ab}\tilde{R}_{ab}\right)\sqrt[3]{\frac{V_0}{V}} + \dots \quad (7)$$

Note that $S_{(0)}$ and $S_{(1)}$ are the only terms with positive dimension. The higher order terms can be obtained straightforwardly but become increasingly more involved because of the non–local terms appearing in the V expansion of $\mathcal{H}_{\text{gl}}$. These solutions, at any given order, provide more information than the usual derivative expansion of S in standard approaches [9, 16] to the AdS/CFT correspondence because they include non–local operators at each order in $\sqrt[3]{V_0/V}$. The local part of Eq. (7) reproduces the counterterm action reported in Sec. III B of [17] but, at each order, contains all of the nonlocal terms ("loop corrections") that contribute at that order. It can be thought of as a resummation of these corrections.

The correspondence suggests an interesting possibility. The volume variable V acts in the field theory as a regulator, and can be interpreted as the parameter of the RG flow out of the fixed point. On the other side, in a CMC foliation of an asymptotically dS spacetime V grows monotonically with time, at an increasingly homogeneous speed. Therefore it represents a natural clock for the gravitational dynamics. This suggests that a 3-dimensional, euclidean CFT might generate the classical dynamics of gravity as the evolution under renormalization group flow. If such a CFT exists, its effective action should depend on the RG parameter V through Eq. (7). From that equation, one can deduce perturbatively the classical dynamics of SD (and therefore GR) as a perturbation of an asymptotically dS spacetime. The fundamental theory would be a timeless euclidean quantum field theory, and time would emerge as evolution under RG flow. The classical dynamics of gravity would be a fundamentally quantum phenomenon.

References

[1] Barbour J and Murchadha N Ó 1999 (*Preprint* gr-qc/9911071)
[2] Anderson E, Barbour J, Foster B Z, Kelleher B and O'Murchadha N 2005 *Class. Quant. Grav.* **22** 1795–1802 (*Preprint* gr-qc/0407104)
[3] Barbour J and O'Murchadha N 2010 (*Preprint* 1009.3559)
[4] Gomes H, Gryb S and Koslowski T 2011 *Class. Quant. Grav.* **28** 045005 (*Preprint* 1010.2481)
[5] Gomes H and Koslowski T 2011 (*Preprint* 1101.5974)
[6] Gomes H, Gryb S, Mercati F and Koslowski T 2011 (*Preprint* 1105.0938)
[7] Maldacena J M 1998 *Adv. Theor. Math. Phys.* **2** 231–252 (*Preprint* hep-th/9711200)
[8] Witten E 1998 *Adv. Theor. Math. Phys.* **2** 253–291 (*Preprint* hep-th/9802150)
[9] de Haro S, Solodukhin S N and Skenderis K 2001 *Commun. Math. Phys.* **217** 595–622 (*Preprint* hep-th/0002230)
[10] Budd T and Koslowski T 2011 (*Preprint* 1107.1287)
[11] Gomes H and Kowslowski T 2011 (*Preprint* 1110.3837)
[12] Gomes H 2011 (*Preprint* 1112.0374)
[13] Arnowitt R, Deser S and Misner C 1959 *Physical Review* **116** 1322–1330
[14] Gomes H 2011 *The Dynamics of Shape* Ph.D. thesis University of Nottingham (*Preprint* 1108.4837)
[15] York J W 1971 *Phys. Rev. Lett.* **26** 1656–1658
[16] de Boer J, Verlinde E P and Verlinde H L 2000 *JHEP* **08** 003
[17] Freidel L 2008 (*Preprint* 0804.0632)

[3] The generalization to the Euclidean AdS case is trivial, requiring the scalar constraint to be expressed as a radial, instead of a time, evolution operator.

AUTHOR INDEX

Agulló, I 189
Alesci, E 193, 213
Ashtekar, A 1
Bahr, B 217
Ballesteros, A 81
Baratin, A 10
Barbosa, R M S 113
Barcaroli, L 294
Beetle, C 197
Beke, D 222
Benedetti, D 85
Bethke, L 20
Bianchi, E 227
Bias, D Martín-De 177
Bilson-Thompson, S 276
Borja, E F 117, 189
Brizuela, D 153
Budd, T G 201
Calcagni, G 157
Caravelli, F 205
Carrozza, S 231
Cianfrani, F 122
Colosi, D 89
Constantinidis, C P 113
Corichi, A 126
Craig, D A 161
Díaz-Polo, J 117, 189
Ding, Y 227
Dittrich, B 29
Dohse, M 89
Dupuis, M 235
Eckert, F C 29
Eichhorn, A 280
Engle, J 197
Fernández-Méndez, M 181
Fleischhack, C 130
Freidel, L 117
G, J F Barbero 189
Gambini, R 39
Garay, I 117
Garay, L J 173
Gielen, S 165
Gomes, H 284
Hazboun, J S 93
Herranz, F J 81
Hinterleitner, F 169
Höhn, P A 97, 239
Kisielowski, M 243
Koslowski, T 289
Livine, E 117, 140, 235
Loret, N 294
Ma, Y 42, 272
Magueijo, J 20
Major, S 169, 298
Martín-Benito, M 173, 177
Marto, J 105
Marugán, G A Mena 173, 177, 181
Mercati, F 302
Meusburger, C 81
Mikovic, A 247
Modesto, L 193
Moniz, P V 105
Noui, K 209
Oeckl, R 89
Olmedo, J 181
Olmo, G J 185
Oporto, Z 113
Oriti, D 10
Perez, A 209
Perini, C 251
Piguet, O 113
Pranzetti, D 209
Puchta, J 255
Pullin, J 39
Raasakka, M 260
Rastgoo, S 39
Reyes, J D 126
Rosati, G 294
Sahlmann, H 52
Sengupta, S 145
Shestakova, T P 101
Singh, P 62, 161
Smerlak, M 264
Speziale, S 85
Steinhaus, S 149
Tambornino, J 140
Tavakoli, Y 105
Varadarajan, M 72
Villaseñor, E J S 189
Vojinovic, M 247
Wheeler, J T 93
Wilson-Ewing, E 173
Wise, D K 109
Zapata, J A 267
Zhang, X 272
Ziaie, A H 105